£7.60

ATOMIC AND NUCLEAR PHYSICS

THE MODERN UNIVERSITY PHYSICS SERIES

This series is intended for readers whose main interest is in physics, or who need the methods of physics in the study of science and technology. Some of the books will provide a sound treatment of topics essential in any physics training, while other, more advanced, volumes will be suitable as preliminary reading for research in the field covered. New titles will be added from time to time.

Clark: *A First Course in Quantum Mechanics*
Littlefield and Thorley: *Atomic and Nuclear Physics*
Lothian: *Optics and Its Uses*
Lovell, Avery and Vernon: *Physical Properties of Materials*
Perina: *Coherence of Light*
Tritton: *Physical Fluid Dynamics*
Wolbarst: *Symmetry and Quantum Systems*

ATOMIC AND NUCLEAR PHYSICS
An Introduction

3rd edition

T. A. LITTLEFIELD

and

N. THORLEY
Senior Lecturer
School of Physics
University of Newcastle upon Tyne

VNR VAN NOSTRAND REINHOLD COMPANY
New York — Cincinnati — Toronto — London — Melbourne

© 1979, Van Nostrand Reinhold Co. Ltd.

First edition 1963
Second edition 1968

All rights reserved. No part of this work covered by the copyright hereon may be reproduced or used in any form or by any means — graphic, electronic, or mechanical, including photocopying, recording, taping, or information storage or retrieval systems — without the written permission of the publishers

**Published by Van Nostrand Reinhold Company Ltd.,
Molly Millars Lane, Wokingham, Berkshire, England**

Published in 1979 *by Van Nostrand Reinhold Company,
A Division of Litton Educational Publishing Inc.,*
135 *West* 50*th Street, New York, N.Y.* 10020, *U.S.A.*

Van Nostrand Reinhold Limited,
1410 *Birchmount Road, Scarborough, Ontario,* M1P 2E7,
Canada

Van Nostrand Reinhold Australia Pty, Limited,
17 *Queen Street, Mitcham, Victoria* 3132, *Australia*

Library of Congress Cataloging in Publication Data

Littlefield, Thomas Albert, 1912–
   Atomic and nuclear physics.

   (The Modern university physics series)
   Bibliography: p.
   Includes index.
   1. Atoms. 2. Nuclear physics. 3. Quantum theory.
I. Thorley, Norman, 1913 — joint author. II. Title.
QC173.L76   1979   539.7   78-31250
ISBN 0-442-30189-8
ISBN 0-442-30190-1 pbk.
ISBN 0-442-30178-2 ELBS

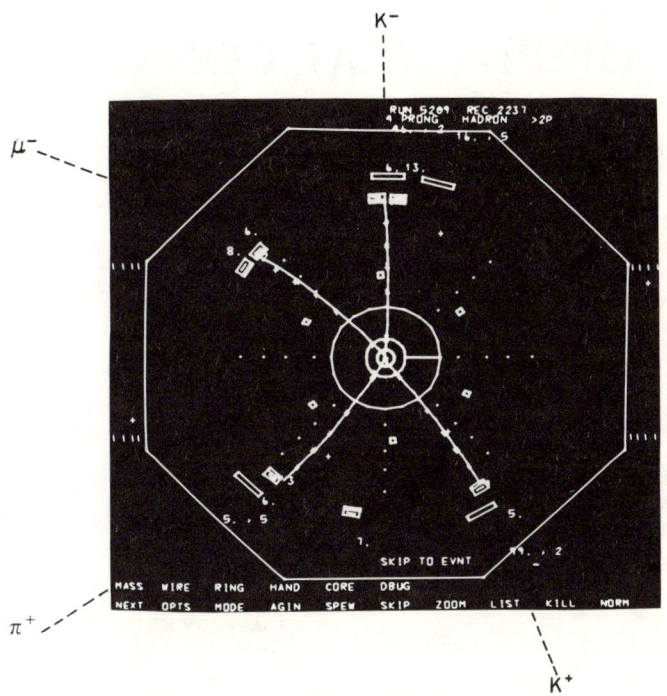

Computer display of the decay of the two charmed particles
$$D^0(c\bar{u}) \to K^-\pi^+ \quad \text{and} \quad D^0(\bar{c}u) \to K^+\mu^-$$
produced from an energetic electron–positron collision. The display shows the cross-section of the beam pipe (centre) and the surrounding spark chambers and scintillation counters of the particle detector arrangement. The tracks of the particles are plotted by the triggered spark chambers. (Photograph taken from *Fundamental Particles with Charm*, Roy F. Schwitters, Scientific American, October 1977, p. 57.)

TABLE OF PHYSICAL CONSTANTS

Speed of light (vacuum) $\qquad c = 2\cdot997\,926 \times 10^8$ m s^{-1}
Elementary charge (proton) $\qquad e = 1\cdot602\,189 \times 10^{-19}$ C
Unified atomic mass constant $\qquad u = 1\cdot660\,565 \times 10^{-27}$ kg
Electron rest mass $\qquad m_e = 9\cdot109\,534 \times 10^{-31}$ kg
$\qquad\qquad = 5\cdot485\,803 \times 10^{-4}$ u

Proton rest mass $\qquad m_p = 1\cdot672\,648 \times 10^{-27}$ kg
$\qquad\qquad = 1\cdot007\,276$ u

Neutron rest mass $\qquad m_n = 1\cdot674\,954 \times 10^{-27}$ kg
$\qquad\qquad = 1\cdot008\,665$ u

Mass of hydrogen atom $\qquad m_H = 1\cdot673\,559 \times 10^{-27}$ kg
$\qquad\qquad = 1\cdot007\,825$ u

Electron charge-to-mass ratio $\qquad e/m_e = 1\cdot758\,804\,5 \times 10^{11}$ C kg^{-1}
Proton-to-electron mass ratio $\qquad m_p/m_e = 1836\cdot151$
Neutron-to-electron mass ratio $\qquad m_n/m_e = 1838\cdot682$
Avogadro constant $\qquad N_A = 6\cdot022\,045 \times 10^{23}$ mol^{-1}
Planck constant $\qquad h = 6\cdot626\,176 \times 10^{-34}$ J s
$\qquad\qquad \hbar = 1\cdot054\,589 \times 10^{-34}$ J s
Faraday constant $\qquad F = 9\cdot648\,456 \times 10^4$ C mol^{-1}
First Bohr radius $\qquad a_0 = 5\cdot291\,771 \times 10^{-11}$ m
Gas constant $\qquad R = 8\cdot314\,41$ J K^{-1} mol^{-1}
Boltzmann constant $\qquad k = 1\cdot380\,662 \times 10^{-23}$ J K^{-1}
Bohr magneton $\qquad \mu_B = 9\cdot274\,078 \times 10^{-24}$ J T^{-1}
Nuclear magneton $\qquad \mu_N = 5\cdot050\,824 \times 10^{-27}$ J T^{-1}
Energy conversion factors $\qquad 1$ eV $= 1\cdot602\,189 \times 10^{-19}$ J
$\qquad\qquad 1$ u $= 931\cdot502$ MeV
Rydberg constant $\qquad R_\infty = 1\cdot097\,373 \times 10^7$ m^{-1}
Magnetic constant (permeability of a vacuum) $\qquad \mu_0 = 4\pi \times 10^{-7}$ H m^{-1}
Electric constant (permittivity of a vacuum) $\qquad \varepsilon_0 = 8\cdot854\,188 \times 10^{-12}$ F m^{-1}
Standard atmosphere $\qquad 1$ atm $= 103\,125$ Pa

These figures are adapted from *Quantities, Units and Symbols*, published by the Symbols Committee of the Royal Society, London, 1975.

PREFACE TO THE THIRD EDITION

After the death of Dr. Littlefield it was decided that I should undertake the revision of the whole of *Atomic and Nuclear Physics: an Introduction* for the third edition, and it was soon apparent that major changes were necessary. I am confident that these changes would have had Dr. Littlefield's approval.

The prime consideration for the present edition has been to modernize at a minimum cost. As much as possible of the second edition has therefore been retained, but where changes have been made they have been fairly drastic. Thus the chapters on fine structure, wave mechanics, the vector model of the atom, Pauli's principle and the Zeeman effect have been completely restructured. The chapters on nuclear models, cosmic rays, fusion systems and fundamental particles have been brought up to date while a new chapter on charm and the latest ideas on quarks has been included. It is hoped that the presentation of the last named will give readers a feeling that physics research can be full of adventure and surprises.

The student targets for the book are the first and second years of an undergraduate course in atomic and nuclear physics at our universities and polytechnics. It will also be useful to those first and second year engineers requiring an atomic and nuclear physics background. The book must not be judged as a final honours text since too many topics are omitted and the mathematical depth is insufficient for this purpose. It is more suitable for all general or ordinary degree students and for first year single honours students. It could also be profitably used by H.N.C. students taking applied physics. The changes made and the new material used are well within the compass of such students.

No references to original papers have been given because it is the writer's experience that only those students who are deeply concerned with the subject in their third years take the trouble to use the library for reference purposes. Furthermore, the addition of references within or at the end of each chapter would have added to the cost of this edition. However, many new problems have been included, some of which (marked N) have been taken from the physics papers of the University of Newcastle upon Tyne, and the writer would like to record his thanks to the Senate of this university for permission to use these questions.

I must also thank my colleagues Dr. I. D. C. Gurney, Dr. E. L. Lewis, Dr. B. Peart and Professor K. T. Dolder for helpful discussions and I am indebted to Professor A. W. Wolfendale for help with the section on cosmic ray research. Needless to say the writer takes full responsibility for any lack of clarity which persists and would be glad to receive comments, corrections and criticisms of the text which might be included in a future printing.

Grateful acknowledgement is made to the Culham Laboratory of the U.K.A.E.A. at Abingdon for the use of Fig. 24.11.

Grateful acknowledgement and thanks are due to Mrs. Dorothy Cooper for the excellent art work she produced from the writer's sketches; to the publishers Van Nostrand Reinhold Co. Ltd.; and most especially to Miss L. J. Ward, the College Editor, for her patience during the preparation of this edition.

Finally, to my wife, Mrs. Joan Thorley, for much encouragement over the last few months and for transforming my scrawl into a typescript.

N.T.
Dec 1978

CONTENTS

Preface   vii

Table of Physical Constants   xvi

Chapter 1   KINETIC THEORY   1
1.1   The Atom in History   1
1.2   Brownian Motion   2
1.3   Basic Assumptions of Kinetic Theory   3
1.4   Pressure of a Gas   4
1.5   Molecular Velocities   6
1.6   Temperature of a Gas: Avogadro's Hypothesis   6
1.7   Mean Free Path   7
1.8   Thermal Conductivity and Viscosity   8
1.9   Specific Heat Capacities   9
1.10   Atomicity   10
1.11   Molar Heat Capacities   12
1.12   Van der Waals' Equation   12
1.13   Molecular Sizes   15
1.14   Summary   16

Chapter 2   THE ELECTRON   19
2.1   Electrical Conduction in Solutions   19
2.2   Conduction in Gases   20
2.3   Properties of Cathode Rays   22
2.4   Thomson's Method for Measuring Charge per Unit Mass (e/m)   22
2.5   Dunnington's Method for e/m 23
2.6   Charge on the Electron   25

Chapter 3   NATURAL RADIOACTIVITY   30
3.1   Introduction   30
3.2 e/m for β-Rays   30
3.3   Bucherer's Method for e/m of β-Rays   32
3.4   The Charge–Mass Ratio (E/M) for α-Rays   33
3.5   Charge on α-Particles   35
3.6   Identification of α-Particles   36
3.7   Early Models of the Atom   37
3.8   The Scattering of α-Particles   38
3.9   Estimates of Nuclear Diameter and Charge   41
3.10   The Neutron   42

Chapter 4 RADIOACTIVE SERIES AND ISOTOPES	46
4.1 Introduction	46
4.2 Equation of Radioactive Decay	46
4.3 Mean Lifetime of Radioactive Substance	47
4.4 Half-Lives of Radioactive Substances	49
4.5 Radioactive Series	49
4.6 Radioactive Equilibrium	50
4.7 Isotopes	52
4.8 The Bainbridge Mass Spectrograph	55

Chapter 5 THE ELECTROMAGNETIC SPECTRUM	59
5.1 Theories of Light	59
5.2 Interference	59
5.3 Diffraction	61
5.4 Spectra	62
5.5 The Electromagnetic Theory	63
5.6 Hertz's Experiment	65
5.7 The Electromagnetic Spectrum	66

Chapter 6 QUANTUM THEORY	69
6.1 The Continuous Spectrum	69
6.2 Planck's Quantum Theory	70
6.3 The Photoelectric Effect	71
6.4 Einstein's Equation	72
6.5 The Discovery of X-Rays	74
6.6 Diffraction of X-Rays	75
6.7 X-Ray Wavelengths	77
6.8 Continuous Spectrum of X-Rays	79
6.9 Compton Effect	80
6.10 Summary	82

CHAPTER 7 SPECTRA	85
7.1 The Hydrogen Spectrum	85
7.2 The Bohr Theory of the Hydrogen Atom	86
7.3 Isotope Effect	90
7.4 The Spectrum of Sodium	93
7.5 Quantum Defects — Interpretation	95
7.6 Selection Rules and the Correspondence Principle	97
7.7 Excitation Potentials	99
7.8 Controlled Excitation of Spectra	101
7.9 X-Ray Spectra	101
7.10 Moseley's Work	102
7.11 The Interpretation of X-Ray Spectra	104

Chapter 8 FINE STRUCTURE AND ELECTRON SPIN	109
8.1 Fine Structure of Alkali–Metal Spectra	109
8.2 Electron Spin	111

8.3	Characteristic X-Rays and Absorption Spectra	114
8.4	Multiplicity of X-Ray Levels	116

Chapter 9 WAVES AND PARTICLES — 120
9.1	The Radiation Dilemma	120
9.2	De Broglie's Theory	120
9.3	Group Velocity	121
9.4	The Davisson and Germer Experiment	122
9.5	The Experiment of Thomson and Reid	124
9.6	The Electron Microscope	125
9.7	Heisenberg's Uncertainty Principle	125
9.8	Born's Statistical Interpretation of Waves and Particles	127

Chapter 10 WAVE MECHANICS — 131
10.1	Some Preliminaries	131
10.2	The Need for Change	132
10.3	The Schrödinger Wave Equation	134
10.4	An Alternative Approach	137
10.5	Solution of the Schrödinger Wave Equation	137
10.6	Simple One-Electron Atom Model	139
10.7	The Hydrogen Atom	143
10.8	Angular Momenta	146
10.9	Summary	147

Chapter 11 THE VECTOR MODEL OF THE ATOM — 150
11.1	Quantum Numbers and Angular Momenta: Summary of Symbols and Notation	150
11.2	Magnetic Moments — Orbital and Spin	151
11.3	The Stern–Gerlach Experiment	153
11.4	Spatial Quantization of Electron Spin	156
11.5	Spin–Orbit Coupling and the Total Angular Momentum **j**	157

Chapter 12 TWO-ELECTRON ATOMS — PAULI PRINCIPLE — 162
12.1	Wave Functions of Two-Electron Atoms	162
12.2	Vector Coupling for Two Electrons	165
12.3	The Helium Spectrum	167
12.4	**jj** Coupling	170
12.5	The Electronic Structure of the Elements and the Periodic Table	172
12.6	The Periodic Table — Some Empirical Rules	176
12.7	Hyperfine Structure and Nuclear Spin Angular Momentum	180

Chapter 13 THE ZEEMAN EFFECT — 183
13.1	Introduction	183
13.2	The Normal Zeeman Effect	183
13.3	Explanation of Zeeman Effect in Terms of Vector Model	184
13.4	Zeeman Effect of Cadmium 643·8 nm Line	187
13.5	The Anomalous Zeeman Effect and the Landé Splitting Factor	188

13.6	Zeeman Splitting in a Strong Magnetic Field: the Paschen–Back Effect	192
13.7	Conclusion	193

Chapter 14 THE STRUCTURE OF THE NUCLEUS — 196
14.1 Introduction — 196
14.2 Nuclear Constituents: Isotopes and Isobars — 197
14.3 The Size of the Nucleus — 200
14.4 Exact Atomic Masses — Mass Excess ΔM — 201
14.5 Binding Energies of Nuclides — Mass Defect — 203
14.6 Stable and Unstable Nuclides — 207
14.7 Derivation of Practical Form of $E = m_0 c^2$ — 208

Chapter 15 PROPERTIES AND USES OF NATURAL RADIOACTIVITY — 212
15.1 The Nature of Radioactivity — 212
15.2 α-Particles and the Geiger–Nuttall Rule — 213
15.3 The Theory of α-Decay — 216
15.4 β-Rays and the Neutrino — 218
15.5 The Absorption and Range of β-Rays — 221
15.6 The Properties of γ-Rays — 223
15.7 Radioactivity as a Measurable Quantity — 226

Chapter 16 NUCLEAR BOMBARDING EXPERIMENTS — 231
16.1 Single α-Particle Scattering — 231
16.2 Nuclear Alchemy — 231
16.3 Cockcroft–Walton Proton Experiments — 235
16.4 The Neutron — 236
16.5 Nuclear Reactions — 236
16.6 Formation of Tritium — 237

Chapter 17 THE MEASUREMENT AND DETECTION OF CHARGED PARTICLES — 240
17.1 The Wilson Cloud Chamber — 240
17.2 The Bubble Chamber — 241
17.3 Ionization Chambers — 244
17.4 The Proportional Counter — 245
17.5 The Geiger–Muller Counter — 248
17.6 Scintillation Counters and Semiconductor Counters — 251
17.7 The Spark Chamber — 252
17.8 The Cerenkov Counter — 254
17.9 Neutron Counting — 255
17.10 The Photographic Plate — 255
17.11 Summary — 255

Chapter 18 ACCELERATING MACHINES AS USED IN NUCLEAR PHYSICS — 258
18.1 Introduction — 258
18.2 The Cockcroft–Walton Proton Accelerator — 258
18.3 The Van de Graaff Electrostatic Generator — 260
18.4 The Linear Accelerator — 261
18.5 The Lawrence Cyclotron — 263
18.6 The Synchrocyclotron — 265
18.7 Electron Accelerating Machines. The Betatron — 265
18.8 Electron Synchrotron — 266
18.9 Proton Synchrotron — 267
18.10 The Alternating-Gradient Synchrotron — 268
18.11 Intersecting Beam Accelerators — 269
18.12 The Growth and Future of Large Accelerating Machines — 269

Chapter 19 NUCLEAR MODELS AND MAGIC NUMBERS — 277
19.1 Introduction — 277
19.2 Neutron Cross-Sections and Nuclear Radii — 277
19.3 The Liquid-Drop Model — 281
19.4 Nuclear Shells and Magic Numbers — 284
19.5 The Theory of the Nuclear Shell Model — 286
19.6 The Collective Model — 290
19.7 Superheavy Elements: Experimental and Theoretical — 291
19.8 Latest Developments — 293
19.9 The Melting of the Moon — 294

Chapter 20 ARTIFICIAL RADIOACTIVITY — 298
20.1 The Discovery of the Positron — 298
20.2 K-Electron Capture — 301
20.3 The Origin of Electrons and Positrons within the Nucleus — 302
20.4 Nuclear Isomerism — 303
20.5 The Production of Radioisotopes — 306
20.6 Some Uses of Radioisotopes — 307

Chapter 21 NEUTRON PHYSICS — 310
21.1 Introduction — 310
21.2 Properties of the Neutron — 310
21.3 Neutron Bombardment Reactions — 313
21.4 Archaeological Dating by the ^{14}C Method — 314
21.5 Tree-Ring Calibration of ^{14}C Dates — 315

Chapter 22 NUCLEAR FISSION AND ITS IMPLICATIONS — 320
22.1 Introduction — 320
22.2 The Theory of Nuclear Fission — 321
22.3 The Energy of Nuclear Fission — 325
22.4 The Distribution of Fission Products — 327
22.5 Characteristics of Fission Neutrons — 328

22.6	The β-Decay Chains of Fission	330
22.7	Controlled Fission–Nuclear Reactors	331
22.8	Nuclear Power Reactors	334
22.9	Nuclear Power Prospects	335

Chapter 23 THE TRANSURANIC ELEMENTS 340

23.1	Neptunium ($Z=93$) to ?	340
23.2	Formation of Transuranic Elements	341
23.3	Neptunium, Np ($Z=93$)	341
23.4	Plutonium, Pu ($Z=94$)	342
23.5	Americium, Am ($Z=95$), and Curium, Cm ($Z=96$)	343
23.6	Berkelium, Bk ($Z=97$), and Californium, Cf ($Z=98$)	343
23.7	Einstein, Es ($Z=99$), and Fermium, Fm ($Z=100$)	344
23.8	Mendelevium, Md ($Z=101$), and Nobelium ($Z=102$)	344
23.9	Lawrencium, Lw ($Z=103$)	344
23.10	Elements with $Z=104$, 105, 106 and 107	345
23.11	The Actinide Series	345

Chapter 24 THERMONUCLEAR REACTIONS AND NUCLEAR FUSION 348

24.1	Introduction	348
24.2	The Source of Stellar Energy	348
24.3	The Plasma	349
24.4	Nuclear Fusion Reactions in the Plasma	350
24.5	Conditions for a Maintained Fusion Reaction	351
24.6	The Possibility of a Fusion Reactor	355
24.7	Tokomak Fusion Systems	357
24.8	Energy in the Future	360

Chapter 25 COSMIC RAYS 364

25.1	Discovery	364
25.2	Nature of Cosmic Rays	364
25.3	The Origin of Cosmic Rays	366
25.4	Geomagnetic Effects	368
25.5	Cosmic Rays at Sea-Level	369
25.6	Extensive Air Showers	371
25.7	The Detection of Cosmic Ray Particles	371
25.8	The Future of Cosmic Ray Research	373

Chapter 26 STABLE AND SEMI-STABLE PARTICLES 377

26.1	Introduction	377
26.2	The Positron: Particles and Antiparticles	379
26.3	Pions, Muons and Kaons	385
26.4	Hyperons	392
26.5	Classification of the Elementary Particles	392
26.6	Mesic Atoms: The Muonium Atom	394

Chapter 27 SHORT-LIVED RESONANCE STATES 401
27.1 Forces and Fields 401
27.2 What is an Elementary Particle? 402
27.3 Short-Lived or Resonance Particles 403
27.4 Conservation Laws: Baryon and Lepton Conservation 409
27.5 Multiplet Structure — Isospin and Hypercharges 410
27.6 Classification of Elementary Particles 415
27.7 Particle Symmetries 416
27.8 Quarks 420
27.9 Conclusions 423

Chapter 28 CHARM AND ALL THAT 426
28.1 The Forces of Nature 426
28.2 The Three-Quirk Trick 430
28.3 The New Quark — Charm 433
28.4 The November Revolution — the J/ψ Particle 435
28.5 Quark Multiplet Representation 436
28.6 Gluons and Colour 438
28.7 The Confinement of Quarks 440
28.8 The Hunting of the Quark 441
28.9 Latest News: New Quarks 442
28.10 Conclusions 443

Appendix A RELATIVITY THEORY 447

Appendix B THE DANGERS OF ATOMIC RADIATIONS 452
B.1 Introduction 452
B.2 Biological Effects of Nuclear and Electromagnetic Radiations 452
B.3 Maximum Permissible Radiation Levels for Safety 455
B.4 Precautions against Radiation Hazards 456

Appendix C COMPLETE LIST OF NUCLIDES OF THE ELEMENTS 458

Index 481

Chapter 1
Kinetic Theory

1.1 The Atom in History

The Greeks speculated whether or not matter could be divided indefinitely into smaller and smaller pieces. Should this be possible, they argued, then matter is continuous, but if not then matter must consist ultimately of very small entities now known as 'atoms'. This situation remained unresolved for many centuries as there was no evidence to support either of these possibilities. In fact, it was not until the beginning of the nineteenth century that the atom became a precise concept based upon the laws of chemical combination and the kinetic theory of gases.

Early in the nineteenth century the quantitative study of chemistry revealed two general laws of chemical combination, the *law of constant composition* and the *law of multiple proportions*. These state, respectively, that a particular chemical compound always contains the same elements combined in the same proportions; and that when one substance unites with another in more than one proportion, these different proportions bear a simple ratio to one another. These were interpreted by Dalton in 1803 to mean that compounds consist of molecules. These molecules are composed of atoms of various elements in definite proportions. There seems to be some doubt which came first, the theory or the experimental results, but there can be no doubt that one inspired and stimulated the other. Soon afterwards, in 1808, Gay-Lussac showed experimentally that simple ratios existed between the volumes of reacting gases. In 1811 Avogadro combined Dalton's atomic theory with Gay-Lussac's observations and suggested that equal volumes of gases in the same conditions of pressure and temperature contain equal numbers of molecules. This is known as *Avogadro's hypothesis* and, since it is found experimentally that the molar volume of many simple gases is the same, viz. 22·4 l at s.t.p. (1 litre = 1 l = 10^{-3} m^3; s.t.p. is 0 °C and 101 325 Pa, see p. xvi), it follows that this particular volume always contains the same number of molecules, irrespective of the gas concerned. This number is the Avogadro constant N_A, which is approximately $6·022 \times 10^{23}$ mol^{-1} and which holds for all those gases approaching ideal gas conditions (see Section 1.3). The ideal gas molar volume of 22·41 l is remarkably close to the values found for real gases and vapours (see Problem 1.20).

These simple ideas led to the formulation of the *atomic theory of matter*, which in turn explained all chemical observations and theory during the next hundred years. Later in the century Mendeleev showed that if the elements were placed in order of atomic weight they displayed a periodicity of behaviour. The atomic

1

theory was quite unable to explain this, which was a clear indication that the atom was not the simple indivisible unit initially conceived by Dalton. The full significance of the periodic table was not apparent until the development of theories of atomic structure in the present century.

As these ideas were taking shape in chemistry, certain rather abstract ideas in physics were beginning to emerge. In particular, experimental evidence began to accumulate which showed first that heat was a form of energy, and later that light, electricity, magnetism and sound were also forms of energy. Physics appeared to be reduced to a study of the interactions of these various forms of energy with matter. It also became clear that in all natural processes energy is converted from one form into another and is never created or destroyed. This is the law of the conservation of energy; apparent exceptions to the law have often been found, but these were almost invariably due to a failure to take all the factors of a situation into consideration. Another conservation law, concerning momentum, applies to both linear and angular momentum, and the study of atomic physics provides many elegant illustrations of this law.

Kinetic theory is based upon the two hypotheses: that matter is composed of molecules and atoms, and that heat is a form of energy. According to the atomic theory, that sensation which we call heat is the external manifestation of the internal kinetic energy of the atoms or molecules due to their random translational motions. It is sometimes called thermal energy.

As early as 1738, and well before the precise formulation of the atomic theory by Dalton, Bernoulli calculated the pressure of a gas from the mechanical properties of molecules striking a boundary. The development of the atomic theory in chemistry was matched by a corresponding refinement of the kinetic theory, especially by Clausius and Clerk Maxwell.

1.2 Brownian Motion

Evidence in support of the existence of molecules comes from observations of Brownian motion, named after Brown who first observed it in 1827. This may be demonstrated by introducing cigarette smoke into a small hollow glass cube of about 10 mm × 10 mm × 10mm (Fig. 1.1) placed on the stage of a microscope and illuminated by a strong horizontal beam of light from one side. When viewed under high magnification, small bright specks of the smoke particles may be seen. These are observed to be continuously agitated in a random fashion. The explanation is that with a heavy body the impacts of individual gas molecules on the surface are relatively too small to displace the body appreciably. Moreover the surface area is so large that the impulses delivered by the numerous molecules balance out. When, however, the size of the body is reduced, the impacts of the molecules all around it are less likely to be balanced and the lighter particle responds more readily to the resultant forces acting upon it. Brownian motion in a liquid can be shown by mixing very dilute solutions of lead acetate and potassium carbonate in a rectangular tank of glass or perspex. When the tank is illuminated by light from an arc lantern and viewed at right angles to the direction of illumination, each crystal platelet shows up as a bright speck when it comes into the position of reflection. The twinkling of these bright specks indicates that the crystals are being continuously agitated in much the same way as the smoke particles. The agitations are due to random molecular collisions.

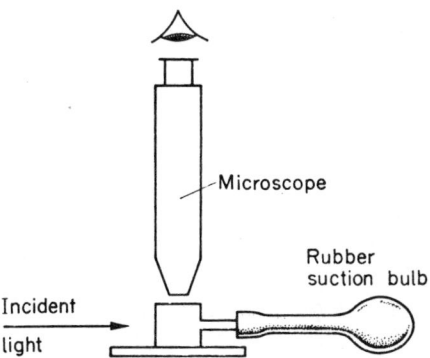

Fig. 1.1 Apparatus to show Brownian motion of smoke particles.

1.3 Basic Assumptions of Kinetic Theory

A gas quickly fills the whole volume in which it is enclosed so that we must imagine the atoms to have great freedom of movement. By contrast, atoms in a solid are fixed in position and are only capable of vibration about a mean position. When we remember that a gas at normal temperature and pressure has a density of about one thousandth that of a solid, it is evident that the separation of the molecules in a gas must be about ten times as great as in a solid. In liquids the molecules are still capable of continuous movement throughout the volume of the liquid, but the speed of movement is very much smaller and the molecules are sufficiently close to be continuously attracted by one another. Thus the molecule acquires greater and greater freedom as we proceed from the solid through the liquid state to the gas.

In the present discussion we shall confine our attention to the gaseous state as certain plausible assumptions are then possible which approximate very closely to the true conditions. These are that the diameter of a molecule is small compared with the distance travelled between two successive collisions; that the velocities of the molecules are so large that many encounters occur in a short time interval; and that the molecules are separated by distances so great that their mutual attractions and repulsions may be neglected. This last assumption implies that the molecules can have no potential energy and that all the energy of the gas must appear in the form of kinetic energy. The molecules are also assumed to be perfectly elastic so that no kinetic energy is lost when a collision takes place. Although all molecules in a gas are assumed to have equal mass, not all, even at a uniform temperature, have the same velocity. There is a velocity distribution so that if we plot n_v, the number of molecules having a particular velocity, against that velocity, we obtain a curve of the form shown in Fig. 1.2. The distribution is not quite symmetrical, and the most probable speed $\hat{v}$ is not quite the same as the mean speed $\bar{v}$. Molecular speeds are assumed to increase with increasing temperature since the molecular kinetic energy represents the heat content of the gas.

If we take a unit cube containing n molecules (Fig. 1.3), each molecule will travel with random motion inside the cube. Since there are a very large number of molecules (about 10^{16} mm^{-3}) and their net motion is entirely random, we may

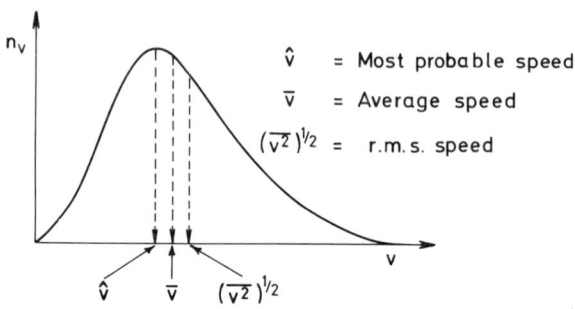

Fig. 1.2 Distribution of molecular speeds in a gas.

assume that at any instant $n/6$ molecules are on the average proceeding towards each of the six faces of the cube. A further simplification can also be made by assuming that all these molecules have the same velocity instead of the distribution described above. This is known as the 'six stream method' of treatment and provides results differing from those of more rigorous treatments only by simple numerical factors.

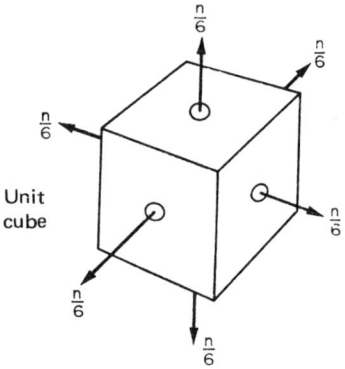

Fig. 1.3 The 'six stream method'

1.4 Pressure of a Gas

To calculate the pressure exerted upon the wall of a container by the impact of molecules we consider a rectangular volume of unit area of cross-section and length v, where v is the velocity of a single molecule (Fig. 1.4). In one second all the molecules travelling towards the right will fall upon this area. The total number of molecules in the volume is nv, n being the number per unit volume. The number which will actually fall upon the unit area of wall will be $nv/6$. If each molecule of mass m and momentum mv is reflected without loss of speed, the momentum change for each molecule on impact is $2mv$, since momentum must be treated as a

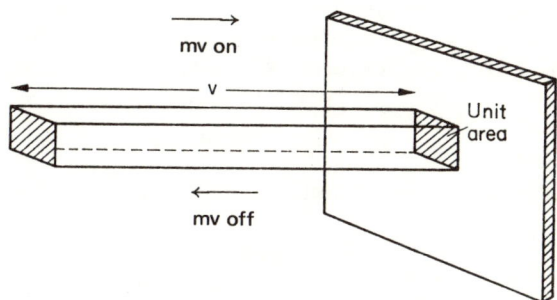

Fig. 1.4 Impact of molecules on wall of container.

vector. Therefore, the total momentum change for all the molecules in one second will be $2nv \cdot nv/6 = mnv^2/3$. This represents the force acting upon the unit area and is therefore the pressure exerted by the gas on the wall of its container, i.e.
$$p = \tfrac{1}{3}mnv^2 = \tfrac{1}{3}\rho v^2,$$
where ρ is the density, or mass of material in unit volume.

For all the particles of the gas, with their different speeds, the speed v in the above equation is the root mean square (r.m.s.) speed $[\overline{v^2}]^{1/2}$, where the mean square speed is defined by

$$\overline{v^2} = \frac{n_1 v_1^2 + n_2 v_2^2 + n_3 v_3^2 + \ldots}{n_1 + n_2 + n_3 + \ldots} = \frac{\sum_i n_i v_i^2}{\sum_i n_i}.$$

Here n_i molecules have speed v_i and the total number of molecules is
$$N = \sum_i n_i.$$
Referring to Fig. 1.2, the relation between the three speeds is
$$\bar{v} \sim 0.9 [\overline{v^2}]^{1/2}$$
and
$$\hat{v} \sim 0.8 [\overline{v^2}]^{1/2}.$$
Throughout this chapter the symbol v is used for the r.m.s. speed and v^2 for the mean square speed. Also 'speed' and 'velocity' are interchangeable.

Returning now to the pressure equation, this may be written as
$$p = \tfrac{1}{3}\rho v^2 = \tfrac{1}{3}\frac{M}{V}v^2,$$
where M is the mass of gas and V is its volume. In Section 1.6 we shall see that if the temperature remains constant no change in the total kinetic energy ($\tfrac{1}{2}Mv^2$) of

the molecule occurs, and $\frac{1}{3}Mv^2 = pV$ is constant. This is Boyle's law, which states that for a given mass of gas pV is a constant at constant temperature. It has been deduced using only the principles of mechanics and certain plausible assumptions about the state of the molecules in a gas.

1.5 Molecular Velocities

A simple calculation using $p = \frac{1}{3}\rho v^2$ enables us to estimate the r.m.s. velocity of, say, hydrogen molecules at s.t.p. Using S.I. units, and remembering that one mole of hydrogen weighs 2·016 g and occupies 22·4 litre at s.t.p., we have

$$101\,325 = \frac{2{\cdot}016 \times 10^{-3} \times v^2}{3 \times 22{\cdot}4 \times 10^{-3}}$$

giving $v = 1838$ m s^{-1}.

Molecular velocities of some common gases are given in Table 1.1.

TABLE 1.1
Molecular Velocities of Some Gases at 0 °C

Gas	Relative molecular mass	v Root mean square velocity (km s^{-1})	$\bar{v} = 0{\cdot}92\,v$ Mean velocity (km s^{-1})
Hydrogen	2·0	1·84	1·69
Helium	4·0	1·31	1·21
Nitrogen	28·0	0·49	0·45
Oxygen	32·0	0·46	0·43
Argon	40·0	0·41	0·38
Krypton	83·7	0·29	0·26

1.6 Temperature of a Gas: Avogadro's Hypothesis

Heating a gas raises its temperature and increases the kinetic energy of the molecules. According to the model we are at present considering, heat energy can only appear as kinetic energy, since potential energy would require mutual forces between the molecules which have been assumed to be negligible. Moreover, when two or more different gases at the same temperature are mixed there is no flow of heat or energy transfer from one to the other. These facts lead us to suppose that temperature is proportional to the kinetic energy of the gas molecules. Furthermore it can be shown that, in a system consisting of different masses $m_1, m_2, m_3 \ldots$ having different velocities, $v_1, v_2, v_3 \ldots$, then $\frac{1}{2}m_1v_1^2 = \frac{1}{2}m_2v_2^2 = \frac{1}{2}m_3v_3^2 \ldots$ after sufficient time has elapsed for equilibrium to be established between the different types of molecules. It is the *average* kinetic energy of a single molecule of each gas which is constant. It seems reasonable therefore to assume that temperature T of a gas is proportional to the average kinetic energy $\frac{1}{2}mv^2$ of a single molecule.

Returning to the pressure equation, we can now write $\rho = mn$ and

$$p = \tfrac{1}{3}mnv^2 = \tfrac{2}{3}n(\tfrac{1}{2}mv^2) = \tfrac{2}{3}\frac{N}{V}(\tfrac{1}{2}mv^2),$$

where N is the number of molecules in an arbitrary volume V. This becomes

$pV = \frac{2}{3}N(\frac{1}{2}mv^2) = sT$, where T is the absolute temperature and s is a constant which takes a value according to the number of molecules N we specify for the system. When $N = N_A$, the Avogadro constant, $s = R$, the gas constant for a mole. Thus we have the well-known law, $pV_m = RT$, for 1 mol of an ideal gas, and substitution of appropriate data gives a value of R equal to $8\cdot31$ J mol^{-1} K^{-1}.

Taking two equal volumes V of different gases at the same temperature and pressure, we get $(pV)_1 = \frac{2}{3}N_1(\frac{1}{2}m_1v_1^2)$ for N_1 molecules of gas 1 in volume V and $(pV)_2 = \frac{2}{3}N_2(\frac{1}{2}m_2v_2^2)$ for N_2 molecules of gas 2 in the same volume V. The pressure is the same, $(pV)_1 = (pV)_2$, and also $\frac{1}{2}m_1v_1^2 = \frac{1}{2}m_2v_2^2$ from our definition of temperature, leaving $N_1 = N_2$. This is Avogadro's hypothesis, which states that equal volumes of different gases at the same temperature and pressure contain equal numbers of molecules. Since the molar volumes V_m of ideal gases are the same, we have $N_1 = N_2 = N_A$ for volume V_m, where, experimentally, $N_A \sim 6\cdot022 \times 10^{23}$ mol^{-1}. Note also that $pV_m = \frac{2}{3}N_A(\frac{1}{2}mv^2) = RT$ so that $E_m = \frac{3}{2}RT$.

1.7 Mean Free Path

We have seen that molecular velocities are about 1 km s^{-1} but it is known that gases diffuse into one another at normal pressure quite slowly. It must be supposed that the molecules are retarded by frequent encounters with other molecules so that their progress is random and irregular. It is also clear that, as the pressure is reduced, the concentration of molecules is smaller and fewer encounters will occur. The average distance travelled between successive encounters has proved a very useful concept in kinetic theory and is called 'mean free path'.

So far we have referred only to an ideal gas consisting of independent dimensionless mass points. In a real gas, however, although the molecules have mutual forces of attraction (see Section 1.12) their speeds are still subject to the distribution of Fig. 1.2. In addition, we must now accept the fact that real molecules have a small but finite size (usually of the order of $0\cdot1$ nm, i.e. 10^{-10} m). It is then possible to find an expression for the mean free path in terms of the molecular diameter.

Consider a molecule A (Fig. 1.5) which we shall assume to be spherical and of diameter d. It will collide with molecules B and C (both assumed stationary) and with any others which lie with their centres within a cylinder of diameter $2d$. In one second, molecule A travels a distance v and sweeps out a collision volume $\pi d^2 v$. The number of molecules it will encounter in this volume is just $\pi d^2 v n$, n

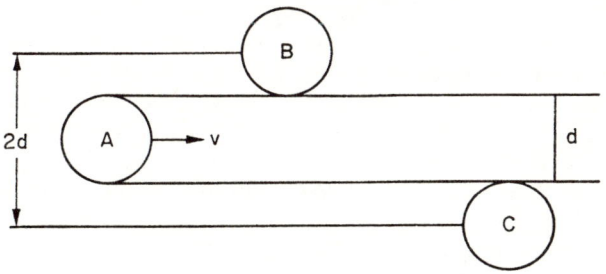

Fig. 1.5 Mean free path.

being the number of molecules in unit volume. This represents the number of collisions it will make in a distance v. The mean distance between two collisions, or the mean free path, L, is given by $v/(\pi d^2 vn) = 1/(\pi d^2 n)$. More rigorous analysis gives $L = 1/(\sqrt{2}\pi d^2 n)$. It will be seen that the mean free path is affected by both the size of the molecule and the density of the gas.

1.8 Thermal Conductivity and Viscosity

The mean free path of a gas can be used to calculate the coefficients of thermal conductivity, viscosity and diffusion. We shall restrict our attention to the first two of these. Consider three layers of a gas of density ρ separated by distances equal to the mean free path, L, and in which a temperature gradient $d\theta/dx$ has been established as shown (Fig. 1.6). Taking the shaded area as unity and using

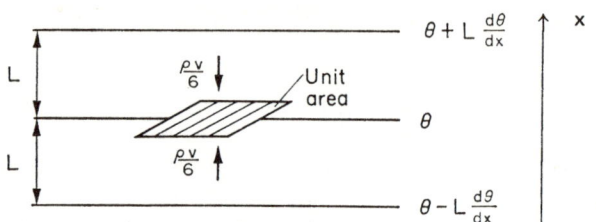

Fig. 1.6 Thermal conductivity.

the six-stream method of treatment, we see that a mass of gas equal to $\rho v/6$ passes across this area in each direction per second. The molecules from above have an average temperature of $\theta + L d\theta/dx$, so that the heat passing downwards per second per unit mass through the area is $(\theta + L\, d\theta/dx)c_V \rho v/6$, where c_V is the specific heat capacity of the gas at constant volume. In like manner, the amount of heat passing upwards per second through the same area by the motion of molecules from the layer at $(\theta - L d\theta/dx)$ is $(\theta - L d\theta/dx)c_V \rho v/6$. The net heat passing downwards per second is therefore equal to the difference, which is equivalent to the rate of transfer of thermal energy, i.e.

$$\frac{dQ}{dt} = c_V \rho \frac{v}{3} \cdot L \frac{d\theta}{dx}.$$

As the coefficient of thermal conductivity, K, is defined by the equation $dQ/dt = KA d\theta/dx$ and we are dealing with unit area, so that $A = 1$, comparison of these equations gives

$$K = \tfrac{1}{3}\rho v L c_V.$$

A very similar argument can be used to derive the coefficient of viscosity of a fluid, again in terms of mean free path. Suppose a velocity gradient du/dx is established in a fluid, and consider three parallel layers, separated by L, the mean free path, as shown in Fig. 1.7. The flow velocity u is very much smaller than the kinetic velocity v of the individual molecules and amounts only to a drift in the direction of flow. A mass of $\rho v/6$ passes each second upwards and downwards through a selected unit area as shown. The average drift momentum in the direction of flow for molecules passing downwards through the unit area is $(u + L du/dx)\rho v/6$; for those passing upwards it is $(u - L du/dx)\rho v/6$. The net momentum change per second associated with the unit area and in the direction of flow is $\frac{1}{3}\rho v du/dx$. From Newton's second law of motion this must represent the force acting upon unit area, so that $F = \frac{1}{3}\rho v L du/dx$. Since viscosity η is defined by the equation $F = \eta A du/dx$ and $A = 1$ in the case discussed, comparison of these equations gives $\eta = \frac{1}{3}\rho v L$. For spherical molecules the formula $\eta = \frac{1}{2}\rho v L$ is obtained by a more rigorous derivation. Combining the formulae for thermal conductivity and viscosity, we have $K = \eta c_V$. Experiment shows that $K = f\eta c_V$, where f is a numerical factor between 1 and 3 depending on the temperature and atomicity of the gas, instead of 1, as given by the simple theory.

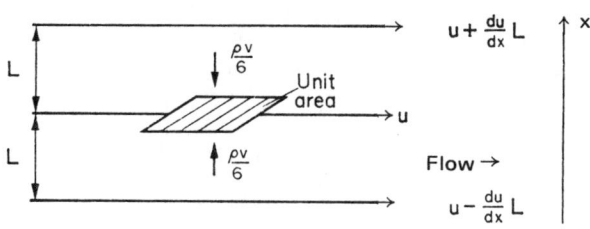

Fig. 1.7 Viscosity.

1.9 Specific Heat Capacities

The kinetic theory has also had quite spectacular success in predicting the ratios of the specific heat capacities of gases. When unit mass of a gas receives a quantity of heat δQ, and its temperature rises by δT, the specific heat capacity is defined by $c = \delta Q/\delta T$. In the case of a gas, the temperature rise depends very largely on the type of change to which the gas is subjected. Two principal specific heat capacities are usually defined, one, c_V, when the heat is added and the volume of the gas kept constant, the other, c_p when the heat is added but the gas is allowed to expand to keep the pressure constant. These two specific heat capacities will thus have different numerical values.

For our purposes the ratio $\gamma = c_p/c_V$ is important because, as we shall see, it is closely connected with the atomicity of a gas — that is, the number of atoms in each molecule of the gas. Thus the atomicities of hydrogen, carbon dioxide and ammonia are two, three and four respectively.

We shall now calculate γ in terms of the molecular energies of a gas. Imagine unit mass of gas enclosed in a cylinder by a movable piston. At constant volume the heat required to raise the temperature by δT is given by $\delta Q_1 = c_V \delta T$. If the gas is now allowed to expand by δV while the heat is being delivered to it, some heat will be needed to do external work $p\delta V$ in pushing the piston against the pressure p of the atmosphere. The specific heat capacity at constant pressure is therefore greater than that at constant volume and is given by $\delta Q_2 = c_p \delta T$ and we can write $\delta Q_2 = \delta Q_1 + p\delta V$. From the atomic point of view, δQ_1 must be regarded as energy supplied to the atoms and molecules of the gas. Suppose the translational energy of the molecules is increased by δE, and the energy of the atoms *within* the molecules is increased by δe, then $\delta Q_1 = \delta E + \delta e$. The ratio of specific heat

capacities now becomes
$$\gamma = \frac{c_p}{c_V} = \frac{\delta Q_2}{\delta Q_1} = \frac{\delta E + \delta e + p\delta V}{\delta E + \delta e}.$$

It has already been shown that $pV = \frac{1}{3}Mv^2$ and $pV = sT$ for an arbitrary gas of mass M and volume V. If the mass is unity and the volume is V_1, then also $pV_1 = rT$, where r is the gas constant per unit mass and $E = \frac{3}{2}rT$, giving $\delta E = \frac{3}{2}r\delta T$ per unit mass. The external work done is $p\delta V_1$, which is $r\delta T$ per unit mass. The ratio of the specific heat capacities may now be written

$$\gamma = \frac{\frac{3}{2}r\delta T + \delta e + r\delta T}{\frac{3}{2}r\delta T + \delta e},$$

where δe depends on how many atoms are contained in the molecule, representing non-translational kinetic energy. Also, from

$$\delta Q_2 = \delta Q_1 + p\delta V,$$

we have

$$\frac{\delta Q_2}{\delta T} = \frac{\delta Q_1}{\delta T} + p\frac{\delta V}{\delta T}$$

or

$$c_p = c_V + r \text{ per unit mass.}$$

For one mole this becomes

$$c_{p,m} = c_{V,m} + R,$$

where R is the gas constant per mole. Hence $R = A_r r$, where A_r is the relative atomic (or molecular) mass and $R = N_A k$, where k is the Boltzmann constant, which is therefore the gas constant for a single molecule of an ideal gas.

1.10 Atomicity

If for the present we limit our considerations to monatomic gases such as helium, neon, argon and krypton, then

$$\delta e = 0 \quad \text{and therefore} \quad \gamma = \frac{\frac{3}{2}r\delta T + r\delta T}{\frac{3}{2}r\delta T} = \frac{5}{3} = 1.67.$$

This is in excellent agreement with the experimental values for the rare gases, as shown for argon and helium in Table 1.2.

In order to predict the ratio of specific heat capacities for diatomic gases, a further principle has to be used, namely the theorem of the equipartition of energy due to Maxwell. The theorem states that the average kinetic energy associated with each degree of freedom of each single molecule is equal to $\frac{1}{2}kT$, k being the

TABLE 1.2
Ratio of Specific Heat Capacities, γ

Gas	α	γ (calc.)	γ (obs.)
Argon	3	1·667	1·666
Helium	3	1·667	1·666
Hydrogen	5	1·400	1·408
Oxygen	5	1·400	1·396
Methane	6	1·333	1·313
Ammonia	6	1·333	1·309
Ethylene	6	1·333	1·255

Boltzmann constant. Since there are N_A molecules in a mole, the total energy associated with each degree of freedom becomes $\frac{1}{2}N_A kT$. The significance of the equipartition of energy is readily appreciated by consideration of the molecules of a monatomic gas. To specify the exact position of such a molecule in space, three coordinates are required (x, y, z in Cartesian, r, θ, φ in polar coordinates). Its position cannot be defined exactly with fewer than three coordinates. We say therefore that it has three degrees of freedom or modes by which energy can be contained or absorbed, and with each of these we must associate energy equal to $\frac{1}{2}kT$.

The kinetic energy of a single gas molecule is $\frac{1}{2}mv^2$ for any direction. Along the axes the kinetic energy has the values $\frac{1}{2}mv_x^2$, $\frac{1}{2}mv_y^2$, $\frac{1}{2}mv_z^2$ since $v^2 = v_x^2 + v_y^2 + v_z^2$. Averaging over long periods of time at constant temperature for all the molecules of the gas, we get

$$\overline{\tfrac{1}{2}mv_x^2} = \overline{\tfrac{1}{2}mv_y^2} = \overline{\tfrac{1}{2}mv_z^2}.$$

Each component corresponds to one translational degree of freedom, carrying with it $\frac{1}{2}kT$ energy. The total translational energy of n molecules per unit *mass* is thus $\frac{3}{2}nkT$ or $\frac{3}{2}rT$, as we have already seen.

Consider now a diatomic molecule, which we shall picture as a dumb-bell of length d (Fig. 1.8). It might at first be supposed that six coordinates are needed to specify its position, but when we remember that

$$d^2 = (x_2 - x_1)^2 + (y_2 - y_1)^2 + (z_2 - z_1)^2$$

it will be realized that only five are essential, the sixth being calculated from this

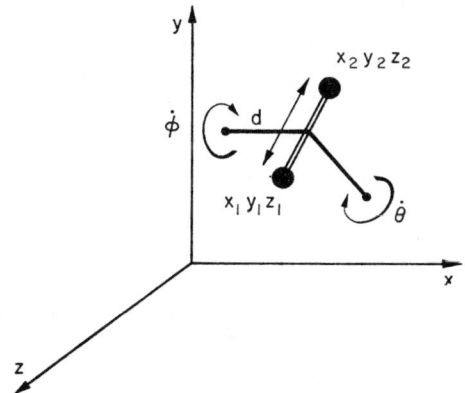

Fig. 1.8 Degrees of freedom of diatomic molecule.

equation. We say therefore that a diatomic molecule has five degrees of freedom, and that we must associate kinetic energy equal to $\frac{5}{2}kT$ with each molecule. This simply means that for unit mass $\delta E = \frac{3}{2}nk\delta T = \frac{3}{2}r\delta T$ and $\delta e = \frac{2}{2}nk\delta T = \frac{2}{2}r\delta T$, so that the ratio of specific heat capacities becomes

$$\gamma = \frac{\frac{3}{2}r\delta T + \frac{2}{2}r\delta T + r\delta T}{\frac{3}{2}r\delta T + \frac{2}{2}r\delta T} = \frac{7}{5} = 1\cdot 4,$$

which again is in excellent agreement with the experimental values for hydrogen and oxygen given in Table 1.2. This may be interpreted in terms of additional kinetic energies, $\tfrac{1}{2}I_\theta \dot\theta^2$ and $\tfrac{1}{2}I_\phi \dot\varphi^2$, arising from rotation of each molecule about two axes mutually perpendicular to the molecular axis d, as shown in Fig. 1.8. These energies are each on average equal to $\tfrac{1}{2}kT$, so that for unit mass $e = \tfrac{1}{2}nI_\theta \dot\theta^2 + \tfrac{1}{2}nI_\phi \dot\varphi^2$, where I_θ, I_ϕ and $\dot\theta$, $\dot\varphi$ are the moments of inertia ana angular velocities about the axes. The energy of rotation about the molecular axis is assumed to be negligible. Moreover, since it is a rigid molecule, no vibrational energies need be associated with this model.

In general, for an atom with α degrees of freedom, the ratio of the specific heat capacities is $\gamma = 1 + 2/\alpha$. As α increases with the number of atoms in the molecule, γ decreases to a theoretical minimum of 1·333. Measurements of γ therefore provide information about the atomicity of a molecule. The observed values of γ are given in Table 1.2, which also shows that the experimental values deviate further from the theoretical values as α increases.

The deviations of the observed values of γ from the theoretical values show that the monatomic and diatomic gases can be treated as 'ideal' gases but that the polyatomic gases and vapours cannot. This indicates the inadequacy of the simple kinetic theory for all but the most simple gases.

1.11 Molar Heat Capacities

Conspicuous success has been achieved by the kinetic theory in predicting the molar heat capacities of gases and even solids. The molar heat capacity is the heat required to raise the temperature of one mole by one kelvin. A very simple calculation enables these to be obtained for both monatomic and diatomic gases. As a monatomic gas has three degrees of freedom, the total molar energy is $\tfrac{3}{2}N_A kT = \tfrac{3}{2}RT$. Recalling that $R = 8\cdot31$ J mol^{-1}, the molar heat capacity becomes $\tfrac{3}{2}R = \tfrac{3}{2} \times 8\cdot31 = 12\cdot46$ J K^{-1}. This is the same as the experimental value of 12·46 J K^{-1} for argon. With a diatomic gas having five degrees of freedom, $E = \tfrac{5}{2}RT$, and the molar heat capacity is $\tfrac{5}{2} \times 8\cdot31 = 20\cdot78$ J K^{-1}. The experimental value for hydrogen is 20·2 J K^{-1}, and again agreement is remarkably good. It is even more surprising to find that this argument can be extended to solids, in which the atoms are packed much more closely together so that each atom is permanently affected by forces due to its neighbours, and its energy has become a function of its position. We must now consider the atoms to have potential energy as well as kinetic energy. The atoms are able to vibrate only about some mean position. They have, therefore, three degrees of freedom with which we must associate kinetic energy equal to $\tfrac{3}{2}RT$. Furthermore, in a vibrating system the average kinetic energy is equal to the average potential energy, so that the total energy is $\tfrac{3}{2}RT + \tfrac{3}{2}RT = 3RT$. The molar heat capacity is then $3R = 24\cdot93$ J K^{-1}. This is, of course, the law of Dulong and Petit.

The above considerations enable us to estimate a value for the Boltzmann constant k, which is given by

$$k = \frac{R}{N_A} = \frac{8\cdot32}{6\cdot02 \times 10^{23}} = 1\cdot38 \times 10^{-23} \text{ J K}^{-1}.$$

1.12 Van der Waals' Equation

The success of the kinetic theory in predicting the gas laws and specific heat capacities leads one to enquire under what conditions it will begin to break down.

Recalling the initial assumptions with regard to the size of the molecules and their lack of mutual attraction, we should expect Boyle's law to break down when the molecules are forced very closely together. This occurs when normal gases are subjected to high pressures, and under such conditions we might expect to find deviations from Boyle's law. Experimental investigations of Boyle's law at very high densities were undertaken for various gases by Regnault, Amagat and Andrews, who showed that the behaviour of a gas could be represented empirically by an equation of the form
$$pV = A + Bp + Cp^2 + \ldots,$$
where B and C are small constants with approximate values of $B = 10^{-3} A$ and $C = 10^{-6} A$. It is clear that the normal ideal gas law is a very close approximation to the true behaviour of a real gas.

Van der Waals introduced corrections into the ideal gas equation to allow for mutual attraction and finite size of molecules. Figure 1.9 shows a given mass of gas molecules condensed as a solid in one corner of the volume V, the whole of which they would normally occupy in the gaseous state. The sizes of the

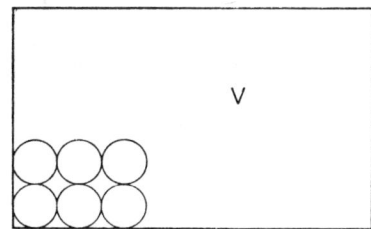

Fig. 1.9 Finite size of molecules.

molecules, which are very greatly exaggerated with respect to the volume, imply that the effective volume in which the molecules move is slightly less than the measured volume. The ideal gas equation becomes $p(V_m - b) = RT$, where V_m is the measured volume of the gas and b is a constant which depends on the volume occupied by the molecules. Precise calculation shows that b is four times the actual volume of the molecules, so that for spherical particles of diameter d the molar value of b is given by
$$b = 4N_A \frac{4\pi}{3} \left(\frac{d}{2}\right)^3 = \frac{2\pi}{3} N_A d^3.$$

Mutual attraction between the molecules leads to a reduction in the pressure exerted by the gas on the boundary walls. A molecule such as A (Fig. 1.10) in the body of the gas experiences forces in all directions, so that on the whole the molecule is not attracted preferentially in any direction. When, however, a molecule approaches the boundary, this balance is no longer preserved, and the molecule is subjected to a retarding force, due to the following molecules, which reduces its momentum. The momentum change on collision with the boundary is less than when this retarding force is ignored, and the pressure exerted is also less. The true molecular pressure is therefore greater than the observed pressure and a small correction term must therefore be added. Detailed calculation shows that this is of the form a/V_m^2, where a is a small constant. Therefore, for 1 mol, van der Waals' equation is

$$\left(p + \frac{a}{V_m^2}\right)(V_m - b) = RT.$$

This equation successfully predicts the general behaviour of many gases over a wide range of conditions, especially under high pressure and near the point of liquefaction. Other equations can be made to fit more accurately over a restricted

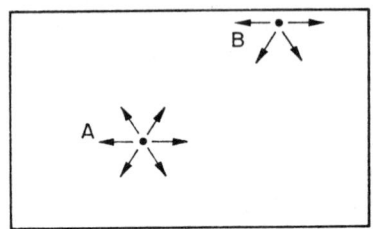

Fig. 1.10 Mutual attraction of molecules.

range of conditions, but none has achieved the general usefulness of van der Waals' equation. The equation may be expanded and rewritten as

$$pV_m = RT + p\left(b - \frac{a}{RT}\right) + p^2 \frac{ab}{R^2 T^2},$$

where it is permissible to use $pV_m = RT$ as a first approximation. Comparing this with the equation determined from experiment, it will be seen that $A = RT$, $B = b - a/RT$ and $C = ab/R^2 T^2$. It will be realized that at a certain temperature T_B the constant $B = 0$, so that the gas obeys Boyle's law very closely indeed. This is known as the Boyle temperature and can be calculated from $T_B = a/Rb$.

One point requires elaboration. We have allowed for mutual attraction of molecules, and yet we depend upon repulsion from the walls of the containing vessel to account for pressure. It is therefore essential to explain how attraction can occur at one time and repulsion at another. From a study of atomic structure, we know that strong repulsion will occur as the outer electron shells of two atoms overlap each other. The force of repulsion is given by β/r^n, where r is the distance between the centres of the two atoms and β is a constant. At the same time, the molecules polarize each other so that each becomes a weak electric dipole. Interaction between such electric dipoles is responsible for the so-called van der Waals' force of attraction, which takes the form α/r^m, α being another constant. The net force of attraction then becomes

$$\frac{\alpha}{r^m} - \frac{\beta}{r^n} \quad \text{where } n \gg m.$$

This expression indicates that for small values of r (on the atomic scale) the repulsive force predominates, whereas for large r the attractive force predominates and at the equilibrium distance the net force is zero. The details of this theory have been worked out by Lennard–Jones, who showed that for many simple molecules $n = 13$ and $m = 7$ in the above expression. These indices refer to

the net force equation
$$F = \frac{\alpha}{r^m} - \frac{\beta}{r^n},$$
and the resulting potential energy diagram is as shown in Fig. 14.3.

1.13 Molecular Sizes

Kinetic theory provides us with valuable indications of the average sizes of atoms and molecules. The mean free path may be calculated from measurements of viscosity or thermal conductivity. In the case of oxygen, the viscosity $\eta = 21 \times 10^{-6}$ N s m^{-2}, the density $\rho = 32 \div 22 \cdot 4$ kg/m^{-3} and the root mean square velocity at 0 °C is 460 m/s^{-1}. The mean free path is given by

$$L = \frac{3\eta}{\rho v} = \frac{3 \times 21 \times 10^{-6}}{(32 \div 22 \cdot 4) \times 460}$$
$$= 96 \times 10^{-9} \text{ m (96 nm)}.$$

Confirmation of this is obtained from thermal conductivity measurements on oxygen gas. The thermal conductivity $K = 23 \cdot 4 \times 10^{-3}$ J m^{-1} s^{-1} K^{-1}, and the specific heat capacity per kilogram $c_V = 650$ J kg^{-1} K^{-1}, so that the mean free path is

$$L = \frac{3K}{\rho v c_V} = \frac{3 \times 23 \cdot 4 \times 10^{-3}}{(32 \div 22 \cdot 4) \times 460 \times 650}$$
$$= 164 \times 10^{-9} \text{ m (164 nm)}.$$

These calculations show that the mean free path may be taken on average as 130 nm. A value for $d^2 n$ may now be obtained using the average of the above two values, so that

$$L = 130 \times 10^{-9} = \frac{1}{\pi d^2 n}$$

therefore

$$d^2 n = \frac{10^9}{\pi \cdot 130} = 2 \cdot 5 \times 10^6 \text{ m}^2.$$

In this formula n is the number of molecules in 1 m^3. Remembering that a mole of any gas occupies 22·4 litres, we may write
$$d^2 N_A = 2 \cdot 5 \times 10^6 \times 22 \cdot 4 \times 10^{-3}$$
$$= 5 \cdot 6 \times 10^4 \text{ m}^2.$$

In order to determine N_A and d separately, a further relationship between them is required. Such a relation is provided by the molecular volume in the solid or liquid state. The value will depend upon the closeness with which the molecules are packed together, but it will suffice here to indicate the order of magnitude involved. If we assume that each molecule occupies a unit cube of side d, then the volume of a mole is given by $d^3 N_A$. As the relative density of liquid oxygen is 1·14, a mole of liquid oxygen will occupy 28×10^{-6} m^3, and assuming the molecules are all touching we have
$$d^3 N_A = 28 \times 10^{-6} \text{ m}^3.$$

Supporting evidence for this value is available from the constant b in van der Waals' equation. For a mole of gaseous oxygen, $b = 31 \cdot 6 \times 10^{-6}$ m^3, which is four times the volume of the molecules. Thus, in 22·4 litres or 0·0224 m^3 of oxygen at s.t.p., the actual volume occupied by the molecules is only about 8×10^{-6} m^3. Assuming the molecules to be spherical we can write

$$\frac{4\pi}{3}\left(\frac{d}{2}\right)^3 N_A = 8 \times 10^{-6}$$

and thus

$$d^3 N_A = 15 \times 10^{-6} \text{ m}^3.$$

This is reasonably good confirmation of the first value, so that when a mean value is adopted we have the second equation $d^3 N_A = 22 \times 10^{-6}$. Comparing this with $d^2 N_A = 5\cdot6 \times 10^4$ obtained earlier, we can calculate the molecular diameter $d = 0\cdot4 \times 10^{-9}$ m or 0·4 nm and $N_A = 3\cdot5 \times 10^{23}$ mol^{-1}. These are simply indications of the magnitudes of d and N_A. Other methods based upon Brownian motion, surface tension and latent heat give similar values for N_A and d. The best estimate of the Avogadro constant is $N_A = 6\cdot022\,045 \times 10^{23}$ mol^{-1}.

1.14 Summary

In this chapter we have seen how the atomic and molecular hypotheses have led to the kinetic theory of gases, which in turn explains a wide range of phenomena, especially in the field of heat. In particular, it has shown that heat itself is a manifestation of kinetic energy of molecules. In the final section, we see how estimates of molecular sizes and the number of atoms and molecules in a given volume can be made from measurements of physical quantities.

Problems

(Those problems marked with an asterisk are solved in full at the end of the section.)

1.1 Calculate the root mean square velocity of nitrogen molecules at s.t.p. The atomic mass number of nitrogen is 14 and the molecule is diatomic. ($4\cdot9 \times 10^2$ m s^{-1})

1.2* Calculate the root mean square velocity of helium molecules at a temperature of $-50\,°$C. The atomic mass number of helium is 4 and it is monatomic. ($11\cdot7 \times 10^2$ m s^{-1})

1.3 The lowest pressure which can readily be attained with a diffusion pump backed by a rotary pump is 10^{-6} mm of mercury. Calculate the number of molecules per cubic millimetre still remaining. For a gas at s.t.p. every cubic millimetre contains $2\cdot7 \times 10^{16}$ molecules. Consider how temperature changes might affect the value you have obtained. ($3\cdot55 \times 10^7$ molecules)

1.4 The equation of a real gas may be written in the form $pV = A + Bp + Cp^2$. Evaluate the constants A, B and C in terms of the constants of van der Waals' equation. Find also the temperature for which the gas most closely obeys Boyle's law. ($A = RT$, $B = b - a/RT$, $C = ab/R^2 T^2$ and $T_B = a/Rb$)

1.5 Define mean free path and thermal conductivity and establish a relationship between these quantities for a gas. The mean free path for oxygen at s.t.p. is 10^{-7} m. Calculate the effective diameter of the oxygen molecule and the number of impacts per second. (0·35 nm, $4\cdot4 \times 10^9$)

1.6 Dulong and Petit's law states that the product of relative atomic mass and specific heat capacity of an element is constant and equal to about 25 J K^{-1}. Show to what extent this can be justified using kinetic theory, explaining carefully the assumptions you make and using mechanical units.

1.7 To what temperature would argon have to be raised so that the molecules would have same root mean square velocity as nitrogen molecules at 0 °C? Argon is monatomic and its atomic mass number is 40. (117 °C)

1.8 Derive an expression for viscosity in terms of mean free path, and show that for a given gas the viscosity is directly proportional to the square root of the absolute temperature.

1.9 Establish a formula for thermal conductivity in terms of mean free path and show that for a given gas the thermal conductivity is directly proportional to the square root of the absolute temperature.

1.10 If the viscosity of oxygen is 20.9×10^{-6} N s m^{-2}, calculate the mean free path of oxygen molecules (a) at s.t.p.; (b) at 0.76 mm of mercury pressure. (95 nm, 95 μm)

1.11* The distance between the electrodes of a discharge tube is 250 mm. Calculate the pressure at which the Crookes dark space will just reach the anode, assuming the tube to be initially filled with air and the mean diameter of the molecules to be 0.1 nm. (3.6×10^{-3} mmHg)

1.12 If the relative density of lead is 11.3 and its atomic mass number 207, calculate the number of atoms in 1000 mm³. (5.5×10^{21})

1.13 Calculate the Avogadro constant and the diameter of a helium atom, given that the viscosity at s.t.p. is 1.9×10^{-5} N s m^{-2} and the relative density of liquid helium is 0.15. (3.5×10^{24}, 92 pm)

1.14 If the relative density of liquid argon is 1.4, compare the separation of the atoms in gaseous argon at s.t.p. with diameter of an argon atom. (9.3)

1.15 Calculate the specific heat capacity for 1 g of argon and, using the value of 1.75×10^{-2} J s^{-1} m^{-1} K^{-1} for the thermal conductivity, estimate the mean free path at s.t.p. (2.3×10^{-4} mm ≡ 0.23 μm)

1.16 Compare the diameters of the xenon and helium atoms if their viscosities are equal and their atomic mass numbers are respectively 131 and 4. (Xenon diameter = 2.4 × helium diameter)

1.17 2.8 mg of oleic acid dropped on to a clean water surface form a circular monomolecular layer 1 m in diameter. Estimate the size of a molecule and the Avogadro constant. The oleic acid molecule is about ten times longer than its width or depth, has a molecular mass of 282 and a relative density of 1.0. (3.5×10^{-7} mm ≡ 0.35 nm, 6.5×10^{23})

1.18 From the data given in Section 1.13, compute the value of the constant f in $K = f\eta c_V$.

1.19 The value of the constant f in the equation $K = f\eta c_V$ is dependent on γ by the equation $f = A\gamma + B$, where A and B are constants. Obtain values of K, η, γ and c_V for various gases from a book of tables and determine the corresponding values of f. Hence find the constants A and B by a graphical method. How does f vary with the molecular weight of the gas? ($A = 2.25$, $B = -1.25$)

1.20 The following are the densities of various gases and vapours in kilograms per cubic metre at s.t.p.:

Hydrogen	0.089
Helium	0.178
Argon	1.78

17

Nitrogen	1·25
Carbon dioxide	1·98
Ammonia	0·771
Methane	0·716
Krypton	3·74
Sulphur dioxide	2·93
Hydrogen sulphide	1·54
Methyl ether	2·18

From your knowledge of their atomic and molecular masses, calculate the molar volume in each case and compare it with the ideal gas value of 0·0224 m³.

Solutions to Problems

1.2 At 101 325 Pa and 0 °C, 4 g of helium occupy 0·0224 m³. The velocity at 0 °C is then given by

$$v_0 = \sqrt{\frac{3p}{\rho_0}} = \sqrt{\frac{3 \times 101\,325}{4 \times 10^{-3} \div 0\cdot 0224}}$$
$$= 1305 \text{ m s}^{-1}$$

Remembering that $v \propto \sqrt{T}$,

$$\frac{v_{-50}}{v_0} = \sqrt{\frac{273-50}{273}},$$

$v_{-50} = 1305\sqrt{0\cdot 82} = 1182$ m s^{-1}.

1.11 The Crookes dark space will just reach the anode when the mean free path becomes equal to the separation of the electrodes. We have therefore to calculate the pressure corresponding to a mean free path of 250 mm in air. At the given pressure the number of molecules per cubic metre n is given by

$$L = \frac{1}{\pi d^2 n},$$

$$0\cdot 25 = \frac{1}{\pi 10^{-20} n},$$

$$n = 1\cdot 27 \times 10^{20}.$$

At normal pressure $N_A = 6\cdot 02 \times 10^{23}$ for 22·4 litres and in 1 m³

$$n' = \frac{6\cdot 02 \times 10^{23}}{0\cdot 0224}$$
$$= 2\cdot 69 \times 10^{25};$$

the required pressure is therefore

$$\frac{n}{n'} = \frac{1\cdot 27 \times 10^{20}}{2\cdot 69 \times 10^{25}}$$
$$= 4\cdot 72 \times 10^{-6} \text{ atm}$$
$$= 0\cdot 478 \text{ Pa}$$

Chapter 2

The Electron

2.1 Electrical Conduction in Solutions

When two copper plates, the electrodes, are placed in a solution of copper sulphate, the electrolyte, and connected to a battery to provide a potential difference across the solution, a current flows in the electrolyte and copper is deposited upon the cathode (Fig. 2.1). This is the process of electrolysis, which takes place whenever electricity passes through a solution of an inorganic salt in water. In this case copper passes into solution at the anode. Various chemical

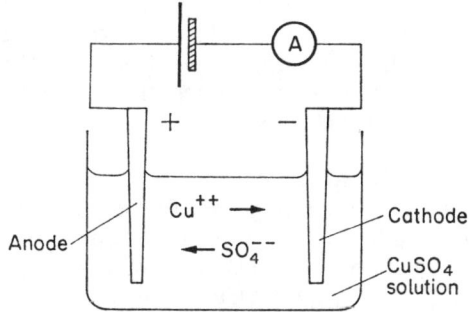

Fig. 2.1 Conduction of electricity through a solution.

actions take place at the electrodes in the electrolytic cell, according to the nature of the electrolyte and the materials employed as electrodes. These reactions are always due to oxidation or reduction of the electrode material and the electrolyte.

Faraday discovered the two laws governing the amount of material deposited during electrolysis. The first law states that the mass m in grams of any substance liberated from the solution is proportional to the quantity of electricity Q coulombs which has passed through the solution. This may be expressed as $m = \varepsilon It$, where ε is a constant called the electrochemical equivalent, measured in kilograms per coulomb, I is the current passed, in amperes, and t is the time for which the current flows, in seconds. Faraday's second law states that the masses of substances liberated by the same quantity of electricity are proportional to their chemical equivalents E, i.e. $m_1/m_2 = E_1/E_2$ for Q coulombs, and for molar quantities $M_1/M_2 = E_1/E_2$ for Q_m coulombs.

This is equivalent to the statement that 1 mol of any monovalent ion is liberated by the same quantity of electricity during electrolysis. This particular quantity is called the Faraday (constant) F, where $F = 9{\cdot}648\,456 \times 10^4$ C mol^{-1}. Faraday's laws are then summarized in the equation

$$n = \frac{It}{Fz}$$

where n is the number of moles liberated, I is the current, in amperes, passing for t seconds, $F = 96\,485$ C and Z is the valence of the ion liberated, i.e. the charge number of the ion.

Even simpler is the statement that it takes a quantity of electricity of 96 485 C to liberate 1 mol of an ion of valence z. By definition, 1 mol contains N_A ions each of charge ze, hence it follows that, for electrons and all monovalent ions, the total charge $N_A e$ is associated with 1 Faraday, i.e.

$$F = N_A e$$

These laws are only applicable to those chemical changes that involve electron transfer, and their importance is that they have been subjected to stringent experimental tests and have so far been found to be exact. The Faraday constant is therefore the same for all ions. It may simply be defined as the charge of 1 mol of electrons.

It is now known that conduction through liquids arises from the presence of ions which are atoms or groups of atoms carrying either a positive or a negative charge of electricity. The current is carried by the movement of ions across the solution under the action of the electric field, as shown in Fig. 2.1. In the case of copper sulphate solution, there are copper ions which carry two positive charges, Cu^{2+}, and sulphate ions which have two negative charges, SO_4^{2-}. In solution copper sulphate becomes ionized thus:

$$CuSO_4 \rightarrow Cu^{2+} + SO_4^{2-}.$$

The Cu^{2+} ions are attracted to the cathode and, on giving up their positive charge, are deposited as metallic copper. The SO_4^{2-} ions, on arriving at the anode, are responsible for a secondary action in which copper from the electrode goes into solution and the anode receives the negative charge

$$SO_4^{2-} + Cu \rightarrow CuSO_4 + 2\ominus.$$

If the electrode is made of material which does not combine with the sulphate radical (e.g. a platinum electrode) then oxygen will be liberated instead according to the equation

$$2SO_4^{2-} + 2H_2O \rightarrow 2H_2SO_4 + O_2\uparrow + 2\ominus.$$

Thus positive and negative electric charges are able to pass across the solution and constitute an electric current. This causes electrolysis of the solution.

2.2 Conduction in Gases

At normal atmospheric pressure air is almost an insulator. An electric field of the order of 3 MV m^{-1} is required to make it conduct electricity. In a discharge tube about 1 m long containing air at a pressure of a few millimetres of mercury, and having a side limb containing charcoal cooled with liquid air, the pressure falls continuously to a very low value as the air is steadily absorbed by the charcoal. If at the same time a potential difference of about 10 kV is applied to the ends of the tube, pink streamers are observed between the electrodes. As the pressure falls, these streamers merge into a continuous band filling the tube. At pressures of about 1 mm of mercury it is possible to distinguish several characteristic regions in the discharge. These are indicated and labelled in Fig. 2.2. As the pressure falls further, the striations spread out and move along the tube, disappearing at the anode. Eventually, at a pressure of about 1 μm of

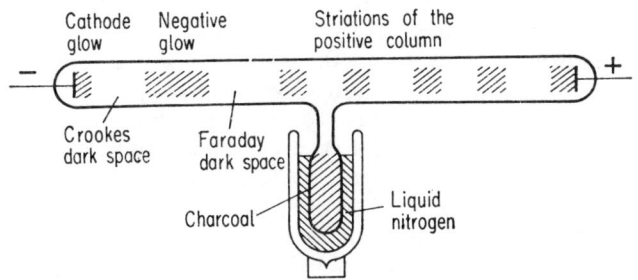

Fig. 2.2   Conduction of electricity through a gas.

mercury, the Crookes dark space (Fig. 2.2) extends the whole length of the tube, leaving only a slight glow on the cathode which is known as the cathode glow.

The mechanism of the discharge is complex, and, in order to get a qualitative appreciation of the process, some knowledge of atomic structure must be anticipated. Some ions are always present in the gas in the tube due to radioactivity, cosmic rays, etc. We now know that such ions consist of negative electrons and positive gaseous ions. Electrons, being much lighter than positive ions, accelerate quickly to the anode, leaving an excess of the less mobile positive ions in the vicinity of the cathode. This concentration of positive ions causes a rapid fall of potential at the cathode. As the accelerated positive ions strike the cathode, more electrons are liberated. These are accelerated in the cathode fall of potential, and each time one encounters a gas molecule much of its energy is transferred to the molecule. The gas molecule is then said to be excited, and on returning to its normal unexcited state loses energy by radiation which is seen as the negative glow. The spectrum of this radiation is characteristic of the gas in the tube. The length of the Crookes dark space is dependent on the mean free path of the electron. After passing through the negative glow the electron is retarded, so that its energy is no longer sufficient to excite the gas molecules, thus giving rise to the Faraday dark space. By the time it has traversed the Faraday dark space the electron has again acquired sufficient energy to excite the gas molecules at the beginning of the positive column.

The striations in the positive column can be attributed to the delaying action of the ionizing process, making it necessary for the electrons to travel a further distance before they again acquire enough energy to excite the gas molecules. This is why the spacing of the striations and the lengths of the Crookes and Faraday dark spaces increase as the pressure falls and the mean free path becomes longer. This state of affairs continues until the pressure is so low that an electron can travel the whole length of the tube without an encounter, which is equivalent to saying that the mean free path is just greater than the separation of the electrodes, and the Crookes dark space fills the whole tube. At this stage a small concentration of positive ions maintains the potential gradient near the cathode. This provides the electrons with most of their energy, after which they travel in straight lines normal to the cathode, and remaining electric fields in the tube have very little influence on them.

2.3 Properties of Cathode Rays

Before the discovery of the electron, these streams of electrons were referred to as 'cathode rays'. The properties of these so-called cathode rays had been studied in the latter half of the nineteenth century, especially in the elegant demonstrations of Crookes and Lenard. Many minerals and glass fluoresce with a characteristic colour when placed in a beam of cathode rays. That these rays travel in straight lines normal to the cathode is shown by placing an object, often in the shape of a Maltese cross, in the path of the cathode rays. The shadow of the cross can be seen on the end of the tube. They also carry energy which can be converted into heat by directing them on to a thin platinum foil which quickly becomes red or even white hot. Cathode rays are also deflected by electric and magnetic fields in a way which clearly indicates that they carry a negative charge. What is perhaps most important of all is that these cathode rays are independent of the material used for the electrodes and the gas filling the discharge tube. They appear to be a common constituent of all matter. They are also able to blacken a photographic plate and to pass through thin sheets of metal.

2.4 Thomson's Method for Measuring Charge per Unit Mass (e/m)

On reviewing the properties of cathode rays in 1897, J. J. Thomson proposed the hypothesis that cathode rays were streams of negatively charged particles produced at the cathode itself and moving with high velocity. He devised an experiment by which the ratio of charge to mass of such particles could be determined. Figure 2.3(a) shows a vacuum tube in which cathode rays streaming from the cathode C fall upon the anode A. In A there is a small hole, so that a pencil of cathode rays passes on to D containing a similar small hole. The narrow pencil of cathode rays can be deflected in the vertical plane by an electric field E between the parallel plates, as shown. It can also be deflected in the same plane by a magnetic field B perpendicular to the paper. The point at which the cathode rays impinge upon the screen S is shown by a fluorescent spot of light, since the screen is coated internally with a fluorescent material such as zinc sulphide. In the analysis which follows, it will be assumed that the electric and magnetic fields are confined to the space between the parallel plates. Although this cannot be achieved in practice, the assumption enables an appreciation of the method to be gained without elaborate mathematics.

Suppose the magnitude of the electric field E and the flux density of the magnetic field B are adjusted so that the pencil of cathode rays is not deflected but falls on S. The force on a particle due to the electric field must be balanced by the force due to the magnetic field, or $Ee = Bev$, where e is charged on the particle and v is its velocity. The velocity is then given by $v = E/B$, the ratio of the electric and magnetic fields. The velocities were found to be very high, up to 10^7 m s^{-1},

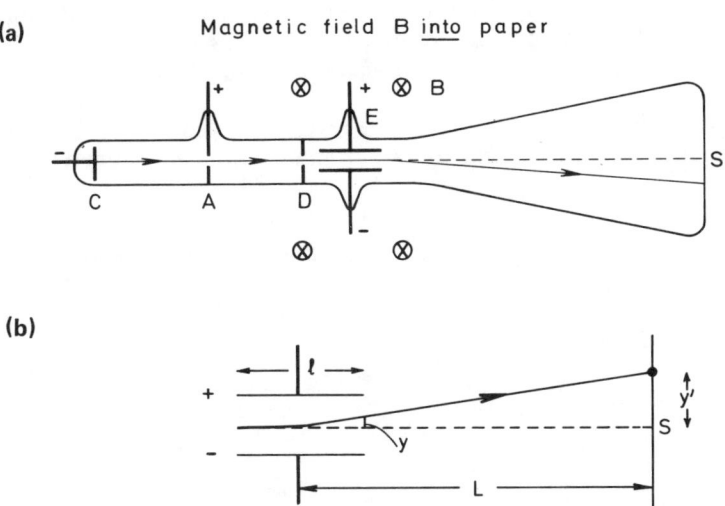

Fig. 2.3 J. J. Thomson's method for measuring e/m for the electron. (a) Deflection with both fields on; (b) deflection by the electric field only.

depending on the potential difference V across the tube, since $Ve = \frac{1}{2}mv^2$.

In the electric field alone (Fig. 2.3(b)) the particle suffers deflection due to the acceleration it receives perpendicular to its direction of motion. Using Newton's second law, $Ee = m\ddot{y} = 2my/t^2$, where $\ddot{y}$ is the acceleration along the y axis, m the mass of the particle, y the vertical displacement when it leaves the plates and t the time it takes to traverse the electric field. Therefore $e/m = 2y/Et^2 = 2yv^2/El^2$, where l is the distance travelled in the electric field. Substitution for the velocity gives $e/m = 2yE/l^2B^2$.

The deflection y of the particle in the uniform electric field may be supposed equivalent to a sudden instantaneous deflection at the centre of the field (Fig. 2.3(b)). It is therefore possible to find y from y' using similar triangles such that $y/\frac{1}{2}l = y'/L$. Hence $e/m = Ey'/B^2Ll$. Note that the length of the deflection plates must be known.

In this experiment Thomson assumed that all particles had the same e/m and the same velocity. This and many other experiments showed that e/m is the same for cathode rays generated in various gases and with several different metals as cathodes, thus confirming Thomson's hypothesis that cathode rays are negatively charged particles common to a large number of elements. It will be apparent too that the apparatus used has the basic features of the oscilloscope so widely used today.

2.5 Dunnington's Method for e/m

Thomson's method provided vital information about e/m at an early stage in the story of the electron, but the method had two main sources of error: the smallness of the deflection and the lack of uniformity in the velocity of the cathode ray particles. These errors have been reduced in later methods, yielding precise values for e/m, the best modern value beint 1.7588×10^{11} C kg^{-1}. A method devised by Dunnington in 1933, which had an error of only 1 in 4500, will be described.

Electrons emitted by the filament F (Fig. 2.4) are accelerated by a potential difference between F and A. Due to the action of a uniform magnetic field of flux

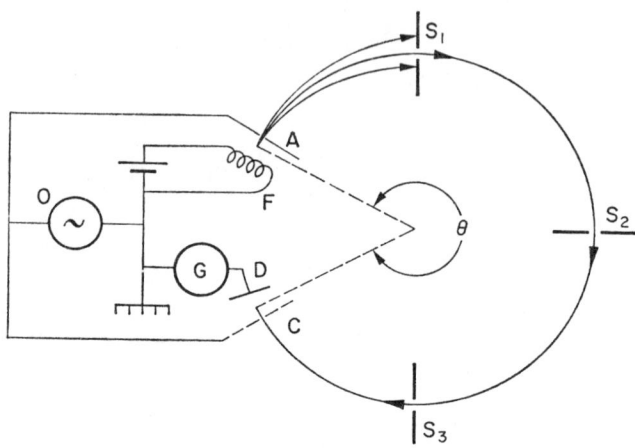

Fig. 2.4 Dunnington's method for the determination of e/m for electron.

density B applied perpendicular to the plane of the paper, the electrons move in a circular path of radius r defined by the slits S_1, S_2 and S_3. Applying Newton's second law of motion to an electron of charge e and mass m moving in a circular orbit of radius r, we get

$$Bev = m\frac{v^2}{r}.$$

Since r is constant, the slits constitute a velocity selector so that only electrons having a velocity given by

$$v = rB\frac{e}{m}$$

can pass through all three slits to C and D. Equal potential differences in phase are applied between A and F and also between C and D by the oscillator O, the frequency of which is controlled by a quartz crystal. C is a grid and D is a Faraday cylinder connected through a galvanometer G to the oscillator O. When A is at a potential $+V$ with respect to F, the electrons leave A with kinetic energy given by $\frac{1}{2}mv^2 = Ve$. These electrons arrive at C with the same kinetic energy. If the period of oscillation T is equal to the time taken for the electron to travel around the circular arc from A to C, the electron will experience a retarding potential between C and D such that will be just unable to reach D, and therefore no current will be recorded by the galvanometer G. This will also happen whenever the time of transit of the electron is equal to nT, where n is an integer. Thus the electron velocity v is given $v = \theta r/nT = \theta rv/n$, where v is the frequency of the oscillator and θ is the angular distance travelled by the electron. The experiment consists in adjusting the value of the magnetic field B until minimum current is recorded by the galvanometer G, in which case the velocity is given by the expressions

$$v = \frac{\theta rv}{n} = rB\frac{e}{m},$$

from which we obtain

$$\frac{e}{m} = \frac{\theta v}{nB},$$

in terms of measurable quantities.

2.6 Charge on the Electron

It has already been shown that the Faraday constant F is given by $F = N_A e$. Early estimates of N_A from observations of Brownian movement enabled values of the electronic charge to be found. Using modern values for these constants, we obtain

$$e = F \div N_A = 9·648\,456 \times 10^4 \div 6·022\,045 \times 10^{23}$$
$$= 1·602\,189 \times 10^{19}\ \text{C}.$$

One of the early estimates of e was due to Planck, but this is not generally recognized. Using the experimental results for black-body radiation, Planck developed his quantum theory (see p. 70) and one of the numerical results he got in the year 1900 was $e = 1·55 \times 10^{-19}$ C.

Direct measurement of e, first attempted by Thomson and others using the Wilson cloud chamber (p. 240), gave values within the range $(1-10) \times 10^{-19}$ C based on measurements on charged droplets. These results were not very consistent. Their methods were refined and developed by Millikan in 1911, who not only proved that each droplet in his experiment carried a charge which was an integral number of electronic charges, but also determined the electronic charge with much greater precision. This in fact was the first direct experimental proof of the atomic nature of electric charge. Millikan's early result was $e = 1·64 \times 10^{-19}$ C. Refinements gave further results, viz. $1·591 \times 10^{-19}$ and $1·602 \times 10^{-19}$ C.

It is worth noting that Planck's early estimate of e was very near to the accepted value and better than all the early direct measurements by Thomson and others.

Millikan's apparatus consisted of two circular brass plates 220 mm in diameter and 16 mm apart forming an air condenser (Fig. 2.5). The upper plate had a

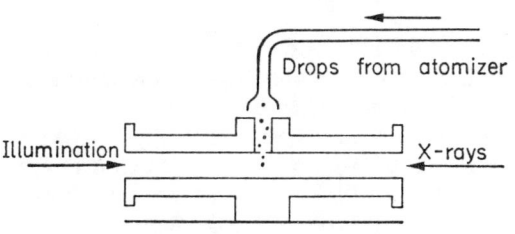

Fig. 2.5   Millikan's apparatus.

minute hole at the centre through which oil drops, formed by a spray in an upper chamber, could pass from time to time. The oil drops were illuminated from the side by a carefully collimated beam of light. This light showed up the oil drops as bright specks against a dark background. The drops were charged by friction or X-ray ionization of the air as they were formed in the spray. They normally fell under the action of gravity, but could be made to rise again by applying an electric field in a suitable direction. The electric field is calculated from the potential difference ($V = 5$ kV) and the separation of the plates ($d = 16$ mm). Thus $E = V/d = 5000/(16 \times 10^{-3})$ V m^{-1}. From observations of the rate of rise and fall of the drop with and without the electric field, the electric charge on an oil drop was found.

When a drop of radius a (Fig. 2.6) falls under the action of gravity alone, its

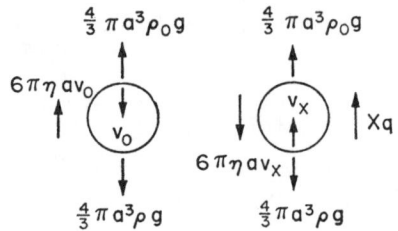

Fig. 2.6 Forces on oil drop.

weight is $\frac{4}{3}\pi a^3 \rho g$, where ρ is the density of the oil and g is the acceleration due to gravity. The upthrust due to the displaced air is, by Archimedes' principle, $\frac{4}{3}\pi a^3 \rho_0 g$, ρ_0 being the air density. The retarding force due to viscous drag as the drop moves through the air is given by Stokes as $6\pi\eta a v_0$, where η is the viscosity of air and v_0 the velocity of the drop. When the velocity becomes uniform, the resultant force on the drop is zero, so that for zero electric field we may write
$$\tfrac{4}{3}\pi a^3(\rho - \rho_0)g = 6\pi\eta a v_0. \tag{2.1}$$
Suppose now that under the action of an electric field X the drop moves upwards with a constant velocity v_X. Again the resultant force is zero, as the velocity is uniform, so that
$$\tfrac{4}{3}\pi a^3(\rho - \rho_0)g + 6\pi\eta a v_X = Xq, \tag{2.2}$$
where q is the charge on the drop. From equations (2.1) and (2.2) we see that
$$6\pi\eta a(v_0 + v_X) = Xq$$
and therefore
$$q = \frac{6\pi\eta a}{X}(v_0 + v_X).$$
Millikan found that values of q for different drops were always multiples of a common value e; that is, $q = ne$, where n is an integer. The value obtained was independent of the manner in which the charge was given to the drop. The highest common factor from a large number of observations was taken to be the electronic charge e, which Millikan found to be $1 \cdot 591 \times 10^{-19}$ C. Later experiments with this method gave $e = 1 \cdot 602 \times 10^{-19}$ C.

Using the values of the constants to four decimal places, the mass m of the

electron is given by
$$m = e \div e/m$$
$$= 1.6022 \times 10^{-19} \div 1.7588 \times 10^{11}$$
$$= 9.1096 \times 10^{-31} \text{ kg}.$$

This number only became significant when compared with the mass M of a single hydrogen atom, which was obtained by dividing the relative atomic mass of hydrogen by N_A. We have
$$M = 1.0078 \times 10^{-3} \div 6.022 \times 10^{23}$$
$$= 1.6735 \times 10^{-27} \text{ kg}.$$

Comparing the masses M and m we get
$$\frac{M}{m} = \frac{1.6735 \times 10^{-27}}{9.1096 \times 10^{-31}} = 1837.$$

Thus we see that the electron is a particle having a mass of just a little more than one two-thousandth of that of hydrogen, the lightest known atom, and carrying a negative charge of electricity equal to 1.602×10^{-19} C.

During the course of these experiments, Millikan observed that the velocity v_X of the drop altered abruptly from time to time. This occurred whenever the drop collided with an ion which changed the number of electrons on the drop. It was most significant that for a given drop the velocity changed by the same amount $\pm \Delta v_X$ each time, or by simple multiples of this amount. As the electronic charge was given by $e = 6\pi \eta a \Delta v_X / X$, this provided conclusive experimental evidence that the charge on the drop always changed by $\pm 1.602 \times 10^{-19}$ C or simple multiples of this amount. This clearly indicated that 1.602×10^{-19} C was the smallest unit of electricity of either sign and might well be regarded as the fundamental charge of electricity, one such negative charge being carried by each electron.

Problems

(Those problems marked with an asterisk are solved in full at the end of the section.)

2.1   A capacitance consisting of two parallel plates 5 mm apart has a potential difference of 300 V. Caluclate (a) the electric field intensity between the plates and (b) the force in a small oil drop carrying a charge of 32×10^{-19} C which is introduced between the two plates. ((a) 60 kV m^{-1}; (b) 0·192 pN)

2.2   In a cathode ray tube the length of the deflector plates is 78 mm and their separation 24 mm, while the distance from the centre of the plates to the screen is 330 mm. No deflection of the electron beam is observed when the potential difference is 2·8 kV and the magnetic field 8.2×10^{-4} T. With only the magnetic field, the deflection is 24 mm on the screen. Calculate e/m. (1.62×10^{11} C kg^{-1})

2.3   An electron emitted from a heated filament is accelerated to the anode by a potential difference of 300 V between the filament and the anode. Calculate (a) the kinetic energy of the electron; (b) its velocity on reaching the anode. ((a) 4·8 $\times 10^{-17}$ J; (b) 1.03×10^7 m s^{-1})

2.4*   An electron accelerated by a p.d. of 5 kV enters a uniform magnetic field of 2×10^{-2} T perpendicular to its direction of motion. Determine the radius of the path of the electron. (12·5 mm)

2.5   Describe how the ratio of the charge to mass of an electron has been determined.

In a cathode tube a potential difference of 3 kV is maintained between the deflector plates whose separation is 20 mm. A magnetic field of 2.5×10^{-3} T at right angles to the electric field gives no deflection of the electron beam, which received an initial acceleration by a potential difference of 10 kV. Calculate the ratio of charge to mass of an electron. (1.8×10^{11} C kg^{-1})

2.6 Describe how the ratio of the charge to the mass of an electron has been measured.

Calculate the mass of an electron using the following data and stating the assumption you make. The Avogadro constant is 6×10^{23} atoms mol^{-1}, $e/m = 1.76 \times 10^{11}$ C kg^{-1} and the Faraday constant is 96 500 C mol^{-1}. (9.1×10^{-31} kg)

2.7* In an oil drop experiment the following data were recorded: plate separation 15 mm, distance of fall 10 mm, potential difference 3.2 kV, viscosity of air 1.82×10^{-5} N s m^{-2}, radius of drop 2.76 μm, successive times of rise of the drop 42 and 78 s. Calculate the change in charge on the drop between the two sets of observations. To how many electrons does this correspond? (4.9×10^{-19} C, 3 electronic charges)

2.8 Describe how the ratio of charge to mass of the electron has been determined. In a cathode ray tube, the length of the deflector plates is 80 mm and their separation 24 mm while the distance from the centre of the plates to the screen is 400 mm. The accelerating potential difference between anode and cathode is 32.5 kV and a magnetic field or 5.6×10^{-4} T produces a displacement of 29.5 mm on the fluorescent screen. Calculate the value of e/m. (1.76×10^{11} C kg^{-1})

2.9 One coulomb liberates 0.001 118 g of silver from solution. Calculate the Avogadro constant assuming that the atomic mass number of silver is 108 and the electronic charge is 1.6×10^{-19} C. (6×10^{23})

2.10 A diode valve has a cylindrical anode of diameter 10 mm with a directly heated filament along its axis. The valve is placed in a solenoid so that the magnetic field is parallel to the axis of the cylinder. Calculate the minimum value of the magnetic flux density required to cut off the anode current when the potential difference between the anode and filament is 50 V. (10^{-2} T)

2.11 Two horizontal condenser plates are each 50 mm long. An electron travelling at 3×10^6 m s^{-1} at 30° to the horizontal enters the space between the plates, just clearing the edge of one of them. Calculate the electric field such that the electron just emerges by the opposite edge of the same plate and the minimum potential difference to which this corresponds. (900 V m^{-1}, 13 V)

Solutions to Problems

2.4 Kinetic energy of an electron is given by $\frac{1}{2}mv^2 = Ve$. Thus
$$\tfrac{1}{2} \times 9.1 \times 10^{-31}\, v^2 = 5000 \times 1.6 \times 10^{-19}\, \text{J}$$
from which $v = 4.2 \times 10^7$ m s^{-1}.

The force on the electron, Bev, gives rise to an acceleration v^2/r in accordance with Newton's second law of motion, so that
$$Bev = m\frac{v^2}{r}.$$

Thus
$$r = \frac{mv}{Be} = \frac{9 \cdot 1 \times 10^{-31} \times 4 \cdot 2 \times 10^{7}}{2 \times 10^{-2} \times 1 \cdot 6 \times 10^{-19}}$$
$$= 1 \cdot 20 \times 10^{-2} \text{ m or } 12 \cdot 0 \text{ mm}.$$

2.7 Applying the equation derived in the text to successive times of rise of the drop corresponding to different changes upon the drop, we get

$$ne = \frac{6\pi\eta a}{X}(v_0 + v_X)$$

$$n'e = \frac{6\pi\eta a}{X}(v_0 + v'_X)$$

$$e(n' - n) = \frac{6\pi\eta a}{X}(v'_X - v_X)$$

$$n' - n = \frac{6\pi\eta a}{eX}(v'_X - v_X),$$

$$= \frac{6\pi \times 1 \cdot 82 \times 10^{-5} \times 2 \cdot 76 \times 10^{-6}}{1 \cdot 6 \times 10^{-19} \times 3200 \div 1 \cdot 5 \times 10^{-2}} \left(\frac{10^{-2}}{42} - \frac{10^{-2}}{78}\right)$$

$$= 3 \text{ (nearly)}$$

Chapter 3

Natural Radioactivity

3.1 Introduction

Whilst investigating the relationship between X-rays (Chapter 6) and fluorescence in 1896, Becquerel accidentally found that a photographic plate was blackened when a uranium compound was placed near it. Careful investigation showed that this property had no relation to fluorescence or to X-rays, but is a property of the element uranium itself. It is now known to be due to radioactivity in which the uranium gives out radiations spontaneously, and these appear to be independent of the physical or chemical condition of the uranium. The rays, like X-rays, can pass through opaque objects and also enable the air to conduct electricity.

The nature of these radioactive radiations was investigated by Rutherford in 1897. He showed there were two kinds, α-radiation which was easily absorbed by thin metal sheets and produced strong ionization, and β-radiation which was more penetrating but produced less ionization. Later Villard discovered a third still more penetrating component which he called γ-radiation. The existence of the three types of radiation may be demonstrated in principle by placing a small amount of radioactive salt at the bottom of a narrow hole drilled in a piece of lead. Lead absorbs all three kinds of rays, so that rays can only escape from the lead block in a narrow pencil. Since some of the rays are absorbed by the air, the whole should be enclosed in a box from which the air can be removed by pumping. If a magnetic field is at right angles to the beam of radiation perpendicular to the plane of the paper, the α- and β-rays are deflected, but the γ-rays are quite unaffected (Fig. 3.1). Similar deflections would occur if an electric field were placed in the plane of the paper (see Fig. 3.1). From the directions of the deflections in relation to the electric and magnetic fields, it appears that the α-rays carry a positive charge, the β-rays a negative charge and the γ-rays are uncharged. The radii of curvature shown in Fig. 3.1 are not to scale.

3.2 e/m for β-Rays

The first measurement of e/m, the charge per unit mass, for β-rays was made by Becquerel in 1900. The results suggested that β-rays were the same as cathode rays, differing only in that the β-electrons were travelling with much higher velocities than the cathode-ray electrons. In the following year this was confirmed by the more refined measurements of e/m for β-rays made by Kaufmann. In

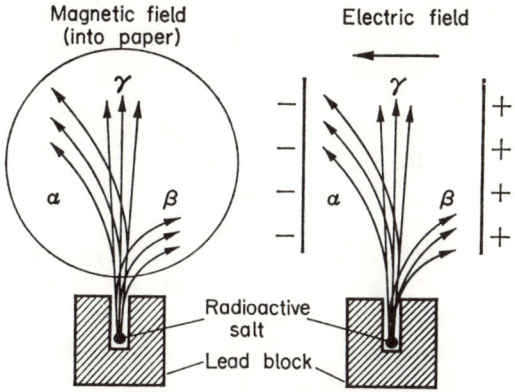

Fig. 3.1 Paths of α-, β-, γ-rays in electric and magnetic fields.

Thomson's experiment with cathode rays (see Section 2.4), the electron velocity was controlled by the potential difference across the discharge tube and was quite well defined. β-rays, on the other hand, have a wide velocity range and Kaufmann was obliged to use parallel electric and magnetic fields rather as Thomson did in his positive ray analysis experiments (Chapter 4). In Kaufmann's apparatus (Fig. 3.2) R was a radioactive source providing a pencil of β-rays which

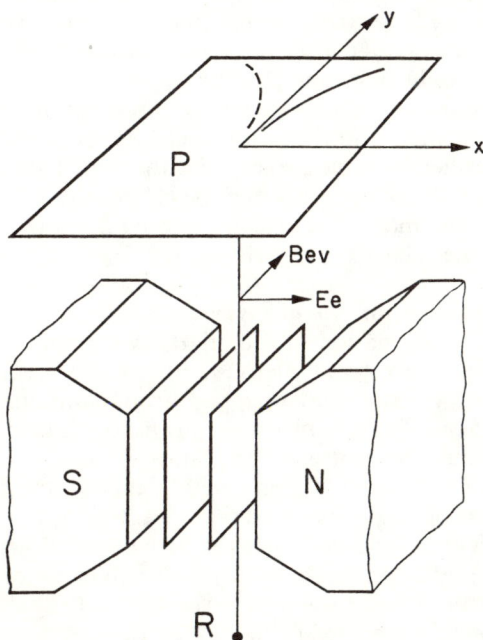

Fig. 3.2 Kaufmann's method for e/m for β-rays.

passed vertically upwards through the electric field E and magnetic field B. The electron experienced two forces Bev and Ee at right angles as shown. Applying Newton's second law we have $Ee = m\ddot{x}$ and $Bev = m\ddot{y}$, where $\ddot{x}$ and $\ddot{y}$ were the accelerations parallel to the x and y axes. Remembering that the displacements were $x = \frac{1}{2}\ddot{x}t^2$ and $y = \frac{1}{2}\ddot{y}t^2$, where t was the time taken by the electron to pass through the electric and magnetic fields, we see that

$$x = \frac{1}{2}\frac{Ee}{m}t^2$$

and

$$y = \frac{1}{2}\frac{Bev}{m}t^2 = \frac{1}{2}\frac{Be}{m}lt,$$

l being the distance travelled by the electron in the electric and magnetic fields.

Eliminating t we get

$$\frac{y^2}{x} = \frac{1}{2}\frac{B^2 e}{Em}l^2 = \text{a constant;}$$

therefore

$$\frac{e}{m} = \frac{2y^2}{x}\frac{E}{B^2}\frac{1}{l^2}.$$

Thus the β-rays, which were recorded upon the photographic plate at P, will lie in a parabola, those from sources with the highest β-energy coming nearest to the origin. Reversal of the electric field produced a mirror image of the parabola in the y axis.

Close examination of the experimental curves shoqed that they were not quite parabolic, especially for electrons with higher velocities. This indicated that the mass of the electron was not constant but increased with velocity, as predicted by the electron theory of Lorentz and later by Einstein's special relativity theory. In Appendix A it is shown how a body of mass m_0 at rest has a mass m given by $m = m_0/(1 - v^2/c^2)^{1/2}$ when in motion with a velocity v, c being the velocity of light. This formula is verified by some earlier measurements by Kaufmann for electron velocities up to $2 \cdot 83 \times 10^8$ m s^{-1} which is 94% of the velocity of light. At this velocity he found the mass of an electron to be 3·1 times its rest mass, m_0, in accordance with the relativity formula.

3.3 Bucherer's Method for e/m of β-rays

A more precise confirmation of the relativity formula was made in 1909 by Bucherer, using a very elegant method. Figure 3.3 represents an evacuated cylinder with photographic film wrapped around inside. A and C were two circular plates 0·25 mm apart, between which an electric field E was applied. Some radium fluoride, S, at the centre of the plates provided a source of β-rays. The whole apparatus was placed in a magnetic field of flux density B which was parallel to the plane of the plates A and C. An electron at P travelling at an angle θ to the magnetic field experienced forces $Bev \sin \theta$, and Ee as shown. When these forces were equal and opposite the electron was able to escape from between the parallel plates, when it had a velocity given by $v = E/B \sin \theta$. Thus the velocity with which the electron emerged from the electric field depended upon the angle θ, so that along the film we have a velocity spectrum. Having left the plates A and

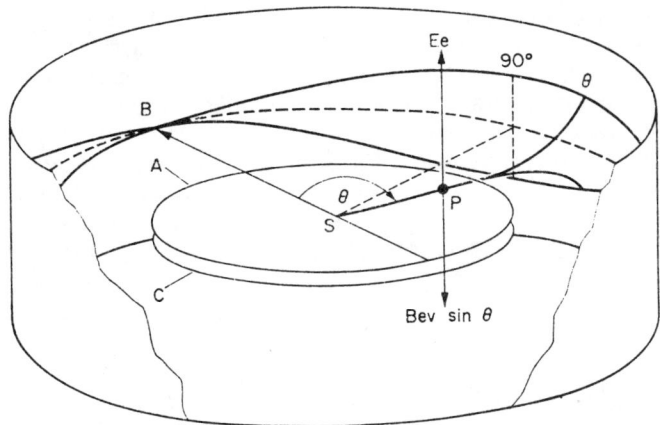

Fig. 3.3 Bucherer's method for the determination of e/m.

C, the electron travelled in the magnetic field only until it struck the film. At first the electron experienced a force $Bev \sin \theta$ and the initial radius of its path was given by $Bev \sin \theta = mv^2/r$. In general the electron will no longer travel in the plane of the plates A and C and will therefore be inclined at some other angle φ to the magnetic field. The force became $Bev \sin \varphi$ in a direction perpendicular to the plane containing B and v. Detailed calculation showed that the electron follows a helical path and this became circular only when $\theta = 90°$. In this case

$$\frac{e}{m} = \frac{v}{r} \frac{1}{B \sin \theta}$$
$$= \frac{1}{r} \frac{E}{B^2}$$

Figure 3.3 shows how the deflection of the electrons is zero when $\theta = 0°$ and $180°$ and maximum when $\theta = \pm 90°$. Reversal of the electric and magnetic fields leads to the symmetrical pattern shown upon the photographic film. In these experiments the pattern near $\theta = 0$ and $180°$ was always missing. This corresponds to an infinite value for the electron velocity. Deflections for the intermediate values of θ enable e/m to be calculated for a wide range of velocities. With this equipment Bucherer showed that the relativity formula for m was obeyed for electron velocities up to 68% of the velocity of light. At this velocity the mass increased by 37%. Whilst Bucherer's experiment did not cover the range of conditions examined by Kaufmann, the precision was higher, so that the work provided a more rigorous test of the relativity formula.

3.4 The Charge–Mass Ratio (E/M) for α-Rays

Having established that β-rays were electrons, it was still necessary to identify the α-rays. The very small deflections they receive in electric and magnetic fields compared with β-rays suggested that they were very much heavier than the electrons which constitute β-rays. Much more precise information was essential before any reasonable attempt at identification was possible. Rutherford and

Robinson measured the ratio of charge to mass, E/M, for α-rays by deflections in electric and magnetic fields.

In the next chapter we shall see that radium disintegrated by α-particle emission into the rare gas radon, which is also radioactive and disintegrates with the emission of further α-particles. In an experiment by Rutherford and Robinson, some radon gas, enclosed in a thin-walled glass tube, was used as the source of α-particles. The velocities of the α-particles were then limited to three values arising from radon and two of its disintegration products, radium A and radium C. In the experimental arrangement shown in Fig. 3.4 the α-rays from the

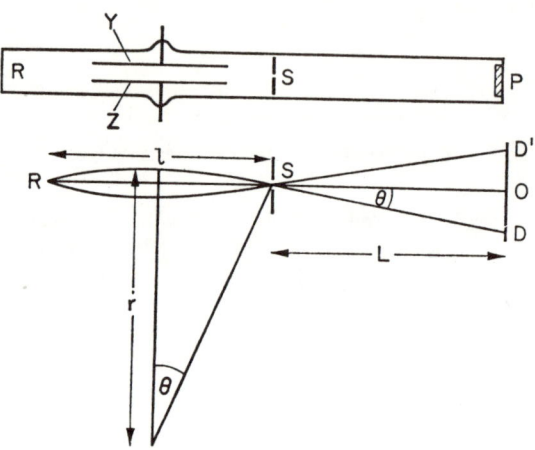

Fig. 3.4 E/M for α-particles.

source R passed between the deflecting plates YZ, which were 350 mm long and 4 mm apart. The rays then passed through a narrow slit S to the photographic plate P ($SP = 500$ mm), wrapped in aluminium foil which protected the plate from light but allowed the α-rays to pass through and be recorded on the plate.

With no electric or magnetic field between the plates YZ, the α-rays passed straight along the tube and were limited by the slit S, so that a sharp line was recorded at O on the photographic plate. When a magnetic field of flux density B was applied perpendicular to the paper in the space between Y and Z, only α-rays which leave R inclined at a small angle to RS were able to pass through the slit S and on to the photographic plate at D. Reversal of the magnetic field gave a symmetrically placed line at D'. The force on the α-particle of charge E, moving with velocity v was given by BEv, and was always perpendicular to the motion. The α-particle therefore moved in an arc of radius r_1, and had an acceleration equal to v^2/r_1. Applying Newton's second law of motion, $BEv = Mv^2/r_1$, where M was the mass of the α-particle.

Deflections were also produced by applying a potential difference of some 2 kV between Y and Z. Here the electric field X and therefore, the force XE upon the charged α-particle was everywhere perpendicular to the line RS. The α-particle therefore moved in a parabolic path. The curvature, however, was so small that it may be assumed to be an arc of a circle of radius r_2. In this case we can again apply Newton's second law of motion and write $XE = Mv^2/r_2$. Elimination of v from these two equations leads to

$$\frac{E}{M} = \frac{X}{B^2} \frac{r_2}{r_1^2}.$$

The radii of the paths can be found from the deflections of the beam. For small θ, $\sin\theta \sim \theta = \frac{1}{2}l/r = OD/L$ so that $r = lL/DD'$. In the actual experiment, three lines were obtained in the place of each of D and D' when either the electric or magnetic fields are applied. These arose from the presence of α-particles having three specific velocities arising from the disintegration of radon, radium A and radium C. The present accepted value for E/M for α-particles is 4.787×10^7 C kg^{-1}.

3.5 Charge on α-Particles

A measurement of the charge E on an α-particle would enable the mass M to be calculated. This measurement was first made in 1908 by Rutherford and Geiger and by Regener. The number of α-particles given off in 1 s from 1 g of radium was determined by placing a very small quantity of radium at a convenient distance from the window of a counter, and counting the number of α-particles arriving in a given interval of time. The counter consisted of an insulated rod W (Fig. 3.5) projecting axially into a metal tube T, at the other end of which there was a mica

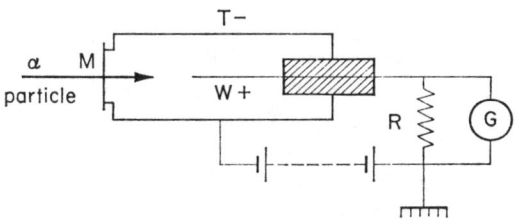

Fig. 3.5   A counter.

window M of known area enabling a gas at a predetermined pressure, usually about 100 mm of mercury, to be enclosed in the tube. A potential difference was applied between T and W such that it was just below the point of discharge. When an α-particle entered the tube it produced local ionization which caused a short discharge and was recorded upon the string galvanometer G. The value of the resistance R was adjusted so that the charge on W discharged quickly in a non-oscillatory manner, enabling α-particles to be recorded in quick succession. The deflections of the string galvanometer were recorded photographically at the rate of 600–900 min^{-1} upon a rapidly moving film. This enabled the total number of α-particles emitted by 1 g of radium in 1 s to be determined. The accepted value is 3.70×10^{10} s^{-1}, or 3.7×10^{10} Bq.†

Crookes had already shown that, when α-particles strike a screen coated with zinc sulphide, the energy of the particle was transformed into visible light. By viewing the screen with a microscope the arrival of each α-particle was observed

† 1 Bq = 1 Becquerel = 1 disintegration per second.

as a scintillation. This method was used by Regener to determine the number of α-particles emitted by 1 g of polonium in 1 s.

About this time, another experiment provided values for the total charge per second carried by α-particles from the above sources. A small known amount of radium C′ in equilibrium with radium C was placed in a small container and covered weth a piece of thin aluminium foil, which absorbed atoms from which the α-particles were emitted and prevented them from escaping, due to their velocity of recoil. The α-particles fell upon the collector plate C (Fig. 3. 6) which

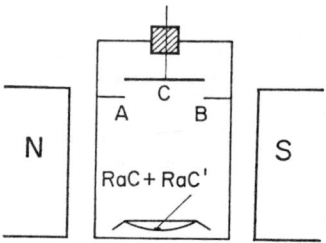

Fig. 3.6   Measurements of the charge on an α-particle.

was connected to a sensitive electrometer. The area of the collector plate used was defined by the diaphragm AB. A strong magnetic field was applied to prevent β-rays emitted from radium C from reaching the collector C, while the α-rays from radium C′ were only slightly deviated. Thus the charge collected per second was found and knowing the number of particles emitted per second, the charge upon one α-particle was determined.

3.6   Identification of α-Particles

The charge on one α-particle was found to be approximately 3.19×10^{-19} C. Comparing this with the electronic charge of 1.60×10^{-19} C, it appeared that the charge on an α-particle was double the charge on an electron; that is $E = 2e$. The mass M_α of the α-particle was given by $M_\alpha = E \div E/M_\alpha = 3.19 \times 10^{-19} \div 4.79 \times 10^{7}$ $= 6.66 \times 10^{-27}$ kg. Comparing this with the mass M_H of a hydrogen atom, $M_\alpha/M_H = 6.66 \times 10^{-27}/1.67 \times 10^{-27} \sim 4$. Thus the α-particle was found to have a mass four times the mass of the hydrogen atom and a positive charge equal to twice the charge on the electron. It seemed very likely that it was the nucleus of the helium atom.

This was confirmed by Rutherford and Royds, who enclosed some radon gas in a thin-walled glass tube A so that particles passed through the walls into D (Fig. 3.7) and were neutralized to helium atoms. After about a week the helium which had collected in B was compressed by the mercury into the fine capillary C, so that on applying a high potential difference between X and Y its spectrum could be examined. The spectrum was identified as that of helium, so that the conclusions drawn from earlier measurements were confirmed.

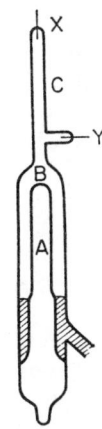

Fig. 3.7 Rutherford and Royds' experiment.

3.7 Early Models of the Atom

Work by Lenard in 1903 on the scattering of swift electrons by thin metallic films suggested that the atom was very largely empty space. Lenard supposed that the mass was concentrated entirely in particles which he called 'dynamids' which were very small compared with the atom itself. A little later, in 1910, J. J. Thomson postulated negatively charged corpuscles embedded in a sphere of positive electricity so that the atom as a whole was neutral. The distribution of positive electricity was uniform throughout the volume of the atom. This was the Thomson 'currant bun' atom.

Very important contributions to our knowledge of atomic structure came from investigations of the scattering of α-particles in thin metallic films by Rutherford, Geiger and Marsden. The α-particles were much more massive than the electrons employed by Lenard. Deflection of the α-particles only occurred when they encountered charged particles of comparable mass, and they were unaffected by any electrons within the atom. A careful experimental study of the scattering of α-particles by Geiger and Marsden provided information about the distribution of mass and charge within the atom and in 1911 led to Rutherford's nuclear model of the atom. In this he supposed that most of the mass of the atom is concentrated in a nucleus at its centre which also carries the positive charge, the diameter being about 10 fm (10^{-14} m). This is surrounded by an electron cloud extending out beyond 0·1 nm and which makes the whole atom electrically neutral. Some details of the evidence upon which Rutherford's nuclear atom is based will now be given.

In the experiment by Geiger and Marsden, α-particles from a source S (Fig. 3.8) were restricted to a narrow pencil by the hole H in the screen and fell upon a piece of metal foil F. The scattered α-particles were detected by a zinc sulphide screen Z, the individual scintillations being observed and counted using the microscope M. Foils of aluminium, copper, silver and gold were used. In all these experiments almost all the α-particles passed straight through the metal foil and were deflected only slightly. This showed that the atom was largely empty space. A small fraction

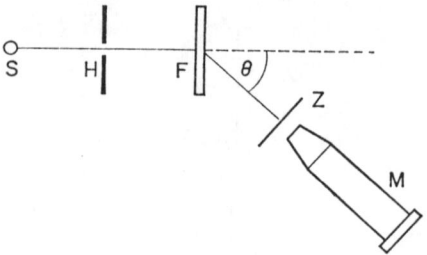

Fig. 3.8 Scattering of α-particles by thin metal films.

was so strongly deviated that the α-particles emerged again on the same side of the foil as they entered, showing that some α-particles were deflected through angles greater than 90°, i.e. 'reflected' back towards the source. This suggested that somewhere within the atom there was a very small massive particle carrying a positive charge so that the charge on the atom as a whole is zero. This small massive particle was called the nucleus.

3.8 The Scattering of α-Particles

In order to appreciate more fully the considerations which led to Rutherford's model of the nuclear atom and to make estimates of the diameter and charge of the nucleus, a detailed study must be made of an encounter between an α-particle and a nucleus. Suppose an α-particle of mass M and initial velocity V approaches a nucleus of charge $+Ze$, where Z is an integer, situated at S (Fig. 3.9) along the

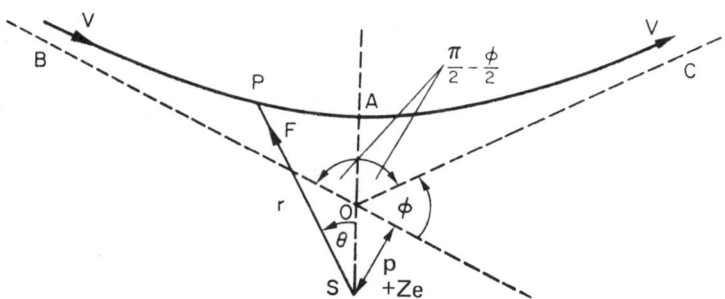

Fig. 3.9 Encounter between α-particle and a nucleus.

path BO. If no interaction occurs it will pass at a distance p from the nucleus. Owing to the Coulomb force between the nucleus and the particle, the latter follows a hyperbola with S as a focus as shown. The lines OB and OC are the asymptotes of the hyperbola and approximate to the initial and final directions of the α-particle when it has passed out of the effective range of the nuclear Coulomb force. The α particle has thus been deflected through an angle φ. All this takes place on a scale much too small for direct observation and only the initial and final directions of the α-particle are observable.

In the following discussion we assume that the nucleus is sufficiently massive to be undisplaced by the encounter with the α-particle. The angular momentum of the system about any point must remain constant. Considering the angular momentum about S we can write

$$MVp = Mr^2\dot{\theta}, \tag{3.1}$$

where V is the initial velocity of approach of the α-particle and $r\dot{\theta}$ is its velocity at any point P in its hyperbolic path (Fig. 3.9). Also the change in linear momentum along the line of symmetry can be set equal to the impulse (the product of force and time) so that

$$2MV\cos\left(\frac{\pi}{2} - \frac{\phi}{2}\right) = 2MV\sin\frac{\phi}{2} = \int_0^\infty F\cos\theta \, dt \tag{3.2}$$

where F is the force along SP. Multiplying equations (3.1) and (3.2) together, we obtain

$$2MV^2 p \sin\frac{\phi}{2} = \int_0^\infty F\cos\theta \, r^2 \dot{\theta} \, dt.$$

When the variable is changed from t to θ, the limits correspond to $t=0$ and ∞ become

$$\theta = -\left(\frac{\pi}{2} - \frac{\phi}{2}\right) \quad \text{and} \quad \theta = \frac{\pi}{2} - \frac{\phi}{2}$$

respectively. Remembering that

$$F = \frac{ZeE}{r^2 4\pi\varepsilon_0},$$

we can now write

$$2MV^2 p \sin\frac{\phi}{2} = \int_{-(\pi/2)+(\phi/2)}^{(\pi/2)-(\phi/2)} \frac{ZeE}{4\pi\varepsilon_0} \cos\theta \, d\theta = \frac{ZeE}{4\pi\varepsilon_0} 2 \cos\frac{\phi}{2};$$

and so

$$\cot\frac{\phi}{2} = \frac{2p}{b}, \tag{3.3}$$

where

$$b = \frac{2ZeE}{4\pi\varepsilon_0 MV^2}$$

and p is the collision distance. (This treatment is given in preference to the more usual one because it does not require a knowledge of the geometry of the hyperbola.)

Suppose Q particles are incident normally upon unit area of a metal surface of thickness t and that q of these particles come within a distance p of a nucleus of a metal atom. The total projected area presented by the nucleii to the α-particles is $\pi p^2 nt$, where n is the number of atoms or nuclei per unit volume. We can therefore write

$$q = Q\pi p^2 nt.$$

Using equation (3.3) we have

$$q = Q\pi nt \frac{b^2}{4} \cot^2 \frac{\phi}{2},$$

which implies that of the Q particles, q will be deflected through angles greater than φ. Further, the number deflected between the angles φ and $\varphi + \Delta\varphi$ is given by

$$\Delta q = Q \frac{\pi}{4} n t b^2 \cot \frac{\varphi}{2} \operatorname{cosec}^2 \frac{\varphi}{2} \Delta\varphi. \tag{3.4}$$

In scattering experiments a count is made of the number of α-particles Δq falling upon an area Δa of a fluorescent screen placed at a distance r normal to the direction of observation. From Fig. 3.10 this is just

$$\frac{\Delta q}{\Delta a} = \frac{\Delta q}{2\pi r^2 \sin \varphi \Delta\varphi} = \frac{\Delta q}{4\pi r^2 \sin\left(\frac{\varphi}{2}\right) \cos\left(\frac{\varphi}{2}\right) \Delta\varphi}.$$

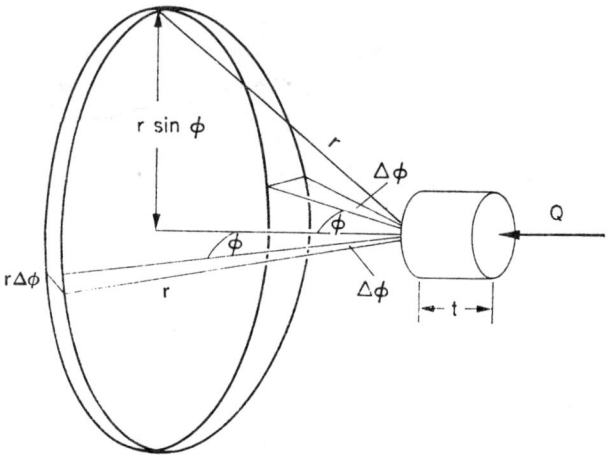

Fig. 3.10 Scattering of α-particles.

Substitution from equation (3.4) gives, in the limit,
$$\frac{dq}{da} = \frac{Qntb^2 \operatorname{cosec}^4 (\varphi/2)}{16r^2}$$
$$= \frac{Qnt}{16r^2} \left(\frac{2ZeE}{4\pi\varepsilon_0 MV^2} \right)^2 \operatorname{cosec}^4 \frac{\varphi}{2}$$
when a substitution is made for b^2.

This is Rutherford's scattering formula and inspection shows that the number of α-particles per unit area dq/da scattered in a given direction φ is proportional to

(1) $\operatorname{cosec}^4 (\varphi/2)$;

(2) the thickness t of the scattering film, provided this is small enough to ensure that second encounters are negligible;

(3) the square of the nuclear charge, i.e. $(Ze)^2$;

(4) the reciprocal of $MV^2)^2$, i.e. $[\text{K.E.}]^2$

These relationships have been confirmed by the experiments of Geiger and

Marsden and provide valuable support for Rutherford's model of the atom. Two arguments lead to the conclusion that large angle scattering of α-particles takes place in a single encounter with a very massive charged particle rather than by successive encounters with a number of smaller particles. The probability of deflections as great as 90° by successive encounters is vanishingly small and the fact that as many as one in 20 000 α-particles are deflected in this way by a gold film 0·4 μm thick, clearly favours single particle scattering. Moreover it can be shown that scattering by successive encounters requires that dq/da shall be directly proportional to $\sqrt{t}$ rather than to t, so that again the experimental evidence is decisively in favour of scattering by a single encounter.

3.9 Estimates of Nuclear Diameter and Charge

We have assumed Coulomb's inverse square law in deriving Rutherford's formula, and the formula has been confirmed by experiment for values of φ between 5° and 150°. It appears therefore that the inverse square law is valid for distances very much less than the diameter of an atom, which we have seen is of the order of 0·1 nm. The closest approach will be made by an α-particle approaching head-on to the nucleus, so that $p=0$ (Fig. 3.9). If d is the distance of closest approach, the potential energy will then be $ZeE/4\pi\varepsilon_0 d$, which may be set equal to the initial kinetic energy of the α-particle $\frac{1}{2}MV^2$ before it enters the electric field of the nucleus. Thus we have

$$\tfrac{1}{2}MV^2 = \frac{ZeE}{4\pi\varepsilon_0 d}.$$

The α-particles for radium C have a velocity of $2\cdot 1 \times 10^7$ m s^{-1} so that for gold (atomic number 79 and atomic mass number 197) we may write

$$\tfrac{1}{2} \times 6\cdot 7 \times 10^{-27} \times 2\cdot 1^2 \times 10^{14} = \frac{79 \times 1\cdot 6 \times 10^{-19} \times 3\cdot 2 \times 10^{-19}}{\frac{1}{9 \times 10^9} d}$$

for which

$$d = \frac{2 \times 79 \times 1\cdot 6 \times 3\cdot 2 \times 10^{-38} \times 10^9 \times 9}{6\cdot 7 \times 2\cdot 1^2 \times 10^{14} \times 10^{-27}}$$
$$= 2\cdot 5 \times 10^{-14} \text{ m}.$$

Support for this value is obtained from the data, already quoted, that only one α-particle in 20 000 is deflected through more than 90° by a gold film 0·4 μm thick of density $19\cdot 3 \times 10^3$ kg m^{-3}. Substitution in the equation $q/Q = \pi p^2 nt$ enables a value for p to be found. Using S.I. units, 1 mol of gold occupies (197×10^{-3}) ÷ $(19\cdot 3 \times 10^3) = 10^{-5}$ m^3, and this we know contains 6×10^{23} atoms. Thus n for 1 m^3 of gold is 6×10^{28} atoms, $t = 4 \times 10^{-7}$ m, $q/Q = 1/20\,000$. Thus

$$\frac{1}{2 \times 10^4} = \pi p^2 (6 \times 10^{28})(4 \times 10^{-7})$$

and so

$$p^2 = 6\cdot 63 \times 10^{-28},$$
$$p = 2\cdot 6 \times 10^{-14} \text{ m or 26 fm}.$$

This value must represent an upper limit for the diameter of the nucleus in so far as we can attach precise meanings to such quantities. For hydrogen it was found that the inverse square law no longer holds for distances as small as 3 fm and Rutherford estimated the nuclear diameter for light atoms to be about 5 fm. In the first chapter we saw that the diameter of an atom is about 0·4 nm, so that the nucleus is only about 10^{-5} of the atomic diameter and occupies only 10^{-15} of the volume of an atom. Its density must be very high indeed — 10^{15} times as great as the densities of the elements we normally measure in the laboratory. Such densities are encountered in some stars, where atoms are stripped of their electrons and the nuclei packed closely together.

Referring again to Rutherford's scattering formula, we see that for a given experiment all the quantities are known except Z, the number of charges upon the nucleus. Chadwick repeated and refined the scattering experiments of Geiger and Marsden and obtained values of Z for copper, silver and platinum. These were respectively 29·3, 46·3, 77·4, and agreed well with the accepted 'atomic numbers' of 29, 47 and 78 which are based upon the numerical order in which these elements appear in the periodic table. It follows that we have to identify 'atomic number' with Z, the net number of charges upon the nucleus. It is also clear that Z rather than the relative atomic mass should be the criterion in deciding the order of elements in the periodic table.

3.10 The Neutron

Hydrogen, being the simplest atom, has only one electron and a nucleus, which is a single particle called a proton. The proton has a positive charge numerically equal to that of the electron, so that the atom as a whole is neutral. We may therefore think of the atomic number of an element Z as being the number of proton charges on the nucleus. If the nucleus were composed entirely of protons, the charge and atomic mass would be the same. Measurements show that the atomic mass is usually nearly twice as large as the atomic number. This is the case with sodium. The atomic mass number is 23 and the atomic number is only 11, implying that the nucleus carries 11 positive charges and these are balanced by 11 negative charges in the electron cloud surrounding it. At first it was supposed that within the nucleus there were 23 protons and 12 electrons which reduced the net change to 11, whilst accounting for the atomic mass number of 23. The discovery of the neutron in 1932 and the accumulation of other evidence, especially in connection with the spin of the nucleus, has led to a revision of this view and it is now supposed that the nucleus consists of protons and neutrons only, collectively called nucleons. The neutron is a new fundamental particle which carries no charge and has a mass very nearly the same as the proton. In the case of sodium therefore, there are 11 protons to account for the atomic number $Z=11$ and 12 neutrons to bring the atomic weight up to the observed value of 23. The atomic mass number is therefore the nucleon number.

At first, it may be surprising that a fundamental particle of the importance of the neutron remained undiscovered until 1932. The fact that the neutron carries no charge made it particularly difficult to detect since it can pass through atoms without deflection and leaves the electron cloud virtually undisturbed. In 1930 Bothe and Becker bombarded light nuclei such as lithium and beryllium with α-particles and observed a very penetrating radiation which they assumed to be very hard γ-rays. Two years later, I. Curie and Joliot found that this radiation could eject protons with very large energies from a layer of paraffin wax. In their experiment α-particles from a polonium source Po (Fig. 3.11) were allowed to fall upon a piece of beryllium Be, beyond which was a block of paraffin wax. Paraffin wax provides a high concentration of hydrogen atoms in a convenient form for such experiments. Curie and Joliot supposed that the energy of the γ-rays was transferred to the protons by the process known as the Compton effect, which will

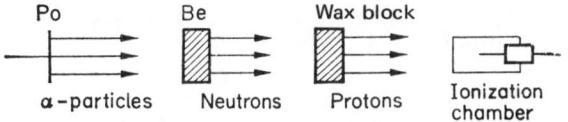

Fig. 3.11 Discovery of the neutron.

be described later. The energy of the incident radiation required to produce protons with the observed energies, was as high as 55 MeV which was greatly in excess of that normally emitted by radioactive substances.

Being dissatisfied with the γ-ray explanation, Chadwick repeated these experiments and later, in collaboration with Feather, he replaced the paraffin block by a small chamber containing gaseous nitrogen. He first measured the range of protons from the paraffin block and then the range of the 'knock-on' nitrogen nuclei in the second series of experiments. In effect he measured the velocities of the protons and the nitrogen nuclei. Chadwick then revived the idea, first suggested by Rutherford in 1920, of a particle having no charge, but of mass comparable with the proton — the neutron. The difficulties, experienced by the earlier workers in accounting for the energy of the protons disappeared when it was assumed that the unknown so-called 'γ-radiation' was composed of neutrons. The problem now became one of simple impact using the classical laws of the conservation of momentum and energy. Restricting our considerations to direct or 'head-on' impact for a neutron of mass M (Fig. 3.12) and velocity V colliding

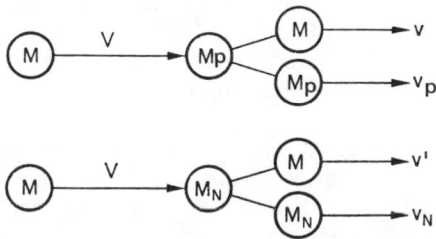

Fig. 3.12 Determination of the mass of a neutron. The collisions are linear.

with a proton of mass M_p at rest, we may write $MV = Mv + M_p v_p$. Using the conservation of energy $\frac{1}{2}MV^2 = \frac{1}{2}Mv^2 + \frac{1}{2}M_p v_p^2$ and eliminating v from these equations yields $v_p = 2VM/(M_p + M)$ and $v = V(M - M_p)/(M + M_p)$. By the same method, an equation may also be obtained for the velocity of the nitrogen nucleus, $v_N = 2VM/(M_N + M)$ from which $v_p/v_N = (M_N + M)/(M_p + M)$. Chadwick measured v_p and v_N obtaining respectively 3.3×10^7 m s^{-1} and 0.47×10^7 m s^{-1}.

Substituting these values, and remembering that the atomic mass numbers of hydrogen and nitrogen are respectively 1 and 14, we have $3.3 \times 10^7/0.47 \times 10^7 = (14+M)/(1+M)$, from which $M = 1.16$. Thus the mass of the neutron is 1·16 times the mass of a proton. This simple calculation serves to show the nature of the calculation used by Chadwick. More recent measurements show the mass to be much more nearly equal to that of the proton, namely 1·008 665 atomic mass units on the ^{12}C scale. (For the definition of the atomic mass unit see p. 56.)

We note that if $M = M_p$ the neutron remains stationary, $v = 0$, while the proton proceeds with the initial neutron velocity, V.

Problems

(*The problems marked with an asterisk are solved in full at the end of the section.*)

Note: 1 MeV is a unit of energy equivalent to 1.6×10^{-13} J.

3.1 A stream of α-particles each carrying a charge of 3.2×10^{-19} C is sent through a uniform magnetic field of 3 T. The velocity of each particle is 1.52×10^7 m s^{-1} and is at right angles to the direction of the magnetic field. Determine the force on each particle. (14·6 pN)

3.2 Calculate in terms of the rest mass m_0, the mass of an electron moving with a velocity of $c/4$, $c/2$ and $3c/4$, where c is the velocity of light. ($1.03\, m_0$, $1.16\, m_0$, $1.51\, m_0$)

3.3* In a Bucherer type experiment the plates are 1 mm apart and have a potential difference of 10 kV. A magnetic field of flux density 2·0 T is applied parallel to the plane of the plates. Calculate the velocity of electrons which are able to escape from between the plates in direction inclined at 30° to the magnetic field (10^7 m s^{-1})

3.4* Calculate the distance of closest approach of 10 MeV α-particles to copper nuclei. The atomic number of copper is 29. (8.4×10^{-15} m ≡ 8·4 fm)

3.5 A beam of α-particles of energy 9 MeV was incident normally upon aluminium foil 1·0 μm thick. A fraction 1 in 10^8 was scattered through an angle of 60° and detected by a scintillation counter of area 25 mm² at a distance of 50 mm. Calculate the atomic number of aluminium, assuming its relative density to be 2·7. (13)

3.6 Establish a relationship between the impact parameter p, the distance of closest approach d and the scattering angle θ. For what scattering angle is the impact parameter equal to the distance of closest approach? ($\theta = 53.2°$)

3.7 What fraction of 7·25 MeV α-particles falling normally upon copper foil of thickness 0·1 μm is scattered through angles greater than 90°? Take the relative density of copper to be 8·9. (1 in 7.3×10^4)

Solution to Problems

3.3 In the Bucherer experiment (see text),

$$v = \frac{E}{B \sin \theta}.$$

Thus
$$v = \frac{10\,000}{10^{-3}} \times \frac{1}{2 \times 0.5}$$
$$= 10^7 \text{ m s}^{-1}.$$

3.4 Energy of α-particle is 10 MeV or $10 \times 1.6 \times 10^{-13}$ J. This is equal to
$$\frac{ZeE}{4\pi\varepsilon_0 d_{min}}$$
(see p. 41), where $Z = 29$, therefore
$$d_{min} = \frac{29 \times (1.6 \times 10^{-19})^2 \times 2}{\frac{1}{9} \times 10^{-9} \times 1.6 \times 10^{-12}}$$
$$= 8.4 \times 10^{-15} \text{ m}$$

Chapter 4

Radioactive Series and Isotopes

4.1 Introduction

We have seen how the atom consists of a very small massive nucleus which contains most of the mass of the atom and carries a positive charge equal to Ze, where Z is the atomic number and e the proton charge. Moreover α- and β-particles have been identified as helium nuclei and electrons respectively. The α-particles, emitted by an atom, can only come from the nucleus and the nucleus must therefore lose two positive charges and some mass equal to that of the helium nucleus. Thus, if Z and A are respectively the atomic number and the atomic mass number of the original atom, the emission of an α-particle gives rise to the following changes:

$Z \to Z-2$ for the proton number,
$A \to A-4$ for the nucleon number.

The change of Z implies that a new element has been created. In the case of radium, for which $Z = 88$ and $A = 226$, a new element, the gas radon, having $Z = 86$ and $A = 222$, is produced.

$$\begin{array}{ccc} A=226 & A=222 & A=4 \\ \text{Radium} \to & \text{Radon} + & \alpha\text{-particle} \\ Z=88 & Z=86 & Z=2 \end{array}$$

In a radioactive change of this type, radium is sometimes referred to as the parent element and radon is known as the daughter. The emission of a β-particle from the nucleus raises the positive nuclear charge by unity whilst the mass remains almost unchanged, so that we have

$$Z \to Z+1.$$

A remains unchanged, but again we have a new element. Such changes are said to be isobaric.

4.2 Equation of Radioactive Decay

When thorium X (radium 224) was first separated chemically from thorium, it was found to be much more active than precipitated thorium. After some time the thorium X gradually lost its activity and the thorium recovered its lost activity. These two effects were exactly reciprocal to each other as shown in Fig. 4.1.

Rutherford and Soddy studied the rates of decay and recovery and showed that they were the same. Thorium itself was not very active but its daughter thorium X was very active. A thorium compound contained both parent thorium and daughter thorium X, and the activity of the whole was largely due to the thorium

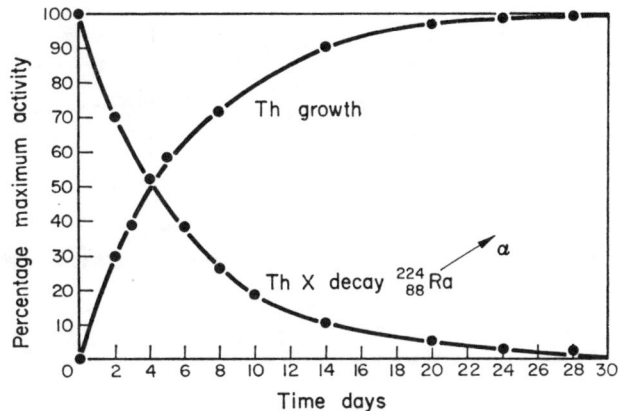

Fig. 4.1 Decay and growth curves of Th X and Th.

X. However, when they were separated chemically the parent thorium had little activity, while the thorium X had a lot. With time the inactive thorium increased its activity due to formation of new thorium X, while the precipitated thorium X lost its activity by natural decay to thorium emanation which was itself a weakly active product. These products were obviously governed by some activity–time law.

Rutherford and Soddy found the experimental curves to be exponential, i.e. the rate of loss of activity was proportional to the amount of activity actually present at the instant when the measurement was made. Thus, $dA/dt = -\lambda A$ (−ve sign because loss occurs), where A is a measure of the activity present.

In terms of atoms this becomes

$$\frac{dN}{dt} = -\lambda N$$

or

$$N = N_0 e^{-\lambda t}$$

where N_0 is the number of atoms at any chosen $t=0$, N is the number of unchanged atoms at time t, and λ is the disintegration constant of reciprocal time, or the fractional disintegration rate.

Since all radioactivity is found to be governed by this law, it follows that the difference between various radioactive atoms lies in the value of λ. Several radioactive atoms may be represented in one diagram, as shown in Fig. 4.2, and this can be replaced by one master curve with the appropriate time scale, as in Fig. 4.3, in which the units of time t, range from microseconds to millions of years. Thus, in order to compare the decay times or life times of radioactive atoms, we must adopt some method of measuring the mean lifetime.

4.3 Mean Lifetime of Radioactive Substance

Soddy showed that the mean lifetime $\bar{T} = 1/\lambda$ by the following argument. Suppose the number of atoms which have survived t seconds (i.e. had a lifetime t) is N, and the number of decaying in the next small time interval Δt is ΔN, the

47

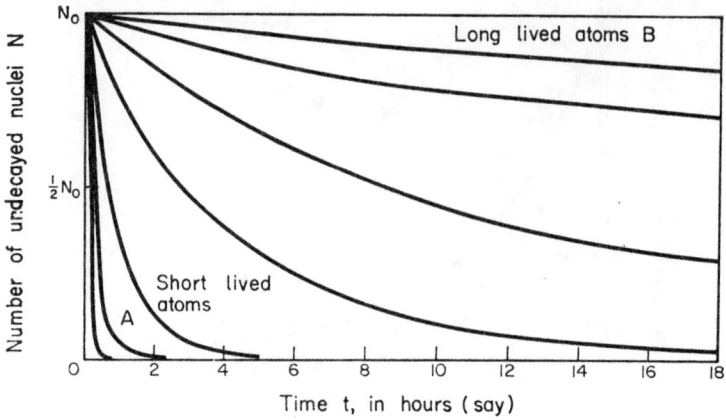

Fig. 4.2 Decay of different nuclides on same time scale.

lifetime of the ΔN decaying atoms is t and the total lifetime of all the N_0 atoms is simply

$$\sum \Delta N t \quad \text{which becomes} \quad \int_{N_0}^{0} t \, dN.$$

The average, or mean, lifetime is then

$$\bar{T} = \int_{N_0}^{0} \frac{t \, dN}{N_0}$$

$$= \frac{1}{N_0} \int_0^\infty t(-\lambda N dt)$$

$$= \frac{1}{N_0} \int_0^\infty -\lambda t N_0 e^{-\lambda t} dt$$

$$= -\lambda \int_0^\infty t e^{-\lambda t} dt$$

$$= -\lambda \left[\frac{t(-e^{-\lambda t})}{\lambda} - \int \frac{(-e^{-\lambda t})}{\lambda} dt \right]_0^\infty$$

$$= -\lambda \left[\frac{-t e^{-\lambda t}}{\lambda} - \frac{e^{-\lambda t}}{\lambda^2} \right]_0^\infty$$

$$= \frac{1}{\lambda}.$$

Hence in Fig. 4.2 the atoms with shortest lives have the largest λ and are at A while the long-lived atoms are situated at B.

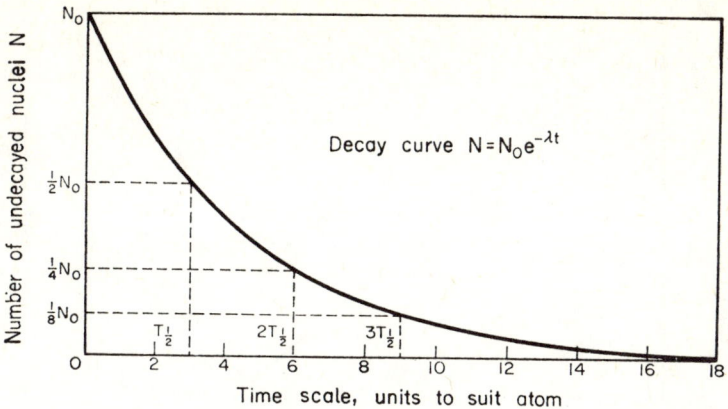

Fig. 4.3 Master exponential decay curve.

4.4 Half-lives of Radioactive Substances

A more usual method of discussing the lifetime of atoms is to consider the half-life $T_{1/2}$. This is the time required for the number of parent atoms to fall from $N = N_0$ to $N = \tfrac{1}{2}N_0$, as shown in Fig. 4.3. Thus in time $T_{1/2}$ one half of the parent atoms have decayed to daughter atoms, so that for the half-life $T_{1/2}$, we have

$$N = \frac{N_0}{2} = N_0 e^{-\lambda T_{1/2}}$$

$$T_{1/2} = \frac{\ln 2}{\lambda} = \frac{0\cdot 693}{\lambda}$$

Therefore
$$T_{1/2} = 0\cdot 693\ \bar{T}.$$

The numerical value of N_0 is taken at any arbitrary zero time.

Half-lives of radioactive substances vary a great deal. Polonium 212 emits α-particles and has a half-life of 3×10^{-7} s, whereas in the same series the precursor atom, thorium, has a half-life of $1\cdot 39 \times 10^{10}$ a, and is therefore almost stable.

From the above we can see that all lifetimes are expressed in terms of the decay constant. This can be measured experimentally since from
$$N_t = N_0 e^{-\lambda t}$$
we get
$$\ln\left(\frac{N_t}{N_0}\right) = -\lambda t$$

where N_0 is the activity at $t=0$ and N_t at time t. N_t/N_0 is the relative activity.

By simple counting techniques described in Chapter 17, counting after known intervals t gives the corresponding N_t values and $\ln(N_t/N_0)$ can then be plotted against t as in Fig. 4.4. The slope of this curve gives λ and hence $T_{1/2}$ or T.

4.5 Radioactive Series

Most of the natural radioactive elements have high atomic numbers. When they are plotted on an A, Z chart, it is found that they fall into three naturally

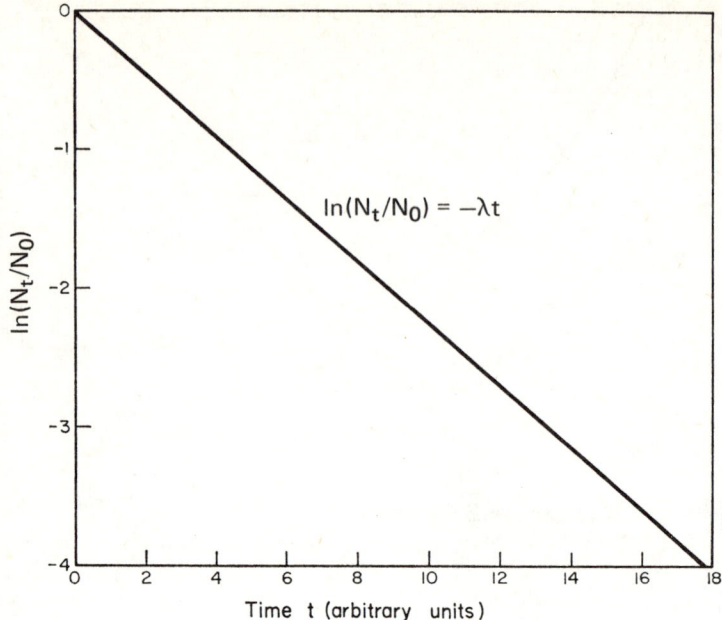

Fig. 4.4 Natural logarithmic plot of Fig. 4.3.

occurring series showing successive transformations. The three series are:

(1) the uranium series beginning with $^{238}_{92}$U and finishing with $^{206}_{82}$Pb (stable), as shown in Fig. 4.5. The atomic mass numbers of the members of this series are given by $4n+2$, n being an integer. It is therefore sometimes known as the $4n+2$ series. Note that a is the S.I. unit for '1 year'.
(2) the $(4n+3)$ or actinium series beginning with $^{235}_{92}$U and finishing with $^{207}_{82}$Pb (stable) as shown in Fig. 4.6.
(3) the $4n$, or thorium series beginning with $^{232}_{90}$Th and finishing with $^{208}_{82}$Pb (stable) as shown in Fig. 4.7.

The discovery of these three naturally occurring radioactive families was very largely the result of the work of Soddy in 1910.

In these diagrams an α-particle emission is given by a diagonal arrow and a β-particle by a horizontal arrow. With the discovery of nuclear fission and the production of the element neptunium ($Z=93$) it has been found that this element is a member of a fourth series, the neptunium or $4n+1$ series, starting at $^{241}_{94}$Pu and having for its stable end product the element $^{209}_{83}$Bi, as shown in Fig. 4.8.

4.6 Radioactive Equilibrium

Figure 4.1 is only true if we have a condition in which the daughter of a radioactive transformation is a solid material or if it cannot escape. Hence the

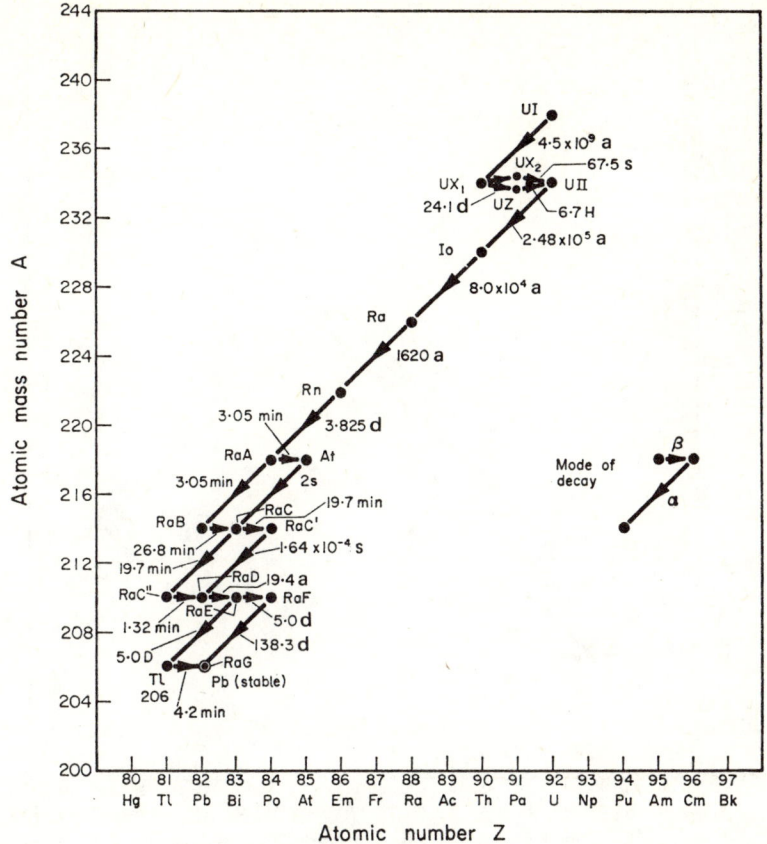

Fig. 4.5 The uranium series ($A = 4n + 2$).

parent and daughter atoms coexist and the two attain equilibrium such that the daughter nuclei disintegrate at the same rate as they are formed from the parent, or

$$\left(\frac{dN}{dt}\right)_{\text{parent}} = \left(\frac{dN}{dt}\right)_{\text{daughter}};$$

therefore $(\lambda N)_{\text{parent}} = (\lambda N)_{\text{daughter}}$ for equilibrium.

If we are dealing with a long series of disintegrations in a family, secular equilibrium is set up, so long as each member remains present, and $\lambda_1 \ll \lambda_2$ so that N_1 remains substantially constant. We therefore have

$$\lambda_1 N_1 = \lambda_2 N_2 = \lambda_3 N_3 \ldots = \lambda_n N_n$$

for n members. In general any two members can be related by

$$\lambda_x N_x = \lambda_y N_y.$$

This is the condition for radioactive equilibrium which is displayed in the four radioactive series just described. It also enables calculations of rates of accumulation on radioactive series from the precursor atom down to the stable

51

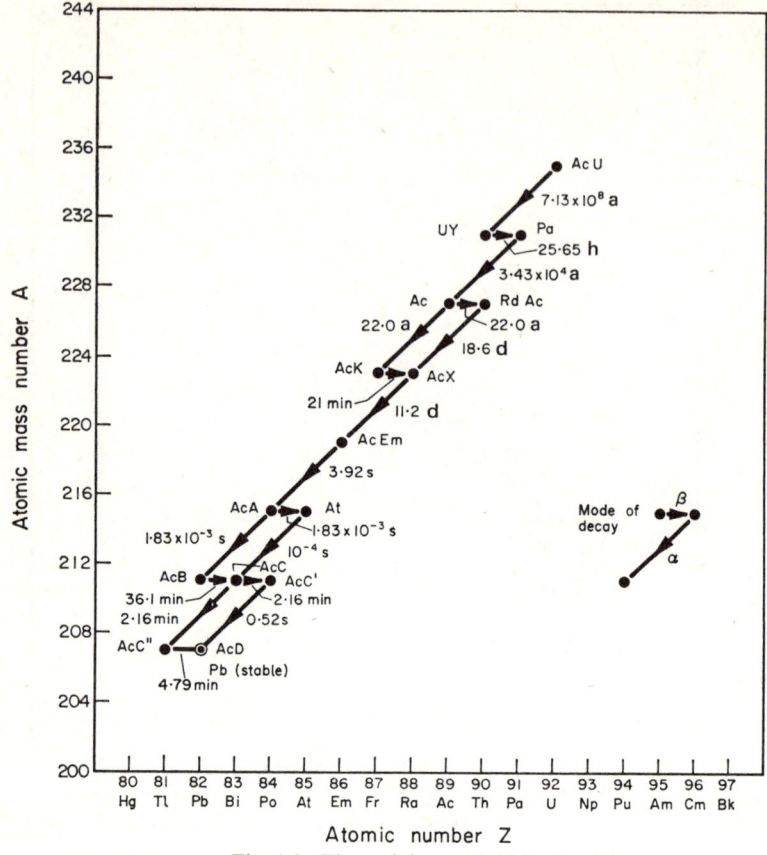

Fig. 4.6 The actinium series ($A = 4n + 3$).

isotope to be made, by applying the appropriate exponential growth and decay equations as required.

4.7 Isotopes

Examination and comparison of the three radioactive series found in nature shows that the same element can have different atomic masses, each of which is almost exactly integral. The name *isotope* was suggested for these by Soddy. For example the last and stable element in the uranium, actinium and thorium series is lead but the atomic mass numbers differ and are respectively 206, 207 and 208. It will be apparent that non-integral relative atomic masses arise from various abundances of isotopes.

In 1910, J. J. Thomson began a search for isotopes among the lighter non-radioactive elements. Neon was the first element to be successfully investigated, the discharge tube being not unlike that which had already been used in his e/m determination for the electron. The polarity of the electrodes of the main discharge tube were reversed and a fine hole, rather less than 1 mm in diameter,

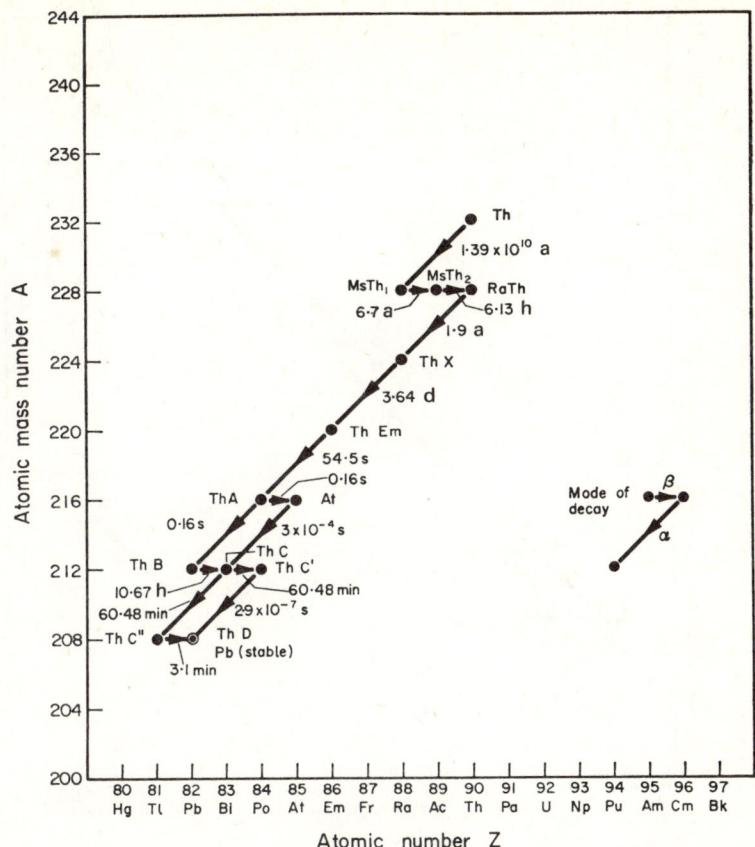

Fig. 4.7 The thorium series ($A = 4n$).

through the cylindrical cathode provided a fine pencil of positive ions instead of electrons. These were subjected to electric and magnetic fields which were parallel (Fig. 4.9). The mechanical forces upon the ions were therefore perpendicular to each other. The wide range of velocity of the ions compelled him to arrange the electric and magnetic fields in this way. The deflection of the pencil of ions was recorded on a photographic plate or fluorescent screen at P.

This arrangement was similar to that used by Kaufmann in his e/m measurements for β-rays described in Chapter 3. Using the same argument it can be shown that ions of charge E and mass M moving along the x axis (Fig. 4.10) will be spread out into a parabola in the y,z plane according to the equation

$$z^2 = \frac{B^2 E}{2YM} L^2 y = k \frac{E}{M} y,$$

k being a constant and L being the distance travelled by the ion in the magnetic and electric fields which were respectively B and Y. Since E was fixed by the degree of ionization, different parabolas were obtained for different values of M,

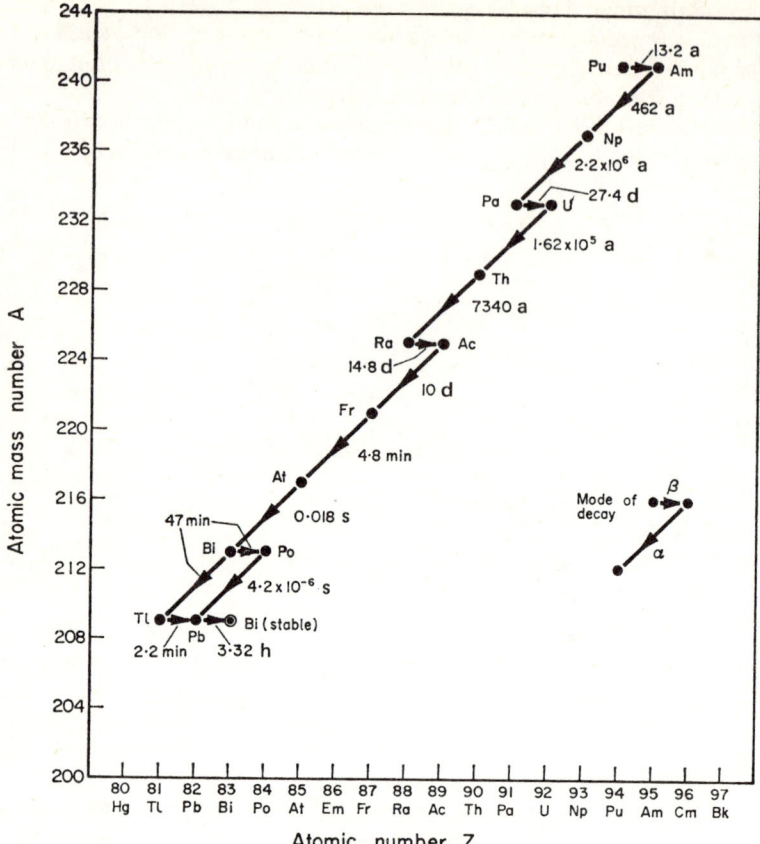

Fig. 4.8 The neptunium series ($A = 4n + 1$).

corresponding to different isotopes. Each parabola corresponds to the velocity distribution of ions of the same E/M produced in the discharge tube. The experiment demonstrated the existence of isotopes in light elements such as neon, but failed to determine their masses with accuracy. In practice a compromise had to be struck between the fineness of the hole in the cathode to ensure a narrow beam for accuracy, and the intensity of the parabola required for visibility or photography.

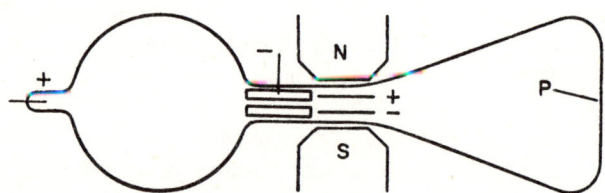

Fig. 4.9 Thomson's positive ray apparatus.

4.8 The Bainbridge Mass Spectrograph

Aston developed and improved Thomson's method, so that positive ions of the same mass were concentrated into a single line on a photographic plate. By this means Aston was able to make precise determinations of mass to one part in 10^3 and later one part in 10^4 in favourable cases. In this form the instrument became known as the mass spectrograph. Rather than give details of Aston's mass

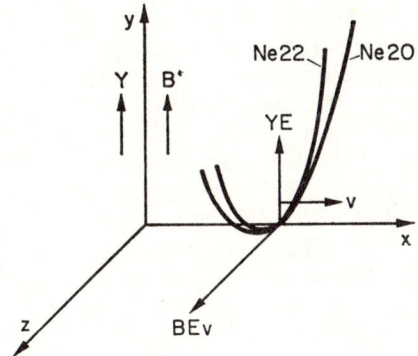

Fig. 4.10   Positive ray parabolas.

spectrograph, a more recent instrument devised by Bainbridge will be described.

This is based upon the deflection of the ions in a magnetic field but since the deflection is also dependent upon the velocity of the ion, a preliminary velocity selector is required to ensure that the velocities of the ions are the same within quite close limits. The beam of ions is restricted to a fine pencil by the slits S_1 and S_2 (Fig. 4.11) after which it is subjected to an electric field X and a magnetic field B perpendicular to each other. The directions are arranged so that the mechanical

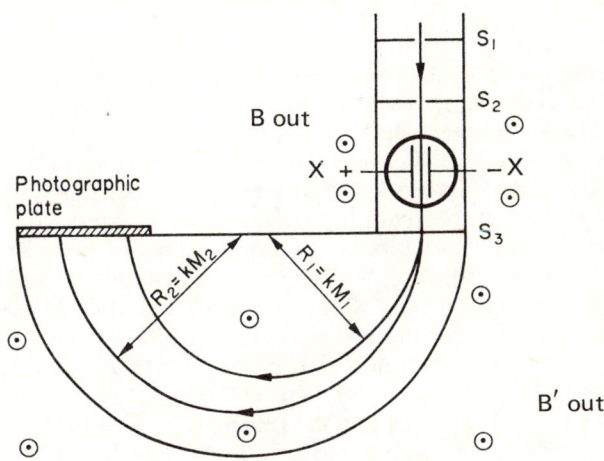

Fig. 4.11   Bainbridge mass spectrograph.

forces upon the ions are equal and opposite for a selected velocity v given by $v = X/B$ since $BEv = XE$. Only ions having this velocity continue in a straight line and pass through the slit S_3 into a region which is exposed to a uniform magnetic field B'. For an ion of charge E and mass M moving in a circular path of radius R, we see that $B'Ev = Mv^2/R$ from which $M = B'ER/v = kR$ when the velocity v is the same for all ions and k is a constant. Isotopes, which have the same charge E but differing mass M will follow paths of different radius and impinge at various points along the photographic plate as shown in Fig. 4.11.

Whilst Thomson's experiment showed that isotopes exist for lighter elements, the later and more refined work of Aston, Bainbridge, Dempster and Nier revealed further very interesting and far-reaching results. It was shown by modern mass spectrometry that exact relative isotopic masses are non-integral. Thus the isotopic masses of the ^{20}Ne, ^{21}Ne and ^{22}Ne isotopes are 19·992 440, 20·993 849 and 21·991 385 respectively. These numbers are relative to the standard mass on the physical scale, which is based on the ^{12}C isotope. The mass of this isotope is undertaken as 12·000 000 u, where u is the unified atomic mass unit which is used throughout this book and is given by

$$1\text{ u} = \tfrac{1}{12} \times \text{mass of 1 atom of the }^{12}\text{C isotope.}$$

By definition we know that 12 g of ^{12}C contain $6·0220 \times 10^{23}$ atoms (N_A). Thus

$$1\text{ u} = \tfrac{1}{12}\left(\frac{12}{N_A}\right)$$
$$= \frac{1}{6·0220 \times 10^{23}}\text{ g}$$
$$= 1·66 \times 10^{-27}\text{ kg.}$$

The modern value is $1\text{ u} = 1·660\ 565 \times 10^{-27}$ kg.

In Chapter 7 we shall see that the difference between the integral mass number and the exact relative isotope mass is related to the mass–energy changes involved in the synthesis of the isotope from its constituent nucleons.

Problems

(*The problems marked with an asterisk are solved in full at the end of the section.*)

4.1 For a radioactive element, define the disintegration constant and the half-life. Establish a relationship between these quantities.

If the half-lives of ^{235}U and ^{238}U are respectively $8·8 \times 10^8$ and $4·5 \times 10^9$ a, calculate the total number of α-particles emitted per second from 1 g of natural uranium. Both isotopes emit α-particles and the abundance of ^{235}U is 0·7%. Assume the atomic mass number of uranium to be 238. ($1·36 \times 10^4 + 1·87 \times 10^2$)

4.2 If the half-life of ^{152}Sm is 10^{12} a, calculate the number of α-particles emitted per second from 1 g of natural samarium. Abundance of ^{152}Sm is 27%, atomic mass number of natural samarium is 150·4. (23·8)

4.3 Outline the experimental evidence which led to the discovery of the neutron. When the nucleus of copper ($A = 65$, $Z = 29$) is bombarded with neutrons, γ-rays are emitted and the resultant nucleus is radioactive, emitting β-particles. If the disintegration constant λ is found to be 0·002 31, calculate the half-life and average life of the radioactive product. ($T_{1/2} = 300$ s, $\overline{T} = 433$ s)

4.4* Calculate the time required for 10% of a sample of thorium to disintegrate. Assume the half-life of thorium to be 1.4×10^{10} a. (2.1×10^9 a)

4.5 If the half-lives of ^{238}U, radium and radon are respectively 4.5×10^9 a, 1620 a and 3·8 d, calculate the relative proportions of these elements in a uranium ore which has attained equilibrium and from which the radon is unable to escape. ($4.7 \times 10^{11} : 1.56 \times 10^5 : 1$)

4.6 The positive ray parabola due to singley ionized ^{20}Ne atoms in a Thomson apparatus is represented by the equation $y = 10z^2$. Find the equation of the parabola due to the other isotope of neon, of atomic mass number 22. ($y = 11z^2$)

4.7 In a Thomson mass spectrometer the plates are 50 mm long and have parallel electric and magnetic fields equal to 15 kV m^{-1} and 0·5 T respectively. Find the equation of the parabola for the singly ionized ^{20}Ne isotope. The mass of a nucleon is 1.67×10^{-27} kg and the electronic charge is 1.6×10^{-19} C. ($y = 10z^2$)

4.8 In the velocity selector of a Bainbridge mass spectrograph the electric field is 20 kV m^{-1} and the magnetic flux density 0·25 T. Calculate the velocity of the ions which pass through undeviated. (8×10^4 m s^{-1})

4.9 Singly ionized magnesium atoms enter a Bainbridge mass spectrograph with a velocity of 3×10^5 m s^{-1}. Calculate the radii of the paths followed by the three most abundant isotopes, of atomic mass numbers 24, 25 and 26, when the magnetic flux density is 0·5 T. The mass of a nucleon is 1.67×10^{-27} kg. (325, 313, 300 mm)

4.10* In a Bainbridge mass spectrograph, singly ionized atoms of ^{20}Ne pass into the deflection chamber with a velocity of 10^5 m s^{-1}. If they are deflected by a magnetic field of flux density 0·08 T, calculate the radius of their path and where ^{22}Ne ions would fall if they had the same initial velocity. ($r = 260$ mm, ^{22}Ne ions at 52 mm beyond ^{20}Ne ions)

Solution to Problems

4.4 It is convenient to work in years. Thus
$$\lambda = \frac{0.693}{1.4 \times 10^{10}} \text{ a}^{-1}$$

From $N = N_0 e^{-\lambda t}$ we have
$$\frac{N}{N_0} = \frac{9}{10} e^{-\lambda t}$$

and so
$$\lambda t = \ln(10/9) = 2.302 \times \log(10/9)$$
$$= 2.302 \times 0.04575$$
$$= 0.1053.$$

Thus
$$t = \frac{0.1053}{0.693} \times 1.4 \times 10^{10}$$
$$= 2.1 \times 10^9 \text{ a}.$$

4.10 For an ion of ^{20}Ne moving in a magnetic field B we can write
$$Bev = \frac{mv^2}{r_{20}}$$
$$r_{20} = \frac{mv}{Be} = \frac{20 \times 10^{-3}}{6 \times 10^{23}} \times \frac{10^5}{0\cdot 08 \times 1\cdot 6 \times 10^{-19}}$$
$$= 0\cdot 26 \text{ m.}$$
In this case $r \propto m$ and the radius of the path followed by ^{22}Ne is given by
$$\frac{r_{22}}{r_{20}} = \frac{22}{20}$$
$$r_{22} = \frac{22}{20} \times 0\cdot 26$$
$$= 0\cdot 286 \text{ m.}$$
The ^{22}Ne ions would therefore fall 52 mm beyond the ^{20}Ne ions.

Chapter 5

The Electromagnetic Spectrum

5.1 Theories of Light

Light travels from the sun to the earth, a distance of about 145 000 000 km, through space containing very little material. When absorbed by a surface it is converted into heat, a form of energy. Energy must therefore have arrived from the sun across this immense distance. In fact almost all the energy known to man has been derived from the sun either now or in past ages. Thus the sun's energy, which millions of years ago was responsible for the growth of luxurious vegetation, is now available to us in the form of coal. To account for this transfer of energy over such a large distance we must know something of the nature of light. Energy can pass from one place to another in two ways. The kinetic energy of a moving body which obeys the laws of mechanics is the essential feature of the corpuscular theory as advocated by Newton at the close of the seventeenth century on the basis of the experimental evidence known to him at that time. On the other hand, energy can also pass from one place to another by a wave motion. This was the basis of the wave theory of light supported by Hooke and Huygens. Sound was then known to be a wave motion and the fact that one could hear but was unable to see around corners proved to be a serious obstacle to the acceptance of the wave theory of light for over a hundred years even though it was known that light deviated very slightly from its straight line path on passing close to the edge of an obstacle.

A satisfactory explanation of rectilinear propagation in terms of wave motion was finally given by Fresnel following the discovery of interference by Young in 1802. In the nineteenth century, investigations of interference and later diffraction and polarization phenomena all received satisfactory explanation by the wave theory, which quickly superseded the corpuscular theory of the previous century. A corpuscular theory has again emerged in the twentieth century in which photons are used to explain photoelectricity and the Compton effect (see Chapter 6).

5.2 Interference

Young's experiment consisted of placing two pin-holes S_1, S_2 (Fig. 5.1) at some distance from a single pin-hole S. A series of bright and dark bands were then observed upon a screen placed at P arising from the interference between light passing through S_1 and S_2 respectively. Young's fringes can be readily observed

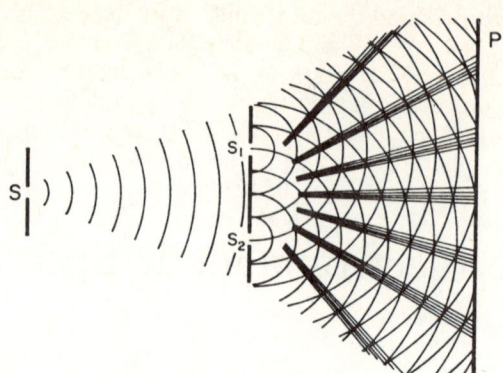

Fig. 5.1 Formation of Young's fringes.

by placing two parallel slits S_1, S_2 (Fig. 5.2) in front of the eye and viewing a single slit S some distance away, when the actual fringes are formed upon the retina of the observer's eye.

Two wave trains can pass through one another without suffering any change. Once they have parted they pursue their respective paths as if no encounter had occurred. Since the waves are independent we must suppose that the displacement of the medium at any point is the algebraic sum of the separate

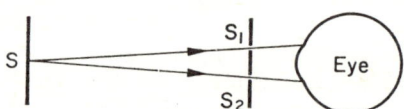

Fig. 5.2 Observation of Young's fringes.

displacements due to each wave. This is a necessary condition so that for two waves of equal amplitude there will be regions of little or no displacement, and also those having up to double the normal amplitude. Such regions correspond to the bright and dark lines or 'fringes' observed in Young's experiment. To observe them another condition must also be satisfied. A definite phase relationship must be maintained between the two wave trains. Since a source emits light in flashes lasting about 10 ns, and there is no phase relationship between the flashes of even a single source, the above condition can only be satisfied by deriving the two wave trains from the same source. Phase changes in one wave train are therefore always accompanied by corresponding phase changes in the other. We sat that the wave trains are coherent.

Returning now to Young's experiment we must imagine two wave systems emerging from S_1 and S_2 respectively. The crests of such waves are drawn with continuous lines (Fig. 5.1). In regions where the crests overlap, corresponding to a path difference of a whole number of wavelengths between the two wave trains, we can expect brightness and the directions along which this occurs are shaded. Between each crest lies a trough and where such a trough coincides with the crest of the other wave system there is little or no displacement, giving darkness. It will be observed that these occur in areas between the bright fringes already marked. The illumination at a given point P can be predicted by counting the number of waves m in the path difference $S_2P - S_1P$. Brightness occurs when the waves are in phase or in step so that $S_2P - S_1P = m\lambda$, λ being the wavelength. Darkness occurs when $S_2P - S_1P = (m + \frac{1}{2})\lambda$, so that the waves are exactly out of step. Calculations along these lines enabled Young to make the first estimates of the wavelength of light.

5.3 Diffraction

The wave theory of light must also be used to explain the phenomenon of diffraction. A simple description of diffraction at a narrow slit is given as it will be required later. If a series of plane waves corresponding to a parallel beam of light fall upon an aperture, edges of the wave fronts beyond the aperture become curved as shown in Fig. 5.3. As the slit is made narrower the effect becomes more

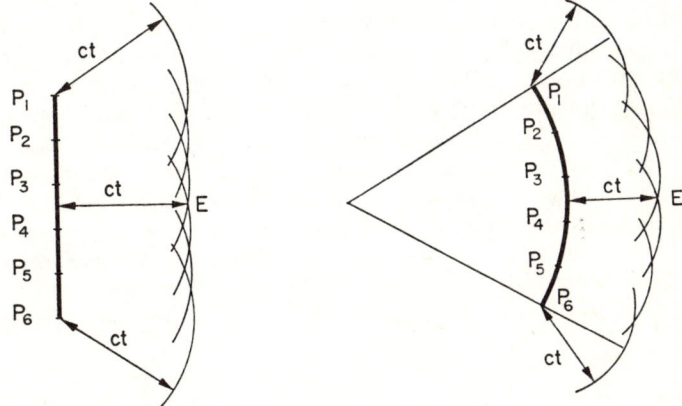

Fig. 5.3   Huygens' principle.

marked and the beam spreads out. Water waves in a ripple tank can be used to demonstrate this phenomenon which is known as diffraction. To understand diffraction we must first describe Huygens' principle. Any one crest of the waves shown between S and S_1, S_2 in Fig. 5.1 can be regarded as the locus of points having the same phase. This is also true of each trough between the crests and for all intermediate stages between these two extremes. A line joining all points of similar phase is known as a wave-front. Huygens' principle enables the position of such a wave-front to be calculated after a given interval of time t. To do this we suppose each point $P_1 P_2 P_3$ (Fig. 5.3) in the wave-front to be a source of so called 'secondary wavelets', and the wavelets from each of these points lie on circles of radius ct, c being the velocity of light. The envelope E to these numerous circles constitutes the new wave-front. In this simple picture of wave propagation we see how the wavelets travelling to the sides become increasingly important as the aperture is reduced. The angle at which the intensity of the spreading beam just falls to zero can be calculated from the width of the aperture a, and the wavelength of the light λ, using the following simple argument. Textbooks on optics should be consulted for more complete and rigorous treatments.

AB is a wave-front (Fig. 5.4) proceeding through the aperture and along which

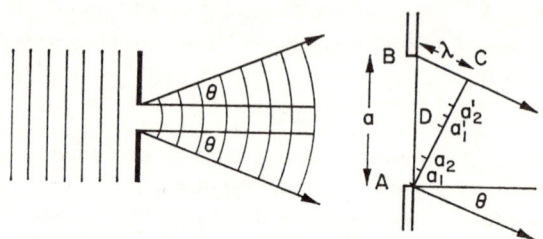

Fig. 5.4 Diffraction at a narrow aperture.

all the displacements are in the same phase. For the direction θ, when $BC = \lambda$, there will be a path difference of one wavelength across the wave-front AC, which is proceeding at an angle θ to the direct wave. The wave-front AC can now be divided into two halves, AD and DC, such that for each point a_1, a_2 in the lower half there is a corresponding point a'_1, a'_2 in the upper half. The disturbances from these points will have a path difference of exactly $\lambda/2$ and will cancel on being brought to a focus by a lens or a mirror. The net effect for the whole wave-front, proceeding in a direction θ and having a path difference of λ across it, is zero displacement of the medium, corresponding to zero intensity of the light beam. The angle θ is therefore given by $\sin\theta = \lambda/a$. It is instructive to examine the values taken by θ as the width of the aperture is reduced below 100λ and we must remember that even a slit 100λ wide is only 50 μm for green light (wavelength 500 nm).

a	100λ	10λ	5λ	2λ	1λ
θ	0·57°	5·75°	11·5°	30°	90°

The spacing of the diffraction grating ruled by Rowland in 1881 was about 3λ.

5.4 Spectra

The nineteenth century also saw a great development in our knowledge of spectra. In 1672 Newton had already shown that the white light from the sun is a mixture of all colours which are revealed when a narrow pencil of sunlight is passed through a triangular glass prism. The colours extend from red through the various rainbow colours to violet. In 1800 Herschel showed, by placing a thermometer beyond the red, that there was a heating effect due to radiation, which we now call infra-red. In the following year Ritter discovered the ultra-violet by showing that silver chloride was blackened even when placed beyond the violet end of the spectrum. This was, of course, before photography as we know it today was developed.

Young's original interference experiment paved the way for all subsequent wavelength measurements and showed clearly the connection between colour and wavelength. The visible spectrum extends roughly from violet light of wavelength 400 nm to red light at 700 nm. Later, in 1814, Fraunhofer extended Young's experiment to many parallel slits producing the first diffraction gratings, with which he made much more reliable wavelength measurements. It was not until 1859 that Kirchhoff stated explicitly that each element radiates a characteristic spectrum, and together with Bunsen laid the foundations of spectrochemical analysis. Several new elements were discovered by this means, the most notable being helium, which was discovered by observations on the sun's spectrum at a time when this element was still unknown on the earth.

Spectra may be divided into three types, according to their appearance and origin. Line spectra, consisting of discrete wavelengths distributed throughout the spectrum and characteristic of the element concerned, arise from vapour in a flame, an arc, a spark, or from the passage of electricity through a gas or vapour at low pressure in a vacuum tube. The lines arise from energy transitions within the atom of the gas or vapour. Continuous spectra arise from incandescent solids such as heated lamp filaments, the positive crater of an arc, or the mouth of a heated furnace. The distribution of energy in such a spectrum depends only upon the temperature of the source and has provided vital information about the interaction between matter and energy as we shall see in the next chapter. We must imagine that the atoms in a solid lie very close together and exert considerable forces upon one another, so that they are no longer able to emit their characteristic wavelength spectra. The third type is called band spectra and when observed with a simple low dispersion spectroscope have a characteristic fluted appearance which readily distinguishes them from line spectra. Several bands normally occur, each having a sharp edge called a 'head', and shading off gradually towards the red or violet. When examined with higher dispersion, such as that available with a large diffraction grating, each band is seen to be composed of many fine lines becoming closer and closer towards the head of the band. Such spectra arise from molecules, in which the very fine lines correspond to energy changes between various possible molecular rotations while each band corresponds to a change in the energy of molecular vibration. Molecular spectra lie outside the scope of this book, but continuous and line spectra will be discussed in more detail later.

5.5 The Electromagnetic Theory

The wave theory of light thus became firmly established in the first half of the nineteenth century. That the waves were transverse was required by the properties of polarized light. The nature of the waves, however, remained a mystery. At first they were supposed to be waves in an elastic medium called the ether, which extended throughout space. Great difficulties were experienced in accepting the mechanical properties required of such a medium. At about the same time Faraday used the concept of electrical tubes of force, which he visualized as strains in a surrounding medium to account for the attraction and repulsion of electric charges. It was Maxwell, in 1864, who linked the electrical medium and the luminiferous ether in his electromagnetic theory of light, to be brilliantly supported by the experimental work of Hertz 24 years later.

Maxwell assumed that when an electrical strain was being established in a medium, a momentary current, called a displacement current, flowed in the medium. This, he supposed, had the properties of a normal current, and was therefore accompanied by a momentary magnetic field. The changing magnetic field in its turn produced a further momentary displacement current, and so the process continued, energy being transferred between the electric and magnetic fields. Maxwell showed that such energy was transferred from one place to another in free space with the velocity of light. The frequency with which the energy was transferred between the electric and magnetic form corresponds to the frequency of the radiation. Thus our concept of a vibrating ether has been replaced by one in which the light wave consists of periodic changes in electric and magnetic fields.

Figure 5.5 shows how an electromagnetic wave travelling along the x axis may

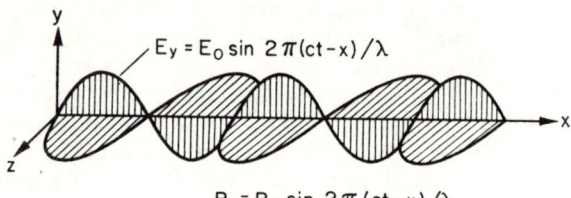

Fig. 5.5 An electromagnetic wave.

be visualized as two sinusoidal waves given by $E_y = E_0 \sin 2\pi(ct-x)/\lambda$ and $B_z = B_0 \sin 2\pi(ct-x)/\lambda$, where E_y and B_z are respectively the electric and magnetic vectors at the point x for a wave of wavelength λ travelling with a velocity c. Differentiation of the first of these equations with respect to x and t gives

$$\frac{d^2 E_y}{dt^2} = -\frac{4\pi^2}{\lambda^2} c^2 E_0 \sin 2\pi(ct-x)/\lambda$$

and

$$\frac{d^2 E_y}{dx^2} = -\frac{4\pi^2}{\lambda^2} E_0 \sin 2\pi(ct-x)/\lambda,$$

from which

$$\frac{d^2 E_y}{dt^2} = c^2 \frac{d^2 E_y}{dx^2};$$

similarly, for the magnetic wave we may write

$$\frac{d^2 B_z}{dt^2} = c^2 \frac{d^2 B_z}{dx^2}.$$

These are differential equations representing electric and magnetic waves travelling along the axis of x with a velocity c. The wave may take the form of either a sine or a cosine wave or any combination of the two.

Maxwell used Faraday's induced e.m.f. law and Ampere's theorem together with his new concept of a displacement current, and showed that

$$\frac{d^2 E_y}{dt^2} = \frac{1}{\mu_0 \varepsilon_0} \frac{d^2 E_y}{dx^2}$$

and
$$\frac{d^2 B_z}{dt^2} = \frac{1}{\mu_0 \varepsilon_0} \frac{d^2 B_z}{dx^2},$$
where μ_0 and ε_0 are respectively the permeability and permittivity of free space. We have just seen that these are differential equations representing electric and magnetic waves travelling along the x axis with a velocity given by
$$c = \frac{1}{\sqrt{\mu_0 \varepsilon_0}}.$$
The values for μ_0 and ε_0 in S.I. units are given by
$$\mu_0 = 4\pi/10^7 \text{ N C}^{-2} \text{ s}^2 \quad (\text{H m}^{-1})$$
and
$$4\pi\varepsilon_0 = 1/(9 \times 10^9) \text{ N}^{-1} \text{ C}^2 \text{ m}^{-2} \quad (\text{F m}^{-1})$$
Substitution gives
$$c = \frac{1}{\sqrt{\mu_0 \varepsilon_0}} = \sqrt{\frac{10^7}{4\pi} \, 4\pi \times 9 \times 10^9} = 3 \times 10^8 \text{ m s}^{-1}.$$
Thus we see that c not only has a numerical value equal to the velocity of light in free space, but also has units (metres per second) which correspond to velocity. The derivation of the velocity of light from constants determined solely by electrical means, represents a remarkable triumph for the electromagnetic theory of light. It also suggested that it should be possible to generate similar waves of different wavelength by purely electrical means, and that these too would have the same velocity in free space.

5.6 Hertz's Experiment

Experiments to produce and detect such waves were carried out by Hertz in 1888 using an induction coil connected to two plates as shown (Fig. 5.6). Sparks

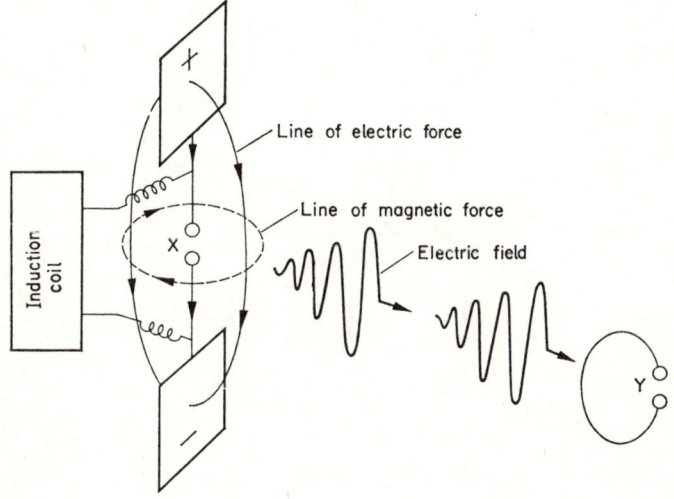

Fig. 5.6   Hertz's experiment.

passed across the spark gap X when the potential difference was raised sufficiently by the induction coil to ionize the air. The places discharged across the conducting path in an oscillatory manner, at frequency governed by the inductance and capacitance of the circuit. Since these were small, the frequency was high (about 10^9 Hz or 1 GHz) and the wavelength was only about 0·3 m. These waves were detected by a loop of wire, Y, some distance away, the length of the loop being adjusted so that the currents induced in it by the waves were at resonance. The presence of the waves was detected by sparks which passed between the balls when the potential in the loop circuit became sufficiently high. That the electric and magnetic fields are perpendicular to each other can be seen from a consideration of the lines of force near the spark gap. When the upper plate carried a positive charge and the lower one a negative charge, the electric field may be represented by lines of electric force passing from the upper to the lower plate. When the current flows there will also be a magnetic field represented by lines of magnetic force in circles about the path of the electric current. Thus we see how the electric and magnetic fields are at right angles.

5.7 The Electromagnetic Spectrum

Hertz showed that these waves had the same velocity as light waves and differed only in wavelength and frequency. Experimental evidence steadily

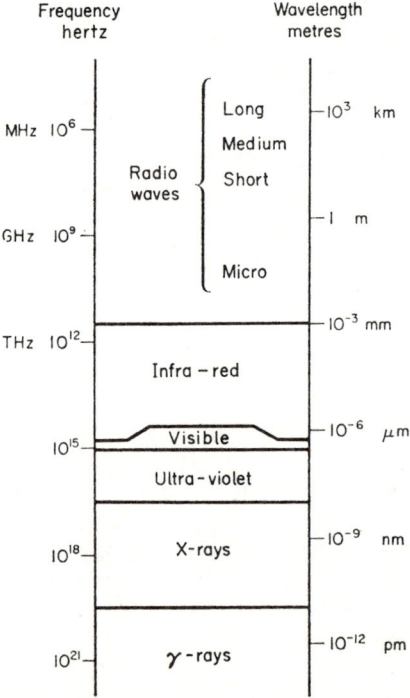

Fig. 5.7 The electromagnetic spectrum.

accumulated which showed that the wavelength and frequency range of electromagnetic waves was very wide indeed, and that such waves displayed widely different properties according to their wavelength. Figure 5.7 shows that the visible light waves occupy a very limited portion of the full electromagnetic spectrum which extends upwards through the infra-red to radio waves having wavelengths of several thousands of metres. Extension to shorter wavelengths goes through the ultra-violet to X-rays and γ-rays of which we shall hear more in later chapters. It must also be appreciated that the boundaries between the various regions of the spectrum are not as clearly defined as Fig. 5.7 suggests and in most cases there is a considerable region of overlap.

Problems
(*The problem marked with an asterisk is solved in full at the end of the section.*)

5.1 Describe the type of motion represented by an equation of the form $y = y_0 \sin 2\pi(t/T - x/\lambda)$, where x, y are Cartesian coordinates and t is the time. Identify the constants y_0, T and λ.

5.2 Show how two similar wavetrains travelling in opposite directions can give rise to stationary waves.

5.3 Show how the law of reflection of light by a plane mirror may be derived from the wave theory of light.

5.4 Assuming refractive index to be the ratio of the velocity of light in vacuum to the velocity in a given medium, use the wave theory of light to establish a formula for refractive index in terms of the angles of incidence and refraction.

5.5 Use the wave theory of light to prove that the radius of curvature of a spherical mirror is twice its focal length.

5.6 Use the wave theory to derive a formula for the focal length of a thin lens in terms of refractive index and the radii of curvature of its surfaces.

5.7 A single slit, illuminated with sodium light of wavelength 600 nm, is viewed at a distance of 5 m through a pair of fine parallel slits 0·1 mm apart, held close to the eye. Calculate the separation of the fringes as seen against a metre rule held in the same plane as the single slit. (30 mm)

5.8 A parallel beam of light of wavelength 546·1 nm from a mercury arc falls normally upon two fine slits 0·5 mm apart. Calculate the linear separation of interference fringes observed at a distance of 2 m beyond the slits. (2·18 mm)

5.9 Ten fringes are observed in a distance of 3·2 mm at 0·5 m from a double slit illuminated by a parallel beam of monochromatic light. Calculate the wavelength of the light used, if the slits are separated by a distance of 1·0 mm. (640 nm)

5.10* Calculate the size of the pin-hole of a pin-hole camera which will yield the best definition of an object at infinity. ($\sqrt{2\lambda d}$, where d is distance from pin-hole to screen)

5.11 What are the frequencies of (*a*) a radio wave of length 100 m, (*b*) a microwave of length 10 mm, (*c*) a light wave of length 600 nm, (*d*) an X-ray wave of length 0·6 nm? (3×10^6, 3×10^{10}, 5×10^{14}, 5×10^{17} Hz)

5.12 A plane surface divides two media having different refractive indices. Show that the path followed by a light ray from a given point in one medium to another point in the other medium takes the least time. (This is known as Fermat's principle.)

Solution to Problem

5.10 Without diffraction a point object such as a star would give rise to a patch of light on the screen equal in diameter a to the pin-hole. Diffraction will cause an angular spread of the light all around the above patch equal to λ/a. If d is the distance of the screen from the pin-hole the linear spread of the light around the original patch is $d.\lambda/a$. The total width D of the light patch is therefore

$$D = a + \frac{2\lambda d}{a}.$$

The minimum value of D is given by

$$\frac{dD}{da} = 1 - \frac{2\lambda d}{a^2} = 0$$

from which

$$a = \sqrt{2\lambda d}.$$

Chapter 6

Quantum Theory

6.1 The Continuous Spectrum

The quantum theory arose in the first instance out of attempts to explain the distribution of energy in the continuous spectrum of an incandescent body. As a piece of metal is heated to incandescence it first becomes red at about 850 K. Later, as the temperature rises further, it becomes yellow and then white at about 3000 K, when all the visible spectrum is being radiated. Experimental investigation of the energy distribution for various temperatures, yields curves of the form shown in Fig. 6.1. It is at once apparent that, as the temperature rises, not only does the energy increase, but the wavelength of maximum energy λ_{max} moves to the region of shorter wavelength. Analysis of the curves shows that $\lambda_{max} T$ = constant. This formula can be derived using the thermodynamical reasoning of classical physics and is known as Wien's displacement law. Wien also showed that the form of the curve could be represented empirically by a formula $E_\lambda = C_1 \lambda^{-5} \exp(-C_2/\lambda T)$, where C_1, C_2 are constants and E_λ is the energy radiated

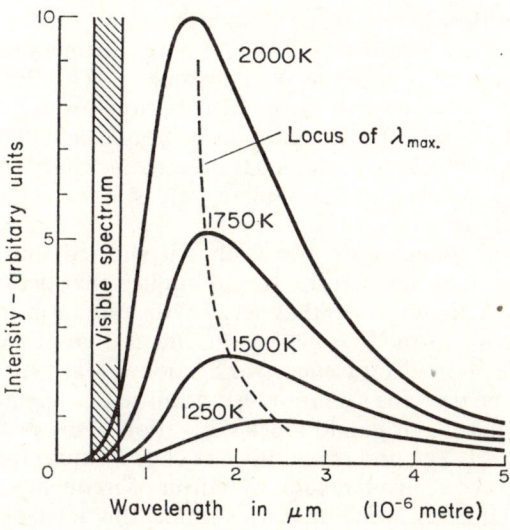

Fig. 6.1 Energy distribution in spectrum of an incandescent solid at various temperatures.

at wavelength λ. This formula is in very close agreement with experiment for small values of T in the visible spectrum corresponding to temperatures up to 2000 K, but diverges for larger values of λT. It also gave the displacement law $\lambda_{\max}T = $ constant. (See Problem 6.14.)

When Rayleigh and Jeans attempted to derive a formula of this type using the concepts of classical physics, especially the law of equipartition of energy, described in Chapter 1, they obtained $E_\lambda = 8\pi k\lambda^{-5}(\lambda T)$, where k $(=R/N_A)$ is Boltzmann's constant, and R and N_A are respectively the molar gas constant and the Avogadro constant. Although this gave agreement for very large values of λT, it failed to give curves resembling those found experimentally. Even more serious, it predicted that the total radiation from a body at a finite temperature should be infinite, since as $\lambda \to 0$, $E \to \infty$. Thus classical physics was unable to explain the facts of temperature radiation.

6.2 Planck's Quantum Theory

In 1901 Planck showed that a successful theory of radiation was possible by making a revolutionary assumption regarding the way in which radiation is emitted or absorbed by atoms. He supposed that the black-body radiator consisted of a set of N resonators, or oscillators, of total energy U_N, such that $U_N = N\varepsilon$, where N is a large integer and ε is a discrete unit of energy. Thus using Planck's theory, the energy is not a continuous infinitely divisible quantity but is made up of a large but finite number of equal units ε. Planck then proved that ε was proportional to the frequency v of the emitted radiation, i.e. $\varepsilon = hv$, where h is the constant of proportionately we now call the Planck constant. This assumption led to the formula

$$E_\lambda = \frac{8\pi hc\lambda^{-5}}{\exp(ch/k\lambda T) - 1}$$

for the energy *density*, where k is the Boltzmann constant.

Planck's new formula agreed very well with the experimental curves of Fig. 6.1 for *all* values of λ and T. It gave the Wien formula as $\lambda T \to 0$ and the Rayleigh–Jeans formula as $\lambda T \to \infty$, as well as the distribution law $\lambda_{\max}T = $ constant and Stefan's fourth power law. The magnitude of h as calculated by Planck was $h = 6.55 \times 10^{-34}$ J s and he also evaluated the electronic charge as 1.55×10^{-19} C, both values being very near to the modern values, a fact not recognized until many years afterwards.

The true value of Planck's work was that it showed that energy must be regarded as having an atomic nature. It also implied that because the energy of the resonators was in discrete units the energy of the radiation emitted was *also* in discrete units in the form $E = nhv$, i.e. in integral multiples of hv. In his investigation of the photoelectric effect (see Section 6.4) in 1905, Einstein proved that this could not be true, the radiation being emitted in *single* localized units of hv with the result that the original Planck derivation of the distribution law was not absolutely correct. The unit of energy hv is called a 'quantum' of energy, or a 'photon' if referred to electromagnetic radiation of frequency v.

At first Planck and his contemporaries found this idea of a discontinuous emission of energy very hard to accept, but it soon became the only explanation of several other phenomena in physics and was recognized as the foundation of modern atomic physics.

It is worthwhile considering how this assumption accounts for the general form of the energy distribution before proceeding to the other evidence for quantum theory. In a solid we have to suppose that the atoms are fixed relative to one another and are only capable of oscillation about a mean position. The thermal energy of the body will be distributed among the atoms as kinetic and potential energy, in much the same way as we supposed in our discussion of the kinetic theory of gases. Not all atoms will have the same energy, but the mean kinetic energy is dependent upon the temperature. A few atoms will have energies greatly in excess of the mean while the energies of others will be very much smaller than the mean. At low frequencies the quantum of energy is small and the process of radiation very nearly approximates to the continuous process visualized by classical physics. It was for this reason that Rayleigh and Jeans were able to predict successfully the energy distribution for large values of λT. As we proceed to higher frequencies, the quantum hv increases until it exceeds the mean energy of the atoms in the solid. Beyond this value fewer and fewer atoms will have sufficient energy available to radiate the necessary quantum hv. Thus the amount of energy rises to a maximum and falls away again as we proceed to higher and higher frequencies. Moreover, as we raise the temperature of the body, the average kinetic energy of the atoms is raised, so that this energy corresponds to a larger quantum hv and a correspondingly larger frequency. At a higher temperature there will be a greater probability of an atom acquiring sufficient energy to radiate a really large quantum of energy hv. Thus we can see in a general way how raising the temperature of a body must lead to shift of the radiation maximum towards the region of higher frequency (i.e. shorter wavelength).

6.3 The Photoelectric Effect

The photoelectric effect, which provides one of the most striking confirmations of quantum theory, was discovered by Hertz in 1887. He showed that when a spark gap was illuminated with ultra-violet light, the electricity could discharge across it more readily. In the following year Hallwachs showed that ultra-violet light, falling upon a negatively charged zinc surface, caused the negative charge to leak away. When, however, the zinc surface was positively charged there was no leakage. This implied that the negative charge was able to escape from an insulated surface, but that a positive charge was unable to do so. Later it was shown that the negative charge was carried by electrons which were able to escape from the surface, whereas there was no corresponding mechanism for the positive charge.

In 1899, from measurements of e/m for the carriers of the negative charge, Lenard established that they were electrons. Ultra-violet light entered through the quartz window Q (Fig. 6.2) and fell upon the clean metal surface at A. The electrons which escaped from A were accelerated to E by a large potential difference V, so that they acquired a velocity v given by $Ve = \frac{1}{2}mv^2$, where e was the electronic charge and m was the electronic mass. In E there was a small hole, so that a pencil of electrons continued to the electrode D. A uniform magnetic field B, perpendicular to the plane of the paper (Fig. 6.2), deflected the electron beam on to a second electrode C, the velocity or magnetic field being adjusted so that the beam just reached C. From the geometry of the apparatus, the radius R of the path of the electron was known, and was given by the equation $Bev = mv^2/R$. From these equations, $e/m = 2V/B^2R^2$ and $v = 2V/BR$.

In this experiment the initial velocity with which the electrons escaped from the metal was ignored. It was, in fact, small compared with the velocity acquired by

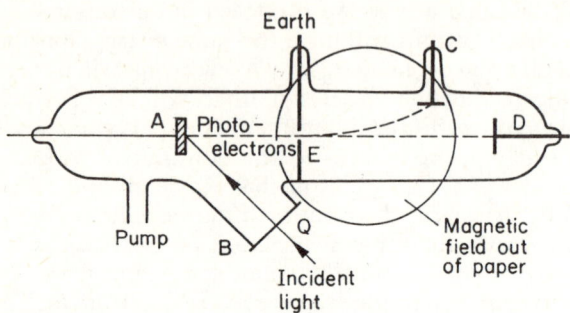

Fig. 6.2 Lenard's apparatus.

the electron in the electric field between A and E. Lenard investigated the velocity with which the electrons were ejected from A by making A positive with respect to E, which was maintained at earth potential. He measured the potential V_0 required to return all the escaping electrons to the metal A, so that $V_0 e = \frac{1}{2} mv_0^2$, where v_0 was the maximum velocity with which electrons escaped from the metal.

These experiments led Lenard to a most important discovery. The velocity v_0, and therefore the maximum energy of the escaping electron, did not depend upon the intensity (i.e. energy) of the incident beam of light, as might well have been supposed from classical physics and the electromagnetic theory of light. The velocity of the electron appeared to be determined solely by the frequency of the light, so that velocity increased weth frequency and the nature of the metal surface illuminated. The intensity only affected the *number* of escaping electrons, as indicated by the current across the tube, and not the energy of the electrons.

6.4 Einstein's Equation

The explanation of Lenard's observations was given by Einstein in 1905, using the quantum theory of radiation initiated by Planck a few years earlier. It gave new emphasis to the fundamental nature of Planck's assumptions in quantum theory. Lenard's work was of such basic importance to quantum theory that it was later repeated and confirmed by Millikan, who used substantial refinements. Using the idea that radiant energy was atomic in nature, and consisted of photons or quanta of frequency v and energy hv, Einstein supposed that the whole energy of a quantum could be transferred to a single electron within a metal. This energy enabled the electron to escape from the metal, but in doing so a certain amount of energy W was used up in bringing the electron to the surface, W being known as the work function of the metal. The remainder of the energy appeared as kinetic energy of the emitted electron. Thus the energy equation can be written $hv = W$

$+\frac{1}{2}mv^2$ and it can be seen how the energy of the electron increases with the frequency of the incident light.

Later, in 1916, Millikan used the alkali metals lithium, sodium and potassium which display the photoelectric effect very strongly with visible light as well as with ultra-violet. He was thus able to test the above relationship over a much wider range of conditions than in earlier work. Reliable results were only possible when clean metal surfaces were available, and Millikan, by the ingenious design of his apparatus, was able to cut clean surfaces in a vacuum.

Light from a mercury source passed through a spectrometer in which the telescope eyepiece was replaced by a second slit so that the instrument became a monochromator enabling light of a selected frequency to enter the vacuum chamber through the quartz window (Fig. 6.3). Quartz components were used

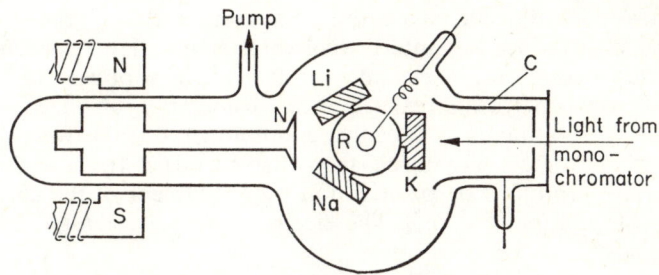

Fig. 6.3 Millikan's photoelectric experiment.

when observations on wavelengths less than 300 nm (i.e. frequencies greater than 10^{15} Hz) were required. The light then fell upon a freshly prepared metal surface. The metals were mounted upon a drum R and were prepared by cutting a thin section with the knife N. Rotation about R enabled the surface to be placed in a position to receive the light. Electrons were collected upon the cylinder C and detected by a sensitive electroscope. The metal surface was given a positive potential with respect to the cylinder C. This potential was adjusted until electrons were just prevented from escaping from the metal surface. Thus the maximum velocity v with which electrons of charge e escape from the surface was given by $Ve = \frac{1}{2}mv^2$, where V is the 'stopping potential'. The maximum energy $\frac{1}{2}mv^2$ of escaping electrons could therefore be found for a range of frequencies v and plotting these gave the straight line shown in Fig. 6.4. The equation of this line was represented by $Ve = hv - hv_0$ where v_0 was known as the threshold frequency. Comparing this with Einstein's photoelectric equation, the work function was given by $W = hv_0$. The gradient of the straight line was equal to the Planck constant h and the method proved one of the most reliable methods of determining this basic constant. The value obtained for h was $6 \cdot 57 \times 10^{-34}$ J s and confirmed Planck's estimate of $6 \cdot 55 \times 10^{-34}$ J s from the distribution of energy in the continuous spectrum of a black body. Note that Millikan had to use the value of e from his oil-drop experiment.

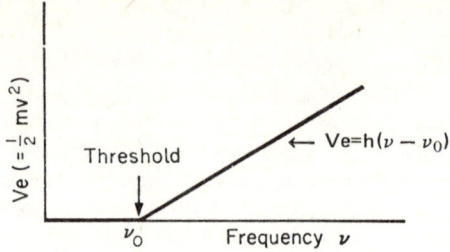

Fig. 6.4 Energies of photo-electrons.

6.5 The Discovery of X-Rays

In 1895 Rontgen observed the fluorescence of crystals of barium platinocyanide at some distance from a cathode ray tube. Although the intensity of the fluorescence was reduced by interposing various materials between the crystals and the tube, it could not be cut off entirely. This he supposed to be due to the emission of very penetrating rays from the cathode ray tube. He called them X-rays, the X indicating that they were unknown. X-rays were found to be produced whenever fast-moving electrons strike a target. In the early X-ray tubes electrons were produced by positive ion bombardment of the cathode as in a simple discharge tube (Fig. 6.5). The electrons were accelerated by a potential

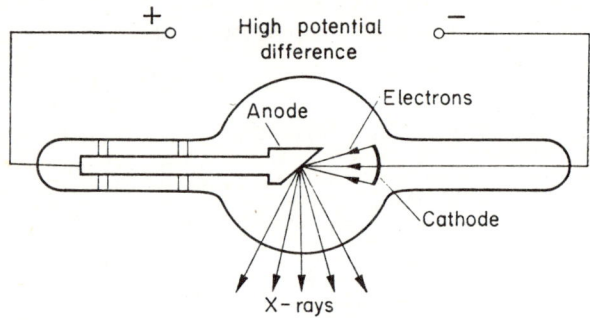

Fig. 6.5 An early X-ray tube.

difference between the cathode and the target which could also be the anode. The cathode was concave so that the electrons fell upon a very limited area of the target. Thus the X-rays originated effectively from a point source so that well-defined shadows of opaque objects could be photographed or observed on a fluorescent screen. A great improvement was achieved by Coolidge in 1913 when he generated a much greater supply of electrons from a heated filament (Fig. 6.6) and obtained a far greater intensity of X-rays than had previously been possible with the cold cathode. Potential differences up to 100 kV are frequently used between cathode and target in crystallographic X-ray units, which are often continuously evacuated with demountable water-cooled targets.

Apart from their great power of penetration, it was found that X-rays

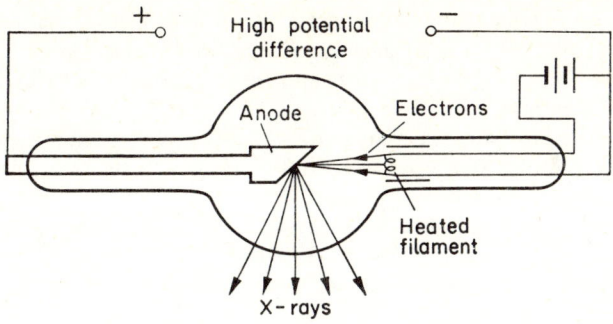

Fig. 6.6 The Coolidge X-ray tube.

blackened photographic plates, thus enabling X-ray photographs to be taken. X-rays were capable of ionizing a gas through which they passed. They did not appear to be reflected or refracted to the same extent as ordinary light waves. Like ultra-violet radiation they also produced skin burns due to ionization, but, owing to their great penetration, X-ray damage could also take place at a much greater depth as in deep X-ray therapy. They were unaffected by electric or magnetic fields, and therefore carried no charge.

6.6 Diffraction of X-Rays

At first there was considerable speculation about the nature of X-rays. Some held that they were very high-speed particles like cathode rays but uncharged, whilst others supposed they were electromagnetic waves of very low wavelengths. A decisive experiment was carried out by Friedrich and Knipping, at the suggestion of van Laue, in 1912. If these waves were of very small wavelength, then an ordinary diffraction grating having 600 lines mm^{-1} would produce no observable effect. Diffraction would only be possible if the spacing of the lines were comparable with the wavelength of the radiation. Von Laue suggested that the ordered arrangement of atoms within a crystal might provide a diffraction grating of sufficiently small spacing (0·1 nm).

In Friedrich and Knipping's experiment, X-rays were limited to a narrow pencil by two small holes in two lead screens S_1, S_2 (Fig. 6.7). The X-rays then fell upon a crystal of zinc sulphide beyond which was placed a photographic plate P. On development of the photograph, a diffraction pattern (now known as a Laue pattern) was observed, indicating that X-rays were wave-like in nature and that

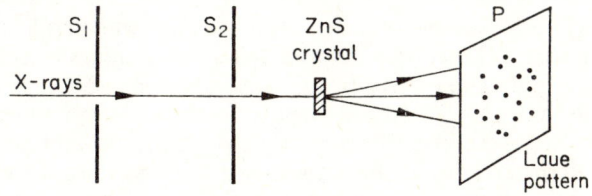

Fig. 6.7 Friedrich and Knipping's experiment.

the wavelength was of the same order as the spacing of the atomic planes of the zinc sulphide crystal (i.e. about 0·1 nm). Each spot on the Laue pattern arose from the reflection of X-rays of certain wavelength from one of the many possible planes within the crystal, as shown in Fig. 6.8.

The condition for reflection is not quite so simple as the above diagram suggests. Consider first a plane wave-front AB (Fig. 6.9) incident upon a plane

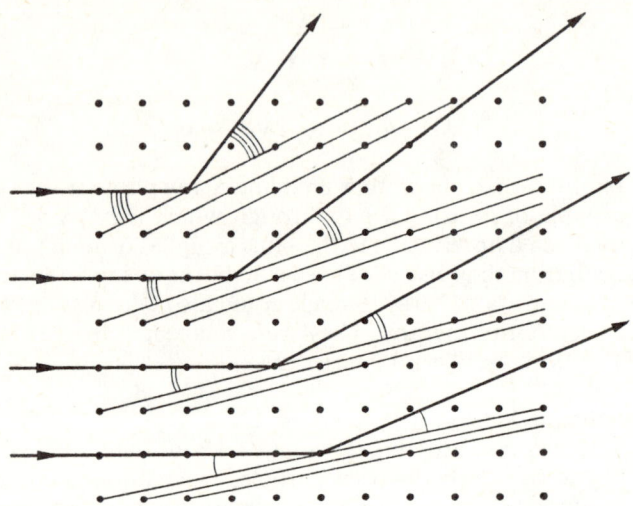

Fig. 6.8 Reflection of X-rays by the various crystal planes.

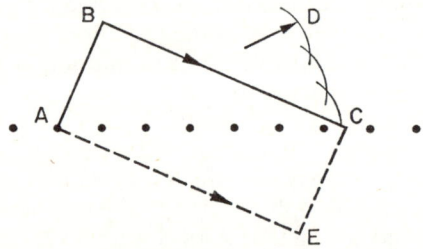

Fig. 6.9 Reflection from a single surface.

AC, containing atoms as shown. Each atom will scatter some of the X-rays from AB and each may be regarded as a centre of secondary wavelets. Huygens' principle tells us that a new wave-front CD will be formed, which is the envelope of the secondary wavelets and corresponds to normal optical reflection for which the angle of incidence is equal to the angle of reflection and is independent of wavelength, or the spacing of the atoms in the plane. Moreover, much of the energy passes straight through undeviated, to form a new wave-front at CE, as well as the reflected wave-front at CD. When we consider a set of such planes all parallel to one another and equispaced at a distance d apart, reflection is only possible at certain discrete angles given by the Bragg equation, as shown below. Consider two rays AC, BD incident upon two successive planes (Fig. 6.10). In accordance with Huygens' principle they will be reflected as CA', DB' such that the angles of incidence and reflection are equal. The additional condition requires that the rays CA', DB' shall be in phase. The incident wave-front is CX and the

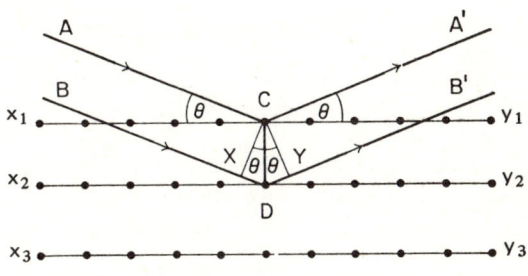

Fig. 6.10 Reflection from successive crystal planes.

reflected wave-front is CY. The rays will be in phase when the distance XDY is equal to a whole number m of wavelengths λ; that is $XD + DY = m\lambda$, and, as $XD = DY$, then $2d \sin \theta = m\lambda$. This is known as the Bragg equation. The Bragg condition together with the law of reflection limits the number of directions in which a pencil of X-rays can emerge from a crystal and so gives rise to a Laue pattern described earlier.

It must be emphasized that the above is a simplified explanation of the so-called 'reflected' beam. When the X-ray photon strikes an atom at a lattice site it is scattered in all directions by the electrons of the atom, generally without any phase relation. However, if rays scattered from the surface $X_1 Y_1$ in Fig. 6.10 are chosen such that they obey the optical reflection condition, then they all 'reflect' in phase since the path difference is zero. This holds for *all* the horizontal layers $X_1 Y_1$, $X_2 Y_2$, $X_3 Y_3$ etc., in Fig. 6.10. The Bragg law $m\lambda = 2d \sin \theta$, where m is integral, means that these 'reflected rays' all scatter in phase *between* the layers $X_1 Y_1$, $X_2 Y_2$ and $X_3 Y_3$. These two rules, i.e. the surface 'reflection condition' and the depth Bragg law, ensure that *all* the 'reflected' rays from the bulk of the crystal are in phase when m is integral. This corresponds to a 'bright fringe' in optical interference and here is often called a 'strong reflection'.

6.7 X-Ray Wavelengths

As Laue patterns were difficult to interpret, Bragg set up an X-ray spectrometer using a rock-salt crystal as a reflecting grating and enabling X-ray wavelengths to be calculated by the above equations. Lead slits S_1, S_2 (Fig. 6.11) restricted the X-rays to a fine pencil which was incident upon the crystal C at the centre of a spectrometer table. The reflected beam was detected by its ability to ionize a gas in an ionization chamber I. The ionization chamber rotated through twice the angle of the crystal turntable so that it was always in a position to receive the reflected X-ray beam from the crystal. Knowing the angle for which a particular

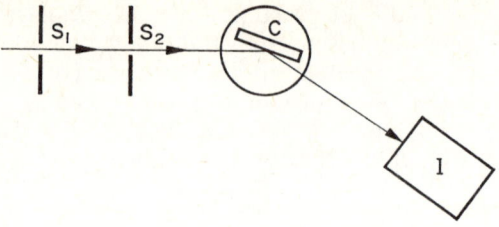

Fig. 6.11 Bragg's X-ray spectrometer.

wavelength was reflected, the wavelength could be calculated from the Bragg equation provided that d, the spacing of the atomic planes in the crystal, was known.

Bragg found that in a crystal of rock-salt the sodium and chlorine ions are arranged in a face-centred cubic structure as shown in Fig. 6.12. Consider a cube

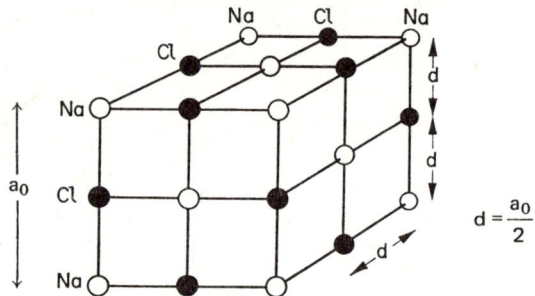

Fig. 6.12 Ions in a rock-salt unit cell.

of side d, of which there are eight in the diagram. Associated with each cube we have four sodium ions, but as each ion is shared by eight unit cubes, each cube contains only half a sodium ion. Likewise it also contains only half a chlorine ion, so that altogether one may only associate half a sodium chloride molecule with each unit cube. The volume of a mole is M/ρ, where M is the relative molecular mass and ρ is the density of sodium chloride. The volume associated with half a molecule is therefore $M/2N_A\rho$, where N_A is the Avogadro constant. As this is the volume of a cube, we may now write $d^3 = M/2N_A\rho$ from which d, the grating spacing, may be calculated. For sodium chloride $\rho = 2 \cdot 16 \times 10^3$ kg m^{-3}; therefore

$$d = \sqrt[3]{\frac{58 \times 10^{-3}}{2 \cdot 16 \times 10^3 \times 2 \times 6 \times 10^{23}}} = 2 \cdot 81 \times 10^{-10} \text{ m or } 0 \cdot 28 \text{ nm.}$$

Thus X-ray wavelengths of wavelength about 10^{-10} m (0·1 nm) may be readily measured by this method, using Bragg's formula $m\lambda = 2d \sin \theta$.

Note that the unit cell is the *whole* of Fig. 6.12 and the unit cell parameter (the repeat distance) is $a_0 = 2d$, the distance between *like* atoms in the lattice.

Bragg used the equation $m\lambda = 2d \sin \theta$ *twice*. In the first place he deduced, from the spectrometer results, that sodium chloride has a crystal *structure* consisting of a face-centred cube of sodium and chlorine ions, as in Fig. 6.12. From this observation Bragg was able to *calculate* the d spacing from the density of sodium chloride as outlined above. This calculation is independent of the diffraction equation. Second, Bragg used this calculated value of d to derive λ from the diffraction equation $m\lambda = 2d \sin \theta$. Typical values were Cu K_α $\lambda = 0.154$ nm and Mo K_α $\lambda = 0.071$ nm.

All diffraction equations, whether for light or X-rays, measure an angle θ and depend on a wavelength λ and a spacing d. In optics, Rowland ruled a diffraction grating mechanically and *counted* the number of lines, giving d independently and λ from the diffraction equation. Likewise Bragg measured d independently for sodium chloride and λ from his diffraction equation.

6.8 Continuous Spectrum of X-Rays

By 1913 wavelengths in the spectrum from a target in an X-ray tube could be measured with the Bragg X-ray spectrometer. It was found to consist of a continuous spectrum or white radiation upon which was superposed a line spectrum, named the KLM series, which was characteristic of the element used as target. At this stage we shall confine our attention to the continuous spectrum. The distribution of energy in the continuous spectrum was found to depend upon the potential difference across the X-ray tube, and not upon the material of the target. Figure 6.13 shows the continuous X radiation from a tungsten target at tube voltages too low to excite the characteristic K lines (see p. 102).

However, the K lines of molybdenum, of longer wavelength than the corresponding tungsten lines, are easily excited by a potential difference of only 40 kV.

Figure 6.13 not only shows that more energy appeared in the tungsten spectrum as the potential difference was increased, but also that the energy maximum moved towards the region of shorter wavelengths. The most interesting feature was that for each potential difference there was a discrete lower wavelength limit, and, as the potential difference increased, the lower wavelength limit fell. The wavelength limit appeared to be independent of the metal used as target in the X-ray tube. In terms of frequency, the maximum frequency v_{max} rose as the potential difference V increased. Experiment showed that they were in fact proportional, i.e. $V \propto v_{max}$. The energy of an electron Ve as it struck the target was also proportional to the frequency. The ratio Ve/v_{max} was therefore constant and found to be h, the Planck constant connecting frequency and energy. On rewriting, the equation became $\frac{1}{2}mv^2 = Ve = hv_{max}$, which gave yet another method of measuring h. The X-ray value for h is 6.56×10^{-34} J s.

The value of λ_{min} is given by

$$hv_{max} = \frac{hc}{\lambda_{min}} = Ve$$

when all the kinetic energy of the electron becomes an X-ray photon. Therefore

$$\lambda_{min} = \frac{hc}{Ve}$$
$$= \frac{6.6 \times 10^{-34} \times 3 \times 10^8}{1.6 \times 10^{-19} \times V}$$
$$= \frac{12\,375}{V} \times 10^{-10} \text{ m}$$

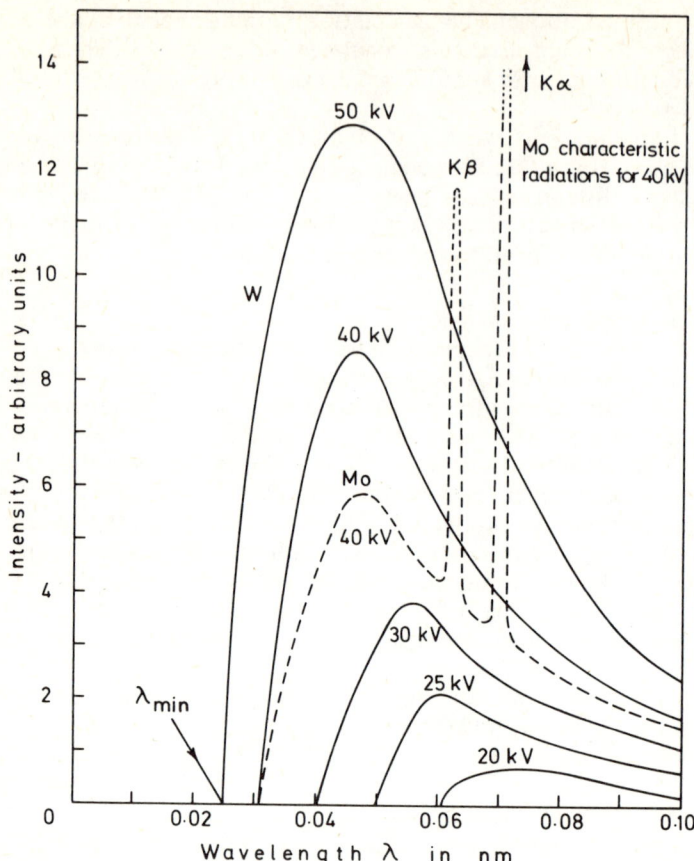

Fig. 6.13 Distribution of energy in X-ray spectrum for various potential differences.

or, approximately, $\lambda_{min} = 1234/V$ nm when V is in volts. This formula, which gives the short wavelength cut-off in Fig. 6.13, is easily remembered.

6.9 Compton Effect

Soon after their discovery, J. J. Thomson used the scattering of X-rays by atoms in his study of atomic structure. With the X-rays available at that time it was found that scattering occurred without change of wavelength in much the same way as light was scattered by the atmosphere. With the development of X-ray tubes giving shorter wavelengths of less than 0·1 nm, later workers found that a wavelength change occurred when scattering took place from the lighter elements. No explanation of this seemed possible using classical physics. Compton and Debye provided the explanation in 1923 by treating the radiation as a stream of individual photons each of which could interact with a single electron.

In Compton's celebrated experiment, X-rays were allowed to fall upon a block

of graphite C (Fig. 6.14) which was a convenient form of loosely bound electrons in carbon atoms. The scattered X-rays were reduced to a fine pencil by a number of lead slits and analysed by a Bragg X-ray crystal spectrometer. The wavelength λ' of the scattered X-rays depended upon the angle of scattering φ and was given by the equation
$$\lambda' - \lambda = (1 - \cos \varphi) h / m_0 c,$$
where λ was the wavelength of the original X-rays and m_0 the rest mass of the electron.

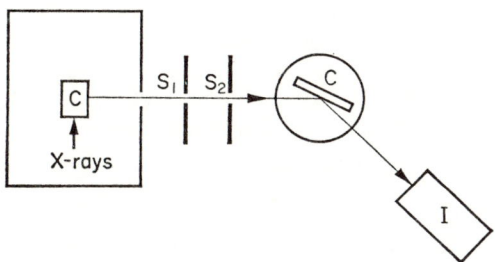

Fig. 6.14 The Compton effect experiment.

From Einstein's work we know that the energy of a photon is hv and the theory of relativity (Appendix A) requires that we associate an energy mc^2 with a mass m. Linking these two concepts, we may put $hv = mc^2$, which implies that a photon has momentum $mc = hv/c$. The interaction between the photon and the electron may now be treated as a simple collision problem in mechanics. The initial momentum vector hv/c (Fig. 6.15) of the X-rays is equal to the two vectors mv and

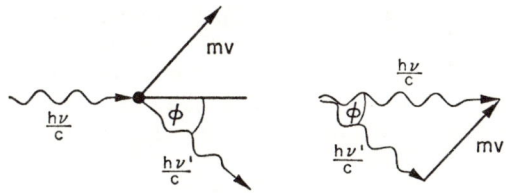

Fig. 6.15 Vector diagram for the Compton effect.

hv'/c, where hv'/c is the momentum associated with the scattered X-rays and mv is the momentum of recoil of the electron. The vector triangle gives the equation
$$m^2 v^2 c^2 = (hv')^2 + (hv)^2 - 2h^2 vv' \cos \varphi. \tag{6.1}$$
The conservation of energy requires that
$$hv = m_0 c^2 = hv' + mc^2, \tag{6.2}$$
where m_0 is the rest mass of the electron. Relativity gives the relation
$$m^2 (1 - v^2/c^2) = m_0^2. \tag{6.3}$$
From equations (6.1) and (6.3) we get $m^2 c^4 - m_0^2 c^4 = (hv)^2 + (hv')^2 - 2h^2 vv' \cos \varphi$. Substituting for $m^2 c^4$ from equation (6.2) yields
$$[h(v - v') + m_0 c^2]^2 - m_0^2 c^4 = (hv)^2 + (hv')^2 - 2h^2 vv' \cos \varphi.$$

On simplification, this gives $m_0c^2(v-v') = hvv'(1-\cos\varphi)$, which becomes $\lambda' - \lambda = (1-\cos\varphi)h/m_0c$, and inserting the numerical values we get $\lambda' - \lambda = 2\cdot4$ pm when $\varphi = 90°$, which is independent of wavelength and becomes increasingly important at shorter wavelengths. This equation was confirmed for all angles of φ up to 150°, showing that a photon hypothesis is required to account for the Compton effect. It also means, that having demonstrated that the photon can be used as a *collision* particle obeying the laws of conservation of momentum and energy we can regard radiation as *corpuscular*, which seems to take us back to Newton's theory of light. The photon of radiation therefore has the mechanical properties of a particle when the conditions are correct for this interpretation. We shall see later than the electron, which we have hitherto regarded as a particle, can also take on wave-like properties when the conditions of interaction favour this. This wave and particle duality of matter and radiation implies that the concepts of 'wave' and 'particle' are merely complementary ways of describing the same process.

6.10 Summary

We have seen how attempts to explain the distribution of energy in the continuous spectrum from a black body led Planck to the concept of the quantum of energy or the photon and that emission of radiation takes place in quanta of energy hv. The connection between energy and frequency is again apparent when we come to study the distribution of energy in the continuous X-ray spectrum. Here the maximum frequency is determined by the energy of the electrons in an X-ray tube showing that the whole energy of an electron may be converted to an X-ray photon. Furthermore Einstein was only able to interpret the photoelectric effect by supposing that the whole energy of a single light photon is transferred to a single electron within a metal. Thus when radiation interacts with matter, it does so in quanta or photons equal to hv, but phenomena such as interference, diffraction and polarization still require that it behaves as a transverse wave. The particle nature of radiation is even more marked when we come to study the Compton effect. Not only does the radiation have energy hv but also momentum hv/c, and its behaviour when interacting with an electron can only be interpreted in terms of the collision of two particles. The relationship between the wave and particle aspects of radiation will be examined in more detail later.

Problems

(*Those problems marked with an asterisk are solved in full at the end of the section.*)

6.1 When a copper surface is illuminated by radiation of wavelength 253·7 nm from a mercury arc, the value of the stopping potential is found to be 0·24 V. Calculate (*a*) the wavelength of the threshold frequency for copper; (*b*) the work done by the electron in escaping through the surface of the copper. ((*a*) 266·5 nm; (*b*) 4·65 eV)

6.2* The wavelength of the photoelectric threshold of tungsten is 230 nm. Determine the energy of the electrons ejected from the surface by ultra-violet light of wavelength 180 nm. (1·49 eV)

6.3 Calculate the grating space of a cubic crystal using the following data: relative molecular mass 100·1, relative density 2·71. (0·313 nm)

6.4 The radiation from an X-ray tube operated at 40 kV is analysed with a Bragg X-ray spectrometer using a calcite crystal with the same spacing as the previous problem. Calculate (a) the short wavelength limit of the X-ray spectrum coming from this tube; (b) the smallest angle between the crystal planes and the X-ray beam at which this wavelength can be detected. ((a) 30·9 pm; (b) 2° 50′)

6.5 The K radiation from a molybdenum target ($\lambda = 70\cdot 8$ pm) is scattered from a block of carbon and the radiation scattered through an angle of 90° is analysed with a calcite crystal ($d = 0\cdot 313$ nm) spectrometer. Calculate (a) the change in wavelength produced in the scattering process; (b) the angular separation in the first order between the modified and unmodified lines produced by rotating the crystal through the required angle. ((a) 2·4 pm; (b) 13·3 min of arc)

6.6 Describe how the energy of an electron liberated from a metal by incident radiation has been investigated experimentally. Explain how the results have been interpreted theoretically.

The wavelength of the photoelectric threshold for silver is 325 nm. Determine the energy of electrons ejected from a silver surface by ultra-violet light of wavelength 253·7 nm from a mercury arc. (1·08 eV)

6.7 Describe how X-ray wavelengths may be measured. Derive any formulae used.

In a Bragg X-ray spectrometer using a calcite crystal for which the grating space is 0·3 nm, X-rays are reflected when the angle between the incident and reflected rays is 5° and again when it is 10°. Calculate the mean wavelength of the X-rays. (261·6 nm)

6.8 Describe a method by which X-ray wavelengths have been determined. A Coolidge type of X-ray tube is operated at 66 kV. Calculate the short wavelength limit of the X-ray spectrum coming from the tube. (18·9 pm)

6.8* If the minimum wavelength recorded in an X-ray spectrum of a 50 kV tube is 24·7 pm, calculate the value of the Planck constant. ($6\cdot 59 \times 10^{-34}$ J s)

6.10 An eye can just detect green light of wavelength 500 nm which arrives on the retina at the rate of 2 aW. To how many photons does this correspond? (5)

6.11 A photon has energy equal to 10 MeV. Calculate the wavelength to which this corresponds. (0·136 pm)

6.12 Show that Wien's empirical radiation law $E_\lambda = C_1 \lambda^{-5} \exp(-C_2/\lambda T)$ approximates to Planck's radiation law for large quanta ($ch \gg k\lambda T$).

6.13 Show that Rayleigh–Jeans radiation law $E_\lambda = 8\pi k \lambda^{-5}(\lambda T)$ approximates to Planck's radiation law for small quanta ($ch \ll k\lambda T$).

6.14 By differentiating the Wien expression in question 6.12 with respect to wavelength, show that $\lambda_m T = $ constant, where λ_m is the wavelength at which E_λ is maximum.

16.15 Given the Wien empirical equation for the spectral energy distribution of black-body radiation of density E_λ

$$E_\lambda d\lambda = C_1 \lambda^{-5} \exp(-C_2/\lambda T)d\lambda,$$

show that this is consistent with the experimental law $\lambda_{\max} T = $ constant. The symbols have their usual meanings.

Sketch the functions
$$f(\lambda T)=e^{c\lambda T}, \qquad f(\lambda T)=e^{-c\lambda T}$$
$$f(\lambda T)=e^{c/\lambda T}, \qquad f(\lambda T)=e^{-c/\lambda T}$$
on the same $f(\lambda T)$ versus λ axes, for constant T. Explain why only one of these functions is suitable to reproduce the observed shapes of the experimental spectral curves. [N]

6.16 State the main features of the experimental black-body spectral radiation curve and verify that the Planck radiation formula given below satisfies these observations:
$$E_\lambda = \frac{C_1 \lambda^{-5}}{e^{C_1/\lambda T}-1}$$

Discuss briefly how Planck was led from this equation to the postulate of energy quantization and describe *in words* how this concept explains the values of E_λ for both large and small values of λT. [N]

6.17 Outline briefly the reasoning by which Wien derived the spectral black-body radiation formula
$$E_\lambda d\lambda = C_1 \lambda^{-5} e^{-C_2/\lambda T} d\lambda$$
and hence deduce the displacement law
$$\lambda_{max} T = \text{constant}.$$
Planck later determined the constant C_2 in terms of atomic constants, viz. $C_2 = hc/k$. Show that this is in agreement with the experimental value of the displacement law constant, $2\cdot 88 \times 10^{-3}$ m K.

What conclusions can be reached from this agreement? [N]

Solutions to Problems

6.2 Minimum energy required to eject electrons from a tungsten surface is
$$h\nu_0 = \frac{hc}{\lambda_0} = 6\cdot 6 \times 10^{-34} \frac{3 \times 10^8}{230 \times 10^{-9}}.$$
Energy of the incident radiation is
$$h\nu = \frac{hc}{\lambda} = 6\cdot 6 \times 10^{-34} \frac{3 \times 10^8}{180 \times 10^{-9}}.$$
Kinetic energy of ejected electron in joules is
$$h\nu - h\nu_0 = 6\cdot 6 \times 10^{-34} \times 3 \times 10^8 \left(\frac{1}{180 \times 10^{-9}} - \frac{1}{230 \times 10^{-9}} \right).$$
Converting to electron volts, this becomes
$$h\nu - h\nu_0 = \frac{6\cdot 6 \times 10^{-34} \times 3 \times 10^8}{1\cdot 6 \times 10^{-19}} \left(\frac{1}{180 \times 10^{-9}} - \frac{1}{230 \times 10^{-9}} \right).$$
$$= 1\cdot 49 \text{ eV}$$

6.9 Maximum energy of electrons in X-ray tube $= Ve$; maximum energy of X-rays emitted $= h\nu$. Therefore
$$h = \frac{Ve}{\nu} = Ve\frac{\lambda}{c}$$
$$= \frac{50 \times 10^3 \times 1\cdot 6 \times 10^{-19} \times 24\cdot 7 \times 10^{-12}}{3 \times 10^8}$$
$$= 6\cdot 59 \times 10^{-34} \text{ J s}.$$

Chapter 7

Spectra

7.1 The Hydrogen Spectrum

That each element displays a characteristic spectrum was first pointed out by Kirchhoff in 1859. It was not until thirty years later that the law governing the distribution of lines in a spectrum was discovered by Balmer and Rydberg. The interpretation of this law in terms of atomic structure began with the work of Bohr in 1913. It revealed that a study of spectra could provide a wealth of information about atomic structure. Hydrogen, being the simplest atom, has the simplest spectrum. The study of this spectrum has provided a key to the interpretation of many of the spectra of more complex atoms.

In the visible region the spectrum of hydrogen was known to consist of four main lines, red (656·3 nm), blue (486·1 nm) and in the violet (434·0 and 410·2 nm) while photography showed that there were others getting closer together in the ultra-violet, converging to a limit of 364·6 nm (Fig. 7.1). It was found that these

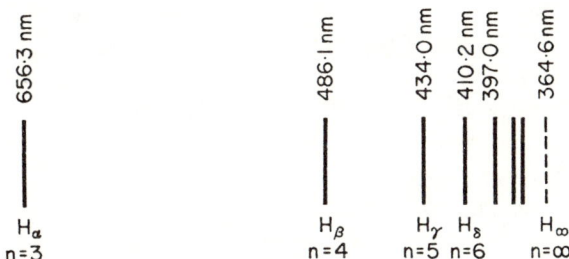

Fig. 7.1 The Balmer series for hydrogen.

could be represented empirically by a formula $1/\lambda = \bar{v} = R(1/2^2 - 1/n^2)$, where R was a constant now known as the Rydberg constant (not to be confused with R, the gas constant used in earlier chapters), n was an integer which can take all values greater than two, and $\bar{v}$ was the wavenumber, i.e. the number of wavelengths in unit length. The formula just quoted was primarily due to the work of Balmer.

The wavelengths of the hydrogen lines in the visible spectrum were brought to Balmer's notice in 1884 by the spectroscopist Angstrom. Balmer looked for a

series relation between the wavelengths of the type
$$\lambda = Kf(n),$$
where K was a constant and n was a running integer whose successive values would give the wavelengths of the lines of the spectrum. He arrived at the formula
$$\lambda = \frac{Kn^2}{n^2 - 4},$$
where $= 3, 4, 5, 6, 7, \ldots$ correspond (Fig. 7.1) to $H_\alpha, H_\beta, H_\gamma, H_\delta$, etc. The value of the constant K was taken from the wavelength of the H_α line, i.e. $\lambda = 656 \cdot 3$ nm experimentally and $n = 3$; hence $K = 364 \cdot 6$ nm. As a result,
$$\lambda_{H_\beta} = \frac{364 \cdot 6 \, n^2}{n^2 - 4},$$
where $n = 4$, giving $\lambda_{H_\beta} = 486 \cdot 1$ nm.

This formula reproduces all the measured wavelengths quoted in Fig. 7.1 to within 1 part in 5000. We conclude that it is an appropriate formula.

Repeating the Balmer formula in terms of the wavenumber $\bar{v} = 1/\lambda$, we get
$$\begin{aligned}
\bar{v} &= \frac{n^2 - 4}{Kn^2} \\
&= \frac{4}{K}\left[\frac{1}{4} - \frac{1}{n^2}\right] \\
&= R\left[\frac{1}{2^2} - \frac{1}{n^2}\right]
\end{aligned}$$

where
$$R = \frac{4}{K} = \frac{4}{364 \cdot 6 \times 10^{-9}} = 10 \cdot 970 \times 10^6 \text{ m}^{-1},$$
which is the Balmer formula value of the Rydberg constant.

It is obvious that this formula must be explained in terms of some theory of atomic structure, both qualitatively and quantitatively.

7.2 The Bohr Theory of the Hydrogen Atom

When Bohr saw the Balmer formula for $\bar{v}$ he was confronted with three facts:

(1) it is a *difference* formula;
(2) it is an *energy* formula if the Planck quantum of energy is incorporated, i.e., $E = hv = hc\bar{v}$;
(3) it can be generalized into
$$\bar{v} = R\left[\frac{1}{n_1^2} - \frac{1}{n_2^2}\right],$$
where n_1 and n_2 are integers somehow defining the energy stages of the atom.

Thus there are many more lines in the hydrogen spectrum that those of the Balmer series seen in the visible region of the spectrum. (See Fig. 7.4(b), p. 91.)

In 1913 Bohr applied the quantum theory to Rutherford's nuclear model of the hydrogen atom using Coulomb's and Newton's laws from classical physics. Since the nucleus was so much heavier than the electron, it was reasonable to suppose that an electron of charge $-e$ (Fig. 7.2) moved in a circle of radius r. By

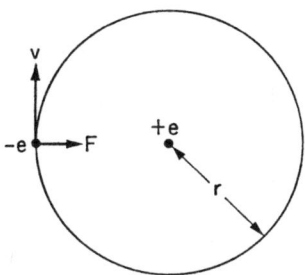

Fig. 7.2 An electron moving in a circular orbit about a proton.

Coulomb's law the force upon the electron was given by $e^2/4\pi\varepsilon_0 r^2$, ε_0 being the permittivity of free space. This was set equal to the product of mass and acceleration so that

$$F = \frac{e^2}{r^2 4\pi\varepsilon_0} = \frac{mv^2}{r},$$

and therefore the kinetic energy was given by

$$\tfrac{1}{2}mv^2 = \frac{e^2}{2r 4\pi\varepsilon_0}.$$

Since the potential energy of the electron was

$$-\frac{e^2}{r 4\pi\varepsilon_0},$$

the total energy (kinetic plus potential) was $-e^2/2r4\pi\varepsilon_0$. The energy of an electron within an atom is always negative since it has been assumed that the potential energy reaches zero when the electron has moved to an infinite distance from the nucleus. Negative energy therefore corresponds to a bound state. According to classical physics an electron moving with an acceleration loses energy by radiation and therefore a revolving electron would quickly spiral into the nucleus. The frequency of the radiation, given classically by the frequency of revolution of the electron in its orbit, would increase continuously giving rise to a continuous spectrum. Instead, a line spectrum corresponding to a number of discrete frequencies was observed as in Fig. 7.1. Some new postulate about the behaviour of an electron within the atom was clearly needed. In his analysis of the problem, using classical physics together with Planck's quantum of energy, Bohr was ultimately forced to the conclusion that the law that regulated the electron orbits was that only those orbits were permissible for which the angular momentum (moment of momentum) was given by $mvr = nh/2\pi$, where n is the quantum number taking values $n = 1, 2, 3, \ldots$ for successive orbits. Whilst an electron remained in one of these orbits, no energy was radiated, and these orbits therefore corresponded to stationary energy states or just stationary states. If we accept these rather strange ideas (justified later by quantum mechanics), the hydrogen calculation is made easier than the method originally due to Bohr. The quantization law $mvr = nh/2\pi$ was a *conclusion* drawn by Bohr in his derivation of the Balmer formula. He did *not* start with it as a postulate.

Bohr argued that when an electron moved from one orbit to another, the energy of the atom as a whole was changed and the energy difference was manifest as radiation. The energy was emitted or absorbed in whole quanta in accordance with quantum theory, such that $E_2 - E_1 = hv$, E_1, E_2 being the energies of the two orbits concerned. Thus, in addition to the laws of classical physics, we have two postulates proposed by Bohr:

(1) electrons in atoms occupy certain discrete stable orbits, of fixed energy, defined by $mvr = nh/2\pi$, and whilst the electron is in one of these orbits no energy is radiated;

(2) when an electron moves from one stable orbit to another of lower energy, a quantum of radiation is emitted, the frequency of this radiation is given by Planck's quantum condition, $hv = E_2 - E_1$.

Eliminating v from the equations $mv^2/r = e^2/r^2 4\pi\varepsilon_0$ and $mvr = nh/2\pi$, the atomic radius became $r = n^2 h^2 4\pi\varepsilon_0 / 4\pi^2 me^2$. Substitution gave 52·9 pm for r when $n = 1$, so that the atom was about 0·1 nm in diameter, agreeing well with kinetic theory estimates. Now, as we have just seen, the total energy E was given by

$$E = -\frac{e^2}{2r 4\pi\varepsilon_0} = -\frac{2\pi^2 me^4}{n^2 h^2 (4\pi\varepsilon_0)^2}.$$

Using the second postulate, the frequency was given by

$$hv = E_2 - E_1 = \frac{2\pi^2 me^4}{h^2 (4\pi\varepsilon_0)^2} \left(\frac{1}{n_1^2} - \frac{1}{n_2^2} \right).$$

The wavenumber ($\bar{v} = 1/\lambda = v/c$) then became

$$\bar{v} = \frac{2\pi^2 me^4}{ch^3 (4\pi\varepsilon_0)^2} \left(\frac{1}{n_1^2} - \frac{1}{n_2^2} \right) = R \left(\frac{1}{n_1^2} - \frac{1}{n_2^2} \right),$$

where R was the Rydberg constant. In all cases $n_2 > n_1$.

It was therefore possible to calculate the Rydberg constant from known atomic constants. Substitution gave $R = 10{\cdot}974 \times 10^6$ m^{-1} or $10{\cdot}974 \times 10^3$ mm^{-1} or $10{\cdot}974$ μm^{-1} in good agreement with experimental values shown in Fig. 7.6, p. 92, and the Balmer value given on p. 86.

This formula also suggested that series other than the Balmer series should exist, corresponding to $n_1 = 1$, $n_1 = 3$, $n_1 = 4$, $n_1 = 5$, etc. and we should expect to find them in regions of the spectrum away from the visible. Such series have since been discovered by Lyman in the far ultra-violet and by Paschen, Brackett and Pfund in the infra-red.

The Bohr model of the atom could then be visualized as consisting of a series of discrete orbits corresponding to constant or stationary energy states for which $n = 1, 2, 3, \ldots$ and the various spectral series were simply electron transitions between them, as shown in Fig. 7.3. As the radii of the orbits would be proportional to n^2, this diagram has not been drawn to scale.

In our considerations of atomic structure we are more concerned with energy changes within the atom, so that the shapes and sizes of the orbits need not be considered at present. We shall just consider the energy level diagrams and, since the wavenumber of a spectral line is given by $\bar{v} = (E_2 - E_1)/hc$ a wavenumber proportional to the energy of the atom may be attached to each energy level. Wavenumbers of spectral lines are obtained by taking differences between the wavenumbers attached to the two levels involved in an electron transition.

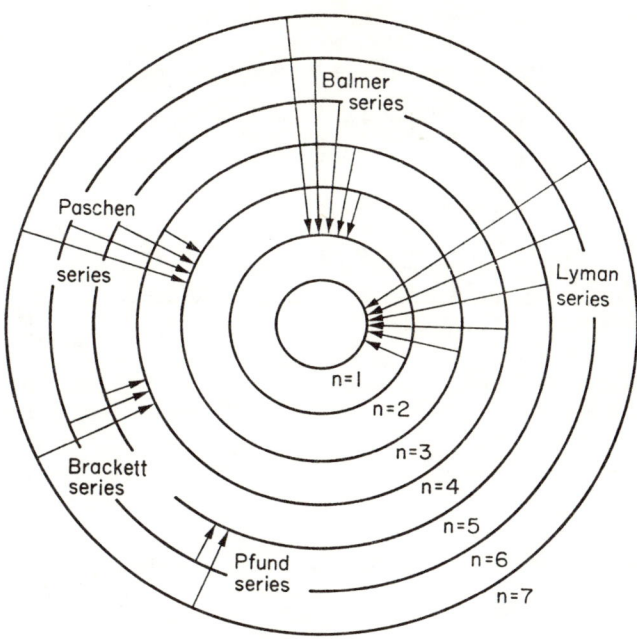

Fig. 7.3 The spectral series of the hydrogen atom.

In the energy level diagram shown in Fig. 7.4(a) horizontal lines are drawn instead of the circular orbits to represent the stationary energy states, $n = \infty$ corresponding to ionization of the atom and zero energy for which the wavenumber is also zero. Energy levels are measured downwards from this and given by R/n^2 so that the deepest level of hydrogen corresponds to $10\,974/1^2 = 10\,974$ mm^{-1} = $10\cdot974$ μm^{-1}. The complete hydrogen spectrum is shown in Fig. 7.4(b).

The success of the Balmer formula and its subsequent analysis by Bohr was due to the fact that the Balmer series lies entirely within the visible and near ultra-violet part of the spectrum, and that there is no overlapping with the other series as there is in the infra-red region where the Paschen, Brackett and Pfund series are found. This is shown in Fig. 7.4(b).

The importance of the Bohr theory in physics was that it was the beginning of theoretical spectroscopy. On the practical side, the measurements of spectral line wavelengths and resolutions became more and more accurate using interferometer methods. This meant that the Bohr theory had to be refined, largely by Sommerfeld, in order to explain the anomalies of wavelength and multiplicity. It also led to the concepts of quanta and the quantization of angular momentum, $mvr = nh/2\pi$, as a general rule.

On the debit side, the Bohr calculation is an odd mixture of classical and quantum methods with no real explanation of radiationless orbital motion or of what happens to the electron in the forbidden zones between orbits. It predicts

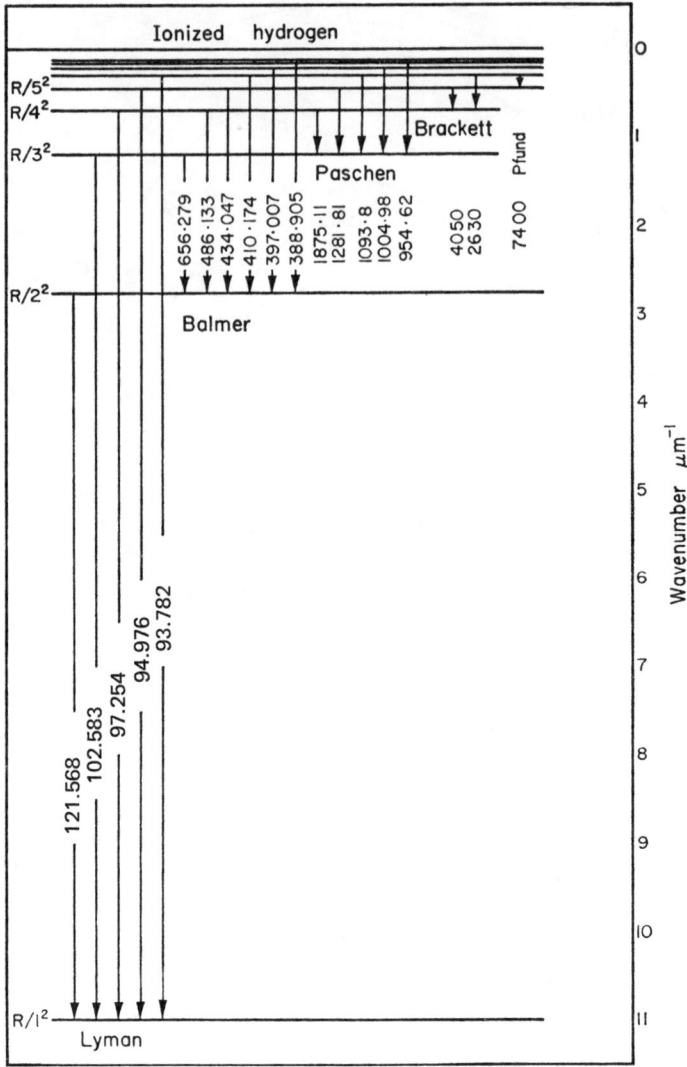

Fig. 7.4(a) Energy levels of the hydrogen atom.

that all lines should be singlets, whereas most spectral lines are multiplets. Simple Bohr theory cannot predict the relative intensities of the lines of a given series. It is only valid for one-electron atoms, i.e. it is successful for He^+ but fails for netural He. However, it does give useful numerical results by simple mathematical methods.

7.3 Isotope Effect

The success of the Bohr theory thus led to many refinements and elaborations to try and account for more and more features of spectra revealed by more careful

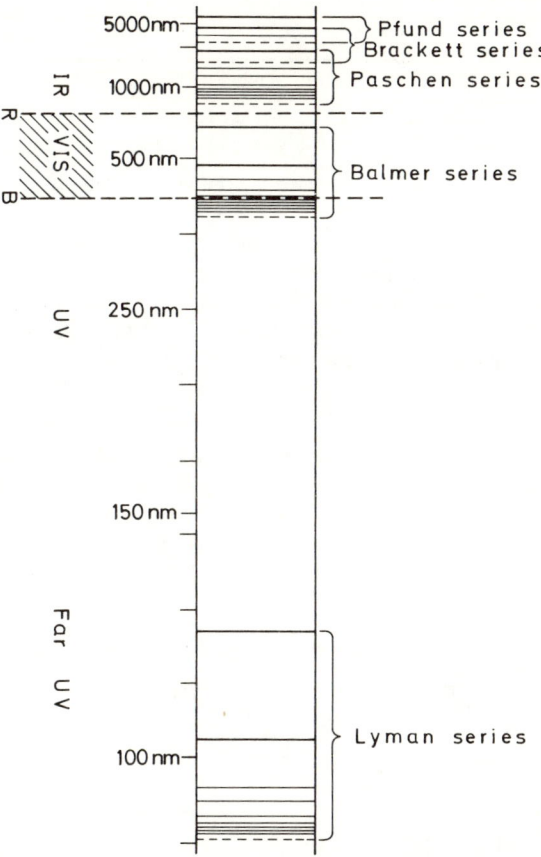

Fig. 7.4(b) Corresponding positions of hydrogen lines.

observation. It has been assumed in the above considerations that the electron moved around a fixed nucleus. This would only be strictly true if the mass of the nucleus were infinite. The fact that it is only some 1836 times as heavy as the electron implies that they each move around a common centre of gravity, as shown in Fig. 7.5.

By moments, we know that $m/M = A/a$, so that $a = Mr/(M+m)$ and $A = mr/(M+m)$, where A, a, are the distances of the nucleus of mass M and electron of mass m from the common centre of gravity and r is the separation of the nucleus and the electron. When ω is the angular velocity of the system the total kinetic energy is given by

$$\tfrac{1}{2}MV^2 + \tfrac{1}{2}mv^2 = \tfrac{1}{2}MA^2\omega^2 + \tfrac{1}{2}ma^2\omega^2$$
$$= \frac{1}{2}\frac{Mm}{M+m}r^2\omega^2$$
$$= \tfrac{1}{2}\mu r^2\omega^2.$$

It thus appears that to allow for the motion of the nucleus we must replace the

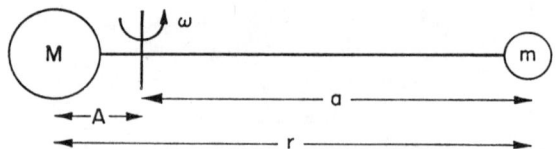

Fig. 7.5 Electron and proton moving around a common centre of gravity.

mass m of the electron by μ, the so called 'reduced mass', where $\mu = m/(1+m/M)$ or $1/\mu = 1/m + 1/M$. From this it can be seen that the reduced mass is equal to the electron mass only when the nuclear mass M is made infinite.

The Rydberg constant then becomes

$$R = \frac{2\pi^2 \mu e^4}{ch^3 (4\pi\varepsilon_0)^2} = \frac{2\pi^2 m e^4}{ch^3 (4\pi\varepsilon_0)^2} \frac{1}{(1+m/M)} = \frac{R_\infty}{1+m/M},$$

where R_∞ is the Rydberg constant which has already been calculated for a nucleus of infinite mass. In general, elements will have Rydberg constants which are slightly less than R_∞, as shown in Fig. 7.6. With increasing M, the Ryberg constant approaches more and more closely to R_∞.

Comparing the experimental values of the Rydberg constants for hydrogen

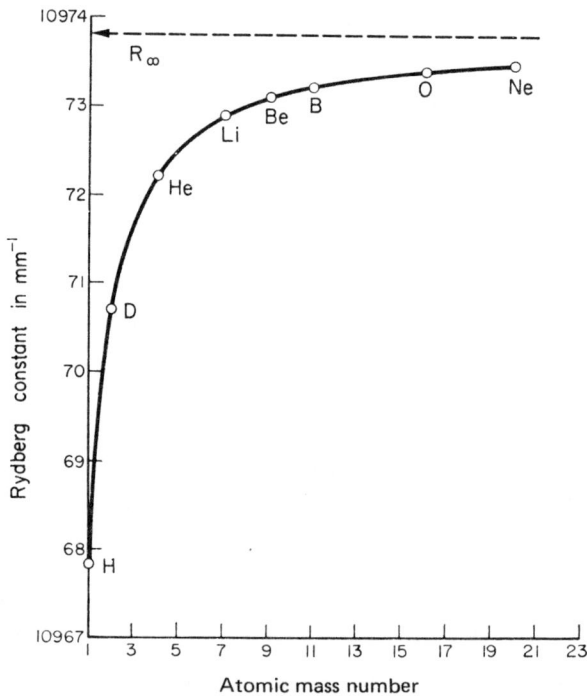

Fig. 7.6 Variation of the Rydberg constant with the atomic mass number.

and helium, we get
$$\frac{R_H}{R_{He}} = \frac{10\,967 \cdot 776}{10\,972 \cdot 243} = \frac{1 + m/M_{He}}{1 + m/M_H}$$
where M_{He} and M_H are here the masses of the helium and hydrogen *nuclei*. Putting $M_{He} = 3 \cdot 9717 M_H$, we get $M_H/m = 1837$, and thus obtain the ratio of the mass of the proton to the mass of the electron by a purely spectroscopic method. This is in excellent agreement with other methods, such as the electrical methods described in Chapter 2.

As we have seen, isotopes of the same element all carry the same nuclear charge, but have different nuclear masses. This gives rise to different values for the Rydberg constant and leads to a splitting of the spectrum lines. It was by this method that Urey, Murphy and Brickwedde discovered deuterium, the isotope of hydrogen of atomic mass 2. The wavelength separation of the first member of the Balmer series will not be calculated. For the hydrogen and deuterium series we may write respectively

$$\bar{v}_H = R_H \left(\frac{1}{2^2} - \frac{1}{n^2} \right) \quad \text{and} \quad \bar{v}_D = R_D \left(\frac{1}{2^2} - \frac{1}{n^2} \right).$$

From these we obtain
$$\frac{\bar{v}_H}{\bar{v}_D} = \frac{R_H}{R_D} = \frac{1 + m/M_D}{1 + m/M_H} = \frac{1 + m/2M_H}{1 + m/M_H}$$
for nuclear masses, and so
$$\frac{\bar{v}_H - \bar{v}_D}{\bar{v}_D} = \frac{1}{2} \frac{1}{(1 + M_H/m)} = \frac{1}{2 \times 1836},$$
therefore
$$\Delta \bar{v}/\bar{v} = \Delta \lambda / \lambda = 1/3672.$$

At $\lambda = 656 \cdot 3$ nm this gives $\Delta \lambda = 0 \cdot 179$ nm.

This was readily measured with a 6·4 m concave grating available to them. Table 7.1 compares the calculated and observed wavelength differences for the first four members of the Balmer series of hydrogen. The agreement is seen to be very satisfactory.

TABLE 7.1
Wavelength Differences for Hydrogen and Deuterium

	α	β	γ	δ
Wavelength λ in nm	656·28	486·13	434·05	410·17
Calculated difference $\Delta\lambda$	0·1793	0·1326	0·1185	0·1119
Observed difference $\Delta\lambda$	0·1791	0·1313	0·1176	0·1088

7.4 The Spectrum of Sodium

The spectra of the alkali metals lithium, sodium, potassium and caesium may be analysed in much the same way as the hydrogen spectrum. The spectral lines may be arranged in various series, but there are many more than for hydrogen. The number and complexity of the series increases as we proceed across the periodic table. For this reason our discussion will first be extended to include the elements in Group I using sodium as our example. In their arrangement of lines into series, the early spectroscopists were guided by the features of the lines, such as intensity, sharpness, diffuseness; the method of production, such as in the electric arc or spark; and the behaviour in electric and magnetic fields. As we shall see when we come to study the Zeeman effect in Chapter 13, the lines of a given series always split up into the same number of components in a magnetic field. Balmer's formula had already indicated that the visible spectrum of hydrogen could be represented by the formula $\bar{v} = R(1/2^2 - 1/n^2)$. Rydberg and Ritz then showed that the whole atomic spectrum of hydrogen could be represented by the formula $\bar{v} = R(1/n_1^2 - 1/n_2^2)$, n_1 and n_2 being integers. Rydberg, in applying this formula to the spectra of the alkali metals, found that it had to be modified to $\bar{v} = R[1/(n_1 - \alpha)^2 - 1/(n_2 - \beta)^2]$, where n_1, n_2 were integers and α, β were fractions now known as quantum defects. The spectra of atoms in Group I consist of four main series known as the principal, diffuse, sharp and fundamental or Bergmann series. Thus, for sodium, assuming a mean wavelength for lines which are double or triple, we have wavelengths associated with the four series shown in Table 7.2, grouped according to their physical characteristics.

TABLE 7.2
Wavelengths of the Four Series of Sodium in nm

Principal	Sharp	Diffuse	Fundamental
589·3	1139·3	818·9	1845·9
330·3	615·8	568·5	1267·8
285·3	515·1	498·1	
268·0	475·0	466·7	
259·4			

These four series may be represented empirically by the formulae:

Principal $\quad \bar{v} = R\left[\dfrac{1}{(3-1\cdot37)^2} - \dfrac{1}{(n-0\cdot88)^2}\right]$

Sharp $\quad \bar{v} = R\left[\dfrac{1}{(3-0\cdot88)^2} - \dfrac{1}{(n-1\cdot37)^2}\right]$

Diffuse $\quad \bar{v} = R\left[\dfrac{1}{(3-0\cdot88)^2} - \dfrac{1}{(n-0\cdot01)^2}\right]$

Fundamental $\quad \bar{v} = R\left[\dfrac{1}{(3-0\cdot01)^2} - \dfrac{1}{(n-0\cdot001)^2}\right]$

The wavenumber of the sodium yellow line may be calculated from the formula for the principal series by putting $n = 3$ and substituting for $R = 10\,973\cdot7$ mm^{-1} so that

$$\bar{v} = 10\,973\cdot7\left[\dfrac{1}{1\cdot63^2} - \dfrac{1}{2\cdot12^2}\right]$$
$$= 1689 \text{ mm}^{-1}.$$

Remembering that there are 10^3 m^{-1} in 1 mm^{-1}, the wavelength is given by
$$\lambda = \frac{10^9}{1689 \times 10^3} = 592 \cdot 1 \text{ nm}.$$
This value is within 0·5% of 589·3 quoted in Table 7.2.

Inspection of these formulae suggests that the quantum defects 1·37, 0·88, 0·01 and 0·001 should be associated respectively with the sharp, principal, diffuse and fundamental levels. If these formulae are now generalized and the quantum defects are denoted by the letters S, P, D and F, the formulae become:

Principal $\quad \bar{v} = \dfrac{R}{(3-S)^2} - \dfrac{R}{(n-P)^2}$

Sharp $\quad \bar{v} = \dfrac{R}{(3-P)^2} - \dfrac{R}{(n-S)^2}$

Diffuse $\quad \bar{v} = \dfrac{R}{(3-P)^2} - \dfrac{R}{(n-D)^2}$

Fundamental $\quad \bar{v} = \dfrac{R}{(3-D)^2} - \dfrac{R}{(n-F)^2}$

To have to write down such formulae each time one wishes to refer to a particular spectrum is laborious, and a shorthand notation has arisen which contains all the essential information, but not in the strict mathematical form given above.

The series then becomes

Principal $\quad \bar{v} = 3S - nP$
Sharp $\quad \bar{v} = 3P - nS$
Diffuse $\quad \bar{v} = 3P - nD$
Fundamental $\quad \bar{v} = 3D - nF$

based on the quantum number $n = 3$.

As in the case of hydrogen these series may be interpreted in terms of an energy-level diagram, as shown in Fig. 7.7. The energy levels for hydrogen are shown on the left-hand side so that they may be compared with the sodium energy levels.

It should be noted that owing to the large quantum defects of the S and P levels, these lie much lower than the corresponding hydrogen levels. With the diffuse and fundamental series, however, the defect has become small and this probably led the early workers to suppose that fundamental series was in some way more important than the others since it almost coincides with the hydrogen Paschen series.

7.5 Quantum Defects — Interpretation

The series found in sodium, and indeed in the alkali metals as a whole, are so like those in hydrogen that some close similarity of atomic structure seems likely. Chemical considerations especially in relation to the periodic table suggest that the sodium atom may be constructed from a neon atom by the addition of a proton to the nucleus and an electron to the outer structure. At this stage we can

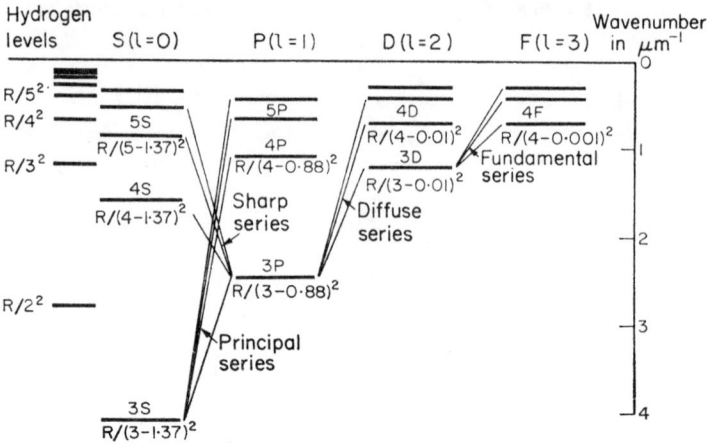

Fig. 7.7 Energy levels of the sodium atom.

ignore the addition of neutrons so that the alkali metals can be regarded as:
(Helium + proton) + outer electron → lithium
(Neon + proton) + outer electron → sodium
(Argon + proton) + outer electron → potassium
and in general
(Rare gas + proton) + outer electron → alkali metal

The electronic structure of a rare gas is most stable and compact and approximates to a spherical distribution of charge. The addition of a proton to the nucleus implies that this stable neon 'atomic' core has attained a single positive charge. The additional outer electron circulates about this neon atomic core, and therefore, while it is outside the core, corresponds closely with the electron of a hydrogen atom. When the electron penetrates the atomic core it is exposed to a very much greater proportion of the nuclear charge and its binding energy is correspondingly greater. The simple Bohr picture of the atom no longer holds, and we must expect substantial departures from the Balmer formula which are revealed by much larger quantum defects. When the energy of binding becomes greater, the energy levels lie lower and the values of the quantum defects increase. Table 7.3 shows the quantum defects for sodium for the four sets of levels and reveals two distinct ranges, the values for the S and P levels being often one hundred times greater than those for the D and F levels. The two ranges of

TABLE 7.3
Quantum Defects for Sodium Atom

	$n=3$	$n=4$	$n=5$	$n=5$
S	1·373	1·357	1·352	1·349
P	0·883	0·867	0·862	0·859
D	0·010	0·011	0·013	0·011
F	–	0·000	0·001	0·008

quantum defects can be attributed to orbits which penetrate or do not penetrate the central atomic core. Figure 7.8 shows the four possible orbits for the principal quantum number $n = 4$.

These orbits are described in terms of a second quantum number l, which is different for each orbit. In the case quoted, l can have four values, 0, 1, 2, of 3. Orbits corresponding to $l = 3$ and 2 giving D and F levels do not penetrate the

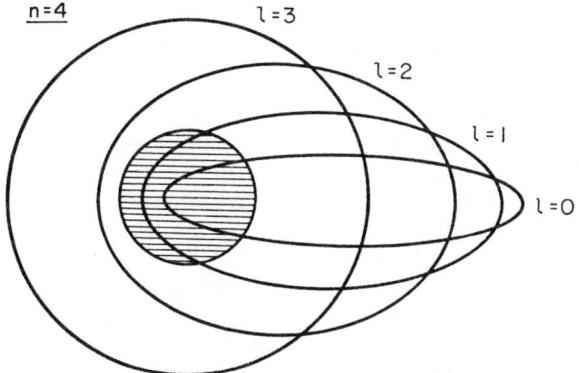

Fig. 7.8 Penetrating and non-penetrating orbits for the quantum number $n = 4$.

central atomic core, so that the structure is similar to hydrogen and the spectrum agrees closely with the Bohr formula. The orbits for which $l = 0$ and 1 corresponding to S and P levels penetrate the atomic core and the spectrum shows wide deviations from the Bohr formula. This picture of the atom is helpful in trying to understand how quantum defects arise, but it should not pursued too far.

7.6 Selection Rules and the Correspondence Principle

Further consideration of the energy level diagram shows that while all lines may be expressed as the difference of two terms, by no means all term differences are observed as spectral lines. Inspection and comparison of a number of spectra led to the discovery of empirical constraints on the possible transitions between energy states. These constraints are called the selection rules for quantum numbers. It appears that when the terms are arranged horizontally, as in Fig. 7.7, transitions can only occur between levels in adjacent sets of levels as shown. This fact may be expressed quantitatively by attaching the quantum number l to each set of levels such that $l = 0, 1, 2, 3$ for the S, P, D and F levels respectively. From an examination of the spectra it became clear that l may change only by ± 1. This is a selection rule, and l is the same quantum number as used above to differentiate between penetrating and non-penetrating orbits, and now called the orbital quantum number. The reason for this will be explained in Chapter 12. We are in fact dealing with quantum number changes in relation to the structure of the atom. A similar selection rule has already been encountered in hydrogen in connection with the Bohr quantum number n, which can only change (for hydrogen) by integral values such that $\Delta n = 1, 2, \ldots$ to infinity. It is found, however, for sodium that Δn can be zero and that l can take all values up to $(n-1)$. Thus for a given value of n, $l = 0, 1, 2, \ldots, (n-1)$ and $\Delta l = \pm 1$. At first these rules were empirical, but they can now be derived as solutions of the wave equation from wave mechanics which we shall discuss in Chapter 10. Thinking in terms of the Bohr model of the atom, we may say that n controls the energy of the electron, and l the angular momentum of the electron about the nucleus. As for the hydrogen atom, the angular momentum is in units of $h/2\pi$ and is equal to $lh/2\pi$. Wave mechanics (see Chapter 10) corrects this to $\sqrt{[l(l+1)]}h/2\pi$. Bearing in mind that the energy of a photon is $h\nu$ and its angular momentum can only be $h/2\pi$, we see that these selection rules arise respectively from the conservation of energy and the conservation of angular momentum.

It is possible to derive a selection rule for n from the Bohr theory. In his 1913 paper, Bohr made the statement that 'the frequency of radiation may be obtained by comparing calculations of the energy radiation in the region of slow vibrations based on the above assumptions with calculations based on ordinary mechanics.' This is the first statement of the much used *Bohr correspondence principle*. This states that, for large orbits, i.e. large quantum numbers, giving 'slow vibrations', quantum physics goes over to classical physics and leads to the same results. This gave Bohr the hint that the radiation emitted by transitions between orbits n_1 and n_2 when both are large integers was essentially the same as that emitted classically by a centripetally accelerated electron revolving in a large orbit, viz. the frequency of rotation. Thus

$$\bar{\nu}_{n_1,n_2} = R\left[\frac{1}{n_1^2} - \frac{1}{n_2^2}\right]$$
$$= R\left[\frac{(n_2 - n_1)(n_2 + n_1)}{n_1^2 n_2^2}\right].$$

Now $n_2 - n_1 = \Delta n$ is small and $n_2 \sim n_1 \sim n$ is large. Thus:

$$\bar{\nu}_{n_1,n_2} = \frac{R 2\Delta n}{n^3}, \quad \text{radiation wave number}$$

or

$$\nu_{n_1,n_2} = Rc\frac{2\Delta n}{n^3}, \quad \text{radiation frequency},$$

for n large. Now in the Bohr calculation, the *orbital* frequency is

$$\omega_n = \frac{4\pi^2 m e^4}{n^3 h^3} = \frac{2Rc}{n^3}$$

for all n. Thus ν_{n_1,n_2} and ω_n will 'correspond' if $\Delta n = 1$, which is therefore a rudimentary selection rule.

While this reveals a reason for selecting allowed transitions, it also reveals a weakness in the Bohr argument. For instance, for an electron in the classical circular orbit, there are possible radiation frequencies of

$$\omega, \quad 2\omega, \quad 3\omega, \quad 4\omega, \quad \text{etc.},$$

i.e. the harmonics generated by one and the same orbital motion, which are not observed. But on the Bohr theory only *one* frequency is generated for each n when $\Delta n = 1$, i.e. between neighbouring orbits, whereas there are others for which $\Delta n = 2$, $\Delta n = 3$, etc., between different but not neighbouring orbits. There are other difficulties, e.g. intensity of lines, multiplicity of lines and polarization, showing that the simple Bohr theory is inadequate.

However, although Bohr initiated the simple correspondence principle in this way, he was able to elaborate it at a later date and it was also helpful to Heisenberg in his formulation of the matrix method of quantum mechanics in 1926.

7.7 Excitation Potentials

Important confirmation of the existence of energy levels within the atom came from a study of the impact of electrons with varying energies upon atoms in a discharge tube. Following the earlier work of Lenard, Franck and Hertz in 1914 established the existence of such levels for the mercury atom, using a tube having three electrodes and containing mercury vapour. Electrons from the filament F (Fig. 7.9) were accelerated through a potential V towards the grid G, after which

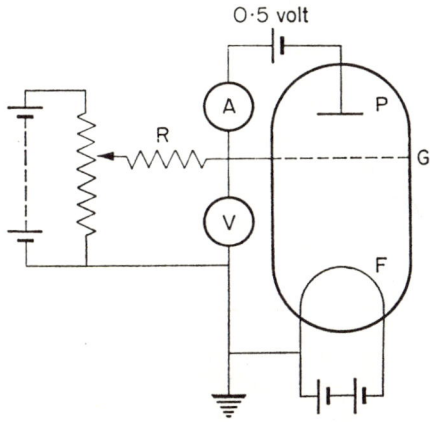

Fig. 7.9   Franck and Hertz apparatus.

there was a small reversed potential of about 0·5 V to the anode P. The distance between the filament and grid was much greater than in a normal triode. A small quantity of mercury was present in the tube and the concentration of mercury atoms was controlled by the temperature of the tube. A pressure of 15 mm was satisfactory and this required a temperature of about 200 °C so that the tube had to be placed in a small oven. As the potential was increased so that G became more positive, the electrons received more and more energy in accordance with the equation $Ve = \frac{1}{2}mv^2$. These soon had sufficient energy to reach the plate against the retarding potential. Collisions with the mercury atoms took place initially in accordance with the law of conservation of energy. When, however the electron energy reached a critical value, the collisions became inelastic and energy was transferred from the electron to the atom of mercury. This occurred when the energy was just sufficient to raise an electron from the lowest orbit to the next within the mercury atom. The impinging electron, having lost most of its energy, was incapable of reaching the anode against the retarding potential. The anode current in consequence showed a sudden drop when the grid potential reached this critical value. When the potential of the grid was increased further the current again increased as the electrons again acquired enough energy to reach the anode. A second critical potential was reached when an electron having excited an atom and lost its energy, was again accelerated so that it was able to excite a second atom. This was revealed by a second peak in the anode current characteristic. This was continued through several more peaks.

The potential difference between successive peaks for mercury is 4·9 V (Fig. 7.10), the value of the first peak being ignored because it includes contact

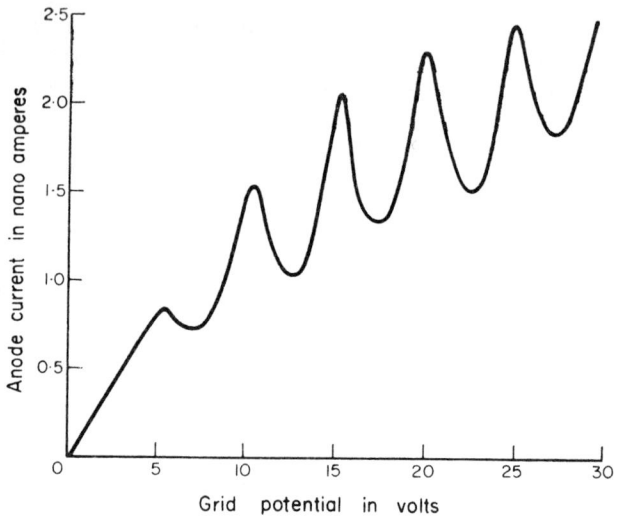

Fig. 7.10 Excitation potentials.

potentials in the circuit. Remembering that the frequency of radiation arising from an energy change in an atom from E_2 to E_1 is given by $E_2 - E_1 = h\nu = hc/\lambda$ and that the critical energy given to the electrons was Ve, we have

$$\lambda = \frac{hc}{Ve}$$

and therefore

$$\lambda = \frac{(6·626 \times 10^{-34}) \times (2·998 \times 10^8)}{4·9 \times (1·602 \times 10^{-19})}$$
$$= 2·531 \times 10^{-7} \text{ m}$$
$$= 253·1 \text{ nm}.$$

This value agrees remarkably well with the line at 253·7 nm in the mercury spectrum which represents a transition from the state of lowest energy, known as the ground state, to the first excited state. Moreover, when the potential is 5·2 V, this line, and only this line, is observed when light from the tube is examined with a spectroscope.

7.8 Controlled Excitation of Spectra

A similar experiment was performed by Newman with sodium vapour in 1925, the whole tube being enclosed in an electric furnace at about 350 °C. Newman showed how each sodium line appeared at a definite potential. At 2·10 V only the yellow sodium D lines appeared at 589·0 and 589·6 nm, corresponding to excitation from the lowest energy state or ground state 3S to 3P. As the potential increased, more and more lines appeared in accordance with the Table 7.4. Reference to the energy level diagram (Fig. 7.7) will show how this comes about, as the 3P, 4S, 3D, etc., levels are energised in turn.

TABLE 7.4
Resonance Potentials of Sodium Atom

Transition	Mean wavelengths of lines (nm)	Potential (V)
3P – 3S	589·3	2·10
4S – 3P	1139·3	3·18
3D – 3P	818·9	3·59
4P – 3S	330·3	3·74
5S – 3P	615·8	4·1
4D – 3P	568·5	4·2
Ionization	All lines	5·12

Here we see that a potential of 5·12 V applied to an atom provides sufficient energy for an electron in the lowest energy state of the sodium atom to be completely removed. This can easily be confirmed from spectroscopic data when we recall that the wavenumber of the lowest term 3S is at $R/(3-1\cdot37)^2$ below the ionization level. Therefore energy equal to $hcR/(3-1\cdot37)^2$ must be supplied and, if this is provided by an electron, we have

$$Ve = \frac{hcR}{(3-1\cdot37)^2}$$

Therefore

$$V = \frac{(6\cdot626 \times 10^{-34}) \times (2\cdot998 \times 10^8) \times (10\cdot97 \times 10^6)}{(1\cdot602 \times 10^{-19}) \times (1\cdot63)^2}$$

$$= 5\cdot12 \text{ V}$$

The agreement with the experimental value of 5·12 V is remarkable.

7.9 X-Ray Spectra

We have mentioned how the spectrum of X-rays consisted of a continuous spectrum or white radiation upon which was superposed a line spectrum. The distribution of energy in the continuous spectrum depended only upon the potential difference across the X-ray tube while the line spectrum was characteristic of the element used as target (see Fig. 7.11). These are called characteristic spectra and were investigated by Moseley in 1913 by making each element in turn the target in an X-ray tube. Thirty-nine elements extending from aluminium to gold were examined in this way. The X-rays were analysed with a Bragg crystal spectrometer, the whole being evacuated to prevent absorption by the air of the

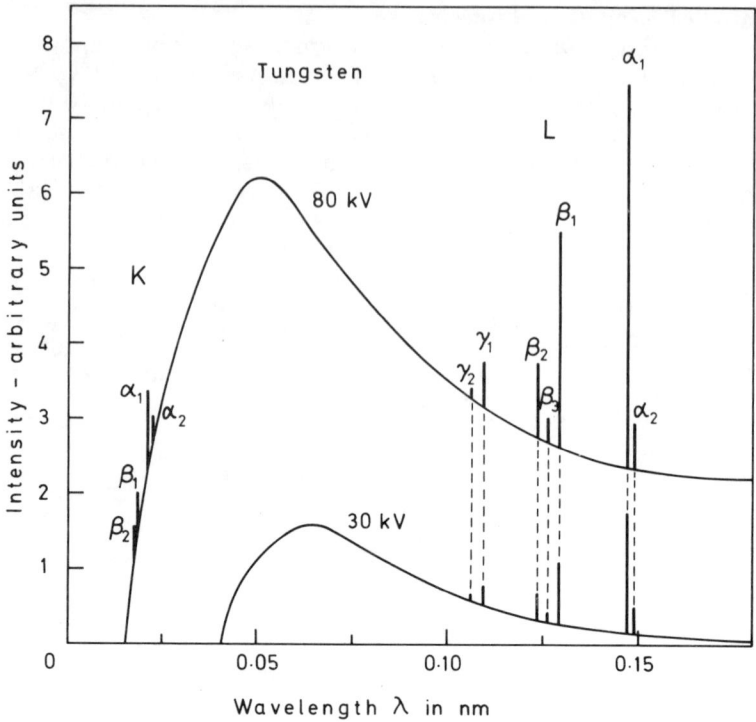

Fig. 7.11 Continuous and characteristic X-ray spectrum of tungsten.

X-rays of longer wavelength. Most elements showed two groups of lines, one generally less than about 0·1 nm called the K series, and another greater than 0·1 nm and called the L series. The wavelengths of the L series were roughly ten times as greater as those of the K series. For elements whose atomic number exceeded 66, further series appeared which were called the M and N series.

7.10 Moseley's Work

Moseley found that the structure or pattern of lines in a given series was similar for most of the elements he examined, but the lines showed a steady decrease in wavelength as the atomic number increased. This is shown in Fig. 7.12. At the time when Moseley carried out this work, the fundamental significance of atomic number had not been realized. Mendeleev, having arranged the elements in order of their atomic masses, found that certain pairs had to be reversed in order to preserve the periodicity of chemical and physical properties. The numerical order in which they were then placed was known as the atomic number. In all, three pairs of elements were reversed in this way, Ar and K, Co and Ni, Te and I. Moseley found that, on plotting the square roots of the frequencies of the K_α and K_β lines against atomic number Z for each pattern, he obtained a straight line which did not quite pass through the origin (Fig. 7.13). It appeared that the X-ray spectrum of each element was characterized by its atomic number, and therefore

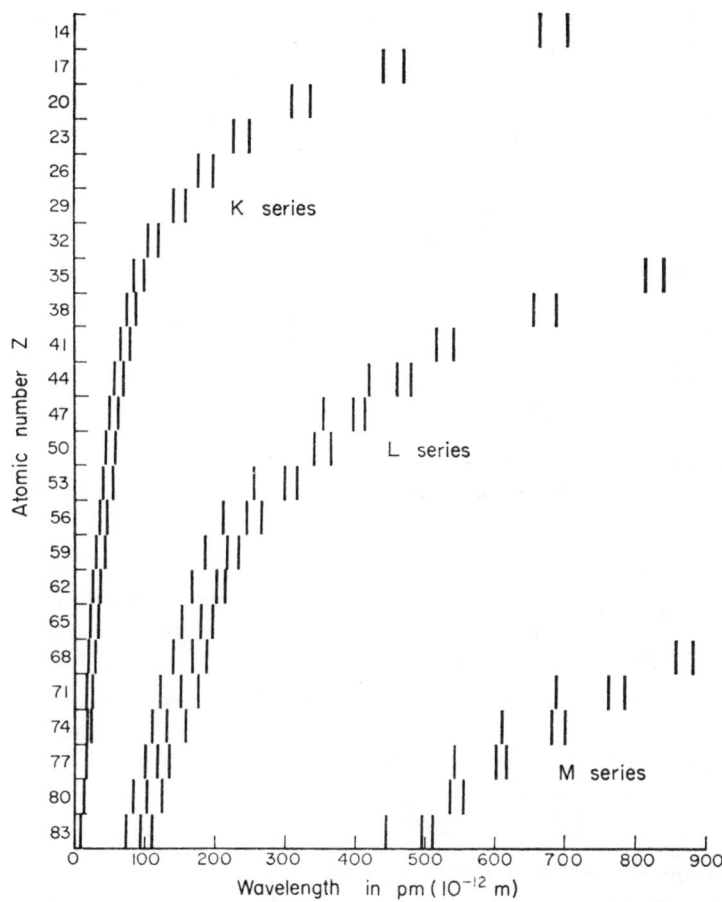

Fig. 7.12 X-ray lines for various elements.

suggested it had more physical significance than just its order in the periodic table. We now know of course that it is the number of positive units of electricity carried by the atomic nucleus, as revealed by Rutherford's α-particle scattering experiments.

It is instructive to give Moseley's own summary of the experiments which led to the above conclusions in 1914. These were:

(1) Every element from aluminium to gold is characterized by an integer N which determines its X-ray spectrum. Every detail in the spectrum of an element can therefore be predicted from the spectra of its neighbours.

(2) This integer N, the atomic number of the element, is identified with the number of positive units of electricity contained in the atomic nucleus.

(3) The atomic numbers for all elements from Al to Au have been tabulated on the assumption that N for Al is 13.

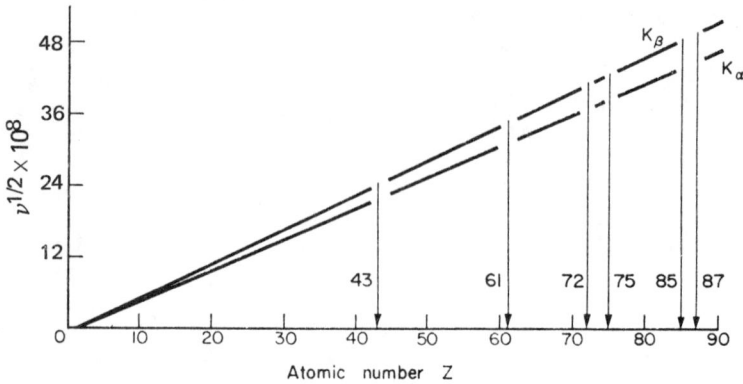

Fig. 7.13 The missing elements.

(4) The order of the atomic numbers is the same as that of the atomic masses, except where the latter disagrees with the order of the chemical properties.

(5) Known elements correspond with all the numbers between 13 and 79 except three. There are here three possible elements still undiscovered.

(6) The frequency of any line in the X-ray spectrum is approximately proportional to $A(N-b)^2$, where A and b are constants.

The symbol N corresponds to our own symbol Z for the atomic number.

Gaps in the Moseley diagram (Fig. 7.13) occurred at $Z = 43$, 61 and 75 corresponding to the (then) undiscovered elements technetium (43), promethium (61) and rhenium (75). Other new elements were later predicted in a similar way, as shown in the diagram.

7.11 The Interpretation of X-Ray Spectra

Moseley's linear relation between (frequency)$^{1/2}$ and atomic number Z can be written in terms of wavenumbers as $\sqrt{\bar{v}} = m(Z-b)$. For the K_α lines, the results gave $m^2 = \frac{3}{4}R$ and $b = 1$, where R is the Rydberg constant, so that the equation becomes

$$\bar{v}_K = R(Z-1)^2 \left(\frac{1}{1^2} - \frac{1}{2^2}\right)$$

which is reminiscent of the Balmer-type formula for hydrogen. This suggested that the X-ray K photon is emitted when an electron changes from the stationary state $n = 2$ to the lower state $n = 1$ from which an electron has been ejected to leave the vacancy. The positive nuclear charge Ze is reduced to $(Z-1)e$ by the presence of another electron in the $n = 1$ orbit. We shall see when we come to study the Pauli exclusion principle that only two electrons are permitted to occupy this orbit in a given atom. For the L_α line the Rydberg formula was again obtained but with different constants such that

$$\bar{v}_L = R(Z-7\cdot 4)^2 \left(\frac{1}{2^2} - \frac{1}{3^2}\right).$$

This indicates that the L_α line arises from an electron transition from $n=3$ to $n=2$. Moreover many more electrons now lie between the excited electron and the nucleus so that the average value of the effective nuclear charge is reduced by 7·4. As in the case of the hydrogen atom, we can represent the various X-ray lines by transitions between the electron orbits within an atom as shown in Fig. 7.14.

It is clear now why the M and N series cannot appear before the atom has reached certain critical sizes and has the required number of stable orbits. As in optical spectra, it is more convenient to use energy level diagrams and Fig. 7.14 shows the energy level diagram corresponding to the electron orbit diagram already discussed.

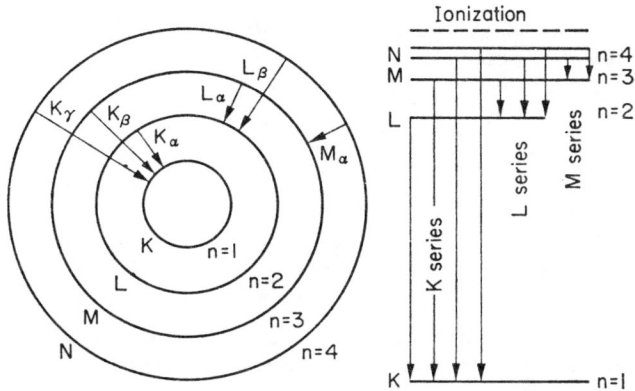

Fig. 7.14 X-ray energy levels.

We conclude this chapter by reminding the reader that all the work of Rutherford and Bohr, Bragg and Moseley we have described was carried out in the years 1911–1914.

Problems

(*Those problems marked with an asterisk are solved in full at the end of the section.*)

7.1 Describe and compare the optical spectra of hydrogen and sodium, showing how they may be represented by energy level diagrams.

If the wavelength limit for the Balmer series for hydrogen is 364·6 nm, calculate the wavelength limits of the sharp $(3P - mS)$ and principal series $(3S - mP)$ of sodium if the quantum defects of the S and P levels are respectively 1·37 and 0·88. Neglect the slight variation of the Rydberg constant with the mass of the nucleus. (409·7 nm, 242·2 nm)

7.2 A series in the spectrum of a gas is given below. It is observed that the series limit and some of the wavelengths are almost identical with those of the Balmer series of hydrogen. Identify the spectrum, account for the extra lines, and also for the small wavelength differences.

Rydberg constant for infinite mass is $10\,973\cdot7$ mm^{-1}.

Gas	Balmer series
656·019	656·280
541·160	
485·940	486·138
454·166	
433·874	434·051
419·990	
410·010	410·178

(Ionized helium, for which

$$\bar{v} = RZ^2\left(\frac{1}{n_1^2} - \frac{1}{n_2^2}\right) = R\left(\frac{2^2}{4^2} - \frac{2^2}{n^2}\right).$$

Small variations arise from variation of R with mass of nucleus)

7.3* The wavenumbers in mm^{-1} obtained from the spectrogram of lithium are classified into three series as follows:

A	B	C
1490·4	1230·5	1638·4
3093·5	2011·3	2172·5
3647·9	2340·1	2420·0
3902·4	2508·9	2554·3
4039·9	2605·4	2635·3

The series limit of A is $4348\cdot6$ mm^{-1} and that of both B and C is $2858\cdot2$ mm^{-1}.

Construct to scale (approximate) an energy level diagram and identify each series. Determine the principal quantum number for the lowest term of each sequence. The Rydberg constant is $10\,973\cdot7$ mm^{-1}. (Principal quantum number is 3 in each case)

7.4 Obtain an expression for the Rydberg constant in terms of atomic constants. Show how it depends upon the mass of the nucleus. The series limits for the Balmer series of hydrogen and for once ionized helium are respectively $2\,741\,950$ and $2\,743\,050$ m^{-1}. Calculate the ratio of the mass of the proton to the mass of the electron. (1865:1)

7.5 Describe the atomic spectrum of hydrogen. Explain how Bohr was able to relate it to the structure of the atom, pointing out carefully the assumptions which were made. Calculate the wavelength limit of the Balmer series. (364·7 nm)

7.6 Explain how the spectra of the alkali metals differ from that of hydrogen. Illustrate your answer by a diagram showing S, P, D, F levels for which the quantum defects are respectively 1·37, 0·88, 0·01 and 0·001. Discuss the interpretation of quantum defects in terms of electron orbits.

7.7 Long exposure photographs of the Balmer series show that each line is accompanied by a second faint line of slightly shorter wavelength. The wavelength differences $\Delta\lambda$ are respectively 0·1791, 0·1313, 0·1176, 0·1088 nm at the wavelengths 656·47, 486·27, 434·17, 410·29. Account in detail for the extra lines.

7.8 Assuming that an amount of hydrogen of mass number 3 sufficient for spectroscopic examination can be introduced into a tube containing ordinary hydrogen, determine the wavelength difference between the H_α lines you would expect to observe. ($\Delta\lambda = 0\cdot238$ nm)

7.9 Find the potential difference through which an electron must be accelerated in order to (a) raise the energy of a hydrogen atom from the ground state to the first excited state; (b) ionize the atom. ((a) 10·2 V; (b) 13·6 V)

7.10 The excitation potential of mercury is 4·9 V. Calculate the wavelength of the radiation which will be emitted as the atoms return to the ground state. (253·0 nm)

7.11 If the series limit of the Lyman series for hydrogen is at 91·2 nm, find the approximate wavelength of the highest energy X-rays emitted by calcium of atomic number 20. (0·255 nm)

7.12* If the series limit of the Balmer series for hydrogen is 364·6 nm, calculate the atomic number of the element which gives X-ray wavelengths down to 0·1 nm. Identify the element. ($Z=31$, Gallium)

7.13 If the first member of the Lyman series is at 121·5 nm, calculate the wavelengths of the first members of the Paschen and Brackett series (1874 nm, 5400 nm)

7.14 Calculate the radius of the electron orbit for (a) hydrogen in the ground state; (b) once ionized helium in the ground state; (c) twice ionized lithium in the ground state. ((a)52·9 pm; (b) 26·5 pm; (c) 17·6 pm)

7.15 The Bohr formula for the hydrogen-like atoms is

$$\bar{v} = RZ^2 \left[\frac{1}{n_1^2} - \frac{1}{n_2^2} \right],$$

where n_1 and n_2 are integers and the other symbols have their usual meanings.

It is found that some lines of the spectrum of ionized helium (He$^+$) correspond almost exactly with the Balmer lines of hydrogen ($n_1=2$) and that their limit is almost the same. Determine the integers n_1 and n_2 in the Bohr formula for He$^+$ and show that n_2 is even for the lines corresponding in wavelength to the Balmer lines.

The series limits are 2 741 944 m^{-1} for hydrogen and 2 743 060 m^{-1} for He$^+$. Obtain the ratio of the mass of the proton to the mass of the electron from these figures. (N)

7.16 What is the physical interpretation of the quantum defect constants in the alkali-metal atomic term formulae?

Given a table of wavenumbers of the spectral lines of an alkali-metal atom explain how you would determine the appropriate quantum defects.

Explain why the wavenumber separation of some of the lines is the same as the separation of certain lines in the atomic hydrogen spectrum. [N]

7.17 'The limit of a spectral series is a term of some other series.' Discuss this statement with reference to the sodium spectrum.

Taking the Rydberg constant for hydrogen as 10 967 mm^{-1} find the approximate numerical values of two diffuse terms of the sodium spectrum, and hence the wavelength of the first member of the diffuse series if the limit of the sharp series is 2 448 mm^{-1}. Show that the quantum defect $P=0.883$. [N]

7.18 Using the Bohr formula, calculate the wavelengths of the first two lines of the Paschen series. Similarly, calculate the wavelengths of the first two lines of the Fundamental series of sodium, taking $D=0.01$ and $F=0.001$ with $R=10974$ mm^{-1}. Comment on the results.

Solutions to Problems

7.3 Since B and C have a common series limit they must be based upon the same energy level. They are probably the sharp and diffuse series but we cannot on this evidence alone distinguish between them. By subtraction we can now find the other energy levels and they are listed below. Lines drawn between the various figures indicate the given series A, B and C

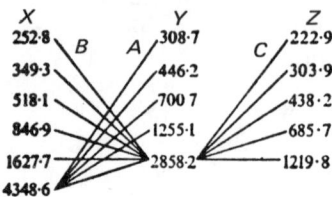

The principal quantum number n is given by the formula

$$\bar{v} = \frac{R}{n^2}:$$

$$n_x = \sqrt{\frac{10 \cdot 937 \times 10^6}{4348 \cdot 6 \times 10^3}} = 1 \cdot 59,$$

$$n_y = \sqrt{\frac{10 \cdot 937 \times 10^6}{2858 \cdot 2 \times 10^3}} = 1 \cdot 96,$$

$$n_z = \sqrt{\frac{10 \cdot 937 \times 10^6}{1219 \cdot 8 \times 10^3}} = 3 \cdot 00.$$

The terms under Z being free from quantum defect must be D terms. Those under X have the largest defect and are therefore S terms while those under Y must be P terms. This is confirmed by the fact that the S terms of an alkali element lie lowest. The three spectral series A, B and C are therefore principal, sharp and diffuse respectively.

7.12 Limit of the Balmer series is given by

$$\bar{v} = \frac{1}{\lambda} = R\left(\frac{1}{2^2} - \frac{1}{\infty^2}\right)$$

so that

$$R = \frac{4}{364 \cdot 6 \times 10^{-9}} \text{ m}^{-1}$$

Wavelengths of the K series are given by

$$\bar{v} = \frac{1}{\lambda} = R(Z-1)^2\left(\frac{1}{1^2} - \frac{1}{n^2}\right).$$

The maximum wavenumber occurs when $n = \infty$ and therefore

$$\bar{v} = \frac{1}{\lambda} = R(Z-1)^2 \frac{1}{1^2},$$

$$(Z-1)^2 = \frac{1}{R\lambda} = \frac{364 \cdot 6 \times 10^{-9}}{4 \times 0 \cdot 1 \times 10^{-9}} = 911,$$

$$Z - 1 = 30 \cdot 2,$$

$$Z = 31, \text{ which corresponds to gallium.}$$

Chapter 8

Fine Structure and Electron Spin

8.1 Fine Structure of Alkali-Metal Spectra
In our first study of the optical spectrum of sodium the fact that many of the lines were double was ignored for the sake of simplicity. It is well known that the first member of the principal series is the sodium yellow line, which, with quite moderate resolving power, is seen to consist of two lines separated by about 0·6 nm. These are the well-known D-lines at 589·0 and 589·6 nm. All lines in the principal, sharp and diffuse series are doublets, so that they are said to have fine structure. It is now necessary to try to interpret this new phenomenon in terms of atomic structure. It is clear that some elaboration of our picture of the atom based upon the Bohr theory is called for. Before attempting this, however, we must examine the experimental facts in more detail and try to devise an appropriate energy level scheme.

In the first place it is apparent that the Bohr theory itself cannot predict a multiplet line structure. Under high resolution, the structure of the H_α line of the Balmer series shows that it is by no means a singlet. Figure 8.1 shows the spectral shape of the H_α line on the Bohr theory and that actually observed spectrophotometrically. Both are therefore line profiles. Figure 8.1(*b*) indicates that the H_α line is at least a triplet. Second, it cannot predict the intensity ratios of the lines of a multiplet. Experimentally, we find that the intensities of the lines of a multiplet are in a simple integral ratio, e.g. $D_1:D_2$ for the components of the D line in the principal series of sodium is exactly 2:1. The simple Bohr theory cannot predict this ratio.

The simple Bohr theory is therefore inadequate to explain the complicated structure of the H_α line. For this reason, Sommerfeld extended the Bohr theory to include relativistic effects, elliptic and three-dimensional orbits showing that energy, linear momentum and angular momentum must all be quantized, so pointing to the necessity of describing the electron orbit in terms of *three* quantum numbers rather than one, as in the simpler Bohr theory. One of these is the quantum number *l* which we have already discussed and which quantizes the orbital angular momentum of the atom. These quantum numbers were not independent and their properties were derived largely by empirical means.

We now consider the structure of the lines of the sodium spectrum. The relevant measurements are in Table 8.1, which shows some of the wavelengths in the sodium spectrum classified into the sharp, principal and diffuse series, as done

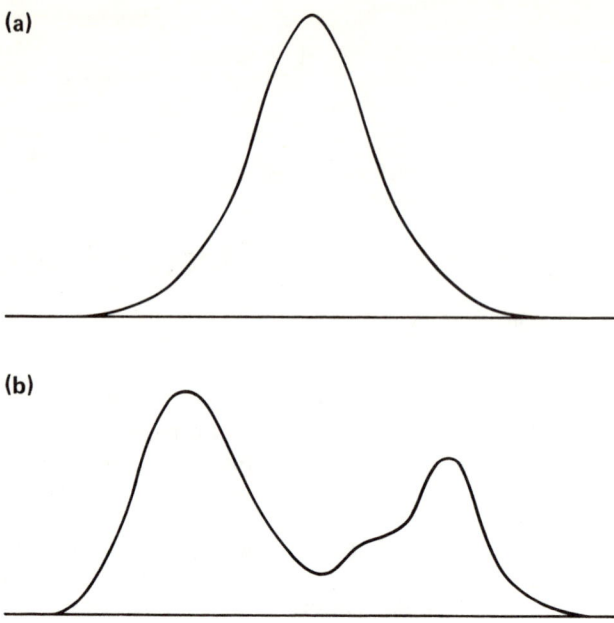

Fig. 8.1 Structure of intensity profile of H_α line. (a) Structure as predicted by Bohr; (b) Structure as found by experiment.

TABLE 8.1
The Sodium Doublets

	Sharp			Principal			Diffuse	
λ (nm)	$\bar{v}$ (mm^{-1})	$\Delta\bar{v}$	λ (nm)	$\bar{v}$ (mm^{-1})	$\Delta\bar{v}$	λ (nm)	$\bar{v}$ (mm^{-1})	$\Delta\bar{v}$
1140·42	876·63	>1·68	588·996	1697·34	>1·72	819·482	1219·95	>1·71
1138·24	878·31		589·593	1695·62		818·330	1221·66	
616·072	1622·74	>1·71	330·234	3027·29	>0·55	568·822	1757·53	>1·72
615·421	1624·45		330·294	3026·74		568·268	1759·25	
515·364	1939·83	>1·72	285·283	3504·26	>0·243	498·287	2006·32	>1·71
514·909	1941·55		285·303	3504·017		497·861	2008·03	
475·189	2103·84	>1·72	268·034	3729·77	>0·15	466·860	2141·37	>1·72
474·802	2105·56		268·044	3729·62		466·486	2143·09	
454·522	2199·50	>1·72	259·383	3854·154	>0·147	449·772	2222·71	>1·72
454·167	2201·22		259·393	3854·007		449·427	2224·43	
442·331	2260·12	>1·72	254·382	3929·920	>0·089	434·345	2275·48	>1·72
441·994	2261·84		254·388	3929·831		439·014	2277·20	

in the previous chapter. The first column of each series gives the wavelength λ of each line, the second column gives the corresponding wavenumber $\bar{v} = 1/\lambda$, while the third column gives the wavenumber separation $\Delta\bar{v}$ of the doublets. Examination of these data reveals that within the experimental error, $\Delta\bar{v}$ is the same for each pair in the sharp and diffuse series. In the principal series, however, $\Delta\bar{v}$ becomes smaller as we proceed to higher members of the series. Moreover for the first member of the principal series, the sodium yellow lines, $\Delta\bar{v}$ is $1\cdot72$ mm^{-1} and is equal to the separation of doublets in the other two series.

Thinking now in terms of the energy level diagram for sodium and remembering that both the sharp and diffuse series involve the 3P term, it is reasonable to suppose that this term consists of two levels with a separation of $1\cdot72$ mm^{-1}. This explains the constant wavenumber separation of the lines in both the sharp and the diffuse series assuming both the S and D levels to be single, although D levels are not always single, as we shall see later (p. 112). To explain how the doublets of the principal series $n\text{P} - 3\text{S}$ close up as we proceed up the series we must suppose that all the P terms are double and that the separation becomes smaller for the higher terms.

With heavier atoms, having more complex electron structures and therefore large central atomic cores, the diffuse series is sometimes triple. As an example the diffuse series of caesium consists of triplets as shown in Table 8.2.

TABLE 8.2
The Diffuse Series of Caesium

Transition	λ (nm)	$\bar{v}$ (mm^{-1})		$\Delta\bar{v}$ (mm^{-1})
6P – 5D	3010·0	332·1		
	3489·2	286·5	⎤	55·4
	3612·7	276·7	⎦ 9·8	
6P – 6D	876·1	1141·1		
	971·2	1090·0	⎤	55·4
	920·8	1085·7	⎦ 4·3	
6P – 7D	672·3	1487·0		
	697·3	1433·7	⎤	55·4
	698·3	1431·6	⎦ 2·1	

The constant splitting of $55\cdot4$ mm^{-1} must, as in the case of sodium, be associated with the 6P term. This has now become very much larger, indicating that the interaction between the electron and the nucleus is much greater, since, the electron, even in the D state, is penetrating the atomic core. For the deepest of these the value is $9\cdot8$ mm^{-1}, but it rapidly closes up as we proceed to higher levels.

8.2 Electron Spin

It is reasonable to assume from the previous section that the principal and diffuse levels of both sodium and caesium are in reality doublets, and give rise to the triplet structure of the diffuse series so readily observed in caesium. For complete doublet level transitions we would expect *four* lines in the diffuse series, as shown in Fig. 8.3 for the caesium 6P–5D transition. In Fig. 8.3, as in similar diagrams, the term separations are not to scale.

The *reason* for the term separations of the *l*-levels is not at once apparent, but it is obvious that some further quantum constraint is needed. Returning to the caesium energy·level diagram of Fig. 8.3, the fact that only *three* transitions are

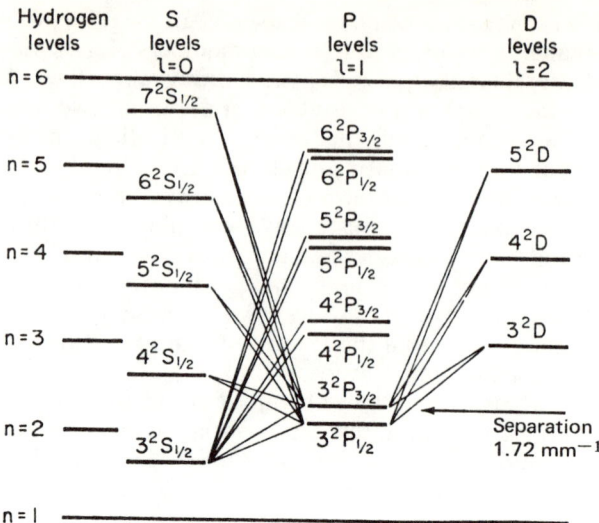

Fig. 8.2 Fine structure of the sodium energy levels.

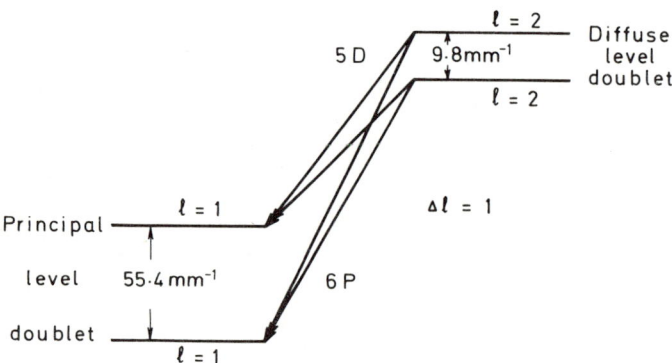

Fig. 8.3 Caesium diffuse series levels showing four possible transitions. Only three are observed. Term separation not to scale.

observed spectroscopically indicates that some new quantum condition operates in addition to the selection rule $\Delta l = \pm 1$, already discussed. Also we note that the S levels ($l=0$) must be single in order to give the principal series *doublets* as found in the sodium D-lines. Thus all but the S levels are doublets. It was found empirically that these difficulties could be resolved by adding algebraically another quantum number $\frac{1}{2}$, of unknown origin, to the orbital quantum l to give two resultant values of the total quantum number j given by $j = l + \frac{1}{2}$ and $j = l - \frac{1}{2}$. This gives *two* different levels for each l except $l=0$. Since the j value of S level can only be $+\frac{1}{2}$ and not $-\frac{1}{2}$.

If we now redraw the energy level diagrams for the two sodium D lines and the caesium 6P–5D diffuse triplet in terms of the new j quantum number, we get Fig. 8.4. We see that the doublets and triplets can arise only when Δj is limited to 0 or 1, but not 2 (and above). This rule was successful in explaining many other complex spectra, and eventually the selection rule was found to be $\Delta j = \pm 1, 0$.

It was this fact, and a search for the understanding of the anomalous Zeeman effect (p. 188), which led Goudsmit and Uhlenbeck (1925) to postulate the

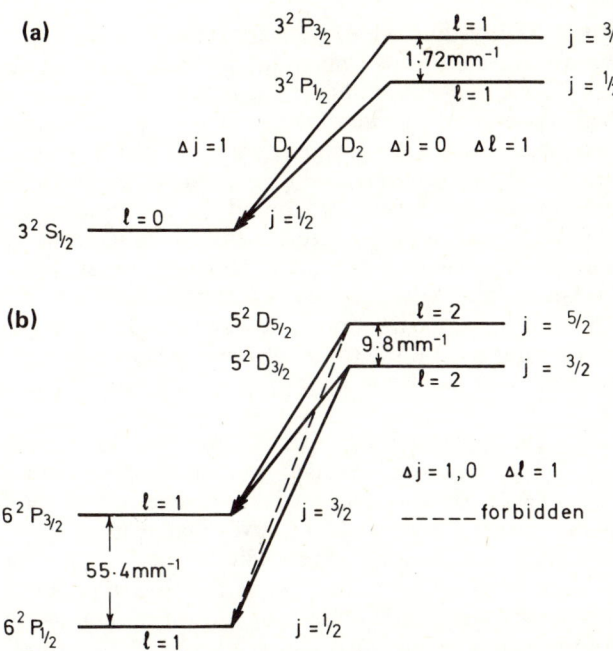

Fig. 8.4 Transitions with selection rules for j. Term separations not to scale. (a) Sodium D-line doublet; (b) caesium diffuse triplet.

hypothesis of the spinning electron with an associated angular momentum of $\frac{1}{2}\hbar$ and a quantum number of $S = \frac{1}{2}$. (The symbol $\hbar$ is the standard abbreviation for the Planck constant h divided by 2π, i.e. $\hbar = h/2\pi$.) The coupling of orbital and spin angular momenta then gave the *total* angular momentum governed by the quantum number rule

$$j = l \pm \tfrac{1}{2}$$

The fine structure of spectral lines could then be described in terms of four quantum numbers n, l, j and s with various empirical relationships between them. The picture of an electron spinning mechanically about its axis must not be taken too literally. In any case it was found that the angular momentum of the spinning electron was not $\frac{1}{2}\hbar$ but $\frac{1}{2}\sqrt{3}\hbar$, as we shall see later.

The spectroscopic notation used to describe the levels in Figs 8.2 and 8.4 is that the term is symbolized by $n^r l_j$, where n, l and j have the meanings already described and r is the term multiplicity. In the alkali metals we have just discussed, we know that $r=2$ and we shall see later that in all cases $r=2s+1$, so that $r=2$, $s=\frac{1}{2}$ for atoms with a single electron. In the case of multi-electron atoms, the resultant spin quantum number can be greater than $\frac{1}{2}$ but the higher multiplicities of terms observed can still be found from the relation $R=2S+1$, where S is the net atomic spin quantum number.

Further proof of these ideas is to be found in X-ray spectroscopy.

8.3 Characteristic X-Rays and Absorption Spectra

X-ray spectra are the results of transitions between inner shell electrons of heavier elements. The K, L shells are made vacant by electron impact and the X-ray frequencies are given by equations of the form $h\nu_K = E_K - E_L$, where E_K, E_L are the ionization energies of the K ($n=1$) and L ($n=2$) shells. The emission spectra are superimposed on the continuous, or 'white', radiation. At low tube voltages only the longer wavelengths are excited, but as the potential difference is increased the N, M, L and K series are emitted in that order. The K series therefore has the shortest wavelength and the highest energy. Figures 6.13 and 7.11 show the relationships between the characteristic and white radiations at various tube voltages for tungsten and molybdenum. The wavelengths of the characteristic X-rays depend on the target element, whereas the white radiation distribution depends only on the tube voltage and is roughly the same for all targets.

With improved techniques for X-ray spectroscopy it was shown therefore that the characteristic X-ray spectrum lines displayed fine structure. This immediately suggests that X-ray levels are multiplets arising from electron spin in the same way as the optical levels. These could be investigated in the same analytical manner as for optical spectra, but more direct methods are available. In particular, X-ray absorption spectra enable the structures of the levels to be found directly. In optical spectra emission and absorption spectra are identical, except in so far as an absorption spectrum is simpler because it is usually limited to transitions from the ground or lowest energy state. This is no longer true with X-rays, which show absorption edges due to ionization rather than absorption lines.

The X-ray absorption spectrum of a metal may be examined by passing a narrow beam of X-rays through the metal in the form of a foil. The spectrum of the incident X-rays should be continuous and the X-rays, after passing through the metal foil, are analysed with a Bragg X-ray spectrometer. Readings with and without the absorber in position at each wavelength are made and a graph of absorption and wavelength plotted.

In general the mass absorption of the incident X-rays varies smoothly as the cube of the wavelength, as shown by the dotted curve in Fig. 8.5, but at certain critical wavelengths, corresponding to the energies of ionization of the K, L, M, etc., electrons, the absorption increases and then falls suddenly. At short wavelengths, with energies greater than the K ionization energy, the incident photons can be totally absorbed by the atom in order to eject a K electron. This requires a photon energy equal to the K ionization energy. The absorption on the short-wavelength side of this wavelength is therefore very high. However, as the wavelength increases the energy is less than the K ionization energy, the

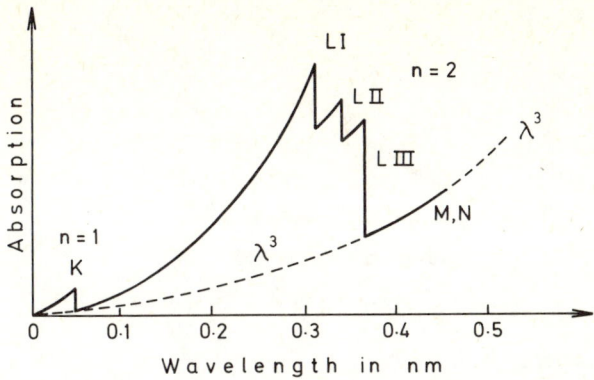

Fig. 8.5 X-ray absorption spectrum of Pd, $Z = 46$.

transmission increases and the absorption falls to the normal value. The K ionization energy wavelength is thus characterized by a sharp, quantum-discrete fall in absorption called the K absorption edge, or just the K edge.

Beyond the K edge the photon has insufficient energy to eject an electron from the K shell, but still enough to eject one from the L shell, $n = 2$. Thus the absorption increases again, corresponding to L ionization and the production of the L edge. This process is repeated for the M, N, ... shells.

The interesting fact which arises from these absorption experiments is that the edges for each principal quantum number above $n = 1$ are multiple, and therefore the inner electron levels of the atom *also* have a multiplet structure, such as we have been discussing for the outer electron levels. The multiplicities are

	K	L	M	N	...	levels
$n =$	1	2	3	4	...	
	1	3	5	7	...	multiplicities

That the edges are sharp is experimental evidence of the discrete value of the energy of each level.

Returning to the K edge, it is found that its wavelength is less than those of the K series, and is given by the *limit* of the K series, as in the Moseley formula

$$\bar{v}_K = R(Z-1)^2 \left(\frac{1}{1^2} - \frac{1}{\infty} \right) = R(Z-1)^2.$$

This is illustrated in Fig. 8.6 for palladium.

The explanation of Fig. 8.6 rests on the fact that X-rays arise from transitions deep within the atom. We shall see later in Chapter 10 how the Pauli principle limits the number of electrons which can occupy successive shells, and, as these are already occupied, the excited electron can only be completely removed from the atom. In effect one cannot observe an X-ray *line* absorption spectrum because the upper energy levels to which the electron would have to be raised are already occupied. The energy required to remove the K electron completely is the ionization energy. Since this is a quantum-determined energy the K absorption edge occurs at a precise wavelength, viz. 50×10^{-12} m from Fig. 8.6. The

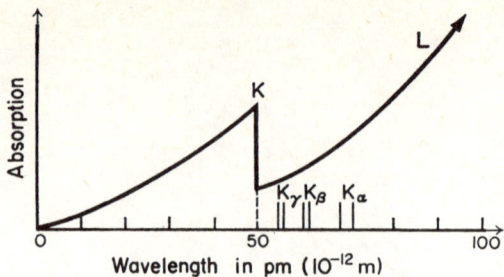

Fig. 8.6 The K absorption edge. (Pd, $Z=46$).

corresponding energy is given by

$$E = h\nu = \frac{hc}{\lambda} = \frac{6\cdot 6 \times 10^{-34} \times 3 \times 10^{8}}{50 \times 10^{-12}}$$
$$= 3\cdot 96 \times 10^{-15} \text{ J}$$
$$= \frac{3\cdot 96 \times 10^{-15}}{1\cdot 6 \times 10^{-19}} \text{ eV}$$
$$= 24\cdot 8 \text{ keV}.$$

The K ionization potential of the Pd atom is therefore 24·8 keV.

8.4 Multiplicity of X-Ray Levels

The multiplicity of the X-ray levels can be explained in terms of the quantum numbers n, l and j which have already been used in the description of optical spectra. Thus for the K shell $n=1$, $l=0$ and $j=0+\tfrac{1}{2}$. In the L shell $n=2$, giving $l=0$ or 1 so that j takes values $j=0+\tfrac{1}{2}$,† $1-\tfrac{1}{2}$ and $1+\tfrac{1}{2}$, explaining the three L levels observed. The M shell for which $n=3$ has $l=0$, 1 or 2. This leads to five j levels as follows: $j=0+\tfrac{1}{2}, j=1\pm\tfrac{1}{2}, j=2\pm\tfrac{1}{2}$. Figure 8.7 shows the various X-ray levels associated with the K, L, M shells and indicates how the various X-ray series of lines arise. In particular it will be seen that the K series are all doubtlets due to the operation of the selection rules $\Delta j = \pm 1$ or 0, and also $\Delta l = \pm 1$. The L and M series are only developed for the larger atoms and have a much more complex structure. In contrast with optical spectra it is interesting to note that all atoms give similar type of X-ray spectra, and the extent to which these are developed is determined by the size of the atom. X-ray spectra are therefore much simpler than optical spectra.

Problems

(*The problem marked with an asterisk is solved in full at the end of the section.*)

8.1 The alkali metal spectra consist only of doublets and triplets. Explain why singlets and other multiplicities are never observed.

8.2* The following series (mm^{-1}) were observed in the spectrum of caesium. Draw an energy level diagram to scale (approximately) showing how these series

† $j=-\tfrac{1}{2}$ is meaningless.

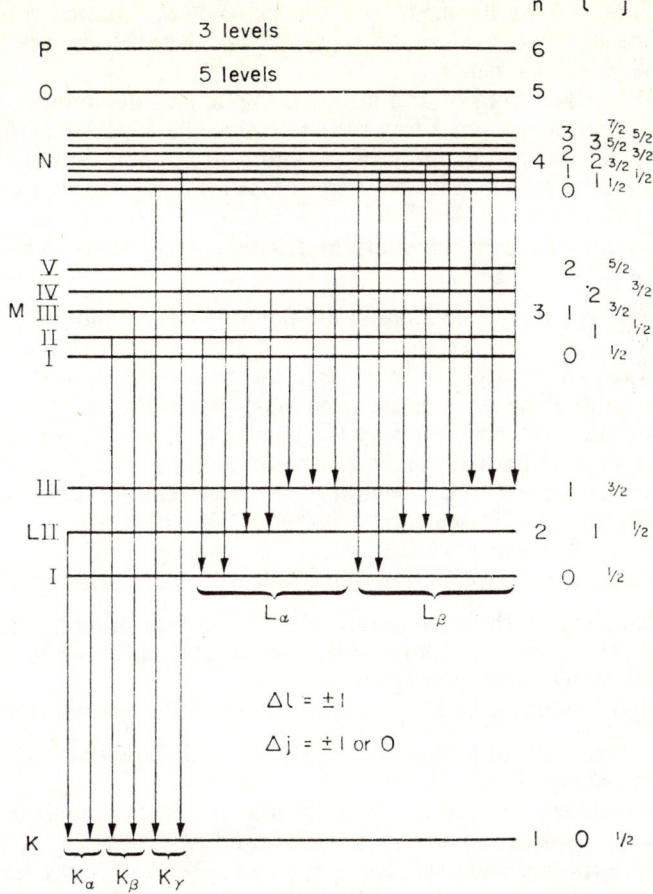

Fig. 8.7 Fine structure of X-ray energy levels.

arise and identifying the levels involved. Assume the deepest term lies at 3140 mm^{-1}.

A	B	C
276·8	680·5	1118·1
286·6	735·9	1173·6
332·2	1258·8	2177·1
1086·0	1314·2	2195·2
1090·3	1518·2	2571·6
1141·4	1573·6	2579·7
1432·0	1657·1	2764·4
1434·1	1712·6	2768·9
1487·4		

(B is sharp, C is principal, A is diffuse series.)

8.3 Interpret the results of Moseley's work on X-ray spectra in terms of the Bohr theory of the atom. Explain the importance of these results in relation to the periodic table of the elements.

If the K absorption limit of uranium is 10·7 pm, find the minimum potential difference required across an X-ray tube to excite the K series. (116 kV)

8.4 If the K absorption limit of platinum is 15 pm, find the minimum potential difference which must be used across an X-ray tube to excite the K series. (90·5 kV)

8.5 Explain briefly how the study of atomic spectra leads to the following conclusions:

(a) The electron has a spin angular momentum corresponding to a quantum number of $\frac{1}{2}$.

(b) The photon resulting from an electron transition between two energy states has an angular momentum of at least one unit of $\hbar$.   [N]

8.6* What do you understand by the 'fine structure' of a spectral line?

The two sodium D lines are given by the transitions $3^2P_{3/2, 1/2} - 3^2S_{1/2}$. The energy required to reverse the direction of the spin, relative to the orbit, of the optically active electron in the atom is 0·002 eV. Calculate the separation of the D lines, taking $\lambda_D = 600$ nm. (0·58 nm)   [N]

8.7 Give a reasoned explanation of the following facts about optical spectra:

(a) the lowest term in the term diagram of the neutral lithium atom is $^2S_{1/2}$;

(b) the highest terms in lithium have almost the same separation as the corresponding terms in the hydrogen atom;

(c) under high resolution the $^1_1H_\alpha$ line of the Balmer series shows a fine structure.

Draw the energy level diagram of the $^1_1H_\alpha$ line and show that there are seven possible components of this line.   [N]

8.9 Draw the energy level diagram of the fine structure of the X-ray spectra of an atom due to transitions between the K, L and M levels.

What is the experimental evidence of the multiplicity of X-ray levels?

An X-ray tube is working at 50 kV. If the lowest wavelength of the X-rays emitted by the tube is 0·0248 nm find the value of the Planck constant.   [N]

Solutions to Problems

8.2 Since the separation between each pair of lines in the B series is constant and equal to 55·4 mm^{-1} we can suppose that these are all based upon an energy level which is double. The other levels involved must each be single. B must therefore correspond to the sharp series. With the C series the separation begins with 55·4 mm^{-1} but rapidly decreases as we proceed down the column. This can best be interpreted by supposing that the lines are based upon a single energy level and that only the upper levels are split, the splitting getting smaller as we go to higher levels. Since the lower level is single it is an S level and the series is the principal series.

A consists of three sets of triplets, the outer members of each triplet have the same separations, 55·4 mm^{-1}, while the first two members become closer as we proceed down the column. These lines arise from transitions between two split levels, the lower one being the same as for B and the upper levels get closer as we go to higher terms. The number of lines observed is limited to three by the j selection rule.

8.6 The fine structure is due to the energy difference between the two P levels. This is
$$0.002 \text{ eV} = 0.002 \times 1.6 \times 10^{-19} \text{ J} = 3.2 \times 10^{-22} \text{ J}$$

Therefore
$$h\Delta v = 3.2 \times 10^{-22} \text{ J}$$

or
$$\Delta v = \frac{3.2 \times 10^{-22}}{6.6 \times 10^{-34}} = 4.85 \times 10^{11} \text{ Hz}$$

Now
$$\frac{\Delta v}{v} = \frac{\Delta \lambda}{\lambda}$$

so that
$$\Delta \lambda = \lambda \frac{\Delta v}{v} = \lambda^2 \frac{\Delta v}{c}$$

Therefore
$$\Delta \lambda = \frac{(600)^2 \times 10^{-18} \times 4.85 \times 10^{11}}{3 \times 10^8} = 0.58 \text{ nm}$$

Chapter 9

Waves and Particles

9.1 The Radiation Dilemma

When electromagnetic radiation interacts with matter it does so in energy quanta equal to $h\nu$. It displays the characteristics of a particle in photoelectricity, in the Compton effect and in the continuous spectrum of X-rays. This is supported and extended by the wealth of experimental data arising from the study of optical and X-ray line spectra. However, the phenomena of interference, diffraction and polarization still require a wave theory for their interpretation. Radiation thus displays a dual character, sometimes behaving as a wave and at other times as a particle. It appears that radiation cannot exhibit its particle and wave properties simultaneously. In Compton's experiment, for example, the X-rays behave as particles on being scattered by the electrons in the graphite. In Bragg's X-ray spectrometer they behave as waves on being diffracted by the crystal but again as particles when they are detected in the ionization chamber or on the photographic plate.

In the previous chapters we have looked at some of the consequences of the quantization of radiation which classically should always show wave properties. We have examined the quantum theory of the atom in terms of its optical and X-ray spectra and we have been led to the idea of discrete quantum numbers, to angular momentum relations, to an understanding of fine structure and to a mysterious idea of a spinning electron whose representative quantum number breaks all the rules by being non-integral. The 'rules', of course, are based on the classical Bohr–Sommerfeld theory of atomic structure, which gave very limited backing to a wealth of empirical facts. The theory is insufficient to deal with these new ideas. The change came in 1924 when de Broglie made the suggestion that complementing the particle (photon) nature of moving particles there was an associated wavelength. Radiation was 'lumpy' by the quantum theory — the new idea was that particles were 'wavy'!

9.2 De Broglie's Theory

Relativity shows that the fundamental law relating energy E, rest mass m_0 and momentum p of a particle is $(E/c)^2 = p^2 + m_0^2 c^2$ (see Appendix A). The rest mass of a photon is zero, so that $m_0 = 0$, and therefore its momentum is $p = E/c$. Setting $E = h\nu$ in accordance with quantum theory, this becomes $p = h\nu/c = h/\lambda$ and we have already seen in Chapter 6 that this is in accordance with the results of the Compton effect. In 1924 de Broglie extended the dualism, already found in radiation, to include material particles such as electrons, protons, atoms and even molecules. He assumed that a wavelength could be *associated* with each particle given by $\lambda = h/p = h/mv$, where m and v were respectively the mass and velocity of the particle. The associated wave velocity ω is given by $\omega = v\lambda$. Substituting for v and λ one obtains

$$\omega = v\lambda = \frac{E}{h}\frac{h}{p} = \frac{mc^2}{mv} = \frac{c^2}{v}.$$

For a material particle, therefore, the wave velocity exceeds the velocity of light, since the particle velocity must be less than c, in accordance with relativity theory (see Appendix A). As this wave velocity does not carry energy and therefore cannot be used to carry a signal or information, the relativity principle remains valid. The limiting case occurs when $\omega = v = c$ for a photon.

The wave velocity is an artificial concept as it cannot be determined experimentally. If one imagines an infinitely extended sine wave, marks must be attached to it in order to determine its velocity. This is only possible by superposing another wave which will provide a convenient measuring mark in the form of an amplitude variation. Experimentally one is only able to measure the velocity of this mark. The velocity with which the mark moves depends upon both waves and is called the group velocity. It is well known in sound that, when two tuning forks of almost the same frequency are sounded, 'beats' occur due to the superposition of the two wavetrains as shown in Fig. 9.1. Here we see how the

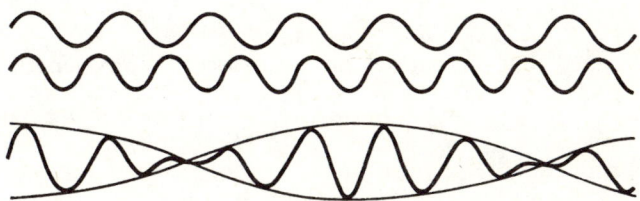

Fig. 9.1 Formation of wave groups.

beats arise from a series of wave groups and the observer hears these as beats when they reach him in regular succession. When a range of wavelengths are superposed only one wave group or pulse is observed. The wave group moves with a velocity which is less than that of the individual waves of which it is composed.

9.3 Group Velocity

Group velocity will now be calculated for the special case of two wavetrains having nearly equal wavelengths although the result is the same as for a larger number of wavetrains. Suppose the wavetrains are of equal amplitude and represented by the equations $y = a \cos 2\pi(vt - x/\lambda)$ and $y' = a \cos 2\pi(v't - x/\lambda')$.

The resultant displacement is given by
$$y+y'=a\cos 2\pi(vt-x/\lambda)+a\cos 2\pi(v't-x/\lambda')$$
$$=2a\cos 2\pi\left[\frac{v-v'}{2}t-\frac{x}{2}\left(\frac{1}{\lambda}-\frac{1}{\lambda'}\right)\right]\cos 2\pi\left[\frac{v+v'}{2}t-\frac{x}{2}\left(\frac{1}{\lambda}+\frac{1}{\lambda'}\right)\right].$$

For nearly equal frequencies and wavelengths, we have $v \sim v'$ and $\lambda \sim \lambda'$ and putting $v-v'=\Delta v$ and $1/\lambda - 1/\lambda' = \Delta(1/\lambda)$ this may be written
$$y+y'=2a\cos 2\pi\left[\frac{\Delta v}{2}t-\frac{x}{2}\Delta\left(\frac{1}{\lambda}\right)\right]\cos 2\pi\left[vt-\frac{x}{\lambda}\right].$$
since
$$2v \sim v+v' \quad \text{and} \quad 2\left(\frac{1}{\lambda}\right) \sim \frac{1}{\lambda}+\frac{1}{\lambda'}$$

The second part, $\cos 2\pi(vt-x/\lambda)$, represents the original wave travelling with mean velocity $v\lambda$. The first part also represents a wave and is the equation of the envelope of Fig. 9.1.; it corresponds to a wave group moving with velocity
$$u=\frac{\Delta v}{2}\div\Delta\left(\frac{1}{2\lambda}\right)=\frac{\Delta v}{\Delta(1/\lambda)}.$$

In the limit as v approaches v' and $1/\lambda$ approaches $1/\lambda'$ this becomes $u=dv/d(1/\lambda)$.

It is now necessary to determine the group velocity for a material particle moving with velocity v. Putting $\beta = v/c$, we may now write
$$v=\frac{E}{h}=\frac{mc^2}{h}=\frac{m_0 c^2}{h}\frac{1}{(1-\beta^2)^{1/2}}$$
(Appendix A). Therefore,
$$\frac{dv}{d\beta}=\frac{m_0 c^2}{h}\frac{\beta}{(1-\beta^2)^{3/2}}.$$

It is also assumed that the particle has an associated wavelength given by
$$\frac{1}{\lambda}=\frac{p}{h}=\frac{mv}{h}=\frac{m_0 c}{h}\frac{\beta}{(1-\beta^2)^{1/2}}.$$
Therefore
$$\frac{d(1/\lambda)}{d\beta}=\frac{m_0 c}{h}\frac{1}{(1-\beta^2)^{3/2}}.$$

The group velocity can be written
$$u=\frac{dv}{d\beta}\div\frac{d(1/\lambda)}{d\beta}=c\beta=v.$$

The group velocity may therefore be identified as the velocity of the particle. It is tempting to try to interpret a particle as a wave group or packet, and reconcile the wave and particle aspects of matter and radiation. This interpretation, however, is unsatisfactory since a wave packet will in time spread out, contrary to the observed stability of material particles such as electrons, protons and neutrons. Further and more detailed consideration of the wave and particle aspects of both radiation and matter will be given later.

9.4 The Davisson and Germer Experiment

The first experimental support for de Broglie's bold hypothesis came in 1927 from experiments by Davisson and Germer. Electrons from a heated filament F

(Fig. 9.2) were accelerated by a small potential difference and allowed to impinge upon the (111) planes of a single crystal of nickel. The intensity of the electrons was measured for various angles of scattering for a range of accelerating potentials from 40 to 68 V. For electrons accelerated through 54 V the reflected intensity was a maximum at an angle of 50° to the incident electron beam. Remembering that the kinetic energy of the electrons is given by $\frac{1}{2}mv^2 = Ve$ and

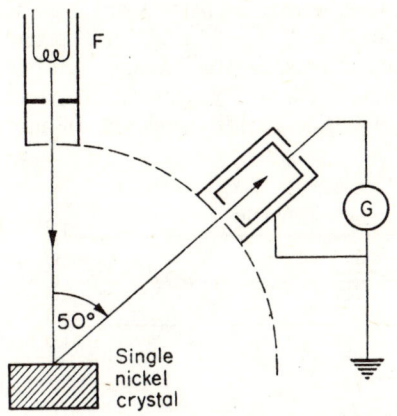

Fig. 9.2 The Davisson and Germer experiment.

that the wavelength associated with them is h/mv, the wavelength is given by $\lambda = h/\sqrt{2Vem}$. Substitution of relevant data gives

$$\lambda = \frac{6 \cdot 6 \times 10^{-34}}{\sqrt{2 \times 54 \times 1 \cdot 6 \times 10^{-19} \times 9 \cdot 1 \times 10^{-31}}} = 1 \cdot 67 \times 10^{-10} \text{ m} = 0 \cdot 167 \text{ nm}.$$

This wavelength is in the X-ray range. From X-ray diffraction we know that the unit cell spacing of nickel is 215 pm, so it is a simple matter to use the Bragg formula to calculate the wavelength of the scattered electrons. As the angle of scattering is 50°, the angles of incidence and reflection must each be 25°. This implies that the reflecting planes in the crystal are inclined at 25° to the top surface of the crystal, as shown in Fig. 9.3, so that the Bragg angle is 65°.

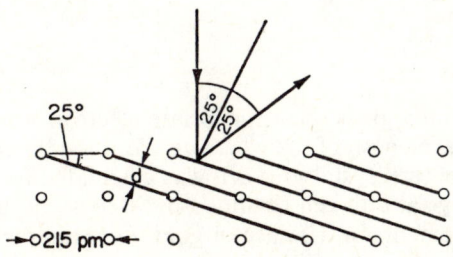

Fig. 9.3 Reflection of electrons from crystal planes in a crystal of nickel.

The distance d between the reflecting (111) crystal planes is $d = 215 \sin 25° = 90.9$ pm. Using the Bragg formula $m\lambda = 2d \sin \theta$ we get $\lambda = 2 \times 90.9 \sin 65° = 0.165$ nm. Agreement between the measured and the predicted wavelength is therefore established.

9.5 The Experiment of Thomson and Reid

The experiment carried out by Davisson and Germer corresponds closely with that used by Bragg for the measurement of X-ray wavelengths. In the following year G. P. Thomson and A. Reid devised an experiment using high-energy electrons produced in a low pressure discharge tube operated at between 10 and 60 kV. These were restricted to a narrow pencil and passed through thin metallic foil, as shown in Fig. 9.4. The metal consisted of many microscopic crystals

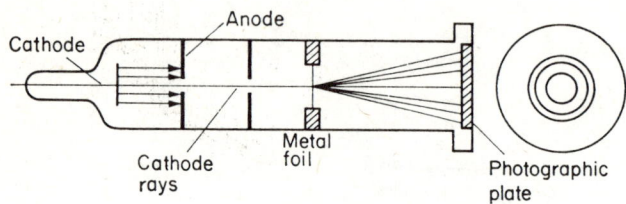

Fig. 9.4 The experiment of Thomson and Reid.

arranged in random fashion so that some were always at the proper angles to give reflection in accordance with the Bragg formula. Copper, aluminium, gold, platinum, lead and iron foils were used and the rings were recorded upon a photographic plate as shown. These interference rings prove the wave nature of the electrons.

If the electrons are accelerated through a potential difference of V volts, then $\tfrac{1}{2}mv^2 = Ve$ with the usual notation. The associated de Broglie wavelength is

$$\lambda = \frac{h}{mv}$$
$$= \frac{h}{\sqrt{2mVe}}$$
$$= \sqrt{\frac{150}{V}} \times 10^{-10} \text{ m}$$
$$= \sqrt{\frac{1.5}{V}} \text{ nm.}$$

This simple formula gives the appropriate electron wavelength immediately.

Knowing the wavelength of the electrons from the potential difference across the tube, the size of the crystal unit cell was calculated from the ring diameters. The sizes of the crystal unit cell obtained by X-rays and by electron diffraction were in close agreement. In the case of gold, for example, the X-ray value was 0.406 nm and the electron diffraction value was 0.408 nm. This agreement amounts to a proof of de Broglie's law.

It is interesting to note that J. J. Thomson was given a Nobel Prize for showing that the electron behaved as a *particle* while his son, G. P. Thomson, was given a Nobel Prize for showing the electron could behave as a *wave*.

9.6 The Electron Microscope

The waves of very small wavelength associated with an electron soon found a very important application in the form of the electron microscope. The resolving power of an optical microscope is limited by the wavelength of visible light used. Some improvement is achieved by using ultra-violet light, but this can hardly exceed a factor of two. The wavelength associated with an electron is governed by its momentum, and potential differences up to 100 kV are readily available. The wavelength associated with an electron accelerated through a potential difference V was shown above to be given by

$$\lambda = \sqrt{\frac{1\cdot 5}{V}} \text{ nm}$$
$$= \sqrt{\frac{1\cdot 5}{10^5}} \text{ nm}$$
$$= 3\cdot 87 \times 10^{-3} \text{ nm}.$$

The electron wavelength, therefore, is very much shorter (10^5 times) than the shortest visible light in the violet at 400 nm. Moreover, the velocity of an electron can be changed by electric and magnetic fields, in much the same way as the velocity of a light wave on passing from one medium to another. These two facts are used in the construction of an electron microscope in which electrons controlled by magnetic coils replace light waves and optical lenses. Electrons of very short wavelengths having passed through a thin object, proceed to form an image upon a photographic plate or a fluorescent screen. It might at first sight be supposed that a resolving power increase of 10^5 should be available in this way. Unfortunately, the electric and magnetic lenses correspond to only simple optical lenses, so that apertures are restricted to about one hundredth of the optical values available. The overall gain, however, is still about 10^3, enabling objects only 1 or 2 nm in size to be observed. Such resolving power is particularly valuable because it enables the larger molecules to be seen and photographed.

9.7 Heisenberg's Uncertainty Principle

Thus we see that the wave–particle dualism goes much further than a study of radiation suggests. It is a feature of the electron and indeed, of all material particles. The electron behaved as a particle when subjected to electric and magnetic fields in J. J. Thomson's e/m experiment, but as a wave in G. P. Thomson's electron diffraction experiment. Like radiation an electron cannot exhibit both particle and wave properties simultaneously. In the electron diffraction experiment an electron behaves as a particle when accelerated in an electron gun at 60 kV, and also when interacting with the silver bromide of the photographic plate. Between these two events it displays the properties of a wave, and is diffracted by the crystals of the metallic foil through which the electrons pass. This is because the electron, when confronted by a periodic potential, i.e. the crystal lattice, responds in a periodic manner i.e. as a wave of wavelength $\lambda = h/p$.

Our conceptual difficulties in accepting the wave–particle dualism arise because our ideas of waves and particles are based upon large-scale observations of large-scale phenomena and experiments. When such ideas are used for atomic phenomena they are on much the same sort of footing as analogies in large-scale physics. It is hardly surprising therefore that difficulties occur. The wave–particle aspects of electrons and photons are closely linked with the uncertainty principle due to Heisenberg. This principle states that *the momentum and position of a particle cannot simultaneously be measured with complete certainty*. It appears to set a definite limit to the amount we can know. The uncertainty principle is readily understood and illustrated by an idealized experiment. Suppose that a beam of electrons of momentum p falls upon a slit of width Δa (Fig. 9.5). This

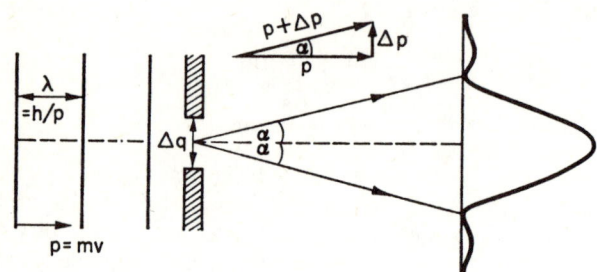

Fig. 9.5 Experiment to illustrate the uncertainty principle.

beam is equivalent to a series of waves, each of wavelength h/p falling upon the slit. Since the electron must pass through the slit Δq is a measure of the precision with which we know its position. The smaller Δq becomes, the more accurately the position of the electron is known. Diffraction will occur, and the diffraction pattern will be recorded upon a screen as shown. Physical optics tells us that the angular width of the central maximum is given by $\sin a = \lambda/\Delta q$ (see Chapter 5). The electron may therefore move anywhere within the angle $\pm \alpha$ as defined by the central maximum. The chance of its lying outside this maximum is small, and given by the size of the subsidiary maxima at each side. Referring now to the vector diagram in Fig. 9.5, we see that this implies an uncertainty Δp in the momentum p given by $\tan \alpha = \Delta p/p$. When α is small, we can now write

$$\tan \alpha = \sin \alpha = \lambda/\Delta q \sim \Delta p/p$$

from which $\Delta p \Delta q = \lambda p = h$. A more rigorous argument gives $\Delta p \Delta q = h/2\pi$, which is usually written $\hbar$ (h cross).

The product of the uncertainty of momentum Δp and the uncertainty of position Δq is equal to the constant $\hbar$. More careful analysis of such problems shows that $\hbar$ is the lower limit of the product $\Delta p \Delta q$, so that the uncertainty principle may be written as $\Delta p \Delta q > h$. It can be seen from diffraction theory that any attempt to define the position of the electron more closely by narrowing the slit width Δq will lead to a broadening of the diffraction pattern. This implies that there will be a corresponding increase in the uncertainty Δp of the momentum of the electron.

Another instructive example concerns the process whereby a particle might be observed in a microscope. Light of frequency v falls upon the particle and is scattered. If it is to be observed at all, the scattered photon must enter the objective of the microscope as shown in Fig. 9.6. There is, therefore, an uncertainty in its momentum given by $\Delta p = p \sin \alpha$ since it can go anywhere within a cone of semi-vertical angle α. The particle at P recoils in accordance with the

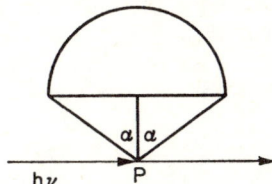

Fig. 9.6   The microscope.

Compton effect. The position of the particle is defined by the resolving power of the microscope such that $\Delta x = \lambda/\sin \alpha$. Again we have $\sin \alpha = \Delta p/p = \lambda/\Delta x$ and therefore $\Delta p \Delta x = \lambda p = h$. To improve the resolving power we can increase the aperture α of the microscope, but in so doing Δp is also increased. Decreasing the wavelength λ would also increase the resolving power, but a correspondingly larger amount of momentum would be transferred to the particle from the photon due to a larger Compton effect. Similar relationships exist for the momenta along each of the Cartesian axes, so that with $\hbar$ written for h we have
$$\Delta p_x \Delta x > \hbar, \quad \Delta p_y \Delta y > \hbar, \quad \Delta p_z \Delta z > \hbar.$$

We may also profitably consider the momenta in terms of wave packets. If Δp_x is very small indeed, then p_x is known precisely and there is only one wavelength associated with it. The wavetrain will extend from $-\infty$ to $+\infty$ along the x axis and the position of the wave packet or group will be correspondingly vague. If, on the other hand, we wish to limit the length of the wave packet in order to define the position of the particle as closely as possible, then a relatively wide range of wavelengths must be employed. This implies that the momentum p_x will not be known with any appreciable precision at all.

The Heisenberg uncertainty principle also applies to the angular momentum L of a body and its angular position φ so that we have $\Delta L \Delta \varphi > \hbar$. That the energy E of a body at a time t is also governed by the same uncertainty principle giving $\Delta E \Delta t > \hbar$ can easily be derived from the above momentum–position relationships.

9.8 Born's Statistical Interpretation of Waves and Particles

As was pointed out earlier, it is tempting to try to identify the particle with the wave packet, but we have seen that there are serious objections to this (p. 122). Born has shown that the wave–particle dualism can best be resolved using a statistical interpretation. It has already been shown in Chapter 5 that a light wave of wavelength λ, travelling along the x axis with a velocity c, may be represented by the equation $A = A_0 \sin 2\pi(ct - x)/\lambda$, where A and A_0 are electric or magnetic vectors. The power of such a wave is proportional to A_0^2, where A_0 is the amplitude of the vibration. The corpuscular theory, on the other hand, represents power as flux, i.e. as the rate of passage of photons through unit area perpendicular to the direction of motion. These are just two ways of saying the same thing, one using the language of the wave theory, and the other the language of particle theory. From this it is apparent that these are complementary views of the same basic phenomena and that A_0^2 is proportional to R, where R is the number of photons passing through unit area in unit time, the area being perpendicular to the direction of flow.

The relationship between A_0^2 and R can be appreciated from a consideration of a simple diffraction experiment. Suppose plane waves W_1, W_2, W_3 (Fig. 9.7) fall

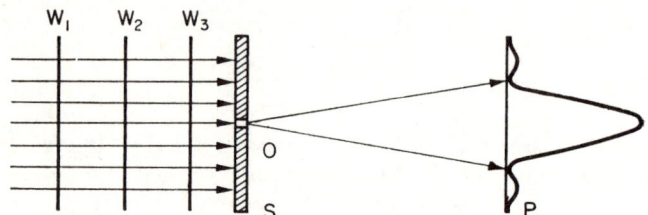

Fig. 9.7 Diffraction at a small aperture.

normally upon a surface S in which there is a small hole O. Behind this is placed a photographic plate P. The photons pass through O at a rate which is governed by the intensity of the light. Each proceeds to the photographic plate and makes a contribution to the blackening. The diffraction pattern which is observed after many thousands of photons have passed represents their statistical distribution. The explanation of the pattern in terms of the wave theory of light is well known. Since the intensity is represented by both R and A_0^2, Born assumed that A_0^2 represented the probability of finding a photon at a particular place. We can now see that as each photon passes through the aperture, A_0^2 governs the probability of finding it at a given place on the photographic plate. When a large number of photons have passed, the familiar diffraction pattern is build up. Fig. 9.8 shows the diffraction pattern built up in this manner from 5000 photons.

The diffraction of electrons in G. P. Thomson's experiment can be treated in much the same way. The pattern observed is explained by associating a wavelength with each electron. If the wave is related to probability in the above manner we are again able to see how the electron pattern would be built up as a large number of electrons arrive at the plate. Thus by associating probability with wave amplitude, Born was able to reconcile the wave and particle theories of both radiation and matter.

Problems

(*The problem marked with an asterisk is solved in full at the end of the section.*)

9.1 Explain how de Broglie was able to extend the wave theory to include material particles. Describe how this has been confirmed experimentally.

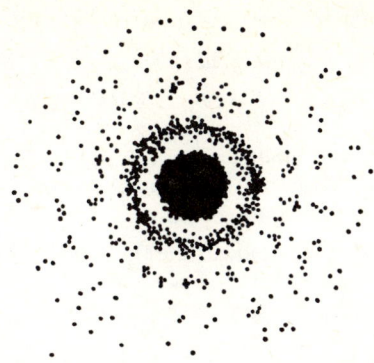

Fig. 9.8 Central disc and first four rings of a diffraction pattern formed by 5000 photons (from *Modern Physics* by M. S. Smith, published by Longmans, London).

Calculate the wavelength associated with an electron after it has been accelerated from rest by a potential difference of 1 MV. (1·22 pm)

9.2 Electrons from a heated filament are accelerated by a potential difference of 10 kV and then passed through a thin sheet of metal for which the spacing of the atomic planes is 40 pm. Calculate the angle of deviation of the first-order diffraction pattern. (17° 16′)

9.3 Calculate the wavelength associated with a proton after it has been accelerated by a potential difference of 1 kV. (0·9 pm)

9.4* 10 kV electrons are passed through a thin film of a metal for which the atomic spacing is 55 pm. What is the angle of deviation of the first-order diffraction maximum? (12° 44′)

9.5 An α-particle has energy equal to 10 MeV. Calculate the wavelength to which this corresponds. (4·2 fm)

9.6 Calculate the length of the wave associated with a body of mass one gram moving with a velocity of 2 m s^{-1}. (3.3×10^{-31} m)

9.7 Electrons from a heated filament are accelerated by a potential difference of 30 kV and passed through a thin sheet of aluminium. Assuming the separation of the atomic planes to be 0·403 nm. Calculate the angle of deviation of the first-order diffraction pattern. ($\varphi = 2\theta = 62$ min of arc)

Solution to Problem

9.4 Velocity of electrons is given by $Ve = \tfrac{1}{2}mv^2$; momentum of electrons is given by $mv = \sqrt{2Vem}$; wavelength of electrons is given by $\lambda = h/\sqrt{2Vem}$. Therefore

$$\lambda = \frac{6 \cdot 6 \times 10^{-34}}{\sqrt{2 \times 10^4 \times 1 \cdot 6 \times 10^{-19} \times 9 \cdot 1 \times 10^{-31}}}$$
$$= 0 \cdot 122 \times 10^{-10} \text{ m} = 12 \cdot 2 \text{ pm}.$$

(Note also that $\lambda = \sqrt{1 \cdot 5/10^4}$ nm = 12·2 pm.)

Applying Bragg's formula for diffraction at the atomic planes, we have
$$m\lambda = 2d \sin \theta$$
$$1 \times 0{\cdot}122 \times 10^{-10} = 2 \times 55 \times 10^{-12} \sin \theta$$
$$\sin \theta = \frac{0{\cdot}122}{1{\cdot}1} = 0{\cdot}111$$
$$\theta = 6° \ 22',$$
which is the Bragg angle. Therefore the angle through which electron is deviated is $2\theta = 12° \ 44'$.

Chapter 10

Wave Mechanics

10.1 Some Preliminaries

Some insight into the possibilities of the wave-mechanical approach to atomic structure is obtained by considering the Bohr quantum conditions. These define stationary states which correspond to definite energy levels of the atom. We have already seen that according to Heisenberg's uncertainty principle $\Delta E \Delta t \geqslant \hbar$, where ΔE, Δt are the uncertainties in measuring energy and time respectively. If the energy is known exactly, then $\Delta E \to 0$ and therefore $\Delta t \to \infty$ implying that the error in measuring time will be very large indeed. It follows that the motion in time will also be unobservable, so that precise electronic orbits become rather meaningless. The whereabouts of an electron at a given instant of time must be replaced by the probability picture given by Born and described in the previous chapter. To do this we must replace the electron in its orbit by a de Broglie wave of wavelength $\lambda = h/mv$, mv being the momentum of the electron. If it is supposed that each circular orbit must contain an integral number of waves as shown in Fig. 10.1, it becomes clear that only orbits of certain radii are possible. Expressed quantitatively this means that $n\lambda = 2\pi r$, where r is the radius of the circle and n is an integer. Remembering that $\lambda = h/mv$, it follows that $mvr = nh/2\pi$. This is just the Bohr condition for a stationary orbit (see Chapter 7) which appeared so arbitrary on the old quantum theory, but now comes quite logically from this model.

We must remember that Fig. 10.1 is only a model. The electron does not travel round the orbit in a wave. The idea that it does seems to give the right answer for the quantization rule.

This simple case is analogous to the vibrations of a stretched string. If the string is unlimited in length, then a vibration can take any form and any wavelength. When, however, the string is stretched between two fixed points, only certain modes of vibration are possible as shown in Fig. 10.2. The corresponding wavelengths are in general given by $\lambda = 2l/n$, l being the length of the string and n the number of loops in it. The positions in the string at which no movement occurs are marked N and are the nodal points or nodes. Similar vibrations occur when a metal plate, clamped at its centre, is set into vibration by bowing its edge. The points are now replaced by lines, which may be revealed by sprinkling a layer of sand upon the plate. The patterns obtained are the well-known Chladni figures. The nodal points of the one-dimensional string and the nodal lines of the two-dimensional plate vibrations must be replaced by nodal surfaces when we come to consider the three-dimensional wave systems associated with an electron

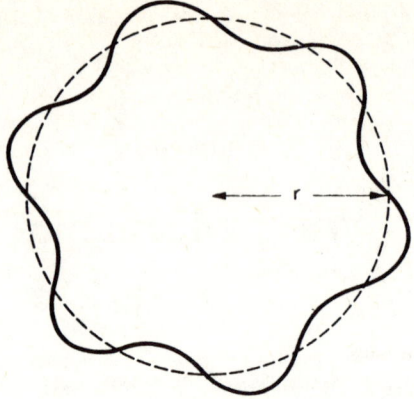

Fig. 10.1 Application of de Broglie wave to circular orbit.

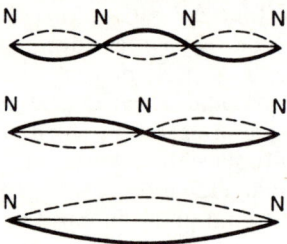

Fig. 10.2 Modes of vibration of stretched string.

within the atom. These nodal surfaces can be either plane, spherical or conical. We have already seen how the fixed ends of the string and the edge of the plate restrict their modes of vibration because certain boundary conditions have been imposed. A *free* electron can take any energy and therefore any wavelength, but when it is attached to an atom the attraction of the nucleus imposes boundary conditions, so that only certain modes of vibration are possible, corresponding to certain allowed or discrete energy states.

10.2 The Need for Change

We have seen that a particle electron of momentum mv can be represented by a wave when confronted by a periodic structure. The associated wavelength $\lambda = h/mv$ or $\lambda = h/p$ therefore represents both the particle and the wave nature of the electron. It is obvious that these ideas are beyond the simple Bohr theory, so that it is desirable to develop a theory which will reflect this dual nature of matter.

In 1926 the limitations of the orbital Bohr–Sommerfeld model of the atom were emphasized by Heisenberg in the uncertainty principle. This tells us that we cannot define precisely and *together* the exact position and momentum (or velocity) of the electron in a Bohr orbit. The complete Bohr theory was a curious mixture of classical mechanics, in which the orbit was defined, and Planck quantum theory, in which the energy of the radiation was defined. The ideas of stationary states and quantization of angular momentum had no real theoretical support. Bohr theory demands that both position and velocity are exactly known at any instant, and this was clearly untenable.

The need to find a microscopic/atomic equation equivalent to the macroscopic second law of motion of Newton became more and more apparent. The new law of motion as applied to atoms would have to include the Planck postulate $E = h\nu$ as well as the de Broglie wave postulate $\lambda = h/mv$. Thus eventually the particle equation of motion of Newton was superseded by the wave equation of motion of Schrödinger. And just as Newton's second law of motion cannot be obtained from first principles — it is judged solely by the results it predicts — so we find that Schrödinger's wave equation cannot be so-derived either: but it has been so successful in solving so many atomic and molecular problems that it is now universally accepted. It has led to an explanation of many empirical relations between quantum numbers, angular momentum coupling relations and selection rules, as well as predicting the relative intensities of spectral lines, a thing not possible with the old quantum theory of the atom.

The new quantum theory dates from 1925 and developed along two lines, one through Heisenberg and the other through Schrödinger. The latter used a wave approach by analogy with wave optics. The former was a more formal approach in terms of operators and matrices. They lead to the same results in those cases where full solutions are possible.

Heisenberg recognized that the fundamental weakness of the Bohr theory was that it was based on unobservables, i.e. electron orbits and electron 'jumps', whereas the only *real* observable quantities are the frequencies and intensities of the radiations and their combination relations. Heisenberg represented the frequencies by a set of matrices and the combination relations by the matrix properties. This led him to replace the classical quantities of position, momentum and energy by complex operators acting appropriately on functions representing the wave nature of the electron or atomic state. This was the beginning of *quantum mechanics*. The wave function, Ψ, was a mathematical concept depicting the behaviour of the electron or the atom depending on both position and time, in the same sort of way that a progressive wave does in classical physics. Thus $\Psi = \Psi(x,y,z,t)$ in general, but in those cases where Ψ is independent of time, as in many atomic and molecular problems, the solution of the wave equation points logically and conclusively to the existence of stationary energy states of the atom and not as an *a priori* postulate as given by Bohr. In quantum mechanics the stationary state is therefore not recognized as a well-defined orbit, as in the old quantum theory of Bohr, but, by virtue of the uncertainty principle, is envisaged as a blurred picture based on a probability concept. But the probability of what?

Born answered this question by using the analogy with intensity in optics (see p. 127). Just as the intensity of the light due to an optical wave is proportional to the squared modulus of the wave amplitude, so Born suggested that the 'intensity' of the electron distribution in an atom was proportional to Ψ^2. Thus the probability of finding an electron in a small volume dv was given by Born as $\Psi^2 dv$

(or, better, $\Psi\Psi^* dv$, where Ψ^* is the conjugate complex of Ψ). Furthermore, since the electron is certainly *somewhere* in space, it then has a probability of 1, so that

$$\int_{-\infty}^{+\infty} \Psi^2 dv = 1,$$

taking the integral over all space. This rule constitutes a condition (called normalization) which must always obtain in real physical problems. For Ψ to be a real solution of the Schrödinger wave equation, we also have the mathematical requirements that Ψ and $d\Psi/dt$ must be continuous and single-valued throughout, and that $\Psi \to 0$ as $x, y, z \to \pm\infty$. In addition to these mathematical requirements there are other limitations to Ψ, known as boundary conditions, set according to the problem. These correspond to the nodes of the vibrating-string problem. Finally, as we have seen, there are the physical requirements that $E = h\nu$ and $\lambda = h/p$, together with the conservation law $E = p^2/2m + V$, where $p^2/2m$ is the kinetic energy and V the potential energy of the system.

10.3 The Schrödinger Wave Equation

The fundamental differential equation for any time-dependent wave motion of velocity v_x is

$$\frac{d^2 y}{dx^2} = \frac{1}{v_x^2} \frac{d^2 y}{dt^2}.$$

This describes a progressive wave moving along the x axis with speed v_x such that the displacement y is a function of x and t, i.e. $y = f(x, t)$. The form of the function which satisfies the differential equation is $y = f(x - v_x t)$ for a wave progressing along $x + $ve or $y = f(x + v_x t)$ for a wave progressing along $x - $ve. *Any* function of $(x \pm v_x t)$ will do, as can be seen by differentiation and substituting in the differential equation.

A commonly used solution for y is the sinusoidal equation

$$y = A \sin \frac{2\pi}{\lambda} (x \pm vt)$$

which can be represented by

$$y = A e^{(2\pi i/\lambda)(x \pm vt)}$$

to include complex solutions. We take only the wave travelling in the $x + $ve direction, so that $v = v_x$. This equation for y is very common in problems of optics and acoustics and other branches of wave physics, but it is important to note that we are actually *starting* with this assumed sinusoidal solution.

If we try to transfer these ideas to the electron, we have to consider some wave property Ψ which is equivalent to y; we must also include the two quantum concepts of $E = h\nu$ and $\lambda = h/p$, as well as the energy condition $E = p^2/2m + V$, as constraints. We see at once that we must somehow get E and p^2 into our wave function equation

$$\Psi = A e^{(2\pi i/\lambda)(x - vt)}.$$

This is a harmonic function, and happily only those wave functions which are harmonic functions of this kind have any real physical significance, so that we can

proceed with this expression for Ψ. We can write
$$\Psi = Ae^{2\pi ix/\lambda}e^{-2\pi ivt/\lambda}$$
$$= Ae^{ipx/\hbar}e^{-iEt/\hbar}$$
if we include the quantum conditions. Thus we have now separated the space and time components of Ψ. We must now extract from this equation a term involving E and one involving p^2 by suitably differentiating Ψ with respect to x and t. Thus we have
$$\frac{\partial \Psi}{\partial t} = Ae^{ipx/\hbar}\left(-\frac{iE}{\hbar}\right)e^{-iEt/\hbar}$$
$$= -\frac{\Psi iE}{\hbar}$$
$$E = -\frac{\hbar}{\Psi i}\frac{\partial \Psi}{\partial t}$$
or
$$E = \frac{\hbar i}{\Psi}\frac{\partial \Psi}{\partial t}.$$
This seems satisfactory for E.

To find p^2 it is obvious that we must differentiate Ψ twice with respect to x. Thus
$$\frac{\partial \Psi}{\partial x} = \frac{Aip}{\hbar}e^{ipx/\hbar}e^{-iEt/\hbar}$$
and
$$\frac{\partial^2 \Psi}{\partial x^2} = -\frac{Ap^2}{\hbar^2}e^{ipx/\hbar}e^{-iEt/\hbar}$$
$$= -\frac{p^2}{\hbar^2}\Psi.$$
Thus
$$p^2 = -\frac{\hbar^2}{\Psi}\frac{\partial^2 \Psi}{\partial x^2},$$
so that from
$$E = \frac{p^2}{2m} + V$$
we get
$$\frac{\hbar i}{\Psi}\frac{\partial \Psi}{\partial t} = -\frac{\hbar^2}{2m\Psi}\frac{\partial^2 \Psi}{\partial x^2} + V,$$
giving
$$\frac{\hbar}{i}\frac{\partial \Psi}{\partial t} = \frac{\hbar^2}{2m}\frac{\partial^2 \Psi}{\partial x^2} - V\Psi.$$

This is the one-dimensional time-dependent Schrödinger wave equation, where $V = V(x, t)$ only, and the energy E is dependent on both x and t. In many physical problems, however, especially in the non-relativistic treatment of electrons and atoms, the wave function Ψ is time-*independent*, so that the potential energy V is $V(x)$ only and the total energy E corresponds to a stationary state, i.e. it is constant for constant x, or better, constant average x.

Thus if Ψ is time-independent we can write $e^{ipx/\hbar}$ as ψ_x, where ψ_x is a time-independent space function now called an eigenfunction.

This gives
$$\Psi = \psi_x A e^{-iEt/\hbar}$$

and we have
$$\frac{\partial \Psi}{\partial t} = \psi_x \left(-\frac{AiE}{\hbar}\right) e^{-iEt/\hbar}$$

Similarly,
$$\frac{\partial^2 \Psi}{\partial x^2} = A e^{-iEt/\hbar} \frac{\partial^2 \psi_x}{\partial x^2}$$

Now substitute in the Schrodinger wave equation we have just derived and we get
$$\frac{\hbar}{i}\left(-\frac{AiE}{\hbar} e^{-iEt/\hbar} \psi_x\right) = \frac{\hbar^2}{2m} A e^{-iEt/\hbar} \frac{\partial^2 \psi_x}{\partial x^2} - V(x) A e^{-iEt/\hbar} \psi_x.$$

Therefore
$$-E\psi_x = \frac{\hbar^2}{2m}\frac{\partial^2 \psi_x}{\partial x^2} - V(x)\psi_x$$

or
$$\frac{\hbar^2}{2m}\frac{\partial^2 \psi_x}{\partial x^2} + [E - V(x)]\psi_x = 0,$$

giving
$$\frac{\partial^2 \psi_x}{\partial x^2} + \frac{2m}{\hbar^2}[E - V(x)]\psi_x = 0.$$

This is the one-dimensional time-*independent* form of the Schrodinger wave equation. It has been 'derived' on the *assumption* that a harmonic solution is possible and therefore can in no way be looked upon as a derivation of the equation from first principles.

In three dimensions the equation becomes
$$\frac{\partial^2 \psi}{\partial x^2} + \frac{\partial^2 \psi}{\partial y^2} + \frac{\partial^2 \psi}{\partial z^2} + \frac{2m}{\hbar^2}(E - V)\psi = 0,$$

where $\psi = \psi(x, y, z)$ and $V = V(x, y, z)$. This is often written
$$\nabla^2 \psi + \frac{2m}{\hbar^2}(E - V)\psi = 0,$$

where ∇^2 is the Laplacian differential operator
$$\nabla^2 = \left(\frac{\partial^2}{\partial x^2} + \frac{\partial^2}{\partial y^2} + \frac{\partial^2}{\partial z^2}\right).$$

The term $(E - V)$ represents the kinetic energy T, and in any problem in which T is a constant of the motion we would have $d^2\psi_x/dx^2 + K\psi_x = 0$ for one dimension, where $K = (2m/\hbar^2)T$. This equation is, of course, the familiar differential equation of simple harmonic motion and leads to the usual sinusoidal solutions.

10.4 An Alternative Approach

We have mentioned previously the Heisenberg method, using operators and matrices. The correct operators are worked out from the observed details of spectral frequencies and their interrelationships and each physical property (e.g. position, momentum, energy, etc.) has its own distinct operator. Now we can write the Schrodinger equation in the form

$$-\frac{\hbar^2}{2m}\nabla^2\psi + V\psi = E\psi$$

or

$$\left\{-\frac{\hbar^2}{2m}\nabla^2 + V\right\}\psi = E\psi.$$

The expression in braces is a differential operator, called the Hamiltonian operator $\hat{H}$. It acts here on the eigenfunction ψ, so that we can write $\hat{H}\psi = E\psi$, which is the simplest designation of the Schrödinger wave equation. The Hamiltonian operator is the quantum-mechanical equivalent of the total energy in Newtonian mechanics, which is also called the Hamiltonian.

In classical mechanics we have $H = T + V$, which is E in the wave equation. Thus the operator $\hat{H}$ is given by

$$H \equiv \left(-\frac{\hbar^2}{2m}\nabla^2 + V\right)$$

In the formal Heisenberg treatment the premises were the observed facts about spectra. From these, Heisenberg was led to state the results in the form of a matrix and out of it came a whole set of operators replacing momentum, energy and position. The rules developed for transferring classical observables into operators then re-awakened the study of the properties of mathematical matrices. The Heisenberg method is probably more rigorous than the Schrödinger method we have outlined, but the final results are the same. In either case it is impossible to say *why* these ideas worked so well — we must be content with the fact that they *do*, so that they must be statements of truth.

In this way we represent the Schrödinger wave equation by the operator equation $\hat{H}\psi = E\psi$. The interpretation of this equation is that we operate on the function ψ with $\hat{H}$ and get a numerical solution E whenever the eigenfunction ψ satisfies the mathematical requirements of continuity and the boundary conditions of a particular problem. These real values of E are the eigenvalues corresponding to those eigenfunctions which are solutions of the equation.

The physical quantities of momentum, position and energy are called dynamic variables and are all replaced by operators in Heisenberg quantum mechanics. Table 10.1 is a table of such operators. Matrix mechanics became a very formal and elegant way of solving the different problems of classical atom mechanics and revealed relationships between the eigenvalues of angular momenta which correspond with observed spectral line properties.

10.5 Solution of the Schrödinger Wave Equation

For simplicity we shall deal only with the one-dimensional case to establish the principles which must obtain for all solutions of wave equation. To begin with we

TABLE 10.1
Some Quantum-Mechanical Operators

Physical quantity Dynamical variable	Symbol	Quantum-mechanical operator
Position	x, y, z, r	x, y, z, r
Linear momentum	p	$\hat{p} \equiv -i\hbar \nabla$
$\quad x$ component	p_x	$\hat{p}_x \equiv -i\hbar \dfrac{\partial}{\partial x}$
$\quad y$ component	p_y	$\hat{p}_y \equiv -i\hbar \dfrac{\partial}{\partial y}$
$\quad z$ component	p_z	$\hat{p}_z \equiv -i\hbar \dfrac{\partial}{\partial z}$
Total energy	E	$\hat{H} \equiv -i\hbar \dfrac{\partial}{\partial t}$
		$\equiv -\left(\dfrac{\hbar}{2m}\right)\nabla^2 + V$
Angular momentum $\quad z$ component	l_z	$\hat{l}_z \equiv -i\hbar \dfrac{\partial}{\partial \varphi}$

$$\nabla \equiv \left(\frac{\partial}{\partial x} + \frac{\partial}{\partial y} + \frac{\partial}{\partial z}\right) \text{ and } \nabla^2 \equiv \left(\frac{\partial^2}{\partial x^2} + \frac{\partial^2}{\partial y^2} + \frac{\partial^2}{\partial z^2}\right)$$

can only solve completely for ψ if we know the potential energy V exactly; its variation with x i.e. $V = V(x)$ must also be completely known.

In these cases a complete solution of the equation will lead to the derivation of the stationary energy states E in the same way that the complete solution of the simple harmonic equation gives us a set of harmonics related integrally, i.e. v_0, $2v_0$, $3v_0$, ..., etc. It is impossible to have frequencies which are, say, half-integral values of the fundamental frequency v_0. Thus, if we take the example of a vibrating string again, we have the mathematical constraints of continuity, etc., as well as the physical constraints or boundary conditions telling us where the nodes and antinodes are. It is essential that all these conditions be fulfilled if we are to get a real observable solution. The problem of the bowed string therefore gives us a set of discrete frequencies. Thus in the case of a quantum problem in which energy and frequency are proportional to each other, we might expect solutions for E of the type E_0, $2E_0$, $3E_0$, ..., etc., where $E_0 = hv_0$. The answer is not quantitatively like this, but still a spectrum of *discrete* energies $E_1, E_2, E_3, \ldots$, is revealed and the classical idea of a continuous distribution of energy is shown to be an extreme case. By analogy with the string case, the surprising result of quantum theory is not that the energies are discrete but that they are not harmonic.

There are relatively few real atomic physics problems in which $V(x)$ or $V(x, y, z)$ is completely known — the hydrogen atom is one of them — so we must illustrate the use of the Schrödinger wave equation by applying it to artificial problems in which various simple forms of $V(x)$ are assumed.

We therefore apply this to the case of a tightly bound electron in an infinitely deep rectangular potential well. The electron is therefore contained within a restricted position, free to vibrate or 'rattle' inside the well. This model corresponds to the bowed string we have mentioned and it demonstrates the technqiues and results which have to be used in all methods of solving the wave equation.

10.6 Simple One-Electron Atom Model

In the case of the vibrating string we have nodes at either end and a standing wave between. So we have our atomic electron in a deep well (Fig. 10.3) from which it cannot escape, with nodes, or boundary conditions of no displacement, at the walls of the well. The width of the well is $2a$ and the potential energy V is

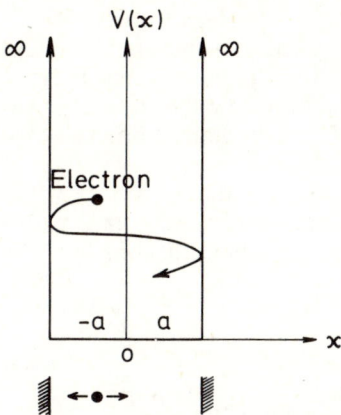

Fig. 10.3 Electron in an infinitely deep potential well.

taken arbitrarily as zero within the walls of the well. The atom has only one electron and the wave function of the atom is taken to be the wave function of the single electron.

The boundary conditions are therefore

$$V(x) = 0 \quad \text{for} \quad x > -a \quad \text{and} \quad x < +a,$$
$$V(x) = \infty \quad \text{for} \quad x < -a \quad \text{and} \quad x > +a,$$
$$\psi(x) = 0 \quad \text{for} \quad x = -a \quad \text{and} \quad x = +a,$$
$$\psi(x) = 0 \quad \text{for} \quad x < -a \quad \text{and} \quad x > +a.$$

We need to find the form of ψ within the potential well where $V = 0$ throughout. In this case the wave equation reduces to

$$\frac{d^2\psi}{dx^2} + \frac{2m}{\hbar^2} E\psi = 0$$

or

$$\frac{d^2\psi}{dx^2} + K^2\psi = 0$$

where $K^2 = 2mE/\hbar^2$. This is the differential equation of classical simple harmonic motion, for which the solutions are

$$\psi_1 = A \sin Kx \quad \text{or} \quad \psi_2 = B \cos Kx,$$

139

where A and B are different constants. Also, of course, $\psi_{12} = \psi_1 + \psi_2$ is a solution. So we can write
$$\psi = A \sin Kx + B \cos x,$$
to which we must apply our boundary conditions, viz.
$$\psi = 0 \quad \text{at} \quad x = a,$$
$$\psi = 0 \quad \text{at} \quad x = -a,$$
giving
$$0 = A \sin Ka + B \cos Ka,$$
$$0 = -A \sin Ka + B \cos Ka.$$
Adding and subtracting gives
$$B \cos Ka = 0,$$
$$A \sin Ka = 0,$$
both of which must be satisfied. It is impossible to find one value of K for which $\sin Ka$ and $\cos Ka$ vanish simultaneously, but they *can* be satisfied separately if we choose K to make $B = 0$ and $A \neq 0$ *or* to make $A = 0$ and $B \neq 0$. It is obvious that A and B cannot be zero together, otherwise $\psi = 0$ throughout, giving a meaningless result.

When $A = 0$, $B \neq 0$, $\cos Ka = 0$ and $Ka = n\pi/2$ with n odd.
When $B = 0$, $A \neq 0$, $\sin Ka = 0$ and $Ka = n\pi/2$ with n even.
There are therefore two possible solutions for ψ:

$$\text{Class 1,} \quad \psi_n(x) = B \cos \frac{n\pi}{2a} x \quad \text{for } n \text{ odd;}$$

$$\text{Class 2,} \quad \psi_n(x) = A \sin \frac{n\pi}{2a} x \quad \text{for } n \text{ even;}$$

giving us odd and even wavefunctions.

These are the eigenfunctions of the electron confined to a deep well and correspond also to the eigenfunctions of the vibrating string with nodes at each end. For both odd or even wavefunctions we have $K = n\pi/2a$, algebraically. Thus

$$\sqrt{\frac{2mE_n}{\hbar^2}} = \frac{n\pi}{2a} \quad \text{and} \quad E_n = \frac{n^2 \pi^2 \hbar^2}{8ma^2},$$

where n is now 1, 2, 3, 4, ..., giving the energies of the odd and even states in turn.

These are the eigenvalues of the problem, i.e. the stationary or quantized energy states. They are quantized through the integer n, which is now formally called a quantum number. The discrete energy values of our model are therefore
$$E_1, 4E_1, 9E_1, 16E_1, \ldots, n^2 E_1, \ldots,$$
where $E_1 = \pi^2 \hbar^2/8ma^2$. There are no others, contrary to the classical model for which the possible energy values are continuous.

The *energy level diagram* is shown in Fig. 10.4. We note that the two classes of solution alternate and that the minimum energy is NOT zero. Actually, E_1 is called the zero-point energy.

The remaining problem is to evaluate the constants A and B to fit our particular boundary conditions. This constitutes the normalization process, in which we have
$$\int_{-\infty}^{\infty} \psi^2 dx = 1.$$

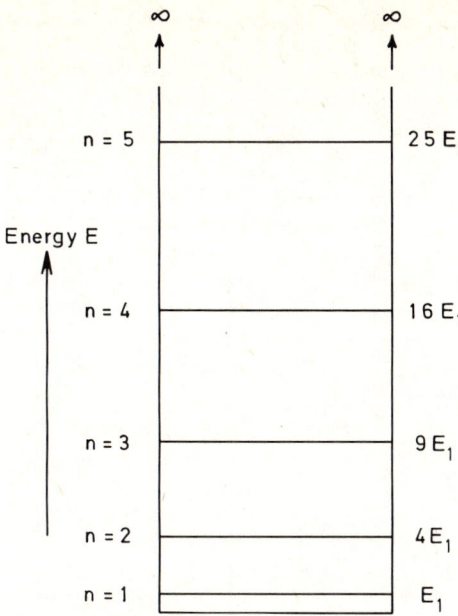

Fig. 10.4 Discrete energy levels of electron in an infinitely deep potential well.

This is a standard procedure in all quantum-mechanical problems.

For the ODD terms we have

$$\int_{-\infty}^{\infty} B^2 \cos^2 \frac{n\pi x}{2a} = 1 \quad \text{for } n \text{ odd,}$$

i.e.

$$\frac{B^2}{2} \int_{-\infty}^{\infty} \left(\cos \frac{n\pi x}{a} + 1 \right) dx = 0$$

so that, on integration,
$$B^2 a = 1 \quad \text{or} \quad B = \pm \sqrt{1/a}.$$

Similarly $A = \pm \sqrt{1/a}$ and, taking the +ve signs, we have

$$\psi_n(x) = \sqrt{\frac{1}{a}} \cos \frac{n\pi}{2a} x \quad \text{for } n \text{ odd}$$

and

$$\psi_n(x) = \sqrt{\frac{1}{a}} \sin \frac{n\pi}{2a} x \quad \text{for } n \text{ even.}$$

These, then, are the final eigenfunctions of the particle.

We show in Fig. 10.5 the first four eigenfunctions and their corresponding ψ^2 values.

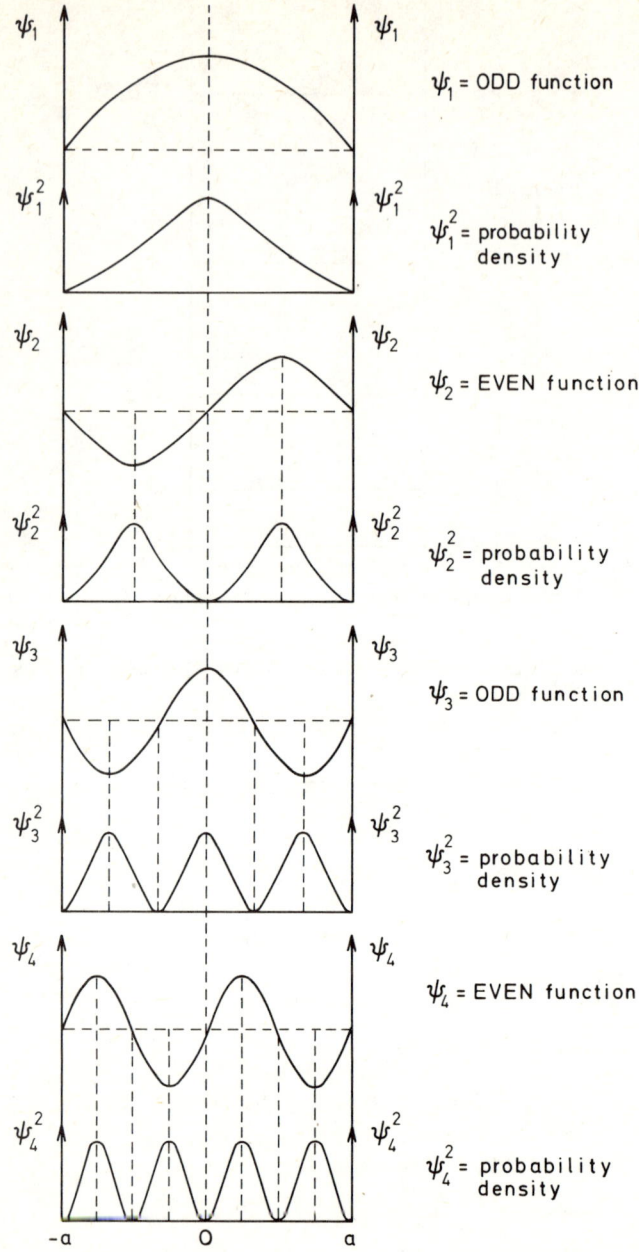

Fig. 10.5 ψ and ψ^2 distributions for first four eigenfunctions of electron of Fig. 10.3.

The objects of this section have been as follows:

(1) to find the quantum-mechanical solution of a simple atom problem with a close classical analogy;
(2) to show that the energy distribution is discrete and not continuous;
(3) to work out the complete eigenfunctions by applying the normalization rule;
(4) to emphasize that this is basically the technique for solving all such problems in atomic physics, the difference being simply one of greater mathematical difficulty.

It is a simple step to extend this section to the real case of an atom in a finite potential well, i.e. $V(x) \neq \infty$ anywhere. The results are very similar to those we have given for the infinite well except that there is now a finite, albeit small, probability that the atom can 'get out', i.e. it is not confined *all* the time to the region $x > -a$ and $x < +a$.

This corresponds to the case when our bowed string is subject to the vibrato of the stopping finger, as in the playing of a violincello note. It is also important in the theory of α-particle emission (see p. 216).

10.7 The Hydrogen Atom

We have said repeatedly that the Bohr theory had its limitations and that the knowledge of quantum numbers, angular momentum selection rules, line intensities, etc., was largely empirical. We have stated that the Schrödinger wave equation is judged on its success in solving various atomic problems in which $V(x, y, z)$ is known completely. The hydrogen atom is one such case since its radial potential $V(r)$ is given by $V(r) = -e^2/4\pi\varepsilon_0 r$ for the proton–electron Coulomb attraction.

The Hamiltonian is
$$H = T + V$$
$$= \tfrac{1}{2}mv^2 - \frac{e^2}{4\pi\varepsilon_0 r}$$

and the wave equation is
$$\frac{\partial^2 \psi}{\partial x^2} + \frac{\partial^2 \psi}{\partial y^2} + \frac{\partial^2 \psi}{\partial z^2} + \frac{2m}{\hbar^2}(E - V)\psi = 0$$

in Cartesian coordinates. Since the potential field is a spherically central field it would seem sensible to transform the Cartesian coordinates to spherical polar coordinates.

The diagram for this is shown in Fig. 10.6.

A particle situated at $P(x, y, z)$ in Cartesian coordinates can be transformed to the r, θ, φ system of the diagram by the transformations
$$x = r \sin\theta \cos\varphi,$$
$$y = r \sin\theta \sin\varphi$$
$$z = r \cos\theta.$$

The effect of this is to transform $\nabla^2 \psi(x, y, z)$ to $\nabla^2 \psi(r, \theta, \varphi)$ so that the

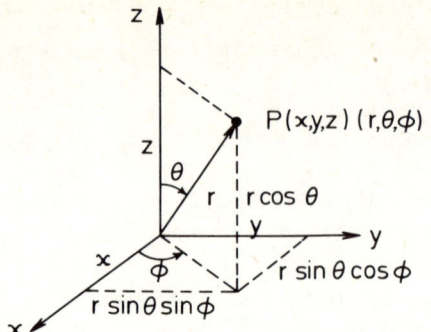

Fig. 10.6 Cartesian and polar coordinate of a point *P*.

Schrödinger equation is now given by

$$\frac{1}{r^2}\frac{\partial}{\partial r}\left(r^2\frac{\partial \psi}{\partial r}\right) + \frac{1}{r^2 \sin\theta}\frac{\partial}{\partial \theta}\left(\sin\theta\frac{\partial \psi}{\partial \theta}\right) + \frac{1}{r^2 \sin^2\theta}\frac{\partial^2 \psi}{\partial \varphi^2} + \frac{2m}{\hbar^2}[E - V(r)]\psi = 0$$

after changing the variables. This, then, is the wave equation we have to solve subject to the boundary conditions of the hydrogen atom.

Just as we did for our model atom, we must now solve this equation to find $\psi = \psi(r, \theta, \varphi)$, and it turns out that $\psi(r, \theta, \varphi)$ can be written in the form

$$\psi(r, \theta, \varphi) = R(r)\Theta(\theta)\Phi(\varphi)$$

where $R(r)$ is a wave function dependent on r only, $\Theta(\theta)$ is a wave function dependent on θ only, $\Phi(\varphi)$ is a wave function dependent on φ only, i.e. the three variables can be written in terms of their own wave functions, independent of the other variables. This technique is known as the method of separated variables; there is no logical reason for doing this, it just works because V is a function of r only, and not of θ and φ.

The solutions obtained are thus three in number, depending on three separate differential equations for three single-valued wave functions. These are found to be

$$\frac{d^2\Phi}{d\varphi^2} + m_l^2 = 0 \quad \text{depending on } \varphi \text{ only;}$$

$$\frac{1}{\sin\theta}\frac{\partial}{\partial \theta}\left(\sin\theta\frac{\partial \Theta}{\partial \theta}\right) + \left[l(l+1) - \frac{m_l^2}{\sin^2\theta}\right]\Theta = 0 \quad \text{depending on } \theta \text{ only;}$$

and

$$\frac{1}{r^2}\frac{\partial}{\partial r}\left(r^2\frac{\partial R}{\partial r}\right) + \frac{2m}{\hbar^2}\left[E + \frac{e^2}{4\pi\varepsilon_0 r} - \frac{\hbar^2 l(l+1)}{2m\,r^2}\right]R = 0 \quad \text{depending on } r \text{ only.}$$

In these equations m_l and l are integers which we shall identify as quantum numbers. The solution of the first equation is trivial. It gives the sinusoidal solution requiring that the integer m_l can only have values $m_l = 0, \pm 1, \pm 2, \ldots$, etc.

The solution to the second equation involves m_l and requires l to be integral also, such that $l = 0, 1, 2, \ldots$, and that for each l there are $2l + 1$ associated values of

m_l, vis. $-l, -(l-1), \ldots, -2, -1, 0, +1, +2, \ldots, +(l-1), +l$ i.e. from $-l$ to $+l$ including 0.

The interpretation of the two quantum numbers l and m_l is that m_l quantizes the rotation, or precession, of the angular momentum vector **l** around the z axis, i.e. keeping r and θ constant but varying the azimuthal angle φ of Fig. 10.6. In this way m_l governs the orientation of **l** in space, allowing only $(2l+1)$ positions of **l** to give the requisite number of m_l values. This is called space quantization and will be dealt with in more detail in the next chapter on the vector model of the atom. For each of these orientations of the angular momentums, the magnitude of the z component of l is given by $l_z = m_l \hbar$ (see Fig. 10.7). (Note that the spherical polar coordinates used here refer to the angular momentum vector and not to the space coordinates of the electron itself.)

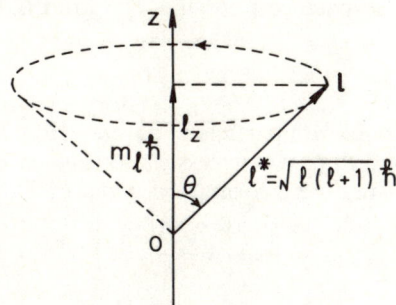

Fig. 10.7 Spatial quantization of orbital angular momentum vector.

The azimuthal quantum number m_l is usually called the magnetic quantum number.

Finally, the solution of the third equation is in terms of r and gives a set of eigenvalues E_n which are subject to discrete set of values given by

$$E_n = -\frac{me^4}{32\pi^2 \varepsilon_0 h^2 n^2}$$

(see p. 88) in exactly the same way that we obtained the eigenvalues E_n for the model atom.

Here n is a positive integer related to l by $l = 0, 1, 2, \ldots, (n-1)$ for each n.

This set of energy values for the hydrogen atom is identical with that obtained by Bohr for his model, but now it has been obtained without resort to classical dynamics or tentative quantum theory. It is the formal result of solving the wave equation for hydrogen.

In this lies the 'proof' that the Schrödinger wave equation is the correct line of approach to these problems — it gives a complete theoretical derivation of facts which we previously offered without explanation.

In this section we have tried to show how the steps used in the complete solution of the single electron atom model of Section 10.6 can be applied to the hydrogen atom. The mathematics is more difficult of course, but the principles are the same. The relationships between the quantum numbers n, l and m_l are then theoretically established and the discrete eigenvalues E_n are theoretically derived, agreeing with the Bohr values.

10.8 Angular Momenta

So far we have only explained the quantum numbers and their derivation from the Schrödinger equation. We know from the old quantum theory that these are closely connected with the magnitudes of the angular momenta of the rotating electron, e.g. orbital momentum $= lh/2\pi = l\hbar$. The Schrödinger equation itself cannot tell us what are the equivalent wave-mechanical equations for the angular momenta. For this we must use the quantum operators equivalent to the classical particle-mechanical expression for the angular momentum. If the angular momentum vector $\mathbf{l}$ has space components l_x, l_y and l_z, then

$$l_x = yp_z - zp_y$$
$$l_y = zp_x - xp_z,$$
$$l_z = xp_y - yp_x,$$

as can be found in textbooks on particle dynamics. Here $f_x = mv_x$, etc. In order to introduce quantum conditions it is necessary to transfer these three equations in their operator forms using the information given in Table 10.1:

$$\hat{l}_x = -i\hbar \left[y\frac{\partial}{\partial z} - z\frac{\partial}{\partial y} \right],$$
$$\hat{l}_y = -i\hbar \left[z\frac{\partial}{\partial x} - x\frac{\partial}{\partial z} \right],$$
$$\hat{l}_z = -i\hbar \left[x\frac{\partial}{\partial y} - y\frac{\partial}{\partial x} \right].$$

In spherical polar coordinates the z-component is

$$\hat{l}_z = -i\hbar \frac{\partial}{\partial \varphi}.$$

So we can write $i\hbar(\partial\psi/\partial\varphi) = m_l\hbar\psi$ since we have seen that $l_z = m_l\hbar$ and the solution is simply

$$\psi = Ae^{im_l\varphi}.$$

Since ψ must be single-valued, it must repeat whenever φ is increased by 2π. Thus we have

$$e^{im_l\varphi} \equiv e^{im_l(\varphi + 2n\pi)},$$

where n is an integer. This cannot be true unless m_l is also an integer or zero, which supports what we have previously deduced about m_l.

Another operator of importance is the square operator $\hat{l}^2$ corresponding to the equation $\hat{l}^2\psi = l(l+1)\hbar^2\psi$, where ψ is an eigenfunctuon of $\hat{l}^2$, l is the quantum number, integral or zero, and the eigenvalues are $l(l+1)\hbar^2$. Since $\hat{l}^2 = l_x^2 + l_y^2 + l_z^2$, we might expect to be able to deduce all three components. A rigorous quantum-mechanical analysis shows that only the square operator and *one* of the components can be known simultaneously. This follows from the Heisenberg uncertainty principle. It is impossible to determine the direction of the angular momentum vector precisely, and only one component can be determined exactly. Conventionally the component $\hat{l}_z$ is taken such that $\hat{l}_z\psi = m_l\hbar\psi$ as before, where m_l is the magnetic quantum number. It has all integral values between $-l$ and $+l$, so that for each eigenvalue of l^2 there are $(2l+1)$ different eigenfunctions, each with a different value of m_l.

The magnitude of **l** is $l^*\hbar$, where $l^* = \sqrt{l(l+1)}$ and, since this is always greater than $l\hbar$, i.e. $|m_l|_{max}\hbar$, the orbital vector **l** cannot lie along the z axis, as was possible in the Bohr theory. Thus, if we know l^* and m_l for a particular eigenfunction, the vector **l** must be thought of as a rotating vector precessing around the z axis. It cannot be fixed in space since $\hat{l}_x$ and $\hat{l}_y$ are unknown. This is shown in Fig. 10.7. Since there are $(2l+1)$ eigenfunctions corresponding to the $(2l+1)$ values of m_l, there will be $(2l+1)$ values of the angle θ, limiting the spatial orientation of **l** to these positions. For each position we have $\cos\theta = m_l/\sqrt{l(l+1)}$, from which the corresponding value of θ can be found.

In a similar way the magnitudes of the other angular momentum vectors, can be shown to be

$$j^*\hbar = \sqrt{j(j+1)}\hbar \quad \text{and} \quad s^*\hbar = \sqrt{s(s+1)}\hbar$$

respectively where j and s are the quantum numbers. Both **j** and **s** are spatially quantized in the appropriate circumstances.

For all these vectors the spatial quantization (i.e. quantum limitations of direction) can only be realized in practice when the vectors are coupled directly with the z direction, i.e. a strong magnetic field direction imposed on the atom as in the Zeeman effect. This gives a *real* z direction in space.

10.9 Summary

The difficulties of the old quantum theory as derived from Bohr theory are overcome by the wave mechanics of Schrödinger and the operator method of Heisenberg. In fact the changes from dynamic variables to operators, as shown in Table 10.1, such as

$$p_x \to -i\hbar\frac{\partial}{\partial x} \quad \text{and} \quad E \to -i\hbar\frac{\partial}{\partial t},$$

are equivalent to postulating the Schrödinger wave equation. We saw that this equation cannot be proved from first principles, nor can it give information about electron spin. It does, however, give a clear meaning to the quantum numbers n, l and m_l and points to the vector model of the atom which we shall discuss in the next chapter.

Problems

(*The problem marked with an asterisk is solved in full at the end of the section.*)

10.1 The wave function for hydrogen in the 1S state for which $n=1$, $l=0$, $m_l = 0$ is given by $\psi = \pi^{-1/2}a^{-3/2}e^{-r/a}$, where a is the Bohr radius and equals 0.53×10^{-10} m. Derive a formula for the probability density and find where the electron is most likely to be. On the same graph plot the wave function and the probability density. (Maximum probability at $r=a$)

10.2 Repeat the above question for hydrogen in the 2S state, where $n=2$, $l=0$,

$m_l=0$ and the wave function is given by
$$\psi = \pi^{-1/2} 2^{-5/2} a^{-3/2}(2-r/a)e^{-r/2a}.$$
(Maxima at $r=0{\cdot}76a$ and $5{\cdot}24a$)

10.3 The wave function for hydrogen in the 2P state ($n=2$, $l=1$, $m_l=0$) is given by $\psi = \pi^{-1}2^{-2}\cos\theta a^{-5/2}e^{-r/2a}$.

Derive a formula for the probability density along the z axis for which $\theta=0$ and find where the electron is most likely to be found. (Maximum probability at $r=2a$)

10.4 Plot the wave function and probability density for the hydrogen in the 3S state ($n=3$, $l=0$, $m_l=0$), the wave function being given by
$$\psi = 2 \times 3^{-3/2} a^{-3/2}\{1 - \tfrac{2}{3}r/a + \tfrac{2}{27}(r/a)^2\}e^{-r/a}.$$

10.5 In the spherical polar coordinate system (r, θ, φ) a particle of mass m which is constrained to move in the circle of constant radius r_0 ($\theta=\pi/2$, $r=r_0$) has a potential energy which is independent of its azimuthal angle φ. Explain why the only relevant term in the operator ∇^2 is
$$\frac{1}{r_0^2}\frac{d^2}{d\varphi^2}.$$

Determine

(a) the possible values of the kinetic energy of the particle,
(b) the possible values of the angular momentum of the particle,
(c) the eigenfunctions of the Schrödinger equation of the particle.

[N]

10.6* Outline briefly how the wave functions and energy levels of the hydrogen atom may be obtained.

The normalized Schrödinger wave function for the ground state of a hydrogen atom is
$$\psi(r) = (\pi a_0^3)^{-1/2}\exp\left(-\frac{r}{a_0}\right),$$
where a_0 is the radius of the first Bohr orbit. Determine the most probable value of r in terms of a_0 and comment on your result. [N]

10.7 A particle moves in a two-dimensional infinitely deep potential well bounded by
$$x=0, a, \quad y=0, b$$

(a) Write down the Schrödinger equation for the motion of the particle.
(b) Write down the boundary conditions of the wave function.
(c) Given that the wave function is of the form $\psi = A\sin k_1 x \sin k_2 y$, find the values of k_1 and k_2.
(d) Substitute the value of ψ in the Schrödinger equation and find the values of E, the energy of the system relative to the bottom of the well.
(e) From the normalizing condition find the value of the constant A.
(f) Hence express the normalized wave function in terms of quantum numbers.

What is the significance of the case when $a=b$? [N]

Solution to Problem

10.6 The most probable value of r is not the mean value but corresponds to the maximum of the probability factor. For electrons between r and $(r+dr)$, this is

$$P(r) = [\psi(r)]^2 dV$$
$$= (\pi a_0^3)^{-1} \exp\left(-\frac{2r}{a_0}\right) 4\pi r^2 dr.$$

For the most probable value of r we put $dP(r)/dr = 0$. Therefore

$$\frac{d}{dr}\left[\frac{4r^2}{a_0^3} \exp\left(-\frac{2r}{a_0}\right)\right] = 0,$$

where $r = \hat{r}$, and so

$$\frac{8r}{a_0^3} \exp\left(-\frac{2r}{a_0}\right) + \frac{4r^2}{a_0^3} \exp\left(-\frac{2r}{a_0}\right)\left(-\frac{2}{a_0}\right) = 0$$

giving

$$8r = 8r^2/a_0$$

or

$$\hat{r} = a_0,$$

which is the radius of the first Bohr orbit.

Chapter 11

The Vector Model of the Atom

11.1 Quantum Numbers and Angular Momenta: Summary of Symbols and Notation

We shall use the following symbols and notation for single-electron atom quantum numbers and angular momenta.

(1) Principal quantum number n. This governs the total energy E_n, where $n = 1, 2, 3, \ldots$, and $E_n \propto 1/n^2$.

(2) Orbital quantum number l. This has values $l = 0, 1, 2, \ldots, (n-1)$ for each n, and governs the orbital angular momentum of magnitude $l^*\hbar$, where $l^* = \sqrt{l(l+1)}$ and l is always a positive integer or zero.

(3) Spin quantum number s. This can only have one value, viz. $s = \frac{1}{2}$ (never put $s = -\frac{1}{2}$), and relates to the intrinsic property of the electron which we call its spin angular momentum, whose magnitude is $s^*\hbar$, where $s^* = \sqrt{s(s+1)}$. Thus $s^*\hbar = \sqrt{\frac{1}{2} \times \frac{3}{2}}\,\hbar = 0.866\,\hbar$ alwaysm, for one electron.

(4) The orbital magnetic quantum number m_l. This relates to the magnitude of the z component of the orbital angular momentum vector **l** through $l_z = m_l\hbar$. The possible values of m_l are dependent on l and are given by $m_l = -l, -(l-1), \ldots, -1, 0, +1, \ldots, +(l-1), +l$, i.e. $(2l+1)$ possible values of m_l for each l.

(5) The spin magnetic quantum number m_s. This depends on the magnitude of the z component of the spin angular momentum **s** through $s_z = m_s\hbar$, where m_s can be either $+\frac{1}{2}$ or $-\frac{1}{2}$.

(6) Angular momentum *vectors* will be indicated by bold-face Roman type, viz. orbital angular momentum vector **l** (magnitude $l^*\hbar$), spin angular momentum vector **s** (magnitude $s^*\hbar$), and total angular momentum vector **j** (magnitude $j^*\hbar$), where $l^* = \sqrt{l(l+1)}$, $s^* = \sqrt{s(s+1)}$ and $j^* = \sqrt{j(j+1)}$.

For a single electron, **l** and **s** couple vectorially in a non-strong field to give **j** = **l** + **s** from the vector triangle made up of the numerical values l^*s^* and j^*. It is to be emphasized that the angular momenta **l** and **s** are NEVER parallel or antiparallel to each other. The total angular momentum is **j**.

(7) The total quantum number of the electron is j, given by $j = l \pm \frac{1}{2}$, i.e. two values for each l in a zero or weak field. This is a numerical relation between quantum numbers and not between angular momenta: in no way does it imply that $s = \pm\frac{1}{2}$. The quantum number j relates to the magnitude of the

total angular momentum through $j^*\hbar = \sqrt{j(j+1)}\hbar$, where j is positive and always half-integral for a single electron. Each l level degenerates into two j levels.

(8) The total magnetic quantum number m_j relates to the component of the total angular momentum vector through $j_z = m_j\hbar$ and is limited to the $(2j+1)$ values given by

$$m_j = -j, -(j-1), \ldots, -1, 0, +1, \ldots, +(j-1), +j.$$

(9) The spectroscopic notation for an electron state depends on l and is denoted by the small letters s, p, d, f, g, ..., etc., when $l = 0, 1, 2, 3, 4, \ldots$, etc.

All necessary information is included in the term level symbol $n^r l_j$, where r is the multiplicity of the term, given by $r = 2s + 1$. Thus $r = 2$ for all single-electron terms except s levels.

(10) The selection rules for allowed transitions are as follows:

n, $\Delta n = (0), 1, 2, 3, \ldots$, etc.
l, $\Delta l = \pm 1$;
s, $\Delta s = 0$;
j, $\Delta j = \pm 1, 0$;

but $j_1 = 0 \to j_2 = 0$ is forbidden.

(11) These quantum rules for the angular momentum vectors are only realized when the appropriate coupling holds, e.g. $j = l \pm \frac{1}{2}$ is true for strong spin–orbit coupling but not for strong spin–field and orbital–field coupling.

11.2 Magnetic Moments — Orbital and Spin

So far, in our discussion of the properties of the atom and of the electron, we have dealt only with mechanical variables. But the electron is a charged body rotating about the nucleus in a closed loop, so that we must now consider the magnetic moment which derives from this. For simplicity we shall confine our discussion to a circular orbit, as shown in Fig. 11.1.

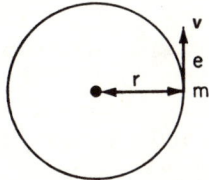

Fig. 11.1 Electron moving in a circular orbit.

An electron moving in such an orbit describes a current loop with an equivalent electric current of $i = e/T$, where T is the period of rotation. The orbital angular momentum p_θ is given by $p_\theta = I\omega = mr^2\omega$, where ω is the angular velocity. Since $T = 2\pi/\omega$, we get $p_\theta = mr^2(2\pi/T) = 2mA/T$, where A is the area of the orbit. Now, from Amperes theorem of magnetic shells, we have $\mu_l = iA$, where μ_l is the equivalent magnetic moment of the current loop, i.e. $\mu_l = (e/T)A = ep_\theta/2m$. Introducing the quantization rule $p_\theta = l^*\hbar = \sqrt{l(l+1)}\hbar$ gives

$$\mu_l = \frac{e\hbar}{2m}\sqrt{l(l+1)}$$

where $l = 0, 1, 2, \ldots$, etc. Since $\sqrt{l(l+1)}$ is a pure number, the units of magnetic moment depend on the expression $e\hbar/2m$. This is, in fact, the atomic unit of magnetic moment and is called the Bohr magneton, denoted by μ_B. Its numerical value is given by

$$\mu_B = \frac{1\cdot 6 \times 10^{-19} \times 1\cdot 05 \times 10^{-34}}{2 \times 9\cdot 1 \times 10^{-31}} = 9\cdot 27 \times 10^{-24} \text{ J T}^{-1}$$

We now have $\mu_l = \mu_B \sqrt{l(l+1)}$ and we can also write

$$\mu_l = \frac{\mu_B}{\hbar} \mathbf{l}.$$

The magnetic moment has a direction which depends on the direction of the current in the loop, and so is a vector. Since it comes from the movement of the negative charge e, the direction of the vector is taken to be *opposite* that of the angular momentum vector **l**. Since μ_l and **l**s are directly proportional to each other, we can draw the vector diagram of these as in Fig. 11.2.

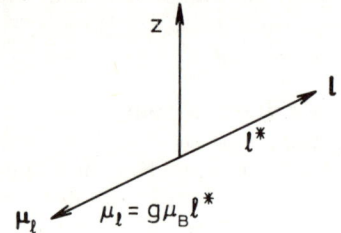

Fig. 11.2 Angular momentum and magnetic moment vectors.

The ratio

$$\left(\frac{\mu_l \text{ in Bohr magnetons}}{l^* \text{ in units of } \hbar}\right)$$

is called the gyromagnetic ratio, g, and in our simple case we see that $g = 1$. This is not true in general, so that we may write $\mu_l = -g\mu_B\sqrt{l(l+1)}$ and $(\mu_l)_z = -g\mu_B m_l$ where $g \geqslant 1$. We shall meet g as the Landé splitting factor when we come to discuss the anomalous Zeeman effect.

Now turning to the model of the spinning electron, we know that the magnitude of its intrinsic angular momentum is $s^*\hbar = \sqrt{s(s+1)}\hbar$, where $s = \tfrac{1}{2}$. This spin constitutes an axially rotating spherical charge giving an equivalent magnetic moment μ_s which is also a vector. By analogy with the orbital case we would expect $\mu_s = -\mu_B\sqrt{s(s+1)}$, but surprisingly it turns out that μ_s is *twice* this value, i.e. $\mu_s = -2 \times \mu_B\sqrt{s(s+1)}$, so that the g factor is 2. This result was proved experimentally by the Stern–Gerlach experiment, as well as spectroscopically from the Zeeman effect. The vector μ_s is taken to be opposite in direction to **s**, in line with the orbital case. The *g* factor of 2 for the spinning electron is an intrinsic property which has to be accepted without explanation.

The atomic electron therefore has associated with it two mechanical and two magnetic vectors, as shown in Fig. 11.3.

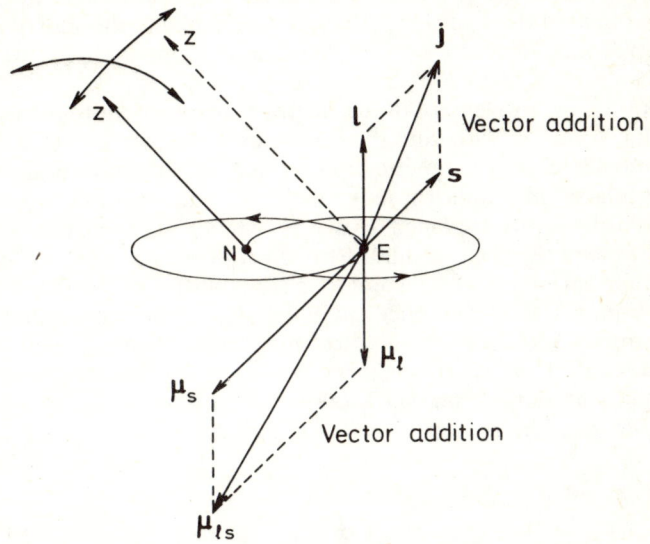

Fig. 11.3 Coupling of mechanical and magnetic vectors.

The electron E travelling round the nucleus N is equivalent to the nucleus travelling round a fixed electron at E. Note from Fig. 11.3 that neither **l** nor **s** is parallel to the axis, nor can they be parallel or antiparallel to each other. The diagram also attempts to indicate that the z axis is *not* fixed in space, so that the atom is completely isotropic in field-free space. Averaged over a fixed time, EZ points in *all* directions. The z axis is used as a convenient mathematical direction and only has a physical reality in the presence of an external field **B**. Conventionally we then take the z axis along the **B** direction and relate all the other vectors to this direction.

For a weak field **B** the **l** and **s** vectors couple to give the total vector **j**, which then precesses around **B** and is spatially quantized. However, as **B** increases the **l** and **s** vectors are uncoupled and precess independently around **B** and are spatially quantized independently. In this case no **j** vector is formed and $m_j = m_l + m_s$.

11.3 The Stern–Gerlach Experiment

Much of what we have discussed so far about atomic and electronic properties, such as space quantization, the existence of magnetic moments and indeed the existence of quantum numbers themselves, has been largely theoretical or, at least, conjectural. But now, in the famous Stern–Gerlach experiment, we have direct experimental proof that an electron has an intrinsic magnetic moment and a spin quantum number of $\frac{1}{2}$.

The experiment measures the magnetic moment of the atom as a whole, and it was not until some years later when the zero orbital momentum of the S state became clear that this magnetic moment was ascribed to the electron. Atoms were chosen which in the ground state had zero orbital momentum. The entire magnetic moment of the atom could then be attributed to the spin of its electrons. Atoms of hydrogen, lithium, sodium, potassium, copper, silver and gold were used.

In such atoms the movement of the electron is equivalent to a current flowing in a circular loop of wire and has corresponding magnetic properties. In particular it behaves as a magnetic dipole which experiences equal but opposite forces when placed in a uniform magnetic field. The resultant force is therefore zero and the atom is not displaced. However, when placed in a non-uniform field, such an atom experiences a resultant force proportional to the gradient of the field. The potential energy of a current loop (Fig. 11.4) of magnetic moment μ with its axis at an angle θ to a magnetic field of flux density B is given by $V = \mu B \cos \theta$. The mechanical force X in a direction x is given by $X = -dV/dx = (-d/dx)(\mu B \cos \theta)$. This is zero when B is constant, in accordance with the above physical argument, but becomes $(dB/dx)(\mu \cos \theta)$ in a non-uniform magnetic field of gradient dB/dx. The maximum value is $\mu dB/dx$.

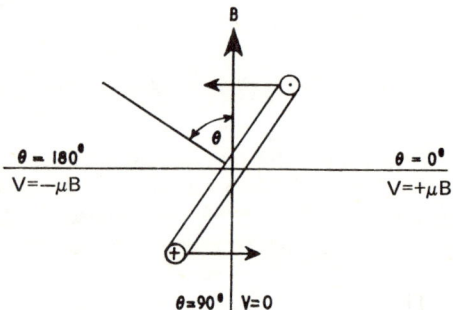

Fig. 11.4 Potential energy V of a current loop of magnetic moment μ in a magnetic field of flux density B.

In the original experiment, silver was heated in an oven and a stream of silver atoms having velocities corresponding to the oven temperature emerged. Two slits S_1, S_2 (Fig. 11.5) limited the stream to a very fine pencil, which was then allowed to pass between the poles of a magnet. The pole pieces were shaped so

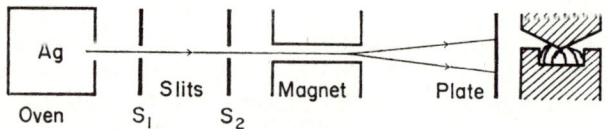

Fig. 11.5 The Stern–Gerlach apparatus.

that a very large magnetic-field gradient dB/dx was obtained. The whole apparatus was evacuated to a pressure sufficiently low to enable a silver atom to traverse the whole length without a collision. The mean free path was therefore greater than the length of the apparatus. Outside the magnetic field the magnetic moments of the silver atoms were orientated in random fashion, as shown in Fig. 11.6. On entering the magnetic field the magnetic moments became orientated parallel and antiparallel to the direction of the field. They were then said to be

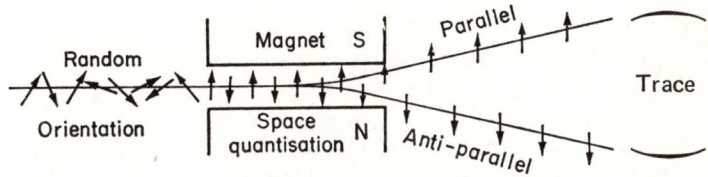

Fig. 11.6 Illustration of Stern–Gerlach experiment.

space-quantized. The atoms also experienced a resultant force due to the large magnetic field gradient, the direction of the force being determined by the direction which the magnetic moment had taken up with respect to the magnetic field. The atoms in the two orientations moved apart under the action of the opposite forces and were recorded as separate deposits upon the plate, as shown in Fig. 11.7(a).

Fig. 11.7 Stern–Gerlach trace pattern. (a) Actual trace of silver atoms; (b) continuous trace expected classically.

Classically, the magnetic moment μ_s could have any value between $-\mu_s$ and $+\mu_s$, so that the traces on the plate should have been continuous, as shown in Fig. 11.7(b).

The fact that *two* discrete, equally-displaced traces were found showed that the magnetic moment had two separate and discrete orientations, one with a component parallel to **B** and the other with a component antiparallel to **B**. These two facts proved conclusively the existence of space quantization for μ_s at least. But there was still the possibility that the inner electrons could be involved orbitally.

This difficulty was resolved by experiments with hydrogen in 1927. These were unambiguous since only one electron was involved, whereas the silver atom contains many electrons. The ground state was $1S_{1/2}$, so that $l=0$, and $m_l=0$ also. If the Stern–Gerlach splitting did not depend on spin, the beam of hydrogen atoms should be undeflected since l of the whole atom was zero.

However, the splitting was exactly as found for the silver atoms, viz. two symmetrical beams due to two magnetic components. Since $m_l=0$ for hydrogen these two components *must* be spin components, i.e. due to μ_s. Thus the intrinsic spin angular momentum **s** has an associated intrinsic spin magnetic moment μ_s, with $\mu_s = -g_s\mu_B s^*$ and $(\mu_s)_z = -g_s\mu_B m_s$, where $s^* = \sqrt{s(s+1)}$, $s_z = m_s\hbar$ and g_s is the spin g factor.

We can now put $\Delta m_s = \pm 1$ just as Δm_l was ± 1 for the orbital case, and the range of m_s is from $-s$ to $+s$, just as the range of m_l was from $-l$ to $+l$. This means that the two values of m_s must be $m_s = -\frac{1}{2}, +\frac{1}{2}$ since s has only *one* value, viz. $s=\frac{1}{2}$.

We saw that the maximum mechanical force X in the x direction was given by $X = \mu_s(dB/dx)$ and, since $(\mu_s)_z = -g_s\mu_B m_s$, we get

$$X = -g_s\mu_B m_s \frac{dB}{dx}.$$

In the actual experiment, dB/dx was easily measured, μ_B is given by $\mu_B = 9{\cdot}27 \times 10^{-27}\,\text{J T}^{-1}$ so that $(\mu_s)_z = g_s\mu_B m_s$ gives $g_s m_s$. The results for hydrogen were $g_s m_s = \pm 1$ and, since $m_s = \pm\frac{1}{2}$ from above, we have $g_s = 2$, confirming our previous assertion that the spin g factor of the electron is *twice* its orbital g factor.

11.4 Spatial Quantization of Electron Spin

We now know that the free electron spin angular momentum **s** vector has an antiparallel vector μ_s associated with it and that the latter gives components $\pm(\mu_s)_z$ along an external magnetic field direction, where
$$(\mu_s)_z = g\mu_B m_s, \qquad m_s = \pm\tfrac{1}{2}.$$
These two positions fix the direction of **s** in space, as shown in Fig. 11.8, where the field direction is taken along the z axis. When **s** is coupled with **B**, it precesses around the field direction at an angle given by

$$\begin{aligned}\cos\theta &= m_s/\sqrt{s(s+1)} \\ &= \pm\tfrac{1}{2}/0{\cdot}866 \\ &= \pm 0{\cdot}5774\end{aligned}$$

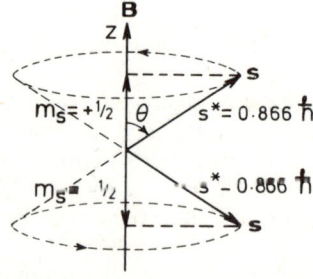

Fig. 11.8   Spatial quantization of electron spin vector showing the two positions.

giving
$$\theta = 54\cdot7° \text{ or } 125\cdot3°.$$
In the absence of any other coupling, these are the only possible positions of **s** relative to **B**, and these are the two space directions revealed by the analysis of the Stern–Gerlach results. Note that $(\mu_s)_z = \mu_B$, the Bohr magneton.

11.5 Spin–Orbit Coupling and the Total Angular Momentum j

The reality of electron spin has now added a fourth quantum number to the total description of the electron state. These are n, l, m_l and m_s and the *total* eigenfunction of such a state will be $\psi_{nlm_l m_s}$. This is the case whenever the orbital and spin angular momenta act independently and do not interact, i.e. it is understood that there are two m_s components $\pm\frac{1}{2}\hbar$ along the z axis.

Thus, if the external field is strong enough, the **l** and **s** vectors precess separately around **B** and the total z component is then $m_l\hbar + m_s\hbar$. In this case m_l and m_s are said to be 'good' quantum numbers. In the absence of a strong external magnetic field **B**, the orbital magnetic field due to μ_l interacts with the spin field due to μ_s. This magnetic coupling results in an interaction between **l** and **s** so that they are no longer independent, and m_l and m_s disappear. The eigenfunction $\psi_{nlm_l m_s}$ is now incorrect. A new vector **j** is formed and both **l** and **s** precess around **j**. The resultant vector **j** is the total angular moment of the electron and represents the electron state *instead* of **l** and **s**, i.e. **j** = **l** + **s**.

By analogy with the orbital and spin momenta we can define the corresponding quantum number j such that the total angular momentum is given by $j^*\hbar = \sqrt{j(j+1)}\hbar$ and its component by $j_z = m_j\hbar$, where m_j is the equivalent total magnetic quantum number which has the $2j+1$ values for each j, i.e.
$$m_j = -j, \;-(j-1), \;\ldots, \;+(j-1), \;+j$$
and $j = (m_j)_{\max} = m_l + m_s$ numerically. It is still true that $j_z = l_z + s_z$, but l_z and s_z are no longer equal to $m_l\hbar$ and $m_s\hbar$.

The new quantum number j is the total quantum number which governs the spatial quantization of **j** relative to the z axis, while the spatial quantization of **l** and **s** becomes redundant. The allowed positions of **j** in space are given by
$$\cos\theta = \frac{m_j}{\sqrt{j(j+1)}}.$$
There will be $2j+1$ values of θ corresponding to the $2j+1$ values of m_j.

In Fig. 11.9 we show the vector model of spin–orbital coupling drawn for the case $l = 2$, $s = \frac{1}{2}$ and $j_+ = \frac{5}{2}$. As can be seen, the components of **l** and **s** oscillate about the extreme positions A and B on the z axis as **l** and **s** precess around **j**. That is why m_l and m_s are not good quantum numbers. On the other hand **j** precesses axially around **B**, z so that j_z is constant and $j_z = m_j\hbar$. That is why j and m_j are the new good quantum numbers. The six spatial positions of **j** are labelled 1–6 in Fig. 11.9.

Since **j** is now quantized along **B**, z while **l** and **s** are not, the quantum numbers m_l and m_s are no longer good quantum numbers and are replaced by j and m_j, so that the correct eigenfunction is now ψ_{nljm_j}. This is always the case when spin–orbit coupling obtains.

Since the scalar addition $m_j = m_l + m_s$ now holds, we have $(m_j)_{\max} = (m_l)_{\max} \pm m_s$ or $j = l \pm \frac{1}{2}$, so that for one-electron systems j is always half-integral. This is the

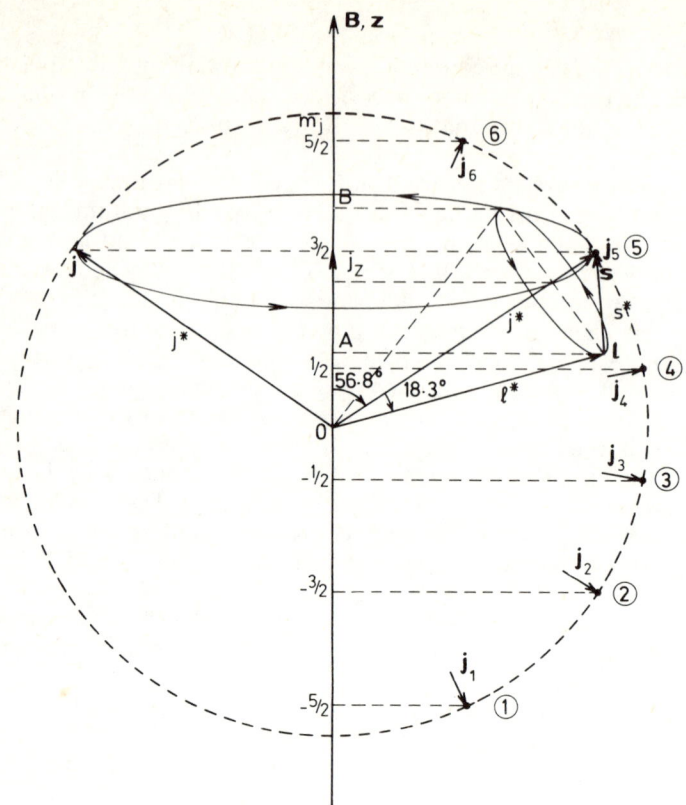

Fig. 11.9 Spatial quantization showing six positions of vector **j** for which $j = \frac{5}{2}$.

result we deduced from the fine structure of the sodium spectrum and used to explain X-ray levels and transitions (see Chapter 8).

11.6 The structure of the Hα line

We can now apply these rules to the structure of the H_α line shown in Fig. 11.10. The H_α line is due to the $n = 3$ to $n = 2$ transition so that, by applying the selection rules for the hydrogen atom as a whole,

$$\Delta n = 1, 2, 3, \ldots;$$
$$\Delta l = \pm 1;$$
$$\Delta s = 0;$$
$$\Delta j = \pm 1, 0.$$

It may seem strange that a transition which does not change the total angular momentum of the atom ($\Delta j = 0$) is able to emit a photon of energy $h\nu$ whose spin angular momentum is 1. This apparent violation of the law of conservation of angular momentum, is due to the fact that the numerical value of j is the same before and after the transition, but the *position* of the vector **j** changes to its antiparallel direction. Thus the total angular momentum changes vectorially and a photon is emitted. From these rules we can construct the complete energy level

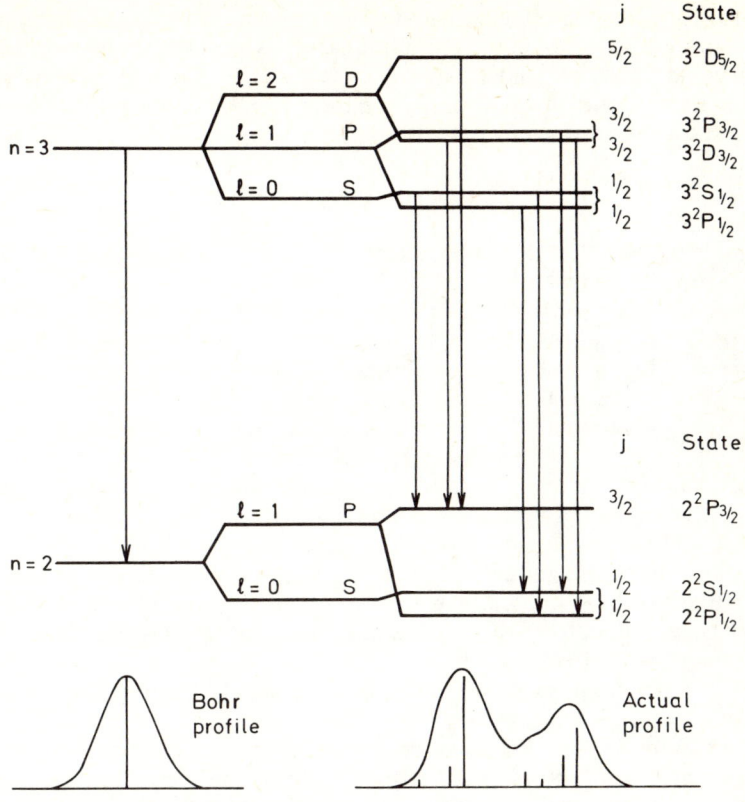

Fig. 11.10 Fine structure of H_α line.

diagram shown in Fig. 11.10. There are *seven* possible transitions. These are not observed as seven spectral lines because, in addition to the insufficient spectrographic resolution, the spectral theories of Sommerfeld and Dirac, including relativistic energy considerations of spin–orbit coupling, gave the same term energy for all those levels with the same n, j but with l differing by one unit only. Thus neighbouring pairs of terms such as

$$nP_j : nS_j, \quad nD_j : nP_j, \quad nF_j : nD_j$$

have the same energy and so give one energy level in the diagram, as in Fig. 11.10, where they are drawn close togehter.

As can be seen from this diagram three pairs of levels coincide, giving two pairs of lines with the same frequency. This reduces the possible number of components in H_α to five. However, the early researches of Lamb and Retherford using microwave techniques showed that the $(D, P)_{3/2}$ and $(P, S)_{1/2}$ terms *were* slightly different and they were able to measure these small term separations. The result is that the term diagram now becomes exactly as shown in Fig. 11.10, as our simple j, l theory predicts, the H_α line now having the seven components originally suggested. These have not yet been resolved experimentally.

The origin of the separation of these terms is in the interaction between the field of the moving electron in the atom and the electromagnetic field corresponding to the zero-point energy (p. 000). The explanation is purely quantum mechanical.

From what has been said it is quite obvious that many of the postulates of quantum theory and of quantum mechanics can be explained and tested by detailed study of the structure of the H_α line.

Problems

11.1 Using the vector model, determine the possible values of the total angular momentum of an f electron. ($j=\frac{5}{2}$ and $\frac{7}{2}$)

11.2 Use the vector model of the atom to determine possible values of the total angular momentum for a sodium atom, when the principal quantum number $n=3$. Draw an energy level diagram and indicate the transitions you would expect to occur. ($^2S_{1/2}$, $^2P_{1/2}$, $^2P_{3/2}$, $^2D_{3/2}$, $^2D_{5/2}$)

11.3* Bearing in mind that the angular momenta associated with l, s and j are respectively $[l(l+1)]^{1/2}h/2\pi$, $[s(s+1)]^{1/2}h/2\pi$ and $[j(j+1)]^{1/2}h/2\pi$, draw vector diagrams to represent the term types predicted in the previous question. Calculate the angles between the **l** and **s** vectors in each case. (Not possible for S term; 35·7°, 114·1°, 45·0°, 118·1°)

11.4 Use the vector model of the atom to explain why the orientations of its total angular momentum vector with respect to the field direction on the 'old' quantum theory differ from those allowed by wave mechanics when the atom is in a weak magnetic field.

What experimental evidence is there of spatial quantization of angular momenta in atoms? [N]

11.5 (a) What experimental evidence is there for the space quantization of the angular momentum vectors of the atom? (b) Draw the vector relations between the orbital, spin and total angular momentum vectors for the $^2D_{3/2}$ and $^2D_{5/2}$ states. (c) Calculate the allowed spin vector directions for the electron of hydrogen in its ground state with respect to a given magnetic field direction. (d) Given that the D-line separation in the sodium spectrum is 60 nm, calculate the magnitude of the magnetic field that splits the P-state, if $\lambda_D = 600$ nm. [N]

11.6 A narrow beam of silver atoms for a furnace at 1500 K pass through a non-uniform magnetic field in a Stern–Gerlach experiment in a direction perpendicular to the maximum field gradient. Calculate the angle through which the beam is deviated in the 100 nm path between the pole pieces if the magnetic flux density varies at the rate of 10 T mm^{-1} and the Boltzmann constant is $1·38 \times 10^{-23}$ J K^{-1}. ($2·5 \times 10^{-2}$ radians)

11.10 Describe and explain the Stern–Gerlach experiment and indicate its importance in atomic physics. Calculate the mechanical force and acceleration of a potassium atom moving in a non-uniform magnetic field and whose magnetic moment is entirely due to the spin of a single electron. The atomic mass number of potassium is 39 and the magnetic flux density varies at the rate of 10 T mm^{-1}. 11.7⁻3 aN, 1·58 Mm s^{-2})

11.8 Calculate the mechanical force and acceleration of a silver atom whose magnetic moment is entirely due to the spin of a single electron. The magnetic flux density varies at the rate of 10 Wb m^{-2} mm^{-1}. ($1·03 \times 10^{-19}$ N, $0·57 \times$ (Mm s^{-2})

Solution to Problem

11.3 Take as an example the $^2D_{5/2}$ term. Here
$$s = \tfrac{1}{2}, \quad l = 2, \quad j = \tfrac{5}{2}$$
and
$$s^* = \sqrt{\tfrac{1}{2} \times \tfrac{3}{2}}, \quad l^* = \sqrt{2 \times 3}, \quad j^* = \sqrt{\tfrac{5}{2} \times \tfrac{7}{2}}$$
$$= 0.866 \qquad\qquad = 2.45 \qquad\qquad = 2.96$$

If θ is the angle between the **l** and **s** vectors, then
$$\cos\theta = \frac{l^{*2} + s^{*2} + j^{*2}}{2 l^* s^*}$$
$$= \frac{6 + \tfrac{3}{4} - 8\tfrac{3}{4}}{2\sqrt{6 \times \tfrac{3}{4}}}$$
$$= -\frac{1}{\sqrt{4.5}}$$
$$= -0.138$$

Whence $\theta = 118.1°$.

Chapter 12

Two-Electron Atoms — Pauli Principle

12.1 Wave Functions of Two-Electron Atoms

In our discussion of the one-electron atom we assumed that the quantum behaviour of the atom was governed by the quantum behaviour of the electron. Ignoring electron spin, this behaviour was analysed by solving the Schrödinger equation for the wave function ψ_{nlm_l}. But what is the appropriate atomic wave function describing the behaviour of atoms with *two* optically active electrons, one of which may be in the ground state and the other in an excited state? To find this new wave function let us suppose the first electron is identified by n_1, l_1 and m_{l_1} and is in the quantum state a, while the other is identified by n_2, l_2 and m_{l_2}, and is in the quantum state b. The wave functions are then quite different in their algebraic form and are

$$\psi_a \equiv \psi_{n_1 l_1 m_{l_1}} \quad \text{and} \quad \psi_b \equiv \psi_{n_2 l_2 m_{l_2}}$$

where a may be the ground state and b an excited state, for example, and $\psi_a \not\equiv \psi_b$.

The electrons are otherwise identical, i.e. they have the same mass and charge and all their quantum information is obtained from the Schrödinger equation, giving the three quantum numbers n, l and m_l.

The wave function of the *atom* is then ψ_{ab} and if the position coordinates of the two electrons are given by x_1 and x_2 we can write $\psi_{ab} \equiv \psi_{ab}(x_1)(x_2)$, i.e. electron 1 is in a at x_1 while electron 2 is in b at x_2.

Using the principle of the separation of variables, this may also be written $\psi_{ab} \equiv \psi_a(x_1)\psi_b(x_2)$, where $\psi_a(x_1)$ is the eigenfunction of electron 1 in state a at x_1 and similarly for 2.

In this equation we have arbitrarily selected electron 1 to be in state a and electron 2 to be in state b. We could, of course, interchange the electrons, since we take them to be physically indistinguishable, i.e. put 1 in b and 2 in a. Thus we get another eigenfunction for the atom:

$$\psi_{ba}(x_1, x_2) \equiv \psi_b(x_1)\psi_a(x_2),$$

which is different from our first eigenfunction since $\psi_a \neq \psi_b$. It is important to remember that $\psi_{ab} \neq \psi_{ba}$. The two atomic eigenfunctions are interchangeable on switching electrons, i.e. $\psi_{ab} \to \psi_{ba} \to \psi_{ab}$ as $1 \to 2 \to 1$, and both satisfy the appropriate Schrödinger equations.

We can test the validity of these eigenfunctions in giving a solution to a real physical problem by choosing some physical property which can easily be deduced from them. The simplest property to consider is the electron density

probability, since this must be the same when the electrons are interchanged, both electrons having the same charge. The probability of finding electron 1 within dx_1 at x_1 in a and 2 within dx_2 at x_2 in b is
$$\psi_{ab}\psi_{ab}^{*}dx_1 dx_2$$
(see p. 000) for the whole atom.

As we noted above, the two electrons are physically identical, so that we may exchange their positions and must leave the probability density unchanged. Therefore, on exchange we have equal probabilities and thus
$$\psi_{ab}\psi_{ab}^{*}dx_1 dx_2 \underset{1 \to 2}{=} \psi_{ba}\psi_{ba}^{*}dx_2 dx_1,$$
giving
$$\psi_{ab}\psi_{ab}^{*}=\psi_{ba}\psi_{ba}^{*} \quad \text{or} \quad \psi_{ab}=\psi_{ba},$$
which, as we have seen, *in not true*. Hence the two eigenfunctions ψ_{ab} and ψ_{ba} cannot both represent an atom of two identical electrons and always give the same probability density, even though they are both solutions of the Schrödinger equation. This is because there is still this arbitrary choice of which electron goes into which state, and we conclude that this holds for any physical problem we could work out. Perhaps the two electrons are *not* indistinguishable.

In quantum mechanics we know that any linear combination of ψ_{ab} and ψ_{ba} is also a solution of the Schrödinger equation, but in atomic physics the only combined eigenfunctions which relate to real physical problems are the *addition* and *subtraction* combinations which then give solutions which are independent of the arbitrary choice of electrons mentioned above. These functions are

$$\psi_S^{+} = \frac{1}{\sqrt{2}}[\psi_{ab}+\psi_{ba}], \qquad \text{the addition combination}$$

and

$$\psi_A^{-} = \frac{1}{\sqrt{2}}[\psi_{ab}-\psi_{ba}], \qquad \text{the subtraction combination,}$$

where $1/\sqrt{2}$ is the normalizing factor.

Now we know that on exchange $\psi_{ab} \to \psi_{ba} \to \psi_{ab}$, etc., so on exchanging the electrons, as before, we get

$$[\psi_S^{+}]_{1 \to 2} = \frac{1}{\sqrt{2}}[\psi_{ba}+\psi_{ab}] = \psi_S^{+}$$

and

$$[\psi_A^{-}]_{1 \to 2} = \frac{1}{\sqrt{2}}[\psi_{ba}-\psi_{ab}] = -\psi_A^{-}$$

Evidently the eigenfunction ψ_S^{+} remains unchanged and is therefore called the symmetric function, while the ψ_A^{-} eigenfunction changes its sign and is called the antisymmetric function.

Let us now test these wave functions against our probability density criterion. First we have to evaluate $\psi_S^{+}\psi_S^{+*}$. This is
$$\tfrac{1}{2}[\psi_{ab}+\psi_{ba}][\psi_{ab}^{*}+\psi_{ba}^{*}] = \tfrac{1}{2}[\psi_{ab}\psi_{ab}^{*}+\psi_{ab}\psi_{ba}^{*}+\psi_{ba}\psi_{ab}^{*}+\psi_{ba}\psi_{ba}^{*}].$$
We now perform $1 \to 2$ switch so that $\psi_S^{+}\psi_S^{+*}$ becomes

$$\tfrac{1}{2}[\psi_{ba}\psi_{ba}^* + \psi_{ba}\psi_{ab}^* + \psi_{ab}\psi_{ba}^* + \psi_{ab}\psi_{ab}^*],$$

exchanging term by term. On inspection we see that this expression is identical with the previous expression (simply because it is symmetrical), so that

$$\psi_S^+ \psi_S^{+*} \underset{1\to 2}{=} \psi_S^+ \psi_S^{+*}$$

and the probability density remains unchanged, as required. Similarly, we find

$$\psi_A^- \psi_A^{-*} \underset{1\to 2}{=} \psi_A^- \psi_A^{-*}$$

i.e. the antisymmetrical wave function *also* remains unchanged. Thus both ψ_S^+ and ψ_A^- are suitable eigenfunctions for an unchanged probability density and are independent of whether we choose electron 1 or 2 to go into state a. Since both state a and state b contain identical electrons, the total energy of the atom will be the same if we exchange the electrons. When we have two different eigenfunctions of the same energy which are acceptable, we say they are degenerate, and this an example of exchange degeneracy.

Although the atom can be represented by *either* of the two functions ψ_S^+ and ψ_A^-, any experiment designed to find out which of these two possibilities obtains for the atom is doomed to failure. During an experiment the atom will spend just as much time in the ψ_S^+ state as in the ψ_A^- state — the time averages are equal. However, if it is not possible to distinguish between ψ_S^+ and ψ_A^- experimentally, there are some interesting conclusions to be drawn from the two cases. Suppose, for instance, that *both* electrons are in the ground state a, say, but 1 is still at x_1, and 2 at x_2. Then

$$\psi_S^+ = \frac{1}{\sqrt{2}}[\psi_{aa}(x_1, x_2) + \psi_{aa}(x_1, x_2)]$$
$$= \sqrt{2}[\psi_{aa}(x_1, x_2)]$$

and

$$\psi_A^- = \frac{1}{\sqrt{2}}[\psi_{aa}(x_1, x_2) - \psi_{aa}(x_1, x_2)]$$
$$= 0.$$

Clearly then, if ψ_A^- is to be a possible eigenfunction, as we have seen it is, the two electrons cannot be in the same quantum state, i.e. they are *not* indistinguishable. Thus if two electrons have the *same* quantum numbers n, l and m_l they must be distinguished by some other quantum condition, different for the two electrons, which makes the two quantum ground states different, i.e. $a_1 \neq a_2$ and $\psi_A^- \neq 0$, as is required. This conclusion is summarized in the *Pauli exclusion principle*, which states that no two electrons can exist in the same quantum state with the same set of quantum numbers. This makes ψ_A^- a possible eigenfunction.

We know spectroscopically, and from the Stern–Gerlach experiment, that the distinguishing quantum number is the spin magnetic quantum number m_s, where $m_s = \pm\tfrac{1}{2}$, corresponding to the two spatial directions of the free spin vector **s**. Electron states can then be described as

$$\psi_{a_1} \equiv \psi_{nlm_l m_s^+} \quad \text{for electron 1}$$

and

$$\psi_{a_2} \equiv \psi_{nlm_l m_s^-} \quad \text{for electron 2,}$$

where $m_s^+ = +\frac{1}{2}$ and $m_s^- = -\frac{1}{2}$, so that $a_1 \neq a_2$ and ψ_A^- is not zero. Then we can write

$$\psi_A^- = \frac{1}{\sqrt{2}} [\psi_{a_1 a_2}(x_1, x_2) - \psi_{a_2 a_1}(x_1, x_2)]$$
$$\neq 0.$$

Formally, the Pauli exclusion principle states that no two electrons can have the same set of four quantum numbers n, l, m_l and m_s. In the case of spin–orbit coupling the four quantum numbers are n, l, j, m_j. The antisymmetric wave function is then non-zero.

In terms of our analysis, this means that the eigenfunctions of a two-electron atom must be antisymmetric and in general we can see that any atomic system containing several optically active electrons must be described by antisymmetric wave functions.

12.2 Vector Coupling for Two Electrons

The simplest atom after hydrogen is helium, with just two electrons. At once we see, from the exclusion principle, that in the ground state the quantum numbers of the two electrons will be

electron 1, $1, 0, 0$ and $+\frac{1}{2}$;
electron 2, $1, 0, 0$ and $-\frac{1}{2}$;

where $n=1$, $l=0$ and $m_l=0$ in both cases, corresponding to an s state for each electron. What is the spectroscopic state of the helium atom as a whole? Suppose, for example, that the atom is in an excited state, one electron in say $n=1$, $l=1$, $m_l=1$ and $m_s=\frac{1}{2}$ (a p state) and the other in $n=2$, $l=2$, $m_l=2$ and $m_s=\frac{1}{2}$ (a d state). How can we find the final angular momentum vectors for the whole atom? Using the capital letters L, S, J, etc., for the atom it is obvious that the final **J** must be compounded somehow of the individual electron values. The electrons couple together either electrostatically or magnetically in the same sort of way that we had spin–orbit coupling, with electrostatic coupling predominating in this case.

There are two possible modes of coupling, **LS** (or Russell–Saunders) coupling, and **jj** coupling, although intermediate cases also arise. We take **LS** coupling first since it predominates in lighter atoms. In **LS** coupling the electrostatic interaction due to the charges is much stronger than the magnetic coupling due to the electron motion. Thus the two electron charges couple strongly to give a resultant spin angular momentum vector **S** and an orbital angular momentum vector **L** such that

$$\mathbf{S} = \mathbf{s}_1 + \mathbf{s}_2 \quad \text{by spin–spin coupling}$$

and

$$\mathbf{L} = \mathbf{l}_1 + \mathbf{l}_2 \quad \text{by orbit–orbit coupling.}$$

The spin quantum number of the atom is S and has values between $(s_1 + s_2)$ and $(s_1 - s_2)$, which is 1 or 0 for the two-electron atom. The *magnitudes* of the corresponding spin vectors are then

$$S^*\hbar = 0$$

and

$$S^*\hbar = \sqrt{1 \times 2}\hbar = \sqrt{2}\hbar$$

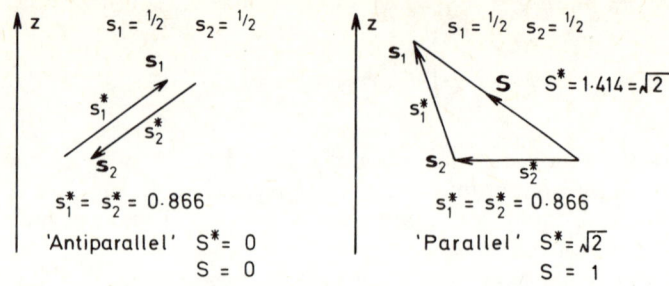

Fig. 12.1 Vector coupling of two spin angular momenta.

with
$$S_z = 0 \quad \text{or} \quad S_z = m_s \hbar$$
where $m_s = \pm 1$ in a strong field.

The vector diagrams for the spin angular momentum are shown in Fig. 12.1. which shows the formation of the final **S** vector for the two electrons of helium by spin–spin coupling. This is true for helium whatever its spectroscopic state.

Let us now turn to the coupling of the orbital angular momenta vectors, i.e. **ll** coupling. Here $\mathbf{L} = \mathbf{l}_1 + \mathbf{l}_2$ and the possible atomic quantum numbers range from $(l_1 + l_2)$ through to $(l_1 - l_2)$ in unit steps. Also $L^*\hbar = \sqrt{L(L+1)}\hbar$ and $L_z = m_L\hbar$, where m_L has the usual $2L+1$ values from $-L$ to $+L$ in a strong field.

We can illustrate **ll** coupling by taking helium in a p, d state, as before:

for the p state, $l_1 = 1$ $l_1^* = \sqrt{2}$;
for the d state, $l_2 = 2$ $l_2^* = \sqrt{6}$.

Quantum numbers for the atom are 3, 2, 1 corresponding to the L^* values of $\sqrt{12}$, $\sqrt{6}$ and $\sqrt{2}$. Figure 12.2 shows the three possible vector positions of **L** relative to a fixed $\mathbf{l}_1$, drawn to scale.

Although the **ss** and **ll** couplings are mainly electrostatic in nature, there is still present a residual, but weaker, magnetic spin–orbit coupling due to the rotation of the electrons. This now gives a final **J** value due to **LS** coupling, via
$$\mathbf{J} = \mathbf{L} + \mathbf{S}$$
and
$$J^*\hbar = \sqrt{J(J+1)}\hbar$$
where J is the total atomic quantum number whose values range from $(L+S)$ down to $(L-S)$ in units. Also $J_z = m_J\hbar$ where m_J has the $2J+1$ values $J, \ldots, -J$ in a strong field.

This is the basis of **LS** or Russell–Saunders coupling.

Returning again to our helium p, d case for which we saw $S = 0, 1$ and $L = 3, 2$ and 1, we see that LS coupling gives two distinct sets of levels:

$S = 0$,   multiplicity $R = 2S + 1 = 1$ (singlets);
$S = 1$,   multiplicity $R = 2S + 1 = 3$ (triplets);

i.e. all levels in helium are *either* singlets or triplets. With each S value we can combine the three L values, so that for the singlet states $L = J$ there would be three

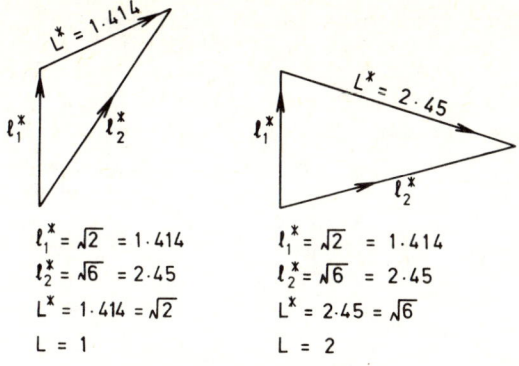

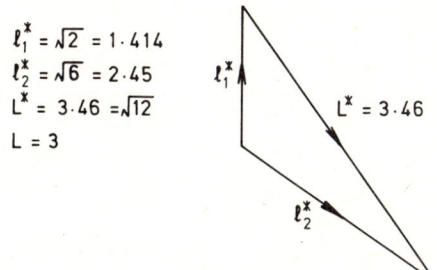

Fig. 12.2 ll coupling for $l_1 = 1$ and $l_2 = 2$. The three cases drawn to scale.

cases $J = 3, 2, 1$ since $S = 0$ with $J^* = \sqrt{J(J+1)}$ in each case. The singlet spectroscopic states of these terms are
$$^1P_1, \quad ^1D_2, \quad ^1F_3.$$
For the triplet states $S = 1$, giving the possible combinations
$$S = 1, \quad L = 1, \quad J = 2, 1, 0,$$
$$S = 1, \quad L = 2, \quad J = 3, 2, 1,$$
$$S = 1, \quad L = 3, \quad J = 4, 3, 2,$$
and the spectroscopic states of these terms are
$$^3P_{2,1,0}, \quad ^3D_{3,2,1}, \quad ^3F_{4,3,2}$$
respectively.

The vector diagrams for the three F states are drawn to scale in Fig. 12.3, drawn to scale

Similar diagrams can be constructed for the D and P states.

Table 12.1 summarizes the quantum notation we have used for two-electron atoms.

12.3 The Helium Spectrum

We have seen that the energy levels of helium are either singlets or triplets, due to Russell–Saunders coupling. Transitions can only take place within these two

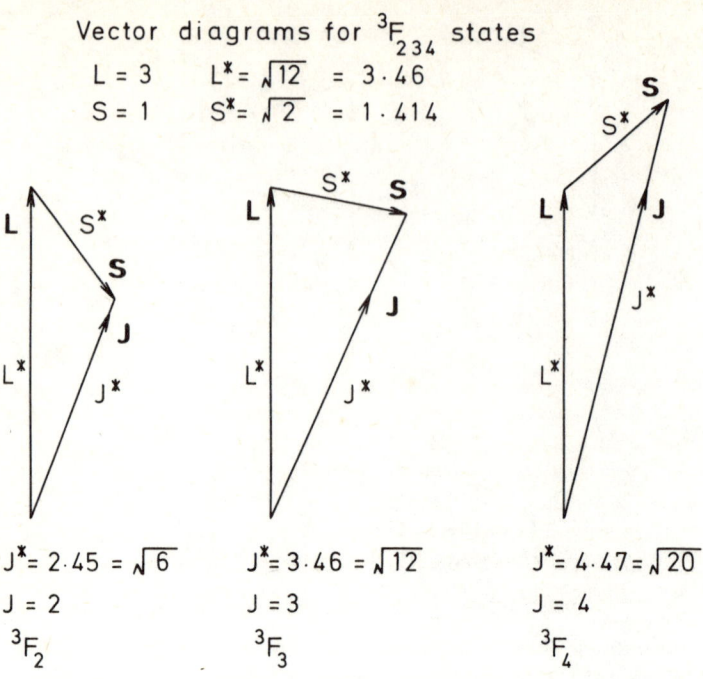

Fig. 12.3 **L, S, J** vector diagram for the ^3F states of helium.

distinct systems, i.e. singlet–singlet, or triplet–triplet, but never between the singlet and triplet systems. Some energy levels and transitions are shown in Fig. 12.4, which has to be compared with Fig. 8.1 for the sodium atom.

The selection rules are

$$\Delta S = 0,$$
$$\Delta L = \pm 1,$$
$$\Delta J = \pm 1 \text{ or } 0 \text{ but not } 0 \rightarrow 0.$$

The spectra of these two systems are quite separate, there being no intercombination lines for helium. However, in heavier two-electron atoms, there are some strong intercombination lines, the best known being that of mercury line at 253·6 nm, which is the 6^1S_0–6^3P_1 transition.

There are no transitions to the 1^1S_1 level, which presumably does not exist. This is a simple 'proof' of the Pauli exclusion principle since 1^1S_1 implies that $n=1, L=0, S=1, J=1$ and this means that the two electrons have the same spin magnetic quantum number $m_s = \frac{1}{2}$ and these are parallel to give $M_s = 1$ and $S = 1$. The two electrons cannot have the same four quantum numbers, hence the complete absence of the 1^1S_1 level.

In Fig. 12.5 we show the energy levels of helium and some typical transitions. All the lines of the spectrum are singlets, triplets or double triplets. The double triplets are due to transitions within the triplet levels which do not involve S

TABLE 12.1
Two-Electron Atoms — Quantum Relations and Russell–Saunders Coupling

Atomic state	Atomic state ≡ Combined electronic states, indicated by upper case letter, e.g. P
Quantum numbers	n, L, J, S, M_L, M_J for atom, where possible values are $L=(l_1+l_2),\ldots,(l_1-l_2),$ $S=(s_1+s_2),\ldots,(s_1-s_2),$ $J=(L+S),\ldots,(L-S),$ for each LS combination, all in integral steps.
Angular momentum vectors and components	**L, S, J** L_x, L_y, L_z
Magnitudes	$L^*\hbar=\sqrt{L(L+1)}\hbar \quad L_z=M_L\hbar$ $S^*\hbar=\sqrt{S(S+1)}\hbar \quad S_z M_S\hbar$ $J^*\hbar\sqrt{J(J+1)}\hbar \quad J_z M_J\hbar$
Vector relationships	$\mathbf{L}=\mathbf{l}_1+\mathbf{l}_2$ using * magnitudes of l_1 and l_2 for vector triangle. $\mathbf{S}=\mathbf{s}_1+\mathbf{s}_2$ using * magnitudes of s_1 and s_2 for vector triangle. Then $\mathbf{J}=\mathbf{L}+\mathbf{S}$ using * magnitudes of **L** and **S** for vector triangle.
Spectroscopic Notation States Terms	S, P, D, F,... $L=0, 1, 2, 3,\ldots$ $n^R L_J$
Multiplicity of terms	$R=2S+1$ $=1$ or 3
Selection rules	$L \quad \Delta L=\pm 1$ $S \quad \Delta S=0$ $J \quad \begin{cases}\Delta J=\pm 1, 0\\ 0 \not\to 0\end{cases}$

The transition $J_1=0 \to J_2=0$ is forbidden because any radiated photon carries away one $\hbar$ unit of angular momentum. The conservation of angular momentum principle thus precludes the possibility of the initial and final J values being identically zero. $\Delta J=0$ is allowed because the **J** vector can rotate through 180° and so emit a photon.

---

levels. However, these fine structure components are often too close together to be resolved. The 587·5 nm double triplet line was found more than a hundred years ago in the solar spectrum near the sodium $D_1 D_2$ absorption lines. It was therefore labelled D_3 and eventually shown to belong to a new element then unknown on earth — appropriately named helium.

The complete helium spectrum has been thoroughly investigated and the many vector and quantization rules we have discussed have been well authenticated.

This is also true of the simple alkaline earth spectra.

The elements Be, Mg, Ca, ..., etc., follow immediately after the alkaline metals Li, Na and K, and each have two optically active valence electrons. Their spectra, therefore, are made up of a singlet system and a triplet system, just as we saw for helium. There are, however, some intercombination lines. The term values within a single series can again be written as a Rydberg–Ritz formula, as for sodium, and

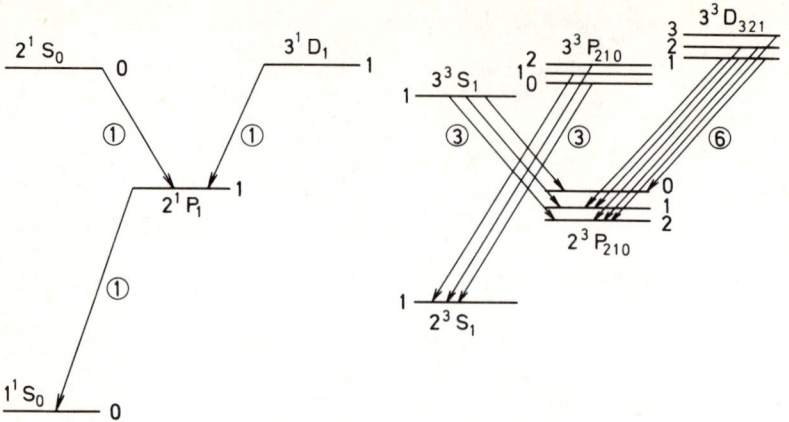

Fig. 12.4 Singlet and triplet transitions in helium applying the rule $\Delta J = \pm 1, 0$. 3D–2P transitions are double triplets.

the lines described as principal, sharp and diffuse series. There are corresponding quantum defects. Coupling is usually by the Russell–Saunders rule.

Proceeding across the periodic table to Group III, we again encounter two independent-term systems, this time doublets ($S = \frac{1}{2}$) and quartets ($S = \frac{3}{2}$) and in Group IV we get singlets ($S = 0$), triplets ($S = 1$) and quintets ($S = 2$) in which **jj** coupling plays an increasingly important role. The vector model of the atom is able to give an interpretation of these complex atomic spectra.

12.4 jj Coupling

We have dealt with the Russell–Saunders coupling scheme because this holds for many of the lighter elements where electrostatic interactions predominate. For the heavier elements, however, the magnetic ineractions also become effective. In the transition from Russell–Saunders coupling, the spin–spin and orbit–orbit couplings are gradually weakened as the *atomic* vectors **L** and **S** are reduced to zero, while the electronic **j**-vectors become more important. Eventually, these **j** vectors couple together to give the final atomic **J** vector for the total atomic angular momentum. This is the basis of **jj** coupling. Thus for Russell-Saunders coupling for two electrons we have

$$\mathbf{L} = \mathbf{l}_1 + \mathbf{l}_2,$$
$$\mathbf{S} = \mathbf{s}_1 + \mathbf{s}_2,$$
$$\mathbf{J} = \mathbf{L} + \mathbf{S}$$

and there are no **j** vectors, whereas for **jj** coupling for two electrons we have

$$\mathbf{j}_1 = \mathbf{l}_1 + \mathbf{s}_1,$$
$$\mathbf{j}_2 = \mathbf{l}_2 + \mathbf{s}_2,$$
$$\mathbf{J} = \mathbf{j}_1 + \mathbf{j}_2$$

and there are no **L** and **S** vectors.

Thus we see that **jj** coupling holds when there is strong spin–orbit coupling for each electron independently. In **jj** coupling the multiplicity rule no longer holds,

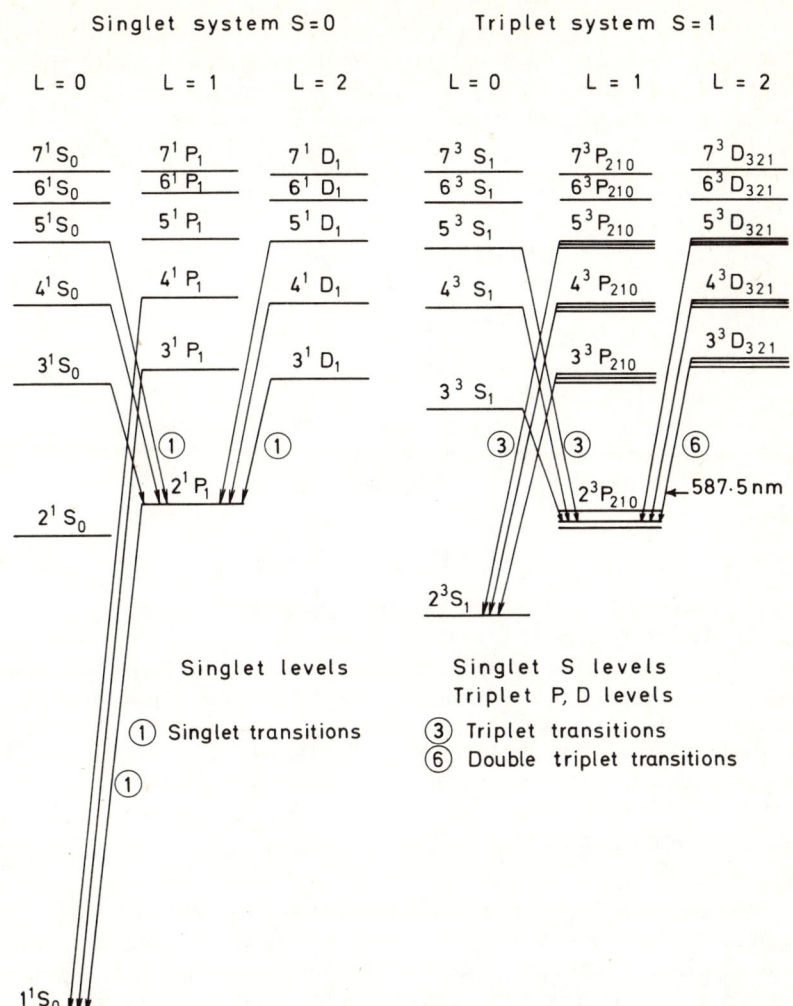

Fig. 12.5 Energy level diagram for helium.

but the quantum *number* rule $j = l \pm \frac{1}{2}$ still applies for each electron. It is found that **jj** coupling is more important in those cases where the s and p levels are inverted. Usually, the s level lies deeper than the p level, but in some of the heavier atoms, e.g. the elements C, Si, Ge and Sn of valency 4, this sequence is inverted and magnetic interactions seem to predominate over electrostatic interactions with the result that **jj** coupling holds.

In the analysis of complex spectra it is sometimes necessary to invoke coupling schemes intermediate between the **LS** and **jj** extremes in order to explain the experimental results.

It is interesting to note that in nuclear spectroscopy the nucleons of the lighter elements interact through **LS** coupling but as the atomic number of the element increases the results can only be explained on the basis of **jj** coupling between the nucleons. We shall deal with this in some detail in our discussion of theoretical nuclear structure, especially the shell model of the nucleus.

For those atoms with many optically active electrons we can extend the previous discussion by simple rules of addition. Thus, for Russell–Saunders coupling,

$$\mathbf{L} = \mathbf{l}_1 + \mathbf{l}_2 + \mathbf{l}_3 + \ldots = \sum_i \mathbf{l}_i,$$

$$\mathbf{S} = \mathbf{s}_1 + \mathbf{s}_2 + \mathbf{s}_3 + \ldots = \sum_i \mathbf{s}_i,$$

$$\mathbf{J} = \mathbf{L} + \mathbf{S},$$

and for **jj** coupling,

$$\mathbf{j}_1 = \mathbf{l}_1 + \mathbf{s}_1,$$
$$\mathbf{j}_2 = \mathbf{l}_2 + \mathbf{s}_2,$$
$$\mathbf{j}_3 = \mathbf{l}_3 + \mathbf{s}_3, \text{ etc.}$$

whence

$$\mathbf{J} = \mathbf{j}_1 + \mathbf{j}_2 + \mathbf{j}_3 + \ldots = \sum_i \mathbf{j}_i.$$

As the number of contributing electrons increases, the number of possible energy states increases, resulting in the complex spectra of the heavier atoms. The justification of these coupling schemes is in the analysis of such spectra for line frequency and intensity. We must also remember that the Pauli principle holds whatever the vector coupling in the atom. It is this fact which leads to an explanation of the periodic table shown in Fig. 12.6. This empirical grouping of elements, showing trends and repetitions of chemical properties, can be explained in terms of the Pauli principle and the electron configuration of the atoms, together with some empirical rules.

12.5 The Electronic Structure of the Elements and the Periodic Table

Beginning with hydrogen, we shall now show how the elements are built up in accordance with the Pauli principle, and how we can account for the periodicity of their properties. Hydrogen consists of a single electron circulating in an orbit about a single proton. For greatest stability this electron will occupy an orbit of least energy. That is $n=1$ and therefore $l=0$, $m_l=0$ and $m_s=+\tfrac{1}{2}$. It is convenient to represent this electron as $1s^1$, where the s indicates that $l=0$ and the superscript shows the number of electrons in this state.

For the next element, helium, we must increase the nuclear charge to two and introduce a second electron. The set of quantum numbers associated with the second electron must not be identical with that of the electron already present. Thus, although we can make $n=1$, $l=0$, $m_l=0$, as before, the spin must be different, so that $m_s=-\tfrac{1}{2}$. According to the Pauli principle no more electrons can now be associated with $n=1$ and the K shell to which this corresponds is now full, and is represented as $1s^2$. We shall see that a completed shell corresponds in each case to an inert gas so that the electron structure must be particularly stable. Lithium may now be formed from helium by increasing the nuclear charge to

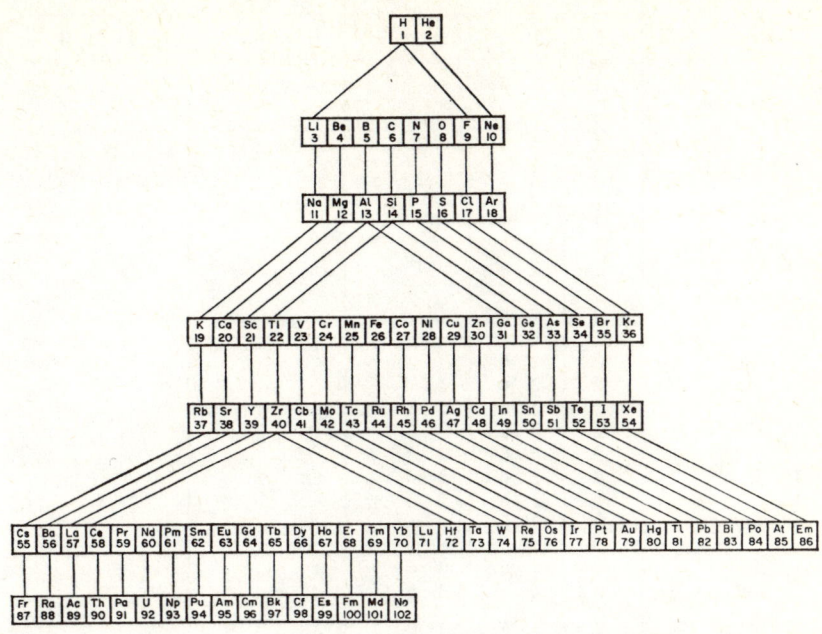

Fig. 12.6 The periodic system (from *Encyclopaedia Britannica*, vol. 17, p. 518, 1962).

three and adding a third electron to make the complete atom electrically neutral. With the third electron we begin to build the second or L shell for which $n=2$. The quantum numbers assigned to this electron in accordance with least energy considerations are $n=2$, $l=1$, $m_l=0$, $m_s=-\frac{1}{2}$, so that the configuration is $1s^22s^1$. The next element, beryllium must arise from the addition of a fourth electron having $n=2$, $l=0$, $m_l=0$, $m_s=+\frac{1}{2}$, and is represented as $1s^22s^2$. Six further elements from boron to neon are obtained as a further six electrons are added, each having $n=2$, $l=1$, but differing in that m_l can take three values ± 1 or 0, and with each of these two m_s values $\pm\frac{1}{2}$ may be associated, making six in all. Thus at neon the L shell is complete, having a total of eight electrons. The electron configuration of neon is therefore $1s^22s^22p^6$, the p indicating electrons for which $l=1$. We see that the complete shell again corresponds to maximum stability, as shown by the fact that the inert gases do not form stable chemical compounds. This process is summarized in Table 12.2. The addition of another electron to neon yields sodium, but this electron has to enter the third shell known as the M shell. The quantum numbers are therefore $n=3$, $l=0$, $m_l=0$, $m_s=-\frac{1}{2}$, and its configuration is $1s^22s^22p^63s^1$. In this fashion eight more electrons are added as we proceed from sodium to argon. Argon is a rare gas and its electron configuration $1s^22s^22p^63s^23p^6$ represents another stable condition. With these simple considerations we have been able to account for the first three periods of the periodic table. The M shell, however, is not full. In accordance with the Pauli principle, there is still room for a further ten electrons in the 3d sub-shell, for

173

TABLE 12.2 Electronic Configurations of the Elements

Element	K	L	M	N	O
	1s	2s 2p	3s 3p 3d	4s 4p 4d 4f	5s 5p 5d 5f 5g
1. H	1				
2. He	2				
3. Li	2	1			
4. Be	2	2			
5. B	2	2 1			
6. C	2	2 2			
7. N	2	2 3			
8. O	2	2 4			
9. F	2	2 5			
10. Ne	2	2 6			
11. Na	2	2 6	1		
12. Mg	2	2 6	2		
13. Al	2	2 6	2 1		
14. Si	2	2 6	2 2		
15. P	2	2 6	2 3		
16. S	2	2 6	2 4		
17. Cl	2	2 6	2 5		
18. Ar	2	2 6	2 6		
19. K	2	2 6	2 6	1	
20. Ca	2	2 6	2 6	2	
21. Sc	2	2 6	2 6 1	2	
22. Ti	2	2 6	2 6 2	2	
23. V	2	2 6	2 6 3	2	
24. Cr	2	2 6	2 6 5	1	
25. Mn	2	2 6	2 6 5	2	
26. Fe	2	2 6	2 6 6	2	
27. Co	2	2 6	2 6 7	2	
28. Ni	2	2 6	2 6 8	2	
29. Cu	2	2 6	2 6 10	1	
30. Zn	2	2 6	2 6 10	2	
31. Ga	2	2 6	2 6 10	2 1	
32. Ge	2	2 6	2 6 10	2 2	
33. As	2	2 6	2 6 10	2 3	
34. Se	2	2 6	2 6 10	2 4	
35. Br	2	2 6	2 6 10	2 5	
36. Kr	2	2 6	2 6 10	2 6	
37. Rb	2	2 6	2 6 10	2 6	1
38. Sr	2	2 6	2 6 10	2 6	2
39. Y	2	2 6	2 6 10	2 6 1	2
40. Zr	2	2 6	2 6 10	2 6 2	2
41. Nb	2	2 6	2 6 10	2 6 4	1
42. Mo	2	2 6	2 6 10	2 6 5	1
43. Tc	2	2 6	2 6 10	2 6 (5)	(2)
44. Ru	2	2 6	2 6 10	2 6 7	1
45. Rh	2	2 6	2 6 10	2 6 8	1
46. Pd	2	2 6	2 6 10	2 6 10	
47. Ag	2	2 6	2 6 10	2 6 10	1
48. Cd	2	2 6	2 6 10	2 6 10	2
49. In	2	2 6	2 6 10	2 6 10	2 1
50. Sn	2	2 6	2 6 10	2 6 10	2 2
51. Sb	2	2 6	2 6 10	2 6 10	2 3
52. Te	2	2 6	2 6 10	2 6 10	2 4
53. I	2	2 6	2 6 10	2 6 10	2 5
54. Xe	2	2 6	2 6 10	2 6 10	2 6
	2	8	18		

Element	K L M	N 4s 4p 4d 4f	O 5s 5p 5d 5f	P 6s 6p 6d	Q 7s
55. Cs	2 8 18	2 6 10	2 6	1	
56. Ba	2 8 18	2 6 10	2 6	2	
57. La	2 8 18	2 6 10	2 6 1	2	
58. Ce	2 8 18	2 6 10 (2)	2 6	(2)	
59. Pr	2 8 18	2 6 10 (3)	2 6	(2)	
60. Nd	2 8 18	2 6 10 (4)	2 6	(2)	
61. Pm	2 8 18	2 6 10 (5)	2 6	(2)	
62. Sm	2 8 18	2 6 10 6	2 6	2	
63. Eu	2 8 18	2 6 10 7	2 6	2	
64. Gd	2 8 18	2 6 10 7	2 6 1	2	
65. Tb	2 8 18	2 6 10 8	2 6 (1)	(2)	
66. Dy	2 8 18	2 6 10 10	2 6 (1)	(2)	
67. Ho	2 8 18	2 6 10 11	2 6 (1)	(2)	
68. Er	2 8 18	2 6 10 12	2 6 (1)	(2)	
69. Tm	2 8 18	2 6 10 13	2 6	2	
70. Yb	2 8 18	2 6 10 14	2 6	2	
71. Lu	2 8 18	2 6 10 14	2 6 1	2	
72. Hf	2 8 18	2 6 10 14	2 6 2	2	
73. Ta	2 8 18	2 6 10 14	2 6 3	2	
74. W	2 8 18	2 6 10 14	2 6 4	2	
75. Re	2 8 18	2 6 10 14	2 6 5	2	
76. Os	2 8 18	2 6 10 14	2 6 6	2	
77. Ir	2 8 18	2 6 10 14	2 6 7	2	
78. Pt	2 8 18	2 6 10 14	2 6 9	1	
79. Au	2 8 18	2 6 10 14	2 6 10	1	
80. Hg	2 8 18	2 6 10 14	2 6 10	2	
81. Tl	2 8 18	2 6 10 14	2 6 10	2 1	
82. Pb	2 8 18	2 6 10 14	2 6 10	2 2	
83. Bi	2 8 18	2 6 10 14	2 6 10	2 3	
84. Po	2 8 18	2 6 10 14	2 6 10	2 4	
85. At	2 8 18	2 6 10 14	2 6 10	2 5	
86. Em	2 8 18	2 6 10 14	2 6 10	2 6	
87. Fr	2 8 18	2 6 10 14	2 6 10	2 6	1
88. Ra	2 8 18	2 6 10 14	2 6 10	2 6	2
89. Ac	2 8 18	2 6 10 14	2 6 10 1	2 6	2
90. Th	2 8 18	2 6 10 14	2 6 10 2	2 6	2
91. Pa	2 8 18	2 6 10 14	2 6 10 3	2 6	2
92. U	2 8 18	2 6 10 14	2 6 10 4	2 6	2
93. Np	2 8 18	2 6 10 14	2 6 10 5	2 6	2
94. Pu	2 8 18	2 6 10 14	2 6 10 6	2 6	2
95. Am	2 8 18	2 6 10 14	2 6 10 7	2 6	2
96. Cm	2 8 18	2 6 10 14	2 6 10 8	2 6	2
97. Bk	2 8 18	2 6 10 14	2 6 10 9	2 6	2
98. Cf	2 8 18	2 6 10 14	2 6 10 10	2 6	2
99. Es	2 8 18	2 6 10 14	2 6 10 11	2 6	2
100. Fm	2 8 18	2 6 10 14	2 6 10 12	2 6	2
101. Md	2 8 18	2 6 10 14	2 6 10 13	2 6	2
102. (No)	2 8 18	2 6 10 14	2 6 10 14	2 6	2
103. Lw	2 8 18	2 6 10 14	2 6 10 14	2 6 1	2
104. Ku(Rf)	2 8 18	2 6 10 14	2 6 10 14	2 6 2	2
105. Ha	2 8 18	2 6 10 14	2 6 10 14	2 6 3	2
106.	2 8 18	2 6 10 14	2 6 10 14	2 6 4	2
107.	2 8 18	2 6 10 14	2 6 10 14	2 6 5	2

which $n=3$, $l=2$. However, the next electron added goes into the N shell with $n=4$ giving potassium, since this represents a condition of smaller energy than placing it in the M shell with $l=2$. With the next element, calcium, the electron also goes into the N shell, giving $1s^2 2s^2 2p^6 3s^2 3p^6 (3d^0) 4s^2$. After this, however, the M shell is completed by the addition of ten further electrons, which brings us to zinc. This is the first interpolated group of the periodic table. After zinc the electron configurations develop in the normal way until we reach the rare gas, krypton, with eight electrons in the N shell.

It now becomes evident that, not only can we account for the existence of eight elements in the first periods of the table, but we can also explain how the interpolated groups arise. Further study of Table 12.2 will show how the next period of eighteen elements from rubidium to xenon develops, with a second interpolated group from yttrium to palladium as further additions to the N shell completes the 4d sub-shell $n=4$, $l=2$. This still leaves the 4f ($n=4$, $l=3$) sub-shell empty and the N shell is finally completed by the rare earth elements from cerium to ytterbium. Thus energy considerations together with quantum theory and the Pauli principle are able to account in quire a remarkable way for the periodic system of the elements.

In general we now see that for each n there are n values of l, viz. $0, 1, 2, \ldots, (n-1)$, so that the total number of substrates for each principal quantum number n is

$$N = \sum_{0}^{n-1} 2(2l+1), \quad n \text{ terms}$$
$$= 2[1+3+5+\ldots+(2n-1)]$$
$$= 2\left[n\left\{\frac{1+(2n-1)}{2}\right\}\right] \quad \text{by arithmetic progression}$$
$$= 2n^2.$$

The first principal quantum state $n=1$ is therefore complete with two electrons, the next $n=2$ closes with eight electrons, the next with 18 and so on. However, this simple rule fails for larger values of n for the energy reasons mentioned above. In many cases outer shells are filled leaving inner shells still incompletely full and the $2n^2$ rule does not extend beyond $n=4$. It is significant that the inert gases after helium all have $s^2 p^6$ outer configurations. Notwithstanding these difficulties, it is clear that the exclusion principle helps to explain the configurations of the lighter elements. For the heavier elements some empirical rules have to be adopted.

12.6 The Periodic Table — Some Empirical Rules

The details of the electron configurations of the elements are governed by two factors:

(1) the most stable configuration for any atom is that for which the total energy is a minimum;
(2) the quantum number rules must be obeyed, including the Pauli principle.

The order in which energy levels appear in atoms as the atomic number is increased can be deduced from the analysis of spectroscopic data. The ascending order, from the deep tightly bound 1s electrons to the more loosely bound 7s electrons can be deduced from the n and l quantum numbers. The level with the lowest $(n+l)$ sum fills first, as electrons are added. Whenever two levels have the *same* $(n+l)$ value, e.g. 4p and 3d, that with the *lower n* value fills first, i.e. 3d before 4p. This arrangement is shown in Table 12.3, which also shows the occupation number of each level.

If we take Sn, $Z=50$, as an example we would expect its configuration on a simple increasing n, l basis to be

$$Z=50, \quad 1s^2 2s^2 2p^6 3s^2 3p^6 3d^{10} 4s^2 4p^6 4d^{10} 4f^4,$$

whereas from the $(n+l)$ sum rule and Table 12.3 we can see that the configuration order is

$$Z=50, \quad 1s^2 2s^2 2p^6 3s^2 2p^6 3d^{10} 4s^2 4p^6 5s^2 4d^{10} 5p^2,$$

as in Table 12.2.

There are other peculiarities, such as the way in which the inner 3d level of the transition elements is gradually filled while the outer 4s level, except chromium $(Z=24)$ is full. Other deviations are found and some of them can be explained by another empirical rule, due to Hund. This states that the distribution of electrons in p, d and f states is always such that as many levels as possible are filled with single electrons before pairing of opposite spins in a level takes place. Thus vanadium $(Z=23)$ has

$$Z=23, \quad 1s^2 2s^2 2p^6 3s^2 3p^6 4s^2 3d^3$$

according to Table 12.2. The spins of the last five electrons are

$$\underset{4s^2}{\uparrow\downarrow} \qquad \underset{3d^3}{\uparrow\ \uparrow\ \uparrow}$$

by the Hund rule. When another electron is added, to give chromium $(Z=24)$, the outer configuration is not $4s^2 3d^4$ as one might expect, i.e. not

$$\underset{4s^2}{\uparrow\downarrow} \qquad \underset{3d^4}{\uparrow\ \uparrow\ \uparrow\ \uparrow}$$

but $4s^1\ 3d^5$ as

$$\underset{4s^1}{\uparrow} \qquad \underset{3d^5}{\uparrow\ \uparrow\ \uparrow\ \uparrow\ \uparrow}$$

to maximize the number of unpaired spins. The next element is manganese $(Z=25)$, for which the 4s level is now restored to $4s^2$ with five 3d electrons still unpaired.

Further study of the complete Table 12.2 shows that the elements can be arrangements in five major subdivisions. These are:

(a) *the inert gases*, which always finish with $s^2 p^6$ electrons, except helium s^2;
(b) the so-called *s-block* elements, which have s^1 or s^2 as their outer levels, the inner levels being completely filled;
(c) the *p-block* elements with outermost p-levels containing 1–5 electrons, with complete inner shells;
(d) the *d-block* elements, which include the transition elements in which the inner d-shells fill up from 1–10 electrons, with the *outer* s shells full (explained by Hund's rule);

TABLE 12.3
Filling Order of Electron Energy Levels

s-States	p-States	d-States	f-States	n	$(n+l)$	Occupation number
		6d		6	8	10
			5f	5	8	14
7s				7	7	2
	6p			6	7	6
		5d		5	7	10
			4f	4	7	14
6s				6	6	2
	5p			5	6	6
		4d		4	6	10
5s				5	5	2
	4p			4	5	6
		3d		3	5	10
4s				4	4	2
	3p			3	4	6
3s				3	3	2
	2p			2	3	6
2s				2	2	2
1s				1	1	2

Energy (-ve) ↓   Filling order ↑

(e) the *f-block* elements, similar to the d-block elements but here the inner f levels fill from 1–14 electrons, with the *outer* s shells again being full. The f-block elements include the rare earths (lanthanides) and their chemical cousins the transuranic elements (actinides).

These blocks are shown in the modern version of the periodic table (see Fig. 12.7).

In this section we have shown the arrangement of the elements in the periodic table can be explained by the theoretical constraint of the Pauli exclusion principle together with a set of empirical rules which are more difficult to explain. It is obvious that the properties of a new element can be predicted from its position in the periodic table, and it was this fact that helped to identify many of the new elements resulting from the fission process. Thus the new transuranic elements lawrencium (Lw, $Z=103$) khurchatovium (Ku, $Z=104$) and hahnium (Ha, $Z=105$) now start a new set of transition elements, as shown in Fig. 12.7, and their chemical properties will be appropriate to their positions in the d-block. The unnamed elements 106 and 107 have also been reported and their chemical properties can be predicted from their positions in the periodic table.

	Inert gases	s^2 s-block			f^{14} f block															d^{10} d-block											p^6 p-block					
			1	2		1	2	3	4	5	6	7	8	9	10	11	12	13	14		1	2	3	4	5	6	7	8	9	10		1	2	3	4	5
H	He	2s	Li	Be																											2p	B	C	N	O	F
	Ne	3s	Na	Mg																											3p	Al	Si	P	S	Cl
	Ar	4s	K	Ca																3d	Sc	Ti	V	Cr	Mn	Fe	Co	Ni	Cu	Zn	4p	Ga	Ge	As	Se	Br
	Kr	5s	Rb	Sr																4d	Y	Zr	Nb	Mo	Tc	Ru	Rh	Pd	Ag	Cd	5p	In	Sn	Sb	Te	I
	Xc	6s	Cs	Ba	4f	La	Ce	Pr	Nd	Pm	Sm	Eu	Gd	Tb	Dy	Ho	Er	Tm	Yb	5d	Lu	Hf	Ta	W	Re	Os	Ir	Pt	Au	Hg	6p	Tl	Pb	Bi	Po	At
	Rn	7s	Fr	Ra	5f	Ac	Th	Pa	U	Np	Pu	Am	Cm	Bk	Cf	Es	Fm	Md	No	6d	Lw	Ku	Ha	?	?											

← Lanthanides / Actinides → ← Transition Elements →

← Metals →

Fig. 12.7 Modern version of periodic table.

12.7 Hyperfine Structure and Nuclear Spin Angular Momentum

In Chapter 7 we saw that the singlet lines predicted by Bohr showed a hyperfine structure due to the presence of the heavy hydrogen isotope, deuterium. This is a simple example of a general isotopic hyperfine structure effect in spectroscopy. However, hyperfine structure can also be found in the spectra of single isotopes of elements, e.g. ^{184}Hg. The wavelength differences in the hyperfine structure of single isotopes are very much smaller than the electron spin fine structure discussed in Chapter 8. Moreover this hyperfine structure does not appear to be related in any systematic manner to the periodic table, so that it seems unlikely that it is associated with outer electron transitions. Pauli suggested that this type of hyperfine structure could be due to the interaction between the atomic angular momentum and a nuclear spin angular momentum via their respective mangetic moments. For nuclear spin the vector is **I** and its magnitude is given by $I^*\hbar = \sqrt{I(I+1)}\hbar$, where I is the corresponding quantum number.

The spin of the nucleus is compounded of the spins of the nucleons (i.e. the protons and neutrons, each of spin quantum number $\frac{1}{2}$), as we shall see when we deal with models of the nucleus in Chapter 19. The symbol **I** refers to the total final angular momentum of the nucleus, called 'nuclear spin' although it can contain an orbital component. The final angular momentum of *the whole atom* is then given by $\mathbf{F}=\mathbf{I}+\mathbf{J}$, where **J** is the electronic contribution. Then $F^*\hbar = \sqrt{F(F+1)}\hbar$, where F is the quantum number and allowed values of F are from $(J+I)$ to $(J-I)$ in unit steps. Each level of given J then splits into $(2I+1)$ hyperfine levels when $i \leqslant J$, or $(2J+1)$ hyperfine levels when $i \geqslant J$. Hence the number of corresponding magnetic quantum numbers M_I, associated with each I, is

$$M_I = I, (I+1), \ldots, -(I-1), -I.$$

Both I and M_I can take half-integral or integral values, but there are always $2I+1$ possible values.

Nuclear spin splitting is very small compared with electron spin splitting since the nuclear magneton is only about $\frac{1}{1840}$ × the Bohr electron magneton. The small wavelength differences involved require spectroscopes of high resolution.

The selection rules for F in hyperfine structure follow the same pattern as those for J in electron fine structure:

$$\Delta F = \pm 1, 0.$$

Each hyperfine structure line is also controlled by the electronic selection rules

$$\Delta J = \pm 1, 0,$$
$$\Delta L = \pm 1,$$

as already discussed.

Following the pioneer work of Stern and Gerlach, the molecular beam method was developed to measure the magnetic moments of nuclei. By analogy with the electron case, the nuclear magnetic moment μ_I is related to the nuclear spin quantum number I by $\mu_I = g\mu_N\sqrt{I(I+1)}$, where g is an appropriate nuclear splitting factor and μ_N is the *nuclear* magneton given by $\mu_N = e\hbar/2m_p$, where m_p is the mass of the proton. Its value is $\mu_N = 5\cdot505\,824 \times 10^{-27}$ J T^{-1} compared with the electron Bohr magnetic $\mu_B = 9\cdot724\,078 \times 10^{-24}$ J T^{-1}. By resonance techniques the molecular beam method determines g, so that if I is known the magnetic moment μ_I is found. The interaction between μ_I and μ_J causes the coupling between **I** and **J** to give the hyperfine structure splitting.

It is interesting to note that $\mu_p/\mu_N = 2{\cdot}792\,84$, $\mu_n/\mu_N = -1{\cdot}913\,15$ and $\mu_d/\mu_N = 0{\cdot}857\,41$, showing that the deuteron is not quite the simple addition of a neutron and a proton.

Problems

(*The problem marked with an asterisk is solved in full at the end of the section*)

12.1 State and explain the Pauli exclusion principle. Use it to determine the maximum number of electrons which can occupy the K, L and M shells in an atom.

Show also how the principle may be used to interpret the interpolated groups of the periodic table of the elements.

12.2 Show how the Pauli principle limits the number of electrons in the nth shell to $2n^2$.

12.3 Write down the electron configurations in the normal state of helium, neon, argon and krypton. ($1s^2$, $1s^2 2s^2 2p^6$, $1s^2 2s^2 2p^6 3s^2 3p^6$, $1s^2 2s^2 2p^6 3s^2 3p^6 3d^{10} 4s^2 4p^6$)

12.4 Use the vector model of the atom to show how the multiplicity in atomic energy levels depends upon the group in the periodic table to which an element belongs.

12.5 Which of the following properties of elements change periodically with atomic number? Explain why: (*a*) atomic volume, (*b*) specific heat, (*c*) atomic heat, (*d*) ionization potential, (*e*) valency.

12.6 Use the Pauli principle to determine the lowest energy state terms of hydrogen, helium, lithium and beryllium. ($^2S_{1/2}$, $^1S_0$, $^2S_{1/2}$, $^1S_0$)

12.7 Use the Pauli principle to show that the resultant angular momenta, associated with the quantum numbers s and j, are each equal to zero for a closed atomic subshell.

12.8 Discuss the meaning of each symbol in the spectroscopic notation ^rL_J used for describing an atomic energy level.

An atomic energy level is designated by $1s^2 2s 2p^{2\,4}P_{3/2}$ in the usual notation. What information about the atom is contained in this designation?

What would you expect to be the ground state of this atom? [N]

12.9* A two-electron atom in an excited state has one electron in a d-state and one in an f-state, coupled according to the LS scheme. Show that there are 20 possible atomic terms which are either singlets or triplets. Write them out.

Determine the orbital quantum number of the atom corresponding to the case when the orbital angular momenta of the electrons are separated by 45°. [N]

Solution to Problem

12.9 From $l_1 = 2$ and $l_2 = 3$ we have the possible L values are 5, 4, 3, 2 and 1 corresponding to H, G, F, D and P states. For two electrons $S = 1$, 0 corresponding to triplet and singlet states. We now find J values from $J = L + S$.

L	S	J	States	S	J=L	States
5	1	6, 5, 4	$^3H_{654}$	0	5	$^1H_5$
4	1	5, 4, 3	$^3G_{543}$	0	4	$^1G_4$
3	1	4, 3, 2	$^3F_{432}$	0	3	$^1F_3$
2	1	3, 2, 1	$^3D_{321}$	0	2	$^1D_2$
1	1	2, 1, 0	$^3P_{210}$	0	1	$^1P_1$
			15 states			5 states

Total 20 states triplets and singlets only.

When l_1 and l_2 are separated by 45° we have

$$L^{*2} = l_1^{*2} + l_2^{*2} - 2l_1^* l_1^* \cos 45$$
$$= 2 \times 3 + 3 \times 4 - 2\sqrt{2 \times 3}\sqrt{3 \times 4}\frac{1}{\sqrt{2}}$$
$$= 6 + 12 - \sqrt{144}$$
$$= 6$$

Now $L^{*2} = L(L+1)$ so that $L = 2$.

Chapter 13

The Zeeman Effect

13.1 Introduction

We have referred to the Zeeman effect several times already. This is because it is of prime importance in confirming experimentally many of the vectorial effects in atom mechanics which are the result of quantum-mechanical theory. Together with the Stern–Gerlach experiment the Zeeman experiemnt confirms the existence of all the magnetic quantum numbers and the space properties of the corresponding vectors. It also gave the earliest value for the charge-to-mass ratio, e/m, of the bound electron. This was a better numerical result than that of J. J. Thomson for the free electron.

In passing it is worth noting that the early work on the Zeeman effect was done long before the advent of the quantum theory of the atom and the identification of quantum numbers. In fact Lorentz was able to give a classical explanation of the normal Zeeman effect in terms of an atom containing a moving charge of unspecified origin.

13.2 The Normal Zeeman Effect

In 1896 Zeeman found that when a sodium flame was placed between the poles of a powerful electromagnet, the spectrum lines were broadened. As we have said, Lorentz explained this in terms of the classical electron theory of matter, by supposing that the periodic motions of the electric charges within the atom were modified by the magnetic field. This in turn led to additional frequencies of radiation so that each spectrum line was split into a number of components. The classical theory of Lorentz predicted that when a source of light S (Fig. 13.1) was placed in a magnetic field of flux density B each spectrum line would, when viewed perpendicular to the field, give rise to frequencies $v + \Delta v$ and $v - \Delta v$ in addition to the original frequency v of the line. When viewed along the direction of the magnetic field the same additional frequencies appeared, but the original frequency v was missing. That the new frequencies were polarized in accordance with the diagram was quickly confirmed by Zeeman.

Lorentz showed that $\Delta v = Be/4\pi m$, where e and m were respectively the electronic charge and mass. Such triplets were soon observed in the singlet spectra of cadmium and zinc and, knowing Δv and B, a value for e/m was calculated. This was the first measurement of e/m and agreed fairly well with the value published soon afterwards by J. J. Thomson. It also indicated that the negatively charged electrons within the atom were responsible for the radiation.

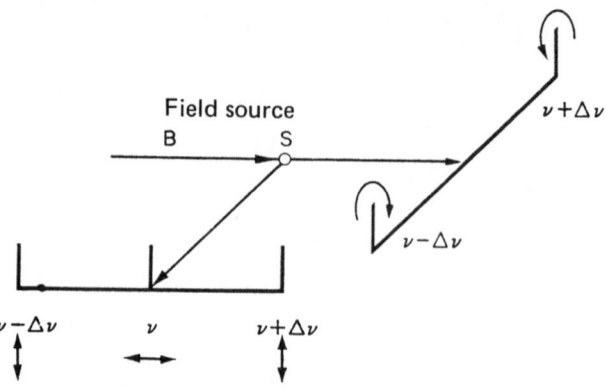

Fig. 13.1 The Zeeman effect, showing polarizations of lines.

It was also found that very few spectrum lines were split into simple triplets in accordance with Lorentz theory. Most lines showed more components and had displacements differing from Δv as calculated above. The sodium lines at 589·0 and 589·6 nm with which the Zeeman effect was first observed were shown to consist of six and four components respectively. This became known as the anomalous Zeeman effect, although it was by far the most common. The explanation had to await the development of quantum theory and the concept of electron spin nearly thirty years later. It is sufficient at this stage to realize that the anomalous Zeeman effect arises because the spin magnetic moment of the electron is $e\hbar/2m$ rather than $\frac{1}{2}e\hbar/2m$, whereas the normal Zeeman effect occurs when the net electron spin is zero. The effect of the field B is the same for all the spectral lines, i.e. equal splitting into three transverse components and two longitudinal components, as in Fig. 13.1.

13.3 Explanation of Zeeman Effect in Terms of Vector Model

An explanation of the simple Zeeman effect will be given in terms of the vector model, rather than the classical Lorentz model. Since it is the simple effect with which we are concerned we need only consider the orbital angular momentum $l^*\hbar$ (Fig. 13.2(a)) and its associated magnetic moment $l^*e\hbar/2m$ drawn anti-parallel because the electronic charge is negative. This magnetic moment interacts with the magnetic field of flux density B and gives rise to a couple C given by

$$C = Bl^*\hbar \frac{e}{2m} \sin\theta.$$

This in turn produces a change in the angular momentum in accordance with Newton's second law of motion, remembering that **l** is a vector such that

$$C = \frac{d}{dt}(l^*\hbar).$$

The momentum change so produced is perpendiculat to **l**, so that the *direction of* the vector is changed rather than its magnitude (Fig. 13.2(b)). This process is continuous, the axis of the couple being always perpendicular to the axis of the angular momentum. The angular momentum vector therefore precesses around

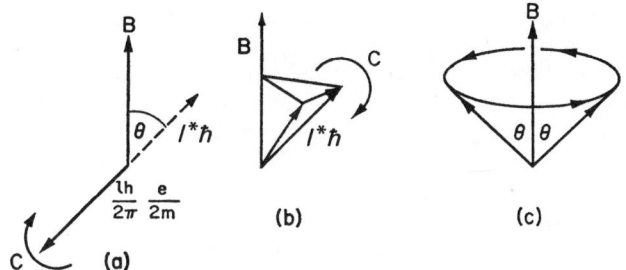

Fig. 13.2 Interaction of magnetic moment with external magnetic field.

the direction of the magnetic field, so that the vector traces out the surface of a cone of semi-vertical angle θ (Fig. 13.2(c)). This is analogous to a gyroscope spinning with its axis at an angle θ to the vertical under the action of the earth's gravitational field. Referring to the vector diagram in Fig. 13.3, we have

$$\delta\varphi = \frac{\delta(l^*\hbar)}{l^*\hbar \sin\theta}$$

The angular velocity of precession ω is given in the limit by

$$\omega = \frac{d\varphi}{dt} = \frac{d(l^*\hbar)/dt}{l^*\hbar \sin\theta}$$

and since

$$C = \frac{d}{dt}(l^*\hbar) = B(l^*\hbar)(e/2m)\sin\theta,$$

from above we have

$$\omega = \frac{B(l^*\hbar)(e/2m)\sin\theta}{(l^*\hbar)\sin\theta} = \frac{Be}{2m},$$

which is the so-called Larmor precessional velocity. This represents an increase of frequency due to B.

Thus we have the angular velocity of precession in terms of the magnetic field and e/m. In effect the angular velocity of the atomic system about the direction of the magnetic field has been changed by an amount $\omega = Be/2m$. This gives rise to

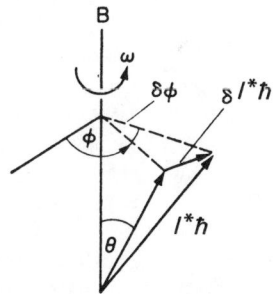

Fig. 13.3 Precession of magnetic moment in an external magnetic field.

185

two energy states from the original single state. The change in energy is given by $\Delta E = \frac{1}{2}I_B(\omega_B^2 - \omega_0^2)$, where ω_B, ω_0 are the angular velocities about the direction of B with and without the presence of the magnetic field and I_B is the moment of inertia about the axis parallel to B. From this we see that $\Delta E = \frac{1}{2}I_B(\omega_B + \omega_0)(\omega_B - \omega_0)$ and, since the change in angular velocity ω is small compared with either ω_B or ω_0, we may write $\omega_B + \omega_0 \simeq 2\omega_0$ and $\omega_B - \omega_0 = \omega$ so that $\Delta E = I_B\omega_0\omega = I_B\omega_0 Be/2m$.

The Stern–Gerlach experiment has shown that an atomic magnet in a magnetic field is subject to space quantization such that it can only set at certain discrete angles with respect to the magnetic field. These are given by $\cos\theta = m_l/l^*$, where $l^* = \sqrt{l(l+1)}$ numerically, and $2l+1$ values of θ are given by $m_l = -l, -(l-1), \ldots, 0, \ldots, +(l-1), +l$. Since the magnitude of the orbital angular momentum is $l^*\hbar$, its component parallel to **B** is given by $l_z = l^*\hbar\cos\theta$, so that

$$l_z = I_B\omega_0 = l^*\hbar\cos\theta \equiv m_l\hbar$$

and the corresponding energy values are increased by

$$\Delta E = l_z\frac{Be}{2m}$$
$$= \mu_B m_l B,$$

where μ_B is the Bohr magneton, $e\hbar/2m$.

We now have to consider two energy states taking part in a transition before and after the application of field B. Let E_0' and E_0'' be the energy states in the absence of the magnetic field and E_B' and E_B'' their values when the field is applied. In each case the magnetic energy $\Delta E = \mu_B m_l B$ is added. Thus

$$E_B' = E_0' + \Delta E'$$
$$= E_0' + \mu_B B m_l'$$

and

$$E_B'' = E_0'' + \Delta E''$$
$$= E_0'' + \mu_B B m_l''.$$

On subtraction,

$$E_B' - E_B'' = E_0' - E_0'' + (m_l' - m_l'')\mu_B B$$

and so

$$\hbar\nu_B = \hbar\nu_0 + \Delta m_l \mu_B B,$$

where ν_B is the frequency of the Zeeman line and Δm_l is the change in the magnetic quantum number between the two Zeeman levels. The effect of the field B is to split the single l level into $2l+1$ magnetic levels of m_l, as required by the Schrödinger theory. Furthermore, the *spacing* of these m_l levels is given by

$$\hbar\nu_B - \hbar\nu_0 = \Delta m_l \mu_B B.$$

Now it is found experimentally that the frequency separation between neighbouring levels from the same l is always the same for constant B. This confirms the quantum mechanical *selection rule* for m_l, viz. $\Delta m_l = 0$ or ± 1, including $0 \to 0$ since neighbouring energy levels are 1 unit of m_l apart.

This is the simple theory of the normal Zeeman effect for spinless atoms, i.e. $S = 0$. We recognize that this can only be tested for two-electron atoms such as Zn or Cd. Figure 13.4 shows the Zeeman transitions for the cadmium 6^1D_2–5^1P_1 line. Note that the Lorentz theory cannot predict the polarization of the lines.

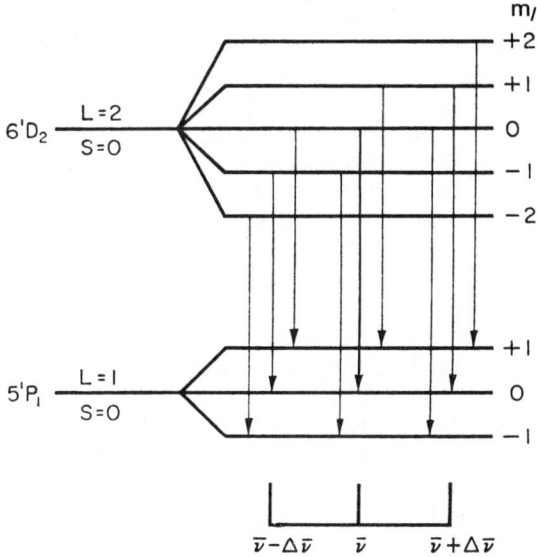

Fig. 13.4 The simple Zeeman effect for the cadmium line at 643·8 nm.

13.4 Zeeman Effect of Cadmium 643·8 nm Line

The way in which a single line becomes a triplet on application of a magnetic field is illustrated by the case of the red line of cadmium at 643·8 nm. For this line the spins of the two electrons are combined so that the resultant spin is zero giving rise to singlet terms. The transition occurs between the states 6^1D_2 and 5^1P_1. For the upper state $L=2$ and m_l takes values $+2, +1, 0, -1, -2$, as shown in Fig. 13.4, while in the lower state $L=1$ and m_l takes values $+1, 0, -1$. The spacing of the levels in wavenumbers is given by the formula

$$\Delta\bar{\nu} = \bar{\nu}_B - \bar{\nu}_0 = \Delta m_l \mu_B B/c\hbar$$
$$= \pm \mu_B B/c\hbar \quad \text{for} \quad \Delta m_l = \pm 1.$$

The levels are therefore equispaced.

By applying the selection rules we see that

(a) the levels are *all* equidistant irrespective of the atomic state, P or D;
(b) since $\Delta m_l = \pm 1, 0$ including $0 \to 0$, there are *nine* possible transitions.

These nine transitions reduce to three *lines*, owing to the equality in spacing just mentioned, whose wave numbers are $\bar{\nu} - \Delta\bar{\nu}$, $\bar{\nu}$ and $\bar{\nu} + \Delta\bar{\nu}$, as in Fig. 13.4. This is an example of the normal Zeeman triplet.

Furthermore, since the wavenumber difference between the inner and outer components is

$$\Delta\bar{\nu} = \frac{1}{c\hbar}\mu_B B$$

$$= \frac{e}{m4\pi c}B,$$

we can see how e/m for the atomic electron in its orbit was first measured. If $\Delta \bar{\nu}$ is measured from the Zeeman effect in a known field **B**, the only unknown is e/m. Zeeman's original value was $1\cdot 6 \times 10^{11}$ C kg^{-1} while that of Thomson for the free electron was 2×10^{11} C kg^{-1} in the single field experiment, and $0\cdot 7 \times 10^{11}$ C kg^{-1} in the zero deflection method described in Chapter 2. The best Zeeman result was $1\cdot 7591 \times 10^{11}$ C kg^{-1}, determined in 1948, and this is equal in value and in accuracy to the best Thomson value. We conclude therefore that the bound electron and the free electron are the same particle.

The present day value is
$$e/m = 1\cdot 758\ 804\ 5 \times 10^{11} \text{ C kg}^{-1}$$
obtained from time-of-flight experiments in which electrons accelerated through a known voltage are timed electronically over a known distance. This gives the velocity of the electrons and the e/m ratio is obtained from the energy equation.

13.5 The Anomalous Zeeman Effect and the Landé Splitting Factor

The early experiments soon showed that the normal Zeeman triplet was the exception rather than the rule. It was called 'normal' because it could easily be explained by the classical Lorentz electron theory without quantum numbers or the notion of electron spin. Thus the Zeeman–Lorentz triplet is always observed when there is no electron spin, as in the singlet lines of many-electron spectra, or when the orbital motion dominates. Most atoms do not show this, so that the majority of Zeeman patterns are not triplets but complex multiplets. Thus the D lines of sodium split into ten lines altogether, with four from D_1 and six from D_2. The Lorentz theory is inadequate in these cases and so the complex Zeeman effect is usually called the *anomalous* Zeeman effect. It is not difficult to see that this is due to an effective electron spin magnetic moment. This immediately reminds us that the peculiar magnetic property of the spinning electron is that its g factor is 2 (p. 156). We now discuss this as the Landé splitting factor.

To account fully for the Zeeman effect it is necessary to assume that the magnetic moment arising from the spin of the electron is given by $\mu_s = 2(\frac{1}{2}\mu_B)$. Assuming **LS** coupling, and applying this to all the electrons involved in an energy change within the atom, we have $\mu_L = L^* \mu_B$ and $\mu_s = 2S^* \mu_B$. Chapter 8 describes how we can account for the fine structure of spectra in terms of the quantum number J which controls the total angular momentum of the electrons. **J** is given by the vector sum of **L** and **S**, so that we may write **J** = **L** + **S**, as shown in the upper part of Fig. 13.5. When, however, we come to study the complex Zeeman effect, it is realized that the pattern and splittings are governed by the total magnetic moment μ_J, rather than the total quantum number J. μ_L may be represented by a vector in the opposite direction to **L** owing to the negative sign of the electronic charge and, for convenience, it is made the same length as L^*, while μ_S can be represented similarly, but its length must be equal to $2S^*$. The resultant of μ_S and μ_L, which may be designated μ_{LS} does not lie in line with **J**. Now μ_J, which we require, is the time average of μ_{LS} over one complete revolution, bearing in mind that the whole system obeys the law of the conservation of angular momentum and therefore precesses about the direction of **J**. It is evident that μ_J is just the projection of μ_{LS} on the direction of the **J** vector as shown, since the component perpendicular to **J** will balance out over one complete revolution. We

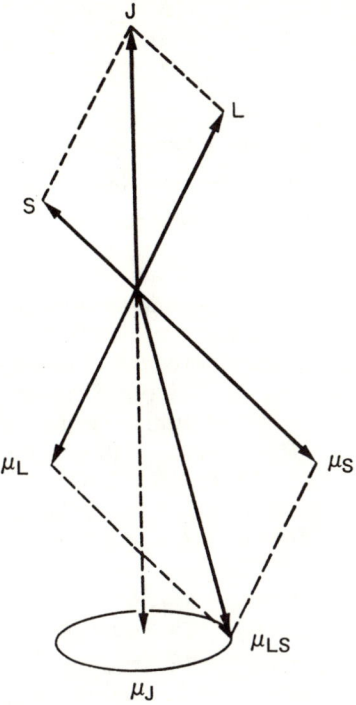

Fig. 13.5 The Landé g factor or splitting factor.

can therefore write
$$\begin{aligned}\mu_J &= \mu_L \cos \widehat{LJ} + \mu_S \cos \widehat{SJ} \\ &= L^* \mu_B \cos \widehat{LJ} + 2S^* \mu_B \cos \widehat{SJ} \\ &= \mu_B(L^* \cos \widehat{LJ} + 2S^* \mu_B \cos \widehat{SJ}).\end{aligned}$$

Using the cosine rule, we may write
$$S^* \cos \widehat{SJ} = \frac{J^{*2} + S^{*2} - L^{*2}}{2J^*}$$

and
$$L^* \cos \widehat{LJ} = \frac{J^{*2} + L^{*2} - S^{*2}}{2J^*}.$$

Hence
$$\mu_J = \mu_B \left(\frac{3J^{*2} + S^{*2} - L^{*2}}{2J^*} \right)$$

or
$$\mu_J = J^* \mu_B \left(\frac{3J^{*2} + S^{*2} - L^{*2}}{2J^{*2}} \right),$$

giving
$$\mu_J = J^*\mu_B\left(1 + \frac{J^{*2} + S^{*2} - L^{*2}}{2J^{*2}}\right)$$
$$= J^*\mu_B g,$$
where g is the Landé splitting factor given by
$$g = 1 + \frac{J(J+1) + S(S+1) - L(L+1)}{2J(J+1)}$$
in terms of quantum numbers.

When we put $S=0$, we see that $J=L$ and therefore g becomes unity. This explains the normal Zeeman effect of spinless atoms, as in Fig. 13.4.

It will now be shown how electron spin leads to the so-called anomalous Zeeman effect and gives rise to many more lines than the simple theory described in Section 13.3. Here it is shown, for an atom whose magnetic moment can be attributed entirely to the orbital momentum of electrons L, the energy levels in a magnetic field of flux density B are given by the expression
$$\Delta E = \mu_B L B \cos\theta = \mu_B B m_L.$$
With the introduction of electron spin, L is replaced by the total quantum number J and its corresponding magnetic moment μ_J. The energy levels are therefore given by
$$\Delta E = g\mu_B J B$$
$$= g\mu_B J \cos\theta B \quad \text{along } z$$
$$= gBm_J\mu_B;$$
m_J being the magnetic quantum number. Following through much the same argument as before, we find that the frequency of the Zeeman component is given by the formula
$$h\nu_B = h\nu_0 + (g'm'_J - g''m''_J)B\mu_B,$$
where g', m'_J and g'', m''_J refer to the upper and lower excited states of the atom concerned. Unless $g' = g''$, making the splitting of the upper and lower states the same, we shall observe more than three components.

Applying this now to the sodium D lines we must first calculate the Landé g factors for the levels concerned ($^2P_{3/2}$, $^2P_{1/2}$, $^2S_{1/2}$).

For $^2P_{3/2}$, $\quad J=\frac{3}{2}, L=1, S=\frac{1}{2}$

and so $\quad g = 1 + \dfrac{\frac{3}{2}\cdot\frac{5}{2} - 1\cdot 2 + \frac{1}{2}\cdot\frac{3}{2}}{2\cdot\frac{3}{2}\cdot\frac{5}{2}} = \dfrac{4}{3}$

for $^2P_{1/2}$, $\quad J=\frac{1}{2}, L=1, S=\frac{1}{2}$

and so $\quad g = 1 + \dfrac{\frac{1}{2}\cdot\frac{3}{2} - 1\cdot 2 + \frac{1}{2}\cdot\frac{3}{2}}{2\cdot\frac{1}{2}\cdot\frac{3}{2}} = \dfrac{2}{3}.$

for $^2S_{1/2}$, $\quad J=\frac{1}{2}, L=0, S=\frac{1}{2}$

and so $\quad g = 1 + \dfrac{\frac{1}{2}\cdot\frac{3}{2} - 0\cdot 1 + \frac{1}{2}\cdot\frac{3}{2}}{2\cdot\frac{1}{2}\cdot\frac{3}{2}} = \dfrac{2}{1}$

It is convenient to tabulate the Landé g factors for some of the doublet levels. They are easy to calculate from the formula. The sodium levels split as shown in Fig. 13.6, the splitting being given by the factor $gm_J\mu_B$. Transitions between the

TABLE 13.1
Some Landé g Factors

L	Term	g	gm_J
0	$^2S_{1/2}$	$\frac{2}{1}$	± 1
1	$^2P_{1/2}$	$\frac{2}{3}$	$\pm\frac{1}{3}$
	$^2P_{3/2}$	$\frac{4}{3}$	$\pm\frac{2}{3}\pm\frac{6}{3}$
2	$^2D_{3/2}$	$\frac{4}{5}$	$\pm\frac{2}{5}\pm\frac{6}{5}$
	$^2D_{5/2}$	$\frac{6}{5}$	$\pm\frac{3}{5}\pm\frac{9}{5}\pm\frac{15}{5}$
3	$^2F_{5/2}$	$\frac{6}{7}$	$\pm\frac{3}{7}\pm\frac{9}{7}\pm\frac{15}{7}$
	$^2F_{7/2}$	$\frac{8}{7}$	$\pm\frac{4}{7}\pm\frac{12}{7}\pm\frac{20}{7}\pm\frac{28}{7}$

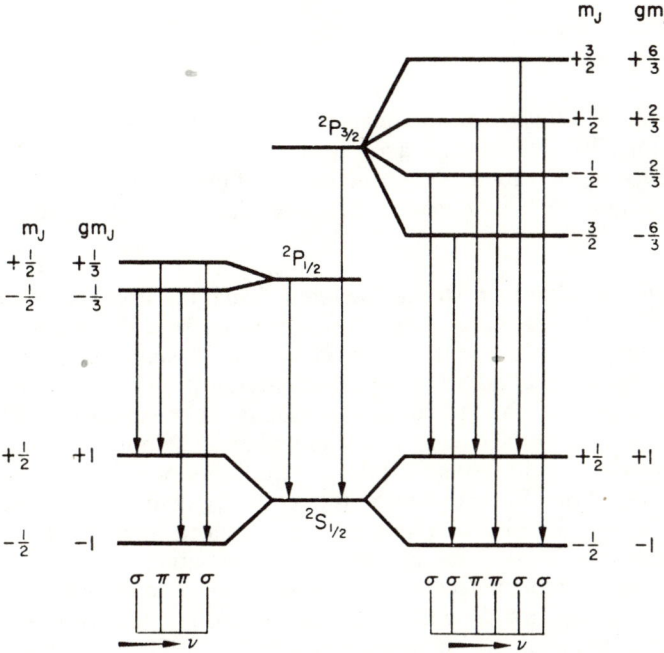

Fig. 13.6 Zeeman effect of the sodium D lines.

levels are determined by the quantum condition $\Delta m_J = \pm 1, 0$. It will be seen how one sodium line gives four components and the other six in contrast with three components when there is no magnetic moment due to electron spin. σ and π indicate the state of the polarization of the component, σ being perpendicular to the magnetic field and given by $\Delta m_J = \pm 1$ and π being parallel to the magnetic field and given by $\Delta m_J = 0$.

This is the energy level-transition diagram for the ten anomalous sodium D

TABLE 13.2
Development of the Paschen–Back Levels

Spectroscopic state			Weak field: Zeeman effect	Allowed m_l, m_s combinations to give m_j		Strong field: Paschen–Back effect	Term Separation
Term	l	j	m_j	m_l	m_s	$(m_l + 2m_s)$	ΔE
$^2P_{3/2}$	1	$\frac{3}{2}$	$+\frac{3}{2}$	$+1$	$+\frac{1}{2}$	2	
$^2P_{3/2}$	1	$\frac{3}{2}$	$+\frac{1}{2}$	0	$+\frac{1}{2}$	1	1
$^2P_{1/2}$	1	$\frac{1}{2}$	$+\frac{1}{2}$	$+1$	$-\frac{1}{2}$	0	1
$^2P_{3/2}$	1	$\frac{3}{2}$	$-\frac{1}{2}$	-1	$+\frac{1}{2}$	0	
$^2P_{1/2}$	1	$\frac{1}{2}$	$-\frac{1}{2}$	0	$-\frac{1}{2}$	-1	1
$^2P_{3/2}$	1	$\frac{3}{2}$	$-\frac{3}{2}$	-1	$-\frac{1}{2}$	-2	1
$^2S_{1/2}$	0	$\frac{1}{2}$	$+\frac{1}{2}$	0	$+\frac{1}{2}$	1	2
$^2S_{1/2}$	0	$\frac{1}{2}$	$-\frac{1}{2}$	0	$-\frac{1}{2}$	-1	

line Zeeman lines. Whereas in the normal Zeeman effect the level separation is always the same, the effect of the g factor is to make the S separation *different* from the P separation so that there is no coincidence of line frequencies. The selection rules are still $\Delta m_J = \pm 1$ or 0 *and* $0 \to 0$ except when $J_1 = J_2 = 0$, although $0 \to 0$ does not signify here.

13.6 Zeeman Splitting in a Strong Magnetic Field: the Paschen–Back Effect

If the field B is increased to about 10 T a very curious thing happens to the anomalous Zeeman pattern. *It reverts to the norman Zeeman triplet.* This is called the Paschen–Back effect and takes place because the coupling between **l** or **s** and **B** is now much stronger than the usual **ls** coupling for each electron. This results in magnetic splitting of the energy levels and when this dominates, i.e. when the fine structure splitting due to **ls** coupling is small, the precessional velocity of **j** about **B** is so rapid that one can neglect the relatively small precessional velocity of **l** and **s** about **j**. The **l** and **s** vectors are therefore uncoupled and precess *independently* about **B** and each is separately space-quantized.

The quantum-mechanical treatment of these strong field effects shows that the energies of the terms are dependent on the quantum number *sum* $(m_l + 2m_s)$, together with a small spin–orbit correction. The factor 2 for m_s is the g factor of the spinning electron. The transition from a weak-field Zeeman effect to a strong-field Paschen–Back effect is illustrated in Table 13.2 for the S and P levels of the sodium atom. The energy level diagram is shown in Fig. 13.7.

From Fig. 13.7 we see that the result of this new quantum rule is to split the two possible P states into five equidistant magnetic states. The selection rules for transitions to the two S states are

$$\Delta m_l = \pm 1, 0 \quad \text{and} \quad \Delta m_s = 0$$

The rigorous derivation of these rules and the details of the allowed levels are purely quantum mechanical.

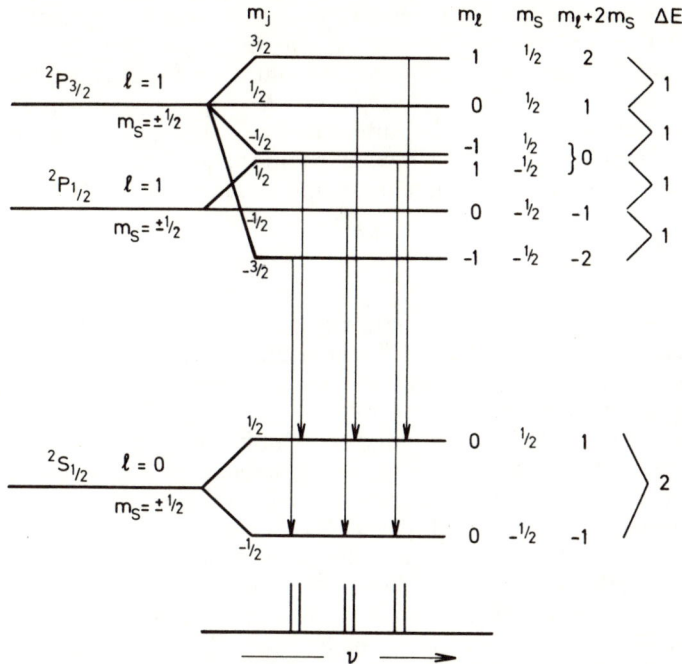

Fig. 13.7 Paschen–Back transitions for sodium D lines showing origin of the three close doublets.

Thus the strong field pattern is a triplet not unlike the classical Lorentz triplet of the normal Zeeman effect but quite different from that of the sodium D lines shown in Fig. 13.6. The change to the simple triplet in very strong fields is the Paschen–Back effect. More rigorous theoretical treatment shows that there is still some residual **ls** coupling energy so that the magnetic separations of the P levels are not all *exactly* the same. Each line of the resulting triplet is therefore a close doublet rather than a singlet. In the *normal* Zeeman effect the term separations are all *exactly* equal, so that the nine transitions form an exact triplet. A comparison between Figs. 13.4, 13.6 and 13.7 shows all the changes mentioned in this section.

The actual Zeeman pattern of an atom changes from the anomalous effect to the simpler Paschen–Back effect as the field increases, but no new information about spatial quantization is revealed. It merely confirms the findings of other magnetic experiments.

The Paschen–Back effect is not readily observed except in light elements such as Li and Be, where the $P_{1/2}$ $P_{3/2}$ splitting is small. It can also be observed in the microwave spectroscopy of hydrogen.

13.7 Conclusion

The effects discussed in this chapter afford ample experimental evidence supporting quantum-mechanical predictions of atomic structure. The wealth of detail in atomic spectroscopy, both in the optical and now in the microwave region of frequencies, has led to much agreement with theory and often to the development of the theories. Work on the hydrogen spectrum and on the hydrogen-like He^+ spectrum still continues but it is a far cry from the simplicity of Balmer and Bohr to the complexity of modern spectroscopy. It all leads to a deeper understanding of the structure of the electronic part of the atom. In the succeeding chapters we shall see to what extent the nuclear structure of the atom can likewise be revealed experimentally and theoretically.

Problems

(*The problems marked with an asterisk are solved in full at the end of the section.*)

13.1 Describe the nature of the Zeeman effect. Explain how the simple effect may be interpreted in terms of circular electron orbits with the magnetic field perpendicular to the plane of the orbit.

The calcium line 422·7 nm, is found to exhibit a simple Zeeman pattern in a magnetic field of 3 T. Calculate (*a*) the difference in frequency between the displaced and undisplaced components; (*b*) the wavelength difference between these components. ((*a*) 42·2 GHz, (*b*) 25·2 pm)

13.2* The red line of cadmium splits into three components separated by 120 MHz when the source is placed in a magnetic field of flux density 8·6 mT, the light being examined in a direction perpendicular to the magnetic field. Calculate the ratio of charge to mass (e/m) of the electron. (1.76×10^{11} C kg^{-1})

13.3 In a magnetic field of flux density 2 T the $^1D_2 - {}^1P_1$ calcium line at $\lambda = 732.6$ nm splits into three components separated by 2.8×10^{10} Hz. Estimate the ratio of charge to mass of an electron. (1.77×10^{11} C kg^{-1})

13.4 In a line which exhibits a simple Zeeman pattern in a magnetic field, calculate the total splitting in wavenumbers for a magnetic field of unit flux density. (0·94 m^{-1} T^{-1})

13.5 Calculate the numerical value of the Bohr magneton. (9.27×10^{-24} J T^{-1})

13.6 The calcium line at 422·6 nm gives a normal Zeeman effect ($J = L$; $S = 0$) showing three components with $\Delta v = 4.22 \times 10^{10}$ Hz in a field of 3 T. Calculate the specific charge of the electron.

Prove any formulae you use. [N]

13.7* In the normal Zeeman effect the separation of the magnetic energy levels is given by

$$\Delta E = M_L \mu_B B,$$

where μ_B is the Bohr magneton and B is the magnetic flux density.

What is the corresponding expression if electron spin is taken into consideration? What is the explanation of this difference?

The fourth member of the principal series of sodium is the doublet
$$6^2P_{3/2,\ 1/2} - 3^2S_{1/2}.$$
Construct a table for these three sodium states showing their values of L, J, S, M_J, g and $M_J g$, where g is the Landé factor

$$g = 1 + \frac{J(J+1) + S(S+1) - L(L+1)}{2J(J+1)}.$$

Hence draw an energy level diagram showing how the $6^2P_{3/2} - 3^2P_{1/2}$ line is split into six equally spaced Zeeman components. [N]

13.8 Distinguish between the *normal* Zeeman effect and the *anomalous* Zeeman effect, illustrating your answer with reference to specific atomic spectra.

Show how the normal Zeeman effect may be deduced in terms of the vector model of the atom.

Explain why it is necessary to introduce the Landé splitting factor g, and also its effect on the relative Zeeman separations of the states of the ^2S and ^2P levels. [N]

13.9 Explain clearly the difference between the normal and the anomalous Zeeman effects.

A spectral line is the result of the $^2P_{3/2} - {}^2S_{1/2}$ transition. Draw an energy level diagram showing the Zeeman splitting in a magnetic field.

Calculate the total splitting in wave numbers in a field of 0·1 T.

Solutions to Problems

13.2 The simple Zeeman splitting of a spectral line in terms of frequency v is given by

$$\Delta v = \frac{\Delta m_l B e}{2\pi \, 2m},$$

where $\Delta m_l = 1$. Therefore

$$\frac{e}{m} = \frac{4\pi \Delta v}{B}$$

$$= \frac{4\pi \times 1 \cdot 2 \times 10^8}{8 \cdot 6 \times 10^{-3}}$$

$$= 1 \cdot 76 \times 10^{11} \text{ C kg}^{-1}$$

13.7 $\Delta E = g M_J \mu_B B$ with electron spin. Calculate g values and tabulate:

State	L	J	S	M_J	g	gM_J	Level separation
$6^2P_{3/2}$	1	$\frac{3}{2}$	$\frac{1}{2}$	$\pm\frac{3}{2}, \pm\frac{1}{2}$	$\frac{4}{3}$	$\pm 2, \pm\frac{2}{3}$	$\left.\begin{array}{c}\frac{4}{3}\end{array}\right\}$
$6^2P_{1/2}$	1	$\frac{1}{2}$	$\frac{1}{2}$	$\pm\frac{1}{2}$	$\frac{2}{3}$	$\pm\frac{1}{3}$	
$3^2S_{1/2}$	0	$\frac{1}{2}$	$\frac{1}{2}$	$\pm\frac{1}{2}$	2	± 1	$\}$ 2

Energy level separation is $\Delta E = \frac{4}{3}\mu_B B$ and $2\mu_B B$ for P and S levels respectively. Applying $\Delta M_J = \pm 1, 0$ gives six components for $6^2P_{3/2} - 3^2S_{1/2}$ each separated by $\frac{4}{3}\mu_B B$.

Chapter 14

The Structure of the Nucleus

14.1 Introduction

Having been so successful in applying quantum laws to the electronic structure of the atom, it was natural that physicists should attempt to apply these laws to the structure of the atomic nucleus. Assuming that the nucleon is a quantum-mechanical particle, that it has an angular momentum and a magnetic moment governed by the rules of coupling and quantization, it should be possible to predict nuclear spectroscopic energy states from collective nucleon properties in much the same way as for electronic states. However, there is no guarantee that the laws of electron interaction will hold for nucleons whose size is only about 10^{-5} that of the whole atom, indeed there is no guarantee that the internucleonic forces will obey the laws of quantum mechanics at all.

From experimental nuclear physics a great deal has been learned about the constituents of the nucleus and of their interchange properties. A wealth of information is therefore available with which to test the quantum-mechanical methods used for the electronic atom. Much of the success of quantum mechanics is due to the extreme simplicity of the two-body hydrogen atom leading to a complete solution of the Schrödinger wave equation. In complex atoms this is not so easy and in the case of the nucleus, with its many nucleons each represented by a wave function, the mathematical difficulties are such that exact solutions are often impossible and we have to resort to nuclear models with built-in properties for which the Schrödinger equation can be solved. Thus quantum mechanics cannot be as successful with the nucleus as with the simpler atom, although the structure of the deuteron has been successfully investigated by these methods.

Research into the structure of the nucleus has led to a knowledge and understanding of those subnuclear particles which are themselves the ultimate structure of matter. The relation between these strange particles their symmetries and order, the possible existence of other particles as the basic building bricks of the physical universe has stimulated world-wide research in high energy nuclear physics and the development of new, and costly, high-energy particle accelerators.

The structure of the nucleus and of its constituents still presents a challenge to both theoretical and experimental physicists.

14.2 Nuclear Constituents: Isotopes and Isobars

In previous chapters we have discussed in detail the properties of the atom based on its electronic structure but have only mentioned in outline some of the properties which depend on its nuclear structure. We saw that the chemical properties of the atom depend on its electronic structure, while its physical properties, its dynamic and kinetic behaviour, depend largely on its mass which is, of course, almost wholly contained in the nucleus. The nucleus is not only the seat of the mass but also the origin of the energy of the atom. The nucleus contains only two types of elementary particle, the proton and the neutron, so that in all there are only three fundamental atomic particles, as shown in Table 14.1. It is true that further particles may be ejected from the nuclear assembly, but they do not exist independently within the nucleus.

TABLE 14.1
Elementary Atomic Particles

Name		Mass	Charge	Symbol
Electron		m_e	$-e$	β^-, e^-
Proton	} Nucleus	$1836 \cdot 1\ m_e$	$+e$	p
Neutron		$1838 \cdot 6\ m_e$	0	n

$m_e = 9 \cdot 109 \times 10^{-31}$ kg; $e = 1 \cdot 602 \times 10^{-19}$ C

The other particles of modern physics are never found as independent entities within the atom. See, however, Chapters 27 and 28.

The nuclear constituents are of roughly equal mass and are referred to collectively as nucleons. It is sometimes useful to take the mass of the proton as unit mass and the charge on the electron as unit charge and use the approximation that the electronic mass is zero. The neutron is radioactive but the proton is stable.

At once we detect here a breakdown of some of our macroscopic laws of electrostatics. The nucleus contains only positive charge — why do not the protons repel each other according to Coulomb's law? If this law is still true there must be further, and stronger, attractive forces at play within the nucleus which exist nowhere else in nature, otherwise we should have found them long ago. Alternatively, Coulomb's law breaks down within the nucleus. At present, as far as we have progressed in this book, we are not able to differentiate between these two possibilities.

Atoms are built up, step by step, by adding neutrons, protons and electrons to the simplest atom of all — hydrogen. This atom is unique in that it is the only atom with less than three particles, and consists of one proton and one electron. It is therefore electrically neutral. Further, it is the only atom which does not contain a neutron. To this atom we can add nucleons and go through the whole gamut of the elements from hydrogen to uranium and beyond. Each time a nucleon is added the atomic mass increases by one unit, but each time a proton is added not only is the atomic mass increased by one unit but the nuclear charge increases by one unit and therefore the element itself changes. Thus an atomic nucleus consists of:

A nucleons ... this is the atomic *mass* number

made up of

Z protons ... this is atomic *number*, or the correct numerical order in the periodic table of the elements.

and

N neutrons

where

$N = A - Z$

If the chemical symbol of the element is X, a particular atom of this element can be wholly described by the notation $^A_Z X$, or sometimes $^A_Z X_N$.

When Z changes the symbol X changes, as indicated above.

Not all combinations of A and Z appear in nature, since energy considerations make some nuclei unstable so that they disintegrate instantaneously. The atoms of the first few elements are given in Table 14.2.

This table can be extended further by adding a proton–electron pair to change the element, or adding neutrons, until all possible atoms have been described.

TABLE 14.2
Light Elements and their Possible Atoms

Element	Protons Z	Electrons Z	Neutrons N	Z	A	Symbol
Hydrogen	1	1	0	1	1	$^1_1$H (p)
	1	1	1	1	2	$^2_1$H (d)
	1	1	2	1	3	$^3_1$H (t)
	1	1	3	Does not exist		
Helium	2	2	1	2	3	$^3_2$He
	2	2	2	2	4	$^4_2$He (α)
	2	2	3	2	5	($^5_2$He)
	2	2	4	2	6	$^6_2$He
	2	2	5	Does not exist		
	2	2	6	2	8	$^8_2$He
Lithium	3	3	3	3	6	$^6_3$Li
	3	3	4	3	7	$^7_3$Li
	3	3	5	3	8	$^8_3$Li
	3	3	6	3	9	($^9_3$Li)
Beryllium	4	4	3	4	7	$^7_4$Be
	4	4	4	4	8	($^8_4$Be)
	4	4	5	4	9	$^9_4$Be
	4	4	6	4	10	$^{10}_4$Be
	4	4	7	4	11	($^{11}_4$Be)

Nuclides in brackets are very unstable.

This table reveals that an element can be represented by different atoms, all with the same chemical properties. Thus hydrogen must always have one electron–proton pair, but it can have 0, 1 or 2 neutrons giving masses of 1, 2 and 3. These atoms of hydrogen are light hydrogen, deuterium (or heavy hydrogen) and tritium, respectively, whose nuclei are the proton (p), the deuteron (d) and the triton (t).

It is evident that since many of the elements can be represented by different nuclei we must use a special name for them. Each of these atoms is called a *nuclide*, i.e. every atom $^A_Z X$ is a nuclide. From Table 14.2 we see that some nuclides have the same atomic number, Z. These nuclides form isotopes of the same element. Isotopes, then, of the same chemical element have different masses. In the Table 14.2 the beryllium isotopes (in which $Z=4$) are:

$$^7_4\text{Be}, \; (^8_4\text{Be}), \; ^9_4\text{Be}, \; ^{10}_4\text{Be}, \text{ and } (^{11}_4\text{Be}),.$$

There are also nuclides having the same atomic mass but with different atomic numbers corresponding therefore to different elements. These are called isobars, and examples are:

^3_1H ^3_2He where $A=3$
^6_2He, ^6_3Li where $A=6$
                   ^7_3Li, ^7_4Be where $A=7$.

Isobaric nuclides can be formed by β^- emission as in the case of tritium:

$$^3_1\text{H} \to {^3_2\text{He}} + {_{-1}^0\text{e}}(\beta^-) + \bar{\nu},$$

where $\bar{\nu}$ is an antineutrino, to be discussed later. Positron ($_{+1}^0\text{e}$, β^+; see p. 218) transformations are also isobaric as in the case of $^{22}_{11}\text{Na}$, for which

$$^{22}_{11}\text{Na} \to {^{22}_{10}\text{Ne}} + {_{+1}^0\text{e}}\,(\beta^+) + \nu,$$

where ν is a neutrino.

Important cases of isobaric groups will be discussed in the chapter on nuclear fission.

A complete list of nuclides (Appendix C') shows that for the lighter elements the number of protons is nearly equal to the number of neutrons. In fact, a plot of N to Z shows a distribution about a smooth curve tending to $N=Z$ at low values, but showing $N>Z$ for the heavier elements. This is shown in Fig. 14.1(a).

A survey of the complete table of stable nuclides shows some interesting features. If we arrange the nuclides according to whether they have odd or even numbers of protons and odd or even numbers of neutrons we get the distribution shown in Table 14.3.

This table shows that the stable nuclides with an even number of protons and an even number of neutrons (even/even nuclides) far outnumber the stable nuclides with an odd number of both protons and neutrons (odd/odd nuclides). The significance of this will be apparent after further discussion on radioactivity, but at this stage it can be said that the nuclear attractive forces referred to earlier in this section must be much stronger for the even/even nuclides than for the odd/odd nuclides.

It is apparent from Fig. 14.1(a) that the number of protons only twice exceeds the number of neutrons and that the number of stable isotopes per element is not constant. For instance, tin, for which $Z=50$, has ten stable isotopes ranging from atomic mass number 112 to atomic mass number 124, whereas nearby caesium,

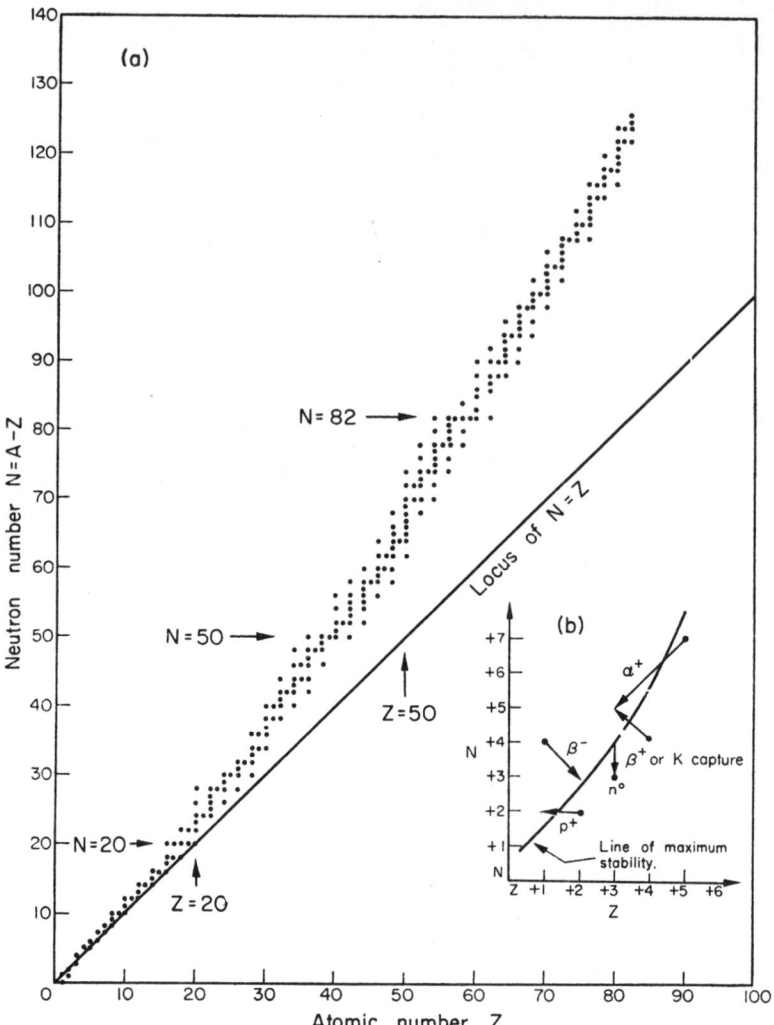

for which $Z = 55$, has only one stable isotope of mass number 133.

Clearly these features are of great significance in nuclear physics and must be accounted for in any theory of nuclear structure.

14.3 The Size of the Nucleus

The atom as a whole is submicroscopic. The 'diameter' of the hydrogen atom as determined spectroscopically and by kinetic theory is about 0·1 nm, but the protonic size is at least four decades lower than this and must be much less than the Rutherford scattering 'size', i.e. less than 26 fm (26 femtometres) for the gold

TABLE 14.3
Nucleon Distribution in Stable Nuclides

Elements Protons Z		Isotopes Neutrons N		Z/N	Distribution	N/Z
Z even	Z odd	N even	N odd	Odd/even	Odd/odd	Even/even
		53		53		
	40		8		8	
		166				166
43			57	57		
Totals		219	65	110	8	166
Total elements	83	Total Isotopes	284			

nucleus, as revealed by the α-particle scattering experiments described in Chapter 3. The results of fast neutron scattering experiments give a nuclear 'radius' R given approximately by the formula:

$$R = R_0 A^{1/3}$$

where $R_0 = 1.4$ fm (see p. 280 for discussion), and A is the atomic mass number. Thus for aluminium, $A = 27$, we have $R = 4.2$ fm. High-energy electron scattering experiments give a similar result.

It is difficult to appreciate this order of magnitude and all one can say is that any nucleus is very small and many times smaller than the atom as a whole. The exact number given to the size of the nucleus depends on the particles used to explore it, together with their energies. The appropriate size unit used for the nucleus is 10^{-15} m $\equiv 1$ fm.

14.4 Exact Atomic Masses — Mass Excess ΔM

We have seen in Chapter 4 that the mass spectrometer can be used for 'weighing' atoms and that the original ideas of integral values have had to be abandoned. Any exact isotopic mass M can be written in terms of the unified atomic mass unit u (see later), $M = A + \Delta M$, where A is an integer = total number of nucleons in nucleus, and ΔM is the mass excess, the sign of which is not always positive because of the choice of ^{12}C as our standard of atomic mass. Some examples are given in Table 14.4, where in all cases, *M is the mass of the neutral atom*, i.e. includes all the electrons, and ΔM is the mass excess $M - A$, given in Appendix C.

The units of this table are unified atomic mass units. This unit is 1 u = $1.660\,565 \times 10^{-27}$ kg, or, more correctly it is $\frac{1}{12} \times M^{12}C$ kg. Note the accuracy in Table 14.4 with which it is possible to measure M with modern mass spectrometer and nuclear reaction methods. It was this order of accuracy that dictated the original

TABLE 14.4
Some Isotopic Masses in u

Nuclide	M	A	ΔM
^{1}H	1·007 825	1	0·007 825
^{4}He	4·002 603	4	0·002 603
^{16}O	15·994 915	16	−0·005 505
^{35}Cl	34·968 851	35	−0·031 149
^{120}Sn	119·902 198	120	−0·097 802

^{12}C = 12·000 000 u

choice of ^{16}O as the physical standard instead of the chemical method of using the natural mixture of oxygen isotopes. Oxygen has three principal isotopes, ^{16}O, ^{17}O, ^{18}O, with abundance ratios of about 500:1 for 16:18 and about 2500:1 for 16:17. The chemical scale of relative atomic masses is based on atmospheric oxygen with all its isotopes. These have masses and abundances as follows, all on the old ^{16}O scale:

$$^{16}O = 16 \cdot 000\ 000 \quad 99 \cdot 759\%,$$
$$^{17}O = 17 \cdot 004\ 534 \quad 0 \cdot 037\%,$$
$$^{18}O = 18 \cdot 004\ 855 \quad 0 \cdot 204\%.$$

If we add these together in the proper proportions, the relative atomic mass of atmospheric oxygen becomes 16·004 453 instead of 16·000 000. To convert the chemical scale to the ^{16}O physical scale we have to multiply by a factor

$$\frac{16 \cdot 004\ 453}{16 \cdot 000\ 000} = 1 \cdot 000\ 278\ 3.$$

This conversion factor is therefore necessary when comparing relative atomic masses obtained chemically with those obtained from the mass spectrometer, in terms of ^{16}O.

Since 1960 isotopic masses have been referred to the mass of ^{12}C as the standard of physical scale, i.e. the unified atomic mass unit is now defined by

$$M^{12}C = 12 \cdot 000\ 000\ u.$$

This new scale supersedes the ^{16}O scale and was accepted by the International Union of Pure and Applied Physics which met in Ottawa in 1960.

The reason for this change is that the conversion factor of 1·000 278 3 from the ^{16}O chemical to the physical scale assumes there is no terrestrial variation in the abundance ratios of the oxygen isotopes. This is now known to be incorrect. Furthermore, carbon has only two stable isotopes as against three for oxygen, and also modern mass spectrometry deals largely with hydrocarbon compounds and M^1H can therefore be determined very accurately in terms of M^{12}C. The absolute value of the atomic mass unit is still the same and the conversion factor is M^{16}O scale = 1·000 317 9 M^{12}C scale. The conversion factor from the chemical scale to the 12C scale is 1·000 275.

Throughout this book the ^{12}C scale is used.

Although we see from Table 14.4 that mass differences ΔM are very small, we know that they are very *accurate* and we shall see later they have an important bearing on the origin of the energy of the atom. Aston suggested the name 'packing fraction' f for the ratio $\Delta M/A$. Since we have put $M = A + \Delta M$, we have $M = A(1+f)$, where f can be positive or negative since ΔM can be positive or negative. The packing fraction curve for the elements is shown in Fig. 14.2. This curve is nearly smooth, but the deviations from it are quite definite and must, therefore, be explicable on any theory of the nucleus. Notice that $f = 0$ for ^{12}C as expected, and that there is a very broad minimum. Regarding this packing fraction curve for the moment as a nuclear potential energy curve, we would expect the elements in the minimum to be the most stable.

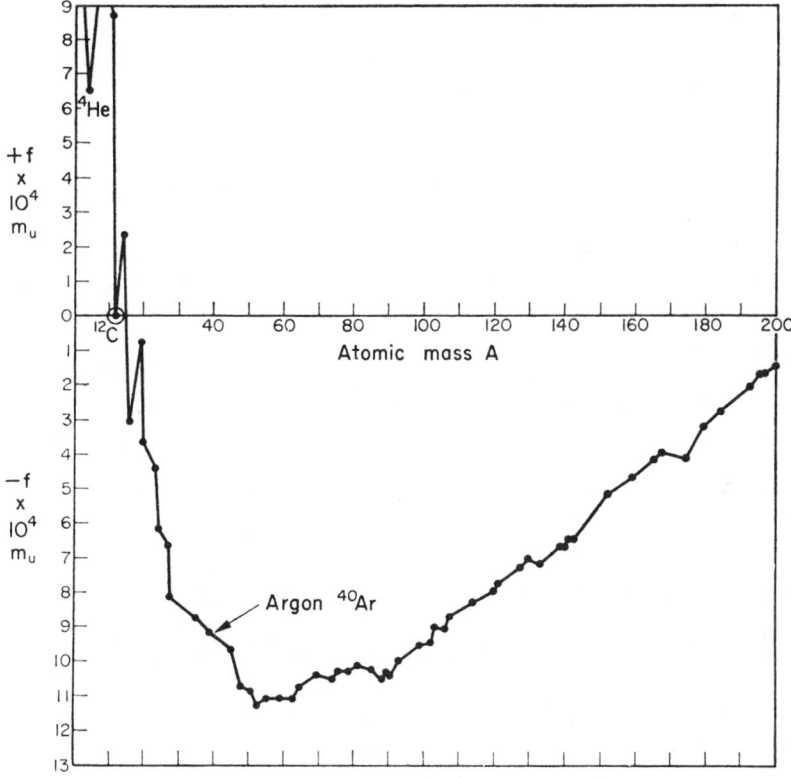

Fig. 14.2 Packing fraction f for the most abundant isotope of each element.

14.5 Binding Energies of Nuclides — Mass Defect $\Delta M_{Z,A}$

Suppose we wish to calculate the mass of the helium atom ($A = 4$) from its nuclear constituents. This we can do quite easily since it contains 2 protons, 2 electrons and 2 neutrons, i.e. the equivalent of 2 hydrogen atoms plus 2 neutrons, so that

$$1 \text{ He atom} = 2 \times (^1_1\text{H} + 1 \text{ neutron}).$$

We would therefore expect the atomic mass of the He atom to be

$$2 \times (M_H + M_n).$$

Working with atomic mass units, u, we have
$$M_H = {}^1_1H = 1 \cdot 007\,825$$
and
$$M_n = {}^1_0n = 1 \cdot 008\,665,$$
so that
$$M_{He} = 2 \times 2 \cdot 016\,490$$
$$= 4 \cdot 032\,980,$$
i.e. the total mass of *all* the constituents of the helium atom is 4·032 980. From Table 14.4 we see that the measured mass of ${}^4_2He$ is 4·002 603 — less than the mass of the nucleons by about 0·03 u. This is rather odd — let us try another example. Take argon ($A = 40$), since this has a *negative* packing fraction, see Fig. 14.2.

This is ${}^{40}_{18}Ar$ so that the full mass is
$$18\,(\text{protons} + \text{electrons}) = 18 \times 1 \cdot 007\,825$$
and
$$22\,\text{neutrons} = 22 \times 1 \cdot 008\,665.$$
This adds up to 40·331 480, whereas the experimental value for ${}^{40}_{18}Ar$ is 39·962 384 — about 0·37 u *less* than the estimated value. Note that the actual mass is again *less* than the 'added' mass, even with a negative packing fraction.

It would appear that the actual mass of a nuclide is never equal to the sum of the masses of its constituents. In any nuclide ${}^A_Z X$ we could write $M_{Z,A}$ for the isotopic mass and $ZM_H + NM_n$ for the total mass of the constituents, where
$$M_H = {}^1_1H = 1 \cdot 007\,825 \quad \text{and} \quad M_n = {}^1_0n = 1 \cdot 008\,665.$$
The difference we have worked out for helium and argon is then
$$\{ZM_H + (A-Z)M_n\} - M_{Z,A},$$
i.e. mass of constituents−experimental mass. If we write $\Delta M_{Z,A}$ for this mass decrease when nuclear constituents join to form a nucleus, we can refer to this as the *mass defect*. (The mass defect must not be confused with the mass excess $M - A$, or ΔM.) This mass disappears on forming a nucleus — where to? One would not expect it to be utterly lost, and it was Einstein who showed that such a loss of mass is equivalent to a gain of energy. From the special theory of relativity (see Appendix A) Einstein showed that there must be a mass–energy equivalence given by the equation
$$E \equiv m_0 c^2$$
for any rest mass m_0,[†] i.e. *a mass of m_0* kg is equivalent to an energy of E joules, where the conversion factor is c^2 (c = velocity of light = 3×10^8 m s^{-1}). Thus, our mass defect $\Delta M_{Z,A}$ appears as an an equivalent amount of energy ΔE on forming a nucleus. It is the energy released, due to the decrease of mass, when nuclei are formed by the fusion together of the requisite number of nucleons; alternatively, it is the energy required to separate the nucleons of the nucleus. It is referred to as the 'binding energy' of the nucleus, B. Thus 'mass' changes are really changes of binding energy — there is no actual destruction of nucleons.

In order to explain the concept of binding energy, let us consider two particles

---

[†] The practical expression for this equation is 1 u = 931 MeV, where 1 MeV is the energy acquired by an electron accelerated by 10^6 V = $1 \cdot 6 \times 10^{-13}$ J. For derivation see the end of this chapter, Section 14.7.

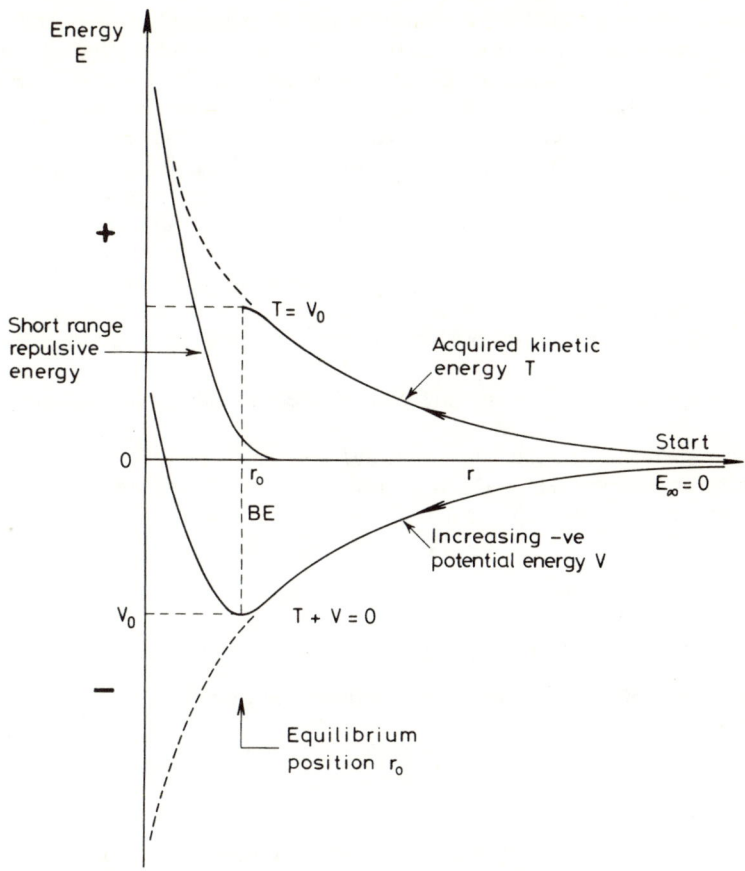

Fig. 14.3 Potential energy diagram showing binding energy V_0.

brought together from infinity under the influence of an attractive force (see Fig. 14.3). At infinity the particles are stationary and the total energy E_∞ is zero. As they approach each other both the kinetic energy T and the potential energy V increase, but V increases *negatively* since work is done *on* the particles by the attractive force between them. At any position $E_\infty = T + V = 0$ by the conservation law, so that $T = -V$ always. At the moment of fusion, when the particles reach their equilibrium position under the influence of the short-range repulsive force, $V = V_0$ and $T = -V_0$. In this bound state the potential energy V_0 is the *binding* energy and the kinetic energy $T = V_0$ is released as radiation energy. This can be converted to other forms of energy on passing through matter.

The minimum potential energy of the bound system is V_0 and this binding energy is equal to the kinetic energy T which is then released. The energy of a bound system such as this is always less than the total rest energy of the constituents of the system when separated.

In the case of a diatomic molecule the binding energy is equivalent to the dissociation energy and for the hydrogen atom it is the ionization energy. When two nuclei are brought together the mass of the compound nucleus is therefore less than the sum of the nuclear rest masses and the binding energy is this mass defect. On fusion the numerical binding energy B is therefore released as T fusion and conversely it is necessary to overcome the binding energy barrier by supplying the equivalent energy T to disintegrate the nucleus completely. The binding energy itself is not released, but that energy which *is* released is numerically equal to it. The two have a different physical entity. Thus in the fusion process (Chapter 24) the energy of fusion can be calculated from the binding energy of the stable system. All systems which move to a stable position of greater ($-$ve) binding energy will give a net release of energy. This explains why the *fission* and *fusion* processes both release available energy. See Chapters 23 and 24.

The relation between binding energy and released energy is illustrated in Fig. 14.3.

We have seen that the packing fraction f is given by $f = (M - A)/A$, where A is the integral mass number or the total number of nucleons, and M is the exact isotopic mass which we have just written as M_{ZA}. Thus

$$\frac{M_{ZA}}{A} = 1 + f.$$

Now

$$\Delta M_{ZA} = \{ZM_H + (A-Z)M_n\} - M_{ZA}$$

is the nuclear binding energy B. The binding energy per nucleon, or $\bar{B}$, is then given by

$$\bar{B} = \frac{B}{A} = \frac{Z}{A} M_H + \left(1 - \frac{Z}{A}\right) M_n - \frac{M_{ZA}}{A}$$

$$= \frac{Z}{A} [M_H - M_n] + M_n - (1+f)$$

$$= -\frac{Z}{A} [0.000\,84] + 0.008\,665 - f \quad \text{in u.}$$

Taking the average value of Z/A as 0.45, we get
$$\bar{B} = -0.000\,38 + 0.008\,665 - f$$
$$= 0.0083 - f,$$

so that the minimum value of f corresponds to maximum value of $\bar{B}$. Note that since f rarely exceeds 10^{-3} u, the value of $\bar{B}$ is roughly constant. Taking an *average* value of $-f$ as 7×10^{-4} u, we find that $\bar{B} = 0.0090$ u about, or approximately 8.4 MeV for most nuclei (1 u = 931 MeV). The reason for this fairly constant value of $\bar{B}$ is that it is made up largely of the neutron mass excess $(M_n - 1)$, as is seen in the above expression for $\bar{B}$. This is really a consequence of the fact that all nuclear forces are short range forces.

Thus, if we plot $\bar{B}$ against A we get the packing fraction curve roughly inverted about the A axis. This is shown in Fig. 14.4, in which we see that the curve has a flat top at about 8.7 MeV., i.e. the addition of a single nucleon to any nucleus in this region of A increases the binding energy by roughly the same amount.

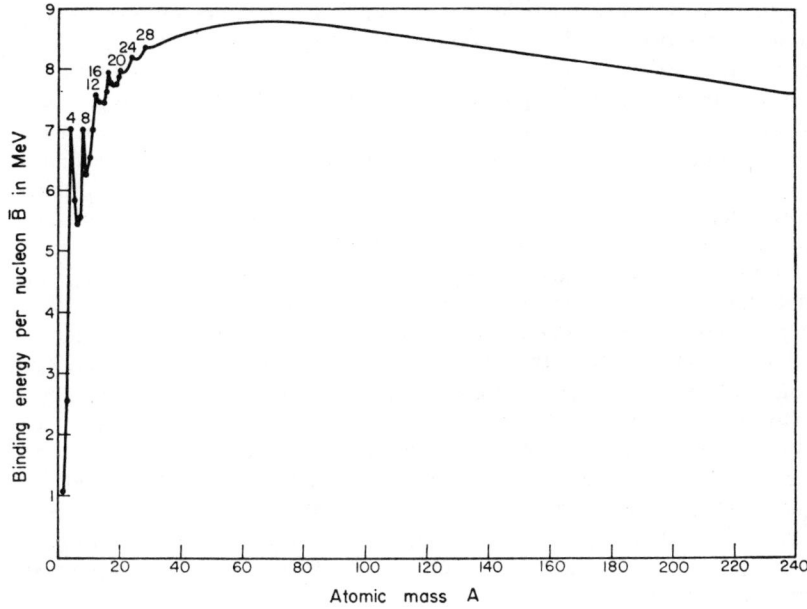

Fig. 14.4 Binding energy per nucleon, showing broad maximum.

This curve gives an idea of the general stability of nuclides. The broad maximum in the middle represents maximum stability, from about $A=40$ to about $A=100$, and the tendency in any nuclear changes is for the resulting nucleus to lie on this portion of the curve. Thus, the *fusion* of light elements tends to produce a single particle nearer the maximum in Fig. 14.4 and so *release* energy, while the splitting (fission) of a heavy element gives two (or more) lighter particles nearer the maximum, again releasing energy. We shall look at these phenomena in greater detail later.

14.6 Stable and Unstable Nuclides

Hitherto our considerations have been restricted to the properties of stable nuclides, in which there is no spontaneous emission of particles. Nuclei can be built up by adding protons and neutrons, systematically altering the masses and the characteristics of the nuclei until the heaviest atoms are realized. From Fig. 14.1(a) we see that as A increases the ratio of neutrons to protons in the nuclei gradually increases. Thus, for oxygen $^{16}_{8}O$ there are 8 neutrons and 8 protons, a ratio of 1:1, whereas for uranium $^{238}_{92}U$ there are 146 neutrons and 92 protons, a ratio of about 3:2.

We can think of the nucleus as a small confined region into which we are forcing nucleons to produce heavier atoms. These nucleons have then an associated binding energy and adjust themselves as far as possible to give a nucleus of minimum potential energy. However, as the number of nucleons increases, there is a tendency for the Coulomb repulsion effect to cause instability so that the nucleus emits particles spontaneously. These need not be nucleons, as we shall see, nor does it follow that *all* such energetic nuclei emit a particle at the same instant of time. The phenomenon is statistical so that the exact time when an individual nucleus emits its particle and increases its stability is governed by chance, i.e. it depends on the instantaneous details of its environment which are always changing. Since the different nuclear states cannot be dealt with individually they must be dealt with collectively. Thus we observe and measure a macroscopic phenomenon, which depends on a large number of microscopic phenomena (nuclear disintegrations) which cannot be observed or measured individually, and we use the macroscopic observations to compare the properties of different nuclei. This macroscopic phenomenon is that of natural radioactivity which has been described in Chapter 3, and will be treated with more detail in the next chapter. We note here that natural radioactivity is almost entirely confined to higher atomic number elements, where the nucleus has become too energetic to retain all its constituents. The nucleus can be regarded as analogous to a drop of liquid near its boiling-point. As the nucleons are added to the nucleus, or, in the analogy, as the temperature of the drop is increased, the particles evaporate from the nucleus in the same way as the molecules evaporate from a drop of liquid. In the case of the nucleus, however, the characteristics of the remaining nucleus differ from the parent, whereas the remaining drop of liquid is simply a smaller edition of its parent.

14.7 Derivation of Practical Form of $E = m_0 c^2$

The energy acquired by an electron of charge e when accelerated by a potential V is given by $E = Ve$. We define units of energy of 1 electron volt (1 eV) as that energy acquired by an electron on being accelerated by 1 V, and 1 MeV (mega-electron volt) when it is accelerated by 10^6 V. Since the charge on the electron is 1.6×10^{-19} C, we have

$$1 \text{ MeV} = 10^6 \times 1.6 \times 10^{-19} \text{ J}.$$

Now $1 \text{ u} \equiv 1.66 \times 10^{-27}$ kg and the energy released when this mass is converted into energy, from Einstein's formula $E = m_0 c^2$, using approximate numerical values, is

$$E = 1.66 \times 10^{-27} \times (3 \times 10^8)^2 \text{ J}$$
$$= \frac{1.66 \times 10^{-27} \times 9 \times 10^{16}}{1.6 \times 10^{-13}} \text{ MeV}$$
$$= 933.7 \text{ MeV}.$$

Thus $1 \text{ u} \equiv 933.7$ MeV and using more accurate values of m, e and c, we find $1 \text{ u} = 931.502$ MeV on the ^{12}C scale. Hence a mass defect of ΔM u in a nuclear assembly corresponds to a release of energy given by $931.502 \Delta M$ MeV. We shall use 931 as the conversion factor in this book.

Problems

(*Those problems marked with an asterisk are solved in full at the end of the section.*)

14.1 From the table of isotopes at the end of the book draw diagrams showing:

(i) the distribution of the number of stable isotopes per element with Z,
(ii) the distribution of the number of stable nuclides against $A - Z$.

List the values at which maxima occur and refer to Chapter 19 for an explanation.

14.2 Draw a suitable histogram of the distribution of stable isotopes in $_{47}$Ag, $_{48}$Cd, $_{49}$In, $_{50}$Sn, $_{51}$Sb, $_{52}$Te, and $_{53}$I.
Explain any systematic features in your diagram.

14.3* Determine which members of the isotopes (^{8}B, ^{12}B); (^{10}C, ^{14}C) and (^{12}N, ^{16}N) are β^- and β^+ emitters. Check your answer from Appendix C.

14.4 If you were making a model of the hydrogen atom on a scale such that a football represented the nucleus, where would the valency electron be found for the atom in its ground state?

14.5 Explain the relation between the mass defect and the packing fraction of a nuclide. Why is it that the latter may be positive or negative whereas the former cannot?

14.6 By taking specific examples discuss the meaning of the term 'binding energy per nucleon'. Why is this approximately constant for all but the lightest elements?

14.7 Three successive krypton isotopes and their isotopic masses are:
$$^{85}_{36}\text{Kr} = 84.913\,523, \quad ^{86}_{36}\text{Kr} = 85.910\,616, \quad ^{87}_{36}\text{Kr} = 86.913\,365.$$
By calculating the binding energy of the last neutron in each case decide which isotope is likely to be unstable to neutrons, given that the isotopic mass for $^{84}_{36}$Kr $= 83.911\,503$. ($^{87}_{36}$Kr)

14.8 It is known that $^{74}_{33}$As decays by β emission with a half-life of 17 days. By considering the binding energies involved determine whether $^{74}_{33}$As decays by positive or negative emission. (β^- and β^+)

14.9* Calculate the binding energies of the following isobars and their binding energies per nucleon:
$$^{64}_{28}\text{Ni} = 63.9280,$$
$$^{64}_{29}\text{Cu} = 63.9298,$$
$$^{64}_{30}\text{Zn} = 63.9292.$$

(577·2 and 9·02 MeV, 575·3 and 8·98 MeV, 574·4 and 8·97 MeV)

Which of these would you expect to be β-active and how would it decay? Why? ($^{64}_{29}$Cu: β^+ and β^-)

14.10 Tritium, $^3_1$H, and helium 3, $^3_2$He, have masses 3·016 050 and 3·016 030 u respectively. Find their binding energies in MeV. What is the cause of the difference between these values? (8·48 MeV, 8·65 MeV)

14.11 A nucleus of mass 240 is broken up into four nuclei each of mass 60. How much energy is released? Hint: use Fig. 14.4. (288 MeV)

14.12 If the nucleus of mass 60 in Problem 14.11 were formed by the fusing together of three nuclei each of mass 20 would there be a release of energy? Calculate this energy and explain why it is not the same as in problem 14.11. (42 MeV)

14.13 The free neutron is not a stable particle. It decays to a proton (stable) by the equation $n^0 \to p^+ + e^- + \bar{\nu} + Q$, where Q is the maximum kinetic energy released and is 780 keV. If M_H is 1·007 825 u and the rest mass of the electron is 0·000 548 u what is the rest mass of the neturon? (1·008 669 u)

14.14 What is the difference in mass between the hydrogen atom and the sum of the proton and electron masses?

14.5 What is meant by the binding energy of an atomic nucleus?

For most nuclei after $Z=10$ the binding energy per nucleon is approximately constant at about 8·4 MeV. Why is this?

Discuss the constancy of the density of the atomic nucleus.

What is the significance of these two constant quantities in formulating a descriptive model of the nucleus? [N]

14.16 Explain the relation between the mass defect and the packing fraction of a nuclide and why the mass defect is always negative whereas the packing fraction can be positive or negative.

By considering the following nuclides and their masses,
$$^{4}_{2}He = 4·002\,603 \text{ u} \quad \text{and} \quad ^{40}_{18}A = 39·962\,384 \text{ u},$$
evaluate:

(1) their packing fractions in mass units,
(2) their mass defects in mass units,
(3) the binding energy per nucleon for each nuclide in MeV units.

The binding energy per nucleon is almost constant at about 8·4 MeV for all but the lightest elements. What is the physical significance of this fact?

(Use the values $M_H = 1·007\,825$ u, $M_n = 1·008\,665$ u.) [N]

Solution to Problems

14.3 Take as an example the carbon isotopes ^{10}C and ^{14}C. If either is a β^- emitter we have, omitting the neutrinos
$$^{10}_{6}C \rightarrow ^{10}_{7}N + _{-1}^{0}e(\beta^-) \quad \text{and} \quad ^{14}_{6}C \rightarrow ^{14}_{7}N + _{-1}^{0}e(\beta^-)$$
or, if β^+ emitters,
$$^{10}_{6}C \rightarrow ^{10}_{5}B + _{+1}^{0}e(\beta^+) \quad \text{and} \quad ^{14}_{6}C \rightarrow ^{14}_{5}B + _{+1}^{0}e(\beta^+).$$
These reactions are always possible if the mass of the final nuclide is less than that of the corresponding carbon nuclide.

Since
$$^{10}C = 10·016\,810, \quad ^{10}N = ?,$$
$$^{14}C = 14·003\,242, \quad ^{14}N = 14·003\,074,$$
$$^{10}B = 10·012\,939, \quad ^{14}B = ?,$$
we deduce ^{10}C is a β^+ emitter and ^{14}C a β^- emitter.

14.9 It is sufficient in calculations of this kind to take
$$M_n = 1·009, \quad M_H = 1·008.$$
Hence for $^{64}_{28}Ni$, $N = 36$ and $Z = 28$. Thus
$$NM_n + ZM_H = 36 \times 1·009 + 28 \times 1·008$$
$$= 36·324 + 28·224$$
$$= 64·548.$$
Therefore
$$\Delta M = 64·548 - 63·928$$
$$= 0·620 \text{ u}.$$
and so
$$\text{B.E.} = 577·2 \text{ MeV} \quad \text{and} \quad \bar{B} = 9·02 \text{ MeV}.$$

For $^{64}_{29}$Cu, $N = 35$ and $Z = 29$. Thus
$$NM_n + ZM_H = 35 \times 1\cdot009 + 29 \times 1\cdot008$$
$$= 64\cdot547.$$
Therefore
$$\Delta M = 64\cdot547 - 63\cdot9429$$
$$= 0\cdot0618 \text{ u}$$
and so
$$\text{B.E.} = 575\cdot3 \text{ MeV} \quad \text{and} \quad \bar{B} = 8\cdot98 \text{ MeV}.$$

For $^{64}_{30}$Zn, $N = 34$ and $Z = 30$. Thus
$$NM_n + ZM_H = 34 \times 1\cdot009 + 30 \times 1\cdot008$$
$$= 34\cdot306 + 30\cdot240$$
$$= 64\cdot546.$$
Therefore
$$\Delta M = 64\cdot546 - 63\cdot929$$
$$= 0\cdot617 \text{ u}$$
and so
$$\text{B.E.} = 574\cdot4 \text{ MeV} \quad \text{and} \quad \bar{B} = 8\cdot97 \text{ MeV}.$$

Chapter 15

Properties of Uses of Natural Radioactivity

15.1 The Nature of Radioactivity

From the discussion in the previous chapter it is apparent that as we progress from light to heavy elements the neutron:proton ratio of the nuclei increases rapidly after calcium, as shown in Table 15.1 for nuclides $^A_Z X_N$.

The maximum ratio appears to be about 3:2. These facts are shown graphically in Fig. 14.1(a) for stable nuclides.

TABLE 15.1
Some Neutron to Proton Ratios

Nuclide	n:p ratio
$^4_2\text{He}_2$	1:1
$^{40}_{20}\text{Ca}_{20}$	1:1
$^{90}_{38}\text{Sr}_{52}$	7:1
$^{134}_{54}\text{Xe}_{80}$	1·48:1
$^{238}_{92}\text{U}_{146}$	1·58:1

As different elements are built up of neutrons and protons to form stable nuclides it appears that on the average rather more neutrons than protons are added. These extra neutrons provide the extra binding energy necessary to overcome the increasing Coulomb repulsion energy of the protons. It is unlikely, therefore, that the properties of a nucleus with a low n:p ratio would be the same as those with a high ratio, and so we find that for very large values of n:p the nucleus tends to be unstable, and spontaneously emits particles in an effort to reduce its potential energy. This is the spontaneous disintegration process we call radioactivity (Chapter 3).

Radioactivity was discovered at the end of the last century and it was soon found that the radiation emitted consisted of three distinct types. These were named α-, β- and γ-rays for simplicity, and were found to be charged helium nuclei, fast electrons and electromagnetic radiation of very short wavelength similar to X-rays, respectively.

The general properties of α-, β- and γ-rays have been discussed in Chapter 4.

15.2 α-Particles and the Geiger–Nuttall Rule

The most important property of α-particles is their ability to ionize any material through which they pass. This property is connected with their range and absorption, and it is found that although they do not penetrate very far into normal materials they cause intense ionization. Thus the range in air for α-particles from $^{210}_{84}$Po (earlier known as radium F) is about 38 mm, and the ionization along the path of the particles increases to a maximum before suddenly decreasing to zero. This was first shown by W. H. Bragg and a typical example of one of his ionization curves for $^{210}_{84}$Po is shown in Fig. 15.1.

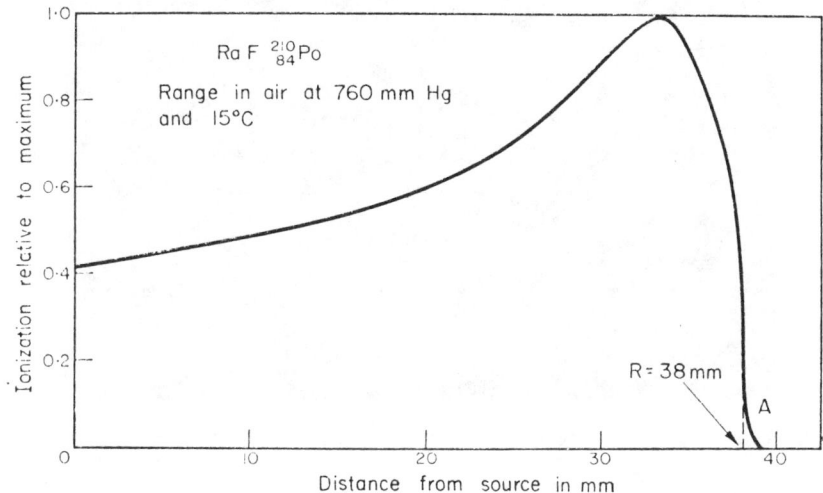

Fig. 15.1 Bragg curve for α-particles from $^{210}_{84}$Po.

As the velocity of the ions is reduced by multiple collisions with the electrons of the gas molecules the ionization efficiency increases until an optimum velocity for ionization is reached. The ionization thereafter decreases rapidly due to ion–electron neutralization, giving a characteristic 'range', i.e. a sharply defined ionization path length. This is best shown in the Wilson cloud chamber pictures given by Blackett (Fig. 15.2). Monoenergetic α-rays have therefore a range R depending on their energy and velocity v, given by the empirical relationship $R = av^3$, where a is a constant. This relationship was originally given by Geiger.

The range as shown in Fig. 15.2 is not always exactly the same for all particles from a given source, due to straggling, or statistical fluctuation in the energy loss process, shown as a slight curvature at the point A on the Bragg curve, Fig. 15.1. The 'range' is found by extrapolating the straight part of the curve to zero ionization, as shown.

Alpha particles are very easily absorbed. A sheet of newspaper will cut off most of them and a postcard will often absorb them completely. Thus from the safety point of view, clothing is sufficient to absorb α-particles and it is the internal hazard which is dangerous (as explained in Appendix B).

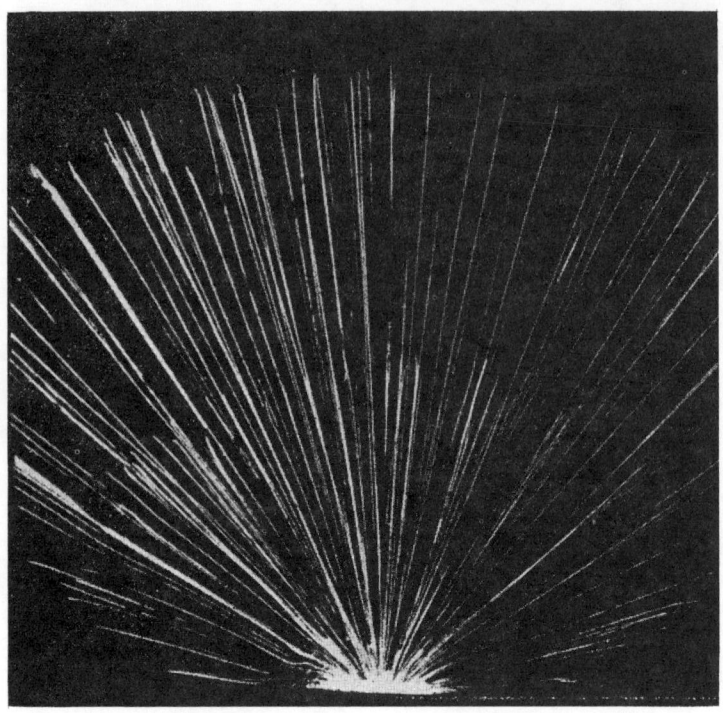

Fig. 15.2 Cloud chamber photograph of α-particle tracks from Th C, C' showing two ranges and straggling effect. (Taken from *Radiations from Radioactive Substances*, Rutherford, Chadwick and Ellis. C.U.P., 1930.)

If we examine the ranges of the common natural α-emitters, together with their respective half-lives, it is apparent that there is a rough reciprocal relation between them. Table 15.2 shows these quantities in detail and it is important to notice the tremendous range of half-life $T_{1/2}$, ranging from 1.4×10^{10} a for $^{232}_{90}$Th which is therefore almost a stable isotope, to 300 ns for $^{212}_{84}$Po which is almost a non-existent isotope. Since these nuclides have the shortest and longest α-particle ranges respectively these figures suggest a reciprocal relation of the form:

$$T_{1/2} \cdot R^m = \text{constant}$$

i.e. $\lambda = AR^m$, where $\lambda = 0.693/T_{1/2}$ (p. 49) and A is a constant. This gives $\log \lambda = m \log R + B$, putting $B = \log A$.

This is the Geiger–Nuttall rule, first discovered in 1911 as the result of a careful survey of the available data. The rule can be verified experimentally but is difficult to explain theoretically.

Plotting the results of Table 15.2 in Fig. 15.3 we find the slope $m = 60$ approximately, and the intercept $B = -44.2$ giving $A = 10^{-84}$, so that

$$\lambda = 10^{-84} \, R^{60}$$

TABLE 15.2
Systematics of the Thorium Series α-Emitters

Radioactive species name	Nuclide symbol	Range in air 760 nm and 15 °C mm	Energy (MeV)	Half-life $T_{1/2}$	Disintegration constant (s^{-1}) λ
Th	$^{232}_{90}$Th	29.0	3.98	1.39×10^{10} a	1.58×10^{-18}
RaTh	$^{228}_{90}$Th	40.2	5.42	1.9 a	1.16×10^{-8}
Th X	$^{224}_{88}$Ra	43.5	5.68	3.64 d	2.20×10^{-6}
Thoron	$^{220}_{86}$Rn	50.6	6.28	54.5 s	1.27×10^{-2}
Th A	$^{216}_{84}$Po	56.8	6.77	0.16 s	4.33
Th C	$^{212}_{83}$Bi	47.9	6.05	60.5 min	1.92×10^{-4}
Th C'	$^{212}_{84}$Po	86.2	8.77	3×10^{-7} s	2.31×10^{6}

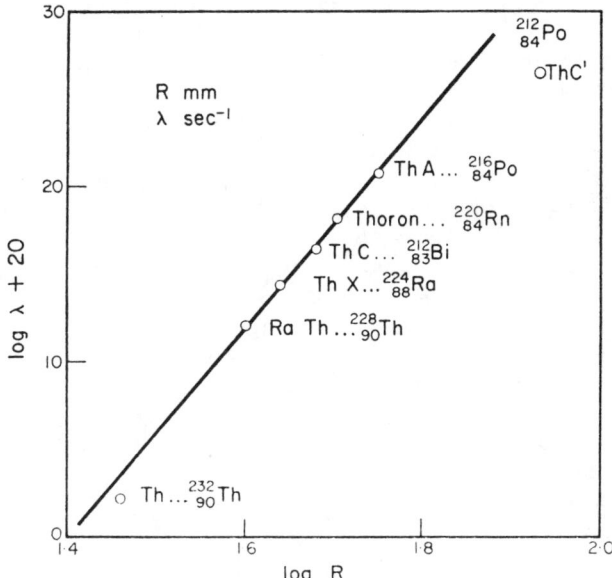

Fig. 15.3 Geiger–Nuttall rule for the thorium series.

or

$$T_{1/2} R^{60} = \frac{0.693}{10^{-84}} = 6.93 \times 10^{83} \text{ s mm}.$$

Different constants would be obtained for the other radioactive series.

Since the range R is connected with the energy of disintegration E by the equations $E = \frac{1}{2}mv^2$ and $R = av^3$, there must be a similar systematic reciprocal connection between half-life and disintegration energy (see Problem 15.1) which has to be explained by any satisfactory theory of the structure of the nucleus.

Furthermore, the α-particle must be a particularly stable combination of protons and neutrons to be ejected as an entity by a radioactive nucleus.

15.3 The Theory of α-decay

Apart from the empirical law of Gieger and Nuttall there are some other puzzling facts about α-particle emission which must be explained. The early α-particle bombarding experiments of Rutherford (see p. 41) showed that the Coulomb law held down to distances of 10^{-14} m with all α-energies of the order of 5 MeV. We are discussing here the naturally radioactive elements of $Z > 80$ and $A > 200$, which means that the range of particle energies is not large. As an example, $^{238}_{92}\text{U}$ is an α-emitter by

$$^{238}_{92}\text{U} \rightarrow {}^{4}_{2}\text{He}\ (\alpha) + {}^{234}_{90}\text{Th} + Q_\alpha$$

where Q_α is the energy of the emitted α-particle, about 4·18 MeV. These α-particles are emitted continuously, although infrequently since $T_{1/2} = 4·5 \times 10^9$ a. On the other hand the $^{235}_{92}\text{U}$ nucleus withstood all attempts to disintegrate it by α-particle bombardment. Firing the swiftest available α-particles of energy 8·3 MeV at $^{238}_{92}\text{U}$ failed to disrupt it. So we have a paradoxical situation where 4·18 MeV α-particles are readily emitted from the $^{238}_{92}\text{U}$ nucleus whereas α-particles of twice this energy fail to penetrate into the nucleus. We are assuming here that the α-particle exists as an entity within the nuclear volume at least just before disintegration. There is some experimental evidence that α-particles exist as 'clusters' within the nucleus for a short time. The energy paradox suggests the existence of an impenetrable wall or potential barrier of potential energy much greater than the α-particle energy within the barrier. This idea of a potential barrier can be understood if we remember the quantum-mechanical discussion of the model atom on p. 139. In that discussion we put the atom in an infinitely deep potential well with infinitely thick walls. In the case of α-emission we are discussing, we have a situation where the wall is finite in both height and width. Inside the well the potential energy is negative, outside the well the potential barrier is subject to the Coulomb law, as shown in Fig. 15.4.

A rough calculation of the maximum energy of the well can be made as follows. The repulsion potential energy is given by

$$V(r) = \frac{2 \times Ze^2}{4\pi\varepsilon_0 r},$$

where Z is the atomic number of the residual nucleus. Now independent experiments show that the effective radius of the uranium nucleus is about $8·6 \times 10^{-15}$ m (see p. 000), so that the maximum potential energy of the barrier is

$$V(r) = \frac{2 \times 90 \times 1·6^2 \times 10^{-38}}{\frac{1}{9} \times 10^{-9} \times 8·6 \times 10^{-15}} \text{ J}$$
$$= 4·82 \times 10^{-12} \text{ J}$$
$$= \frac{4·82 \times 10^{-12}}{1·6 \times 10^{-13}} \text{ MeV}$$
$$= 30 \text{ MeV}$$

This is the peak barrier height, although the actual shape of the distribution is more like the dotted portion of Fig. 15.4. In order to get out of the well, the α-

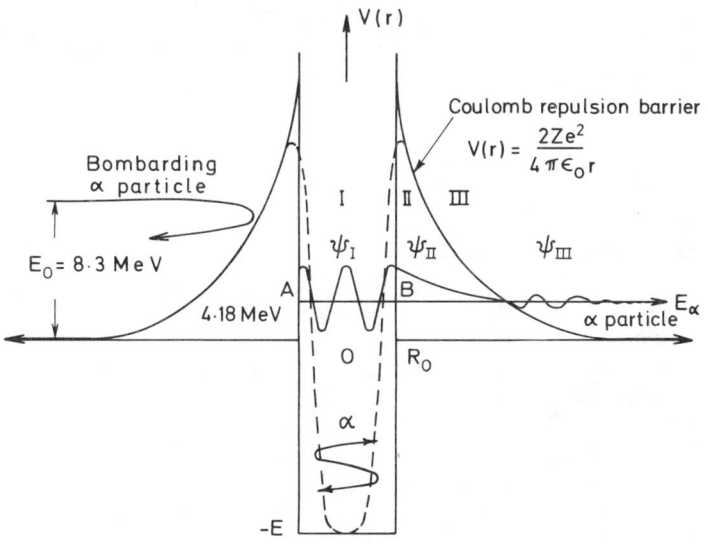

Fig. 15.4 'Tunelling' effect for α-particles.

particle has to surmount this 30 MeV barrier. Classically this is impossible because the α-particle 'rattles' about inside the well and fails to rise above the 4·18 MeV energy level shown as AB in Fig. 15.4. Thus classical physics fails to solve the paradox just as it failed to predict the Geiger–Nuttall law.

At once we see here a case for a quantum-mechanical approach. We saw in Chapter 10 that the atom (although here we are discussing a charged α-particle) behaved as a wave represented by a wave function ψ. Early quantum-mechanical calculations showed that the α-particle had a finite probability of penetrating the barrier from inside the well, and could be represented by three different wave functions as follows, referring to Fig. 15.4:

in region I, ψ_I, which is sinusoidal,
in region II, ψ_II, which is exponential,
in region III, ψ_III, which decreases sinusoidally.

After solving the Schrödinger wave equation for each region and inserting the boundary and continuity conditions, the transmission coefficient from inside can be calculated from the amplitudes of the wave functions. The probability of an α-particle escaping can be calculated and the decay constant is of the form
$$\lambda = A\mathrm{e}^{-B},$$
where A depends on the nuclear radius R and B depends both on Z and on the velocity of emission. Thus we get
$$\log \lambda = a_0 - \frac{b_0}{v_\alpha}$$
(where v_α is the velocity of emission of the α-particle) or
$$\log \lambda = a - \frac{b}{E_\alpha^{1/2}},$$

where a depends on R and b depends on Z. This is not quite the same as the empirical law of Geiger and Nuttall but can be regarded as a theoretical 'Geiger–Nuttall' rule. It is completely borne out by experiment. It can be written as

$$\log T_{1/2} = k_1 + \frac{k_2}{E_\alpha^{1/2}},$$

where k_1 and k_2 are constants, and this formula can easily be verified from Appendix C.

Although an exact replica of the Geiger–Nuttall law has not been found by quantum mechanics, the linear relation between $\log \lambda$ and $E_\alpha^{-1/2}$ has been predicted and verified. It shows that λ is very sensitive to changes in E_α or v_α, as found in the great range of $T_{1/2}$ values. As such it gives great credence to the quantum-mechanical theory of barrier penetration sometimes called the 'tunnelling effect'. A complete account of this can be found in H. Clark, *A First Course in Quantum Mechanics*, Van Nostrand Reinhold, p. 79.

The successful explanation of the experimental results for α-particle systematics represents one of the early justifications for applying quantum-mechanical methods to the problems of nuclear structure. It also enabled the nuclear radius R to be calculated from α-particle emission, and tables of α-particle nuclear radii show that most of the natural α-emitters have nuclear radii of about $(9 \cdot 2 - 9 \cdot 5) \times 10^{-15}$ m. This agrees well with radii measured by other probes (see p. 000).

15.4 β-Rays and the Neutrino

The identification of β^--rays with fast electrons (see Chapter 3) means that all the β^--emitters change their chemical nature and atomic numbers but do not appreciably change their atomic mass numbers, i.e. isobars are different chemical elements. It is possible to measure the energies of nuclear β^--rays by means of the magnetic spectrograph and it is found that the velocities of ejected electrons are not constant, but are spread out in a spectrum as shown in Fig. 15.5.

This continuous velocity spectrum refers to nuclear electrons only and shows that the energy of disintegration of a nucleus is not always given completely to the ejected β^--particle. The most probable value of the energy of an ejected electron is given by E_1 (Fig. 15.5) but the most energetic electrons are actually comparatively few in number. At first sight this suggests that the law of conservation of energy fails with β^--emitters, but Pauli in 1931 suggested informally that there is no violation of the conservation laws if *another* particle as well as the electron is simultaneously emitted by each nucleus. This particle is known as the antineutrino ($\bar{v}$), and the energy balance now becomes

$$E_{\beta^-_{\text{Total}}} = E_{\bar{v}} + E_\beta$$

where both $E_{\bar{v}}$ and E_{β^-} are variable, as shown in Fig. 15.5.

This equation was verified in cloud-chamber experiments with β-rays and careful measurements showed that the conservation of linear momentum was also violated unless another particle was emitted with the β^- particle.

We shall see later that a positron may be emitted from proton rich nuclides. Positrons are positive electrons, β^+, and are accompanied by neutrinos v. They are antiparticles to negative electrons (see p. 000) and their neutrinos are antiparticles to antineutrinos. Experiments on nuclear spins show that all odd-A

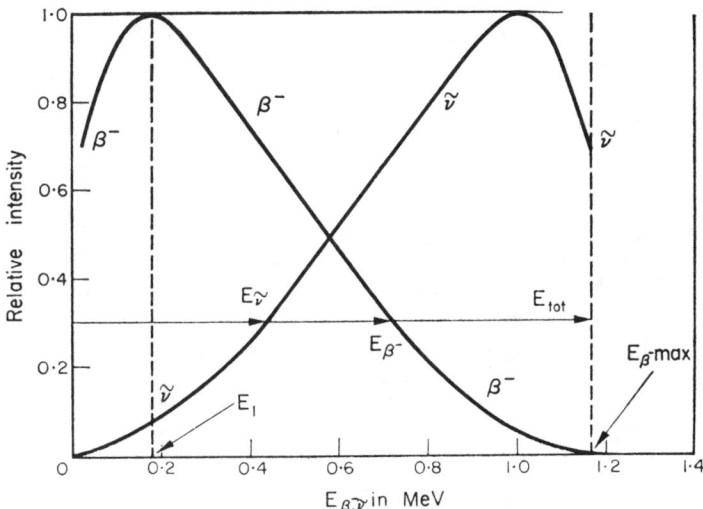

Fig. 15.5 Typical β^-, $\tilde{\nu}$ energy spectrum.

nuclides have half-integral spins, and many have $s = \frac{1}{2}$, so that a decay equation such as

$$^A_Z X \rightarrow {}^A_{Z+1}Y + {}^{0}_{-1}e \; (\beta^-)$$

may have spins

$$\tfrac{1}{2} \rightarrow \tfrac{1}{2} + \tfrac{1}{2}$$

when A is odd. Conservation of angular momentum thus requires another particle to be ejected simultaneously with the β^--particle and with the same spin $\tfrac{1}{2}\hbar$ but opposite in sense. All $\beta^\pm$ decays are isobaric, so that the nucleon number remains unchanged. The angular momentum of the nucleus must therefore remain the same or change by an integral number of $\hbar$ units (all nucleons have spin $\tfrac{1}{2}\hbar$). To conserve angular momentum the spins of the $\tilde{\nu}$ and β^- particles must each be $\tfrac{1}{2}\hbar$ and oriented to give a total change of 0 or 1 units. Thus considerations of energy, linear momentum and angular momentum all point to the existence of a neutrino. The basic equation for all β^--decays is

$$n^0 \rightarrow p^+ + e^- + \tilde{\nu}$$

and it is from this equation that we infer that the spin of the neutrino is $s = \tfrac{1}{2}$.

It was 25 years (after Pauli's original suggestion) before the elusive neutrino and antineutrino became 'respectable', i.e. were accepted as real particles interacting with matter. Reines and Cowan in the mid-1950s tested the inverse equation

$$\tilde{\nu} + p^+ \rightarrow n^0 + \beta^+$$

in a classic experiment using the high flux of antineutrinos from the β^- decays in a nuclear reactor. The neutrinos were fired into a large liquid collision chamber containing an organic liquid or water of high proton density and surrounded by liquid scintillation counters (see Fig. 15.6). By detecting the presence of n^0 and β^+

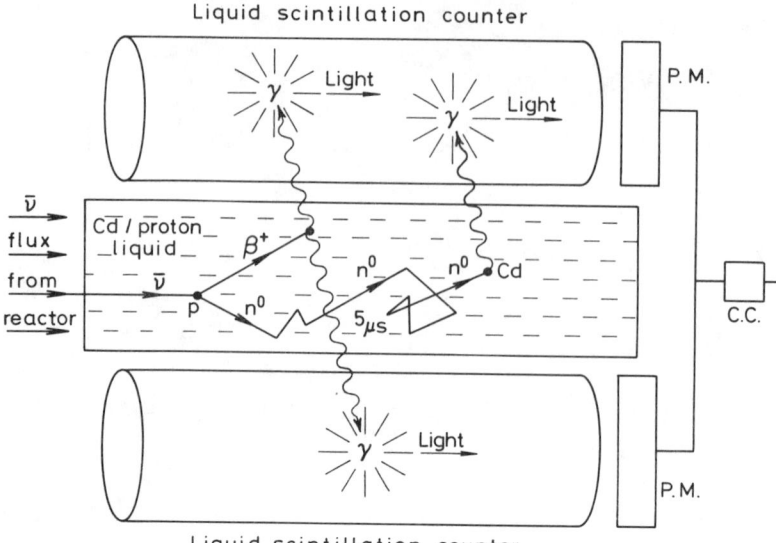

Fig. 15.6 Diagrammatic representation of the Reines and Cowan neutrino experiment.

simultaneously the neutrino equation was confirmed. For detecting the neutrons, the liquid was doped with a cadmium salt. Cadmium has a high neutron capture cross-section through the n, γ reaction. The neutrons travel in a random direction for about 5 μs before producing simultaneous γ-rays from the cadmium collisions. The positrons from the neutrino collisions also produced γ-rays from their decay in flight by the annihilation process. On entering the scintillator tube the γ-rays ionize and excite the liquid molecules. The light radiation from the de-excitation of these molecules causes the 'scintillations' which are then detected by the photomultiplier tubes (P.M. in Fig. 15.6). The calculated time difference between the two γ-pulses was such that the neutron γ's were expected to be recorded about 5 μs after the annihilation γ's. A delayed coincidence counter (C.C. in Fig. 15.6) was therefore set at this figure to detect the existence of neutron- and positron-produced γ's simultaneously. The count rate for these coincidences was 0.41 ± 20 counts per minute or about 2·88 counts per hour. This very low figure agreed well with the anticipated cross-section of only 10^{-48} m^2.

Reines and Cowan proved conclusively that the $\tilde{\nu}p^+ \rightarrow n^0\beta^+$ reaction was real. Checking the energy of the β^+ γ-pulse against that from a known β^+ emitter such as ^{64}Cu identified it unambiguously and removing the Cd from the scintillation liquid removed the second pulse completely, as expected.

The two neutrinos are now well established as free particles in collision experiments. They are fermions with spin $\frac{1}{2}\hbar$, with no charge, no magnetic moment, no rest mass and they do not cause ionization on passing through matter. Like the photon, which is a boson of spin 1, the neutrino has a finite energy and momentum in flight. It travels with the speed of light c and is subject to the laws of relativity. Unlike the photon, the neutrino does not interact with the electromagnetic field, but it does interact with particle matter although with a cross-section of only 10^{-48} m^2. Matter is almost totally transparent to neutrinos.

The shapes of both β^- and β^+ curves such as in Fig. 15.5 can be explained by quantum mechanics, assuming that the neutrino has zero rest mass. Direct experiment shows that the rest mass of the neutrino is certainly less than $0.0005m_e$. The theory was originally due to Fermi.

15.5 The Absorption and Range of β^- rays

The maximum energy of the β^--rays is almost the energy of disintegration and for experimental purposes can be taken as such. The β^-- and $\bar{\nu}$-particles are ejected simultaneously, so that along with the recoil of the product nucleus they conserve linear momentum and together they conserve spin angular momentum.

Beta particles are comparatively easily absorbed by thin sheets of metal, e.g. a sheet of Al 5 mm thick will cut down the intensity of a β^--beam by about 90%. They are thus rather more penetrating than α-rays and in consequence the ionization caused by β^--particles is less than that caused by α-particles of the same energy, due to the high velocity of the β^--particles. The hazard from β^--particles is therefore mainly an internal one by ingestion of contaminated food and drink or by inhalation of airborne radioactive dust.

The absorption of β^--particles by matter follows roughly the exponential law
$$I = I_0 e^{-\mu x}$$
where I_0 is the initial intensity of beam, I is the final intensity of beam after passing through thickness x, μ is the absorption coefficient as measured by this equation.

This equation is rather fortuitous as the β^--rays are not monoenergetic. It is therefore to be treated as an experimental relation. On testing this equation experimentally for pure β^--emitters it is found that the graph of $\log I$ against thickness x is not a straight line as the above equation suggests. Furthermore, the intensity persists even for very thick absorbers and remains constant up to large values of x, as shown in Fig. 15.7. This constant intensity arises from the formation of '*Bremsstrahlung*', or 'braking radiation', by the sudden arrest of the β^--particles within the absorber. This radiation is equivalent to the formation of penetrating X-rays of short wavelength at high values of x, so giving the effect of a beam of β^--particles of constant intensity.

As a single β^--particle passes through an absorbing material it loses energy by multiple collisions until it is no longer able to ionize the atoms of the absorber. It therefore has a fairly well-defined range beyond which it has no ionizing effect. The exponential equation mentioned above which is used to describe this attenuation is purely fortuitous and the range R as shown in Fig. 15.7 then corresponds to those β^--particles in the spectrum with maximum energy $E_{\beta^- \text{max}}$, i.e. the end point of Fig. 15.5. This is also the value of the decay energy. The range of β^--particles can be determined by a simple absorption experiment. Instead of plotting the logarithm of the activity to the thickness x, it is customary to plot it to the area density, in mg cm^{-2} units. The reason for this is twofold. It is much easier to weight a foil and measure its area than to measure its thickness. Also it appears that to a good approximation absorber thicknesses expressed as mg cm^{-2} are nearly independent of the material of the absorber for a given β^- energy. This is

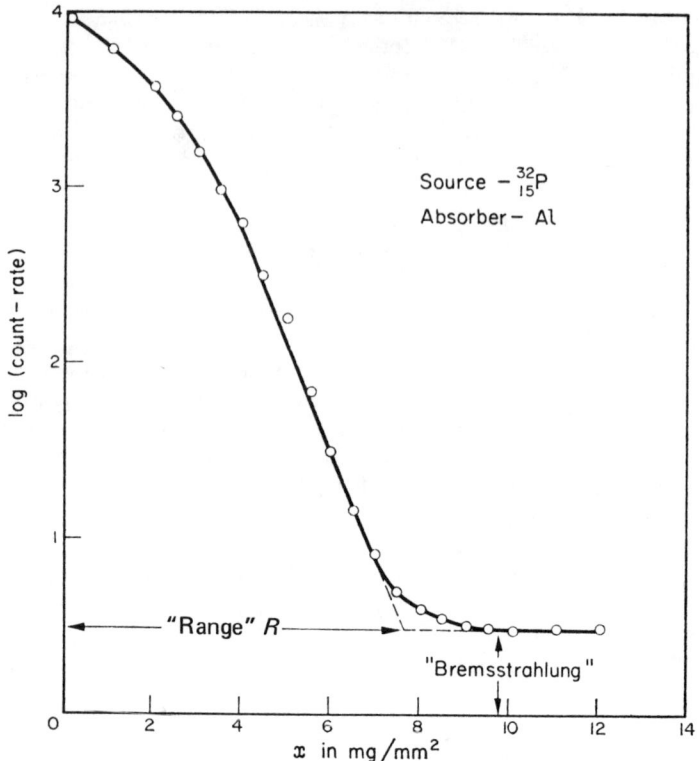

Fig. 15.7 Logarithmic β^--absorption curve.

because the measured intensity of the beam depends on the number of atoms struck, i.e. on the material density, as well as on the initial β^- energy. The ranges of β^--articles in many metal foils are therefore roughly the same, being of the order of 500 mg cm^{-2}. Experimental range–energy curves are therefore much the same for many metals.

A convenient metal absorber is aluminium and Feather has given a formula for the extrapolated range of β^--particles in aluminium in terms of the maximum energy of the β^--particles. This is

$$R = 543 E_{\beta^- \text{max}} - 160$$

with R in mg cm^{-2} and $E_{\beta\text{max}}$ in MeV. From Fig. 15.7 we see that R is 7.7 mg mm^{-2} for $^{32}_{15}\text{P}$ β^--particles, i.e. 770 mg cm^{-2} in aluminium. From the Feather formula, the value of $E_{\beta^- \text{max}}$ for $^{32}_{15}\text{P}$ is given by

$$E_{\beta^- \text{max}} = \frac{770 + 160}{543}$$
$$= 1 \cdot 71 \text{ MeV}$$

This corresponds with the true value as found in Appendix C.

The absorption method outlined above is very convenient for measuring the unknown β^--decay energy of radioactive elements, and many empirical formulae have been used to cover a wide range of energies. It is not as accurate as the magnetic spectrograph method, but is good enough for most purposes.

15.6 The Properties of γ-Rays

Unlike the other types of nuclear radiation, the γ-rays are not corpuscular but consist of short-wavelength electromagnetic radiation in many ways like hard X-rays. Since the energy of the radiation is proportional to the frequency (quantum theory) it is to be expected that the γ-rays, with $\lambda \sim 100$ fm, are correspondingly more energetic than the lower frequency X-rays, for which $\lambda \sim 100$ pm. This is made manifest in the greater penetrating and ionizing properties of γ-rays.

As γ-rays penetrate through matter they are attenuated by three main processes:

(1) Photoelectric effect; the action of the γ-photons emitted by the nucleus on the orbital electrons is exactly the same as the action of an incident ultra-violet photon from outside.
(2) Compton recoils; the loss of energy by a γ-photon by collisions with electrons.
(3) Pair production; the simultaneous creation of a negative and positive electron from a γ-photon.

These three factors can be combined into one single absorption coefficient μ in the exponential law of absorption

$$I = I_0 e^{-\mu x} = I_0 e^{-x/l},$$

where I_0 is the intensity of the beam before passing through a medium of thickness x; I is the intensity after passage through x; μ is the total absorption coefficient; l is the relaxation length $= \mu^{-1}$.

The relative values of different γ-absorbers are usually quoted in half-value thicknesses, $x_{1/2}$, such that $I = \frac{1}{2}I_0$.

We have

$$\ln \frac{I_0}{I} = \mu x$$

and therefore

$$\ln 2 = \mu x_{1/2}$$

or

$$x_{1/2} = \frac{0.693}{\mu}.$$

This is, of course, exactly analogous mathematically to the radioactive half-life $T_{1/2}$. The half-value thickness of lead is about 9 mm for 1 MeV γ-rays.

The absorption coefficient μ can be found experimentally from the exponential equation using monoenergetic γ-rays, thin absorbers and 'good geometry', i.e. narrow collimated beams. Hence $x_{1/2}$ can be calculated. Some typical values of $x_{1/2}$ are shown in Table 15.3 for ^{60}Co γ-rays of average energy 1·25 MeV.

It will be obvious from Table 15.3 that as the relative density ρ of the absorber increases the half-value thickness decreases, i.e. the absorption is proportional to the relative density. In fact it is found experimentally that the mass absorption

TABLE 15.3
$x_{1/2}$ and $x_{1/2}\rho$ for Various Thin Absorbers and ^{60}Co γ-radiation

	Wood	Water	Concrete	Steel	Lead
Relative density, ρ	0.55	1·0	2·3	7·8	11·3
$x_{1/2}$ cm	24	12	5	1·5	1·1
$x_{1/2}\rho$	13·2	12·0	11·5	11·7	12·4

coefficient μ/ρ is nearly constant for most materials. We can write $\mu = \sigma_a N_V$, where σ_a is the absorption coefficient per atom expressed as a cross-section and N_V is the number of absorbing atoms per unit volume, so that the units of $\sigma_a N_V$ are reciprocal length, as required. Thus, $\mu = \sigma_a N_A \rho/A$ where N_A is the Avogadro constant and A is the relative atomic mass number, Putting $\sigma_a = Z\sigma_e$, where σ_e is the effective absorption of a single electron, we finally get

$$\mu = N_A \left(\frac{Z}{A}\right) \rho \sigma_e.$$

Since Z/A is nearly constant for all atoms, we see that $\mu \propto \rho$. Now $x_{1/2}\mu = 0.693$, so that $x_{1/2}\rho$ is also constant. This is true for different materials so long as the same γ-rays are used. Table 15.3 shows that the product $x_{1/2}\rho$ for several thin absorbers for ^{60}Co radiation is nearly constant.

Taking the product $x_{1/2}\rho$ as 12, we can easily find the $x_{1/2}$ value for any material from its relative density. This gives a quick estimate of thin shielding requirements for radiation protection. Thus a thickness of $3x_{1/2}$ will give a transmission of only 12·5% of the incident radiation, for thin shields only.

A more practical concept is that of one-tenth value thickness $x_{1/10}$. This is given by $x_{1/10} = 3.32 x_{1/2}$ for thin shields. In practice, for thick shields, there is an additional factor to account for the multiple scattering of the γ-photons through the shield. This is called the 'build-up' factor', B. The exponential equation is then $I = I_0 B e^{-\mu x}$ so that the incident radiation intensity is enhanced. Real $x_{1/10}$ values are therefore greater than those given by the simple formula. Typical values are given in Table 15.4, for ^{60}Co radiation again.

The product $x_{1/10}\rho$ is still fairly constant at about $x_{1/10}\rho = 70$, except for lead which has a relatively high μ/ρ value at energies of about 1 MeV.

TABLE 15.4
Experimental $x_{1/10}$ Values for Thick Absorbers and ^{60}Co γ-Radiation

	Wood	Water	Concrete	Steel	Lead
Calculated from $3.32\ x_{1/2}$ (cm)	80	40	16·6	5	3·7
Experimental $x_{1/10}$ values (cm)	125	65	30	9·4	4·6
Experimental values of $x_{1/10}\rho$	69	65	69	73	42

This discussion is very relevant to the radiation protection problems outlined in Appendix B.

γ-rays have their origin in the excited states of nuclei after the emission of an α- or β^--particle or both or after nuclear reactions. As one might expect by comparison with the electron energy states of the atom as a whole, the nuclear energy states are also quantized. The γ-photon energy is then a measure of the energy difference between the states involved in the γ-decay, as shown in Fig. 15.8. Thus a complex pattern of γ-ray energies can be associated with nuclear energy states in much the same way that the optical spectra can be associated with atomic energy states, but so far no accurate theoretical predictions of nuclear energy states are more difficult to make.

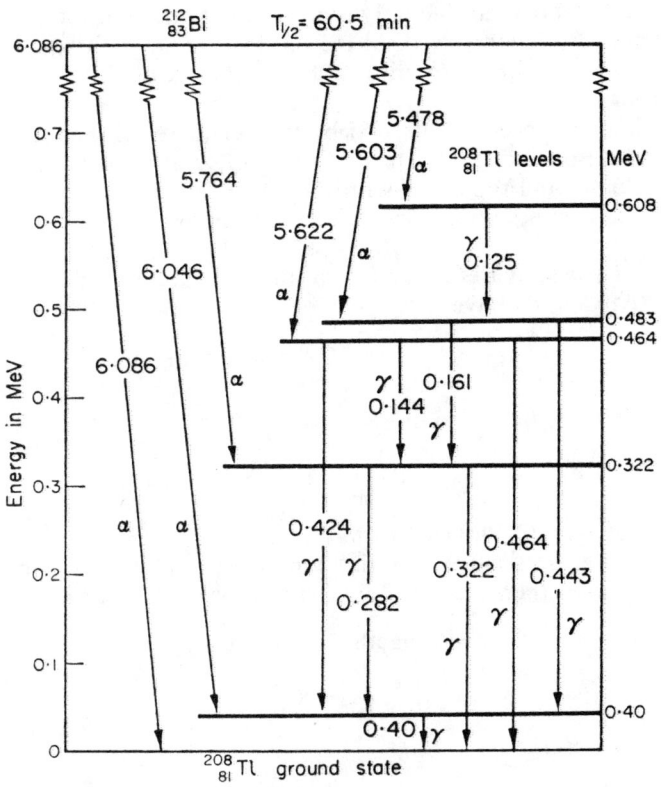

Fig. 15.8 Gamma energy scheme for $^{208}_{81}$Tl after α-decay from $^{212}_{83}$Bi. Recoil energy omitted. All energies in MeV.

γ-rays are the most penetrating of all natural radiations and together with neutron beams constitute the major external hazard to the human body. Radiation dangers to laboratory personnel are mainly neutron and γ-ray hazards and are dealt with in Appendix B.

15.7 Radioactivity as a Measurable Quantity

In Chapter 4 we saw how the law of radioactive decay may be written in the form $N_t = N_0 e^{-\lambda t}$, where N_0 and N_t are respectively the numbers of atoms present at $t=0$ and $t=t$. These numbers are never actually known, so that we must choose a standard unit quantity of radioactivity and measure the N values relative to this quantity. For many years the standard unit of radioactivity was the *curie*, Ci, defined as that quantity of any radioactive substance which has a decay rate of 3.7×10^{10} disintegrations per second. This awkward number is taken from the specific disintegration rate of radium, viz. 1 g of radium disintegrates at the rate of 3.7×10^{10} atoms s^{-1}. The common subunits are:

$$1 \text{ millicurie (1 mCi)} = 3.7 \times 10^7 \text{ dis s}^{-1},$$
$$1 \text{ microcurie}(1\mu\text{Ci}) = 3.7 \times 10^4 \text{ dis s}^{-1}.$$

Substances with a short half-life only require small quantities to give a curie of activity and conversely long-lived substances require large quantities, e.g. $^{238}_{92}\text{U}$, for which $T_{1/2} = 4.5 \times 10^9$ a, requires about 3 tonnes of metal to give a total activity of 1 Ci.

The curie has now been superseded by the becquerel (Bq) as the S.I. unit of activity, where 1 Bq = 1 dis s^{-1}. Thus 1 Ci = 37 GBq. It is intended to phase out the curie gradually and replace it with the becquerel.

The disintegration equation can also be written as
$$A_t = A_0 e^{-\lambda t}$$
in terms of the curie activities A_0 and A_t at time zero and time t respectively. For any curie activity A, we have
$$\frac{dN}{dt} = A \times 3.7 \times 10^{10} \text{ numerically}$$
$$= \lambda N$$
so that
$$A = \frac{\lambda N}{3.7 \times 10^{10}}$$
and therefore $A_t = A_0 e^{-\lambda t}$ follows from $N_t = N_0 e^{-\lambda t}$.

If the mass of a radioactive material consisting of a nuclide of gram atomic mass M is m grams, then the number of nuclei present is given by
$$N = \text{Avogadro constant} \times \frac{m}{M}$$
$$= 6.02 \times 10^{23} \frac{m}{M} \text{ nuclei}$$
and the activity will be
$$A = \frac{\lambda}{3.7 \times 10^{10}} \times 6.02 \times 10^{23} \times \frac{m}{M}$$
$$= 1.63 \times 10^{13} \frac{\lambda m}{M} \text{ Ci}$$

Hence 1 μg of ThX ($T_{1/2} = 3.64$ d, $\lambda = 2.20 \times 10^{-6}$ s^{-1}) will have an activity of
$$1.63 \times 10^{13} \times 2.2 \times 10^{-6} \times \frac{10^{-6}}{222} = 0.154 \text{ Ci}.$$

Note that a nuclide in a radioactive decay chain emitting two particles (e.g. α- and β^--particles) simultaneously will have twice the activity calculated for a single particle.

Problems

(Those problems marked with an asterisk are solved in full at the end of the section.)

15.1 From a table of half-lives and corresponding energies of α-emitters of the same series verify that the Geiger–Nuttall relation may be written
$$\log T_{1/2} = m_1 \log E + B_1$$
and evaluate the constants m_1 and B_1. ($m_1 = -70$, $B_1 = 52$ in a MeV units)

15.2 Examine the manner in which the constants m and B in the Geiger–Nuttall relation vary between the different radioactive series.

15.3* The mass of a moving electron, m, is given by the relativity relation
$$m = \frac{m_0}{\sqrt{1 - v^2/c^2}},$$
where m_0 = rest mass, v velocity and c velocity of light, so that
$$\left(\frac{e}{m}\right)_{\text{measured}} = \left(\frac{e}{m_0}\right)_{\text{rest}} \left(1 - \frac{v^2}{c^2}\right)^{1/2}.$$
The maximum energy of β^--particles from $^{32}_{15}\text{P}$ is 1·71 MeV. What magnetic field perpendicular to a beam of β^--particles from $^{32}_{15}\text{P}$ would bend it to give a radius of 100 mm? (0·072 T)

15.4 It is possible to measure the energy of β^--particles by measuring their absorption in aluminium. The following are some results for $^{32}_{15}\text{P}$, after correcting for background in such an experiment:

Absorber thickness (mg mm^{-2})	Activity (counts min^{-1})
12	3
10	3
9	3
8	4
7	7
6	32
5	161
4	596
3	1493
2	3370
1	5411
0	9023

Using the empirical formula $E = 0.185R + 0.245$, where R is the range in mg mm^{-2} and E is the maximum β^- energy in MeV, determine E and compare your result with Problem 15.3. (1·75 MeV)

15.5 Determine the one-tenth value thickness of aluminium for γ-rays of various energies from the following data of relative intensities:

Aluminium thickness (mm)	2·7 MeV	1·2 MeV	0·8 MeV
300	0·060	0·010	0·005
200	0·150	0·045	0·025
150	0·240	0·095	0·065
100	0·385	0·210	0·160
50	0·620	0·455	0·400
0	1·000	1·000	1·000

Comment on the results obtained. (240 mm, 145 mm, 125 mm)

15.6 Using the radioactive decay formula show that the rate of accumulation of a daughter nucleus is given by

$$\frac{dN_2}{dt} = \lambda_1 N_1 - \lambda_2 N_2$$

and that the total amount of daughter nucleus accumulated in time t is approximately

$$N_2 = \frac{\lambda_1}{\lambda_2} N_1 (1 - e^{-\lambda_2 t}).$$

What is the value of N_2 when secular equilibrium has been reached?

15.7 What is the volume of 1 μCi of radon gas at s.t.p.? Density $= 10$ kg m^{-3}. (10^{-10} m^3 = 0·1 mm^3)

15.8 Calculate the disintegration rate per gram of each member of the uranium series, and the corresponding activities in curies.

15.9* Using the present-day abundances of the two main uranium isotopes and assuming that the abundance ratio could never have been greater than unity, estimate the maximum possible age of the Earth's crust. (6×10^9 a)

15.10 The abundances of $^{238}_{92}U$ and $^{234}_{92}U$ in present-day natural uranium are:

$$^{238}_{92}U = 99·28\%$$
$$^{234}_{92}U = 0·0058\%$$

The half-life of $^{238}_{92}U$ is well established as $4·498 \times 10^9$ a. Calculate the half-life of $^{234}_{92}U$. ($2·63 \times 10^5$ a)

What further data are required to calculate the half-life of the other important uranium isotope, vix. $^{235}_{92}U$, with an abundance of 0·71%?

15.11 Calculate the specific activity in microcuries per gram of (a) $^{238}_{92}U$ and (b) $^{24}_{11}Na$.

15.12 How much lead is produced by a mass M of primaeval $^{238}_{92}U$?

Solutions to Problems

15.3 Relativity kinetic energy is given by

$$E = m_0 c^2 \left[\frac{1}{\sqrt{1-v^2/c^2}} - 1 \right] = E_0 \left[\frac{1}{\sqrt{1-v^2/c^2}} - 1 \right].$$

Thus
$$1.71 \text{ MeV} = 1.71 \times 1.6 \times 10^{-13} \text{ J}$$
$$= 2.74 \times 10^{-13} \text{ J for } E.$$

Hence
$$\frac{2.74 \times 10^{-13}}{9.1 \times 10^{-31} \times 9 \times 10^{16}} = \frac{1}{\sqrt{1-v^2/c^2}} - 1,$$

from which
$$\frac{1}{\sqrt{1-v^2/c^2}} = 4.34$$

giving
$$\frac{v^2}{c^2} = 0.947$$

or $v = 2.92 \times 10^8$ m s^{-1}, i.e. nearly the velocity of light. Now
$$Bev = \frac{mv^2}{r}$$

giving
$$B = \frac{m_0}{\sqrt{1-\frac{v^2}{c^2}}} \frac{v}{er}$$
$$= \left(\frac{m_0}{e}\right) \frac{1}{\sqrt{1-v^2/c^2}} \frac{v}{r}$$
$$= \frac{1}{1.76 \times 10^{11}} \times 4.34 \times \frac{2.92 \times 10^8}{10^{-1}} \quad \text{for} \quad r = 100 \text{ mm}$$
$$= 0.072 \text{ T}.$$

15.9 This problem is due to Rutherford. The uranium isotopes involved are $^{238}_{92}$U and $^{235}_{92}$U with a present-day abundance ratio of 137·8:1.

Now for $^{238}_{92}$U we have $N_8 = N_{08}e^{-\lambda_8 t}$, where N_8 and N_{08} refer to present and original numbers of $^{238}_{92}$U atoms involved, t is measured from $t = 0$, i.e. is the age of the crust, and λ_8 is the radioactive decay constant of $^{238}_{92}$U such that $\lambda_8 = 0.693/T_8$, where T_8 is the corresponding half-life.
$$\frac{N_8}{N_{08}} = e^{-\lambda_8 t} = e^{-0.693 t/T_8}.$$

Similarly
$$\frac{N_5}{N_{05}} = e^{-\lambda_5 t} = e^{-0.693 t/T_5}.$$

Therefore
$$\frac{N_8}{N_5} \frac{N_{05}}{N_{08}} = \exp\left\{0.693 t \left(\frac{1}{T_5} - \frac{1}{T_8}\right)\right\},$$

where $T_5 = 7.13 \times 10^8$ a and $T_8 = 4.5 \times 10^9$ a.

Now $N_8/N_5 = 137.8$ and we assume $N_{05}/N_{08} = 1$, the maximum value with $t = t_{max}$. Thus

$$\log_{10} 137\cdot8 = 0\cdot4343 \times 0\cdot693 \times t_{max} \times (1\cdot18 \times 10^{-9})$$
$$2\cdot1392 = 0\cdot4343 \times 0\cdot693 \times t_{max} \times 1\cdot18 \times 10^{-19},$$
where t is in years, from which
$$t_{max} = 6 \times 10^9 \text{ a}.$$

Chapter 16

Nuclear Bombarding Experiments

16.1 Single α-Particle Scattering

In Chapter 3 we gave an account of the early work of Rutherford on the theory of single α-particle scattering and of the experiments of Geiger and Marsden on α-particle scattering in metals. The work on gold foil was of great importance because it was these experiments that formed the basis of the so-called Rutherford–Bohr atom. Since most of the incident α-particles were not noticeably deflected by the gold foil in Rutherford's experiment, he deduced that the atom was almost 'empty', and since some particles were actually repelled it was concluded that the whole positive charge of the atom was concentrated in a single central point, see Fig. 16.1. It was shown later that the total positive charge on the nucleus was Ze, where Z is the order number of the element in the periodic table, or simply the atomic number. Finally, for the first time, it was possible to deduce from the results of these early scattering experiments that the size of the nucleus was less than about 10 fm, a figure which agrees well with nuclear sizes as measured by modern methods.

16.2 Nuclear Alchemy

Towards the end of the First World War, Rutherford turned his attention to the scattering of α-particles by gases, using the apparatus shown in Fig. 16.2. The range of α-particles from $^{214}_{84}$Po (RaC′) of energy 7·68 MeV is about 70 mm in air at normal pressure, so that using air there would be no scintillations on the screen when $AB > 70$ mm. This apparatus enabled Rutherford to measure the effect of bombarding nuclei of gas atoms and molecules with α-particles. With hydrogen in the chamber the scintillations were due to projected protons as expected, since the action of the α-particles was merely to 'knock-on' the hydrogen nuclei by direct collision. The maximum range of these protons in air was equivalent to about 300 mm, or to about 1·2 m in hydrogen, and their positive charge was demonstrated by magnetic deflection. When oxygen or carbon dioxide were added the number of proton scintillations decreased in accordance with the increased mass absorptive power of the gas. In dry air, however, the number of long-range proton scintillations on the screen actually *increased* and the equivalent range in air was found to be about 400 mm. The possibility of hydrogen contamination providing the protons was eliminated by careful cleaning, and eventually Rutherford showed that exactly the same effect could be

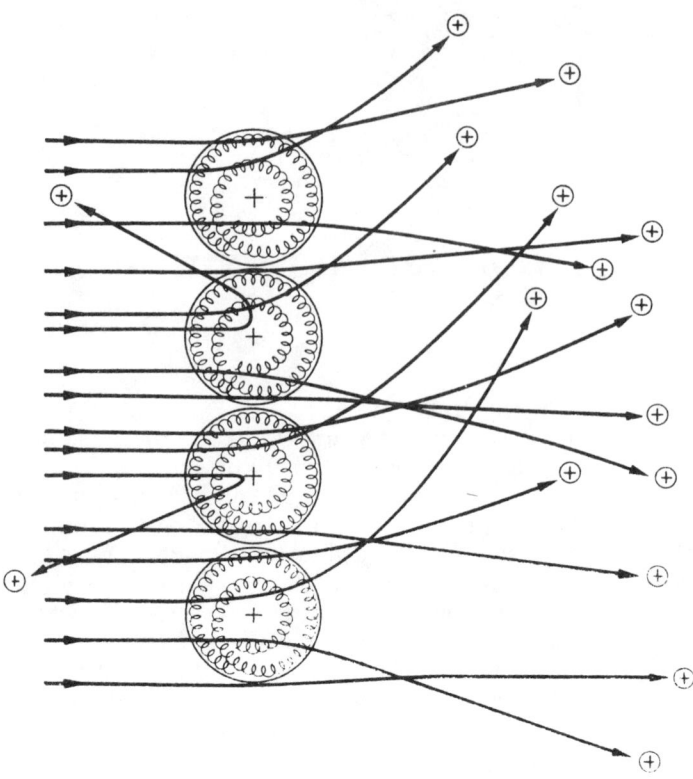

Fig. 16.1 Passage of α-particles through Rutherford–Bohr atom showing forward and backward scatter.

obtained by replacing the air by nitrogen gas. From this it was concluded that fast protons of maximum range equivalent to 400 mm in air were always produced by the α-particle bombardment of air and nitrogen, the nitrogen in the air being responsible for the protons. The most important conclusion from these experiments was that nitrogen gas *always* produced fast protons when bombarded with particles and so Rutherford wrote: '... if this be the case we must conclude that the nitrogen atom is disintegrated under the intense forces developed in a close collision with a swift α-particle.' He suggested that the effect of an α-particle on a nitrogen nucleus leads to the formation of fast protons by the disintegration of the nucleus.

We can thus write the Rutherford nuclear reaction in the same way as a chemical reaction:

$$^4_2He + ^{14}_7N \rightarrow ^{17}_8O + ^1_1H,$$

where both Z and A must balance on both sides of the equation. This equation can be expressed simply by the notation $^{14}_7N\,(\alpha, p)\,^{17}_8O$. It must not be considered that this reaction always takes place for all nitrogen atoms bombarded with α-particles. It only takes place when an α-particle happens by chance to make an

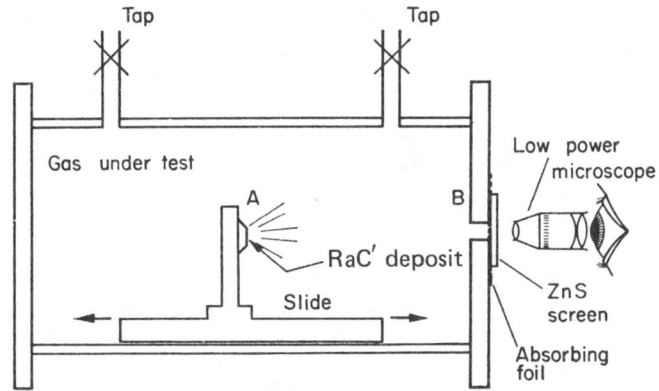

Fig. 16.2 Rutherford's apparatus for α-particle scattering in hydrogen and nitrogen.

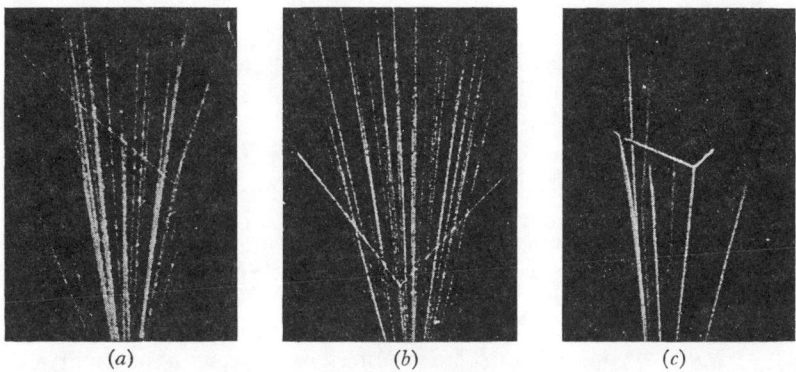

Fig. 16.3 Wilson cloud chamber photographs of single α-particle collisions with (*a*) a hydrogen atom; (*b*) a helium atom; (*c*) an oxygen atom. (Taken from *Radiations from Radioactive Substances* by Rutherford, Chadwick and Ellis, C.U.P., 1930.)

almost 'head-on' collision with a nitrogen nucleus, which occurs about once in 50 000 particles. Sincle α-particle collisions are shown very beautifully by the single fork-like tracks in the Blackett cloud chamber photographs of α-particle tracks in Fig. 16.3. This cloud chamber technique is dealt with more fully in Chapter 17.

Blackett examined some 20 000 cloud-chamber photographs including 400 000 α-particle tracks. Only eight were of the forked type of Fig. 16.3. Stereographic photographs (Fig. 17.2) showed unambiguously that the three tracks of a collision event were coplanar and there was never a fourth track. Thus there were only two product particle tracks, a thin one due to the proton and a thick one due to a much heavier nucleus. This proved that the N (α, p) O interpretation was essentially correct.

In this experiment Rutherford achieved the first artificial transmutation of one element into another, viz. nitrogen into oxygen, by occasional α-particle collisions with nitrogen nuclei, and so laid the foundation of many nuclear experiments carried out during the next decade.

When writing down the equation of a nuclear reaction not only must we remember to balance both Z and A but also to insert the total kinetic energy change Q of the reaction, thus:
$$^4_2He + ^{14}_7N \rightarrow ^{17}_8O + ^1_1H + Q.$$
This nuclear reaction energy Q can be positive or negative. When positive the reaction is said to be exoergic (cf. exothermic in chemistry), and when negative the reaction is endoergic (cf. endothermic in chemistry). In the case of Rutherford's $^{14}_7N$ (α, p) $^{17}_8O$ reaction the value of Q can be calculated from the masses of the constituents. Since Q is the net kinetic energy of all the particles (incident and final) we can calculate this from the masses of the particles as follows:

$$\begin{array}{llll}
\text{L.H.S.} & ^4He = 4\cdot0026 & \text{and R.H.S.} & ^1H = 1\cdot0078 \\
& ^{14}N = 14\cdot0031 & & ^{17}O = 16\cdot9991 \\
& \overline{18\cdot0057} & \longrightarrow & \overline{18\cdot0069}
\end{array}$$

giving an increase of mass of 0·0012 u. Since 1 u ≡ 931 MeV, we have $Q = -0\cdot0012 \times 931 = -1\cdot12$ MeV.

The reaction is therefore endoergic, i.e. 1·12 MeV energy has to be provided before the reaction can take place.

Thus $Q = E_H + E_O - E_\alpha = -1\cdot12$ MeV, assuming the $^{14}_7N$ nucleus is initially at rest. But $E_\alpha = 7\cdot68$ MeV from $^{214}_{84}Po$ and so
$$E_H + E_O = 6\cdot56 \text{ MeV,}$$
which is the energy available for distribution between the products.

It is now assumed in this reaction that the α-particle is captured by the nitrogen nucleus and forms a new or compound nucleus of fluorine, $^{18}_9F$, which immediately emits a proton and an oxygen nucleus in opposite directions. By applying the law of conservation of momentum we have, with the usual notation, $m_H v_H = m_O v_O$ (assuming the compound nucleus of fluorine is momentarily at rest). Therefore
$$\frac{v_H}{v_O} = \frac{m_O}{m_H} = \frac{17}{1}$$
and so
$$\frac{E_H}{E_O} = \frac{\tfrac{1}{2}m_H v_H^2}{\tfrac{1}{2}m_O v_O^2} = \left(\frac{m_H}{m_O}\right)\left(\frac{m_O}{m_H}\right)^2 = \frac{1}{17} \times \frac{17^2}{1} = \frac{17}{1}.$$
Thus
$$\frac{E_H}{E_O + E_H} = \frac{17}{18} \quad \text{or} \quad E_H = \frac{17}{18} \times (\text{Total available energy}).$$
Hence
$$E_H = \frac{17}{18} \times 6\cdot56 \text{ MeV}$$
$$= 6\cdot2 \text{ MeV.}$$

This is the proton energy which corresponds to the observed range of 400 mm in air as already described. (See Problem 16.7 for a more general treatment.)

This theoretical value is confirmed by measurements of the range of the protons ejected. From Geiger's empirical law (see p. 213) connecting the range R and the velocity v, we have $R = av^3$, where a is a constant. For α-particles in air, $R = 70$ mm and $E_\alpha = 7\cdot68$ MeV, so that the constant a can be found. If we now assume that Geiger's law is true for protons as well as α-particles, we can calculate the proton energy E_H as follows:
$$E_\alpha = \tfrac{1}{2}m_\alpha v_\alpha^2 \quad \text{and} \quad E_H = \tfrac{1}{2}m_H v_H^2,$$
so that
$$\frac{E_\alpha}{E_H} = 4\left(\frac{v_\alpha}{v_H}\right)^2 \quad \text{or} \quad \left(\frac{v_\alpha}{v_H}\right)^3 = \frac{1}{8}\left[\frac{E_\alpha}{E_H}\right]^{3/2},$$
giving $R_H = 8R_\alpha[E_H/E_\alpha]^{3/2}$ from Geiger's law. Therefore
$$E_H = \left[\frac{1}{8}\frac{R_H}{R_\alpha}\right]^{2/3} \times E_\alpha = \left[\frac{1}{8}\frac{40}{7}\right]^{2/3} \times 7\cdot68 = 6\cdot14 \text{ MeV},$$
agreeing with the experimental value worked out above.

Thus, Rutherford's α-particle reaction was shown to be a true transmutation and in the 1920s it was followed by the investigation of many more reactions of the same nature. In some of these, as for example, the $^{27}_{13}$Al (α, p) $^{30}_{14}$Si reaction,
$$^{27}_{13}\text{Al} + ^{4}_{2}\text{He (α)} \rightarrow ^{30}_{14}\text{Si} + ^{1}_{1}\text{H (p)} + 2\cdot26 \text{ MeV},$$
the value of Q was positive, and kinetic energy was therefore *created* by the disintegration process, and this was often typical for α-particle reactions with light elements. For heavy elements the energy required for an α-particle reaction was greater than that available from natural radioactive substances due to the high potential barrier of the heavier nuclei, so that the early researches on the (α, p) reactions were limited to elements for which the atomic number was less than 20. This is not surprising in view of what we have said about the barrier potential of heavy elements.

16.3 Cockcroft–Walton Proton Experiments

During the 1920s nuclear research was limited to the study of various (α, p) reactions for many light element targets. The reaction products were studied quantitatively in terms of range and angular distribution in space. In many cases it was established that the protons were emitted in two homogeneous groups, each with its own characteristic range, indicating that energy states might exist within the nucleus.

There was no significant advance in this work until 1930–2, when Cockcroft and Walton developed their machine for accelerating protons and so provided missiles from non-natural sources. These artificially produced fast particles were used to bombard light elements and interesting transmutations of another type were obtained.

The machine used by Cockcroft and Walton was a version of the R–C coupled voltage doubler (see Chapter 18, p. 260) and this was the first important charged particle machine to be used in nuclear physics research. The protons were accelerated to about 0·7 MeV, which was not so high as the natural α-particle energy used by Rutherford but enough to cause transmutations by proton penetration of the potential barriers of light elements.

The first element to be transmuted with protons was lithium, and the resulting tracks were eventually observed in the cloud chamber by Dee and Walton. Two tracks of α-particles were found at 180° to each other and were attributed to the reaction
$$^7_3\text{Li} + ^1_1\text{H}(p) \rightarrow ^4_2\text{He}(\alpha) + ^4_2\text{He}(\alpha).$$
This reaction can be written $^7_3\text{Li}(p, \alpha)^4_2\text{He}$ and in general Cockcroft–Walton reactions were of the (p, α) type. This was confirmed in several cases, as for example in the interesting reaction investigated by Dee, in which three α-particles were observed:
$$^{11}_5\text{B} + ^1_1\text{H}(p) \rightarrow ^8_4\text{Be} + ^4_2\text{He}(\alpha)$$
followed by $^8_4\text{Be} \rightarrow 2^4_2\text{He}(\alpha)$, establishing that the ^8_4Be isotope is unstable. Again the emission of various energy groups of α-particles indicated the possible existence of nuclear energy levels.

16.4 The Neutron

The new Cockcroft–Walton type of reaction was immediately followed by the discovery of the neutron, the identification of which has already been discussed in Chapter 3. It will be recalled that neutrons were produced when α-particles bombarded beryllium or boron and Chadwick suggested the following reactions took place:
$$^9_4\text{Be} + ^4_2\text{He}(\alpha) \rightarrow ^{12}_6\text{C} + ^1_0\text{n} \quad \text{and} \quad ^{11}_5\text{B} + ^4_2\text{He}(\alpha) \rightarrow ^{14}_7\text{N} + ^1_0\text{n}.$$
Maximum energy protons were ejected from the paraffin by occasional direct 'knock-on' neutron collisions in which the neutrons, having nearly equal mass, transferred all their energy to the protons. The proton energy is then a measure of the incident neutron energy. (See Problem 21.2.)

Details have already been given in Chapter 3 of the method by which it was first shown that the neutron had about the same mass as the proton. The accepted value for the mass of the neutron is now 1·008 665 u, so that the masses of the proton and the neutron are nearly the same and, except in precise energy calculations, can be regarded as identical. We shall deal in more detail with this in Chapter 20.

16.5 Nuclear Reactions

In all types of nuclear reaction we have
$$A + B + E_1 \rightarrow C + D + E_2,$$
where A, B, C and D are the reacting nuclei and E_1 and E_2 are the initial and final kinetic energies respectively. The Q value is $E_2 - E_1$, or more conveniently $Q = (M_A + M_B) - (M_C + M_D)$ in terms of *atomic* masses.

Bohr suggested that the collision of the A and B nuclei leads in the first place to the formation of a compound nucleus AB^*. This compound nucleus disintegrates into $C + D$ with a lifetime which may be as long as 0·1 fs. This has to be compared with the basic nuclear time of 10^{-23} s = 10^{-8} fs corresponding to the time taken for a particle travelling with the velocity of light (3×10^8 m s^{-1}) to cross a nuclear diameter (say 3 fm). The lifetime of the compound nucleus is then relatively long.

When an individual compound nucleus is formed it is not always certain into what particles it will disintegrate. Thus α-particle bombardment may produce protons or neutrons, as we have already seen, so that

$$(\alpha, p) \quad \text{or} \quad (\alpha, n)$$

reactions are possible. Similarly, proton bombardment may produce

$$(p, \alpha) \quad \text{or} \quad (p, n)$$

reactions, or even (p, γ) reactions. Occasionally a deuteron is emitted as in

$$^{9}_{4}\text{Be} + ^{1}_{1}\text{H (p)} \rightarrow ^{10}_{5}\text{B}^{*} \rightarrow ^{8}_{4}\text{Be} + ^{2}_{1}\text{H (d)},$$

where the deuteron $^{2}_{1}\text{H}$ is a heavy hydrogen nucleus and can itself be used as a bombarding particle.

In addition to the above, we must therefore include deuteron reactions in our brief survey of possible nuclear reactions. Some examples are

$$(d, \alpha), \quad (d, p), \quad (d, n), \quad (d, 2n)$$

reactions. The list of possible bombarding particles now becomes α, p, d, n and γ (photons).

In the last case, we refer to the γ-bombardment as a photoreaction, for example the reaction

$$^{9}_{4}\text{Be} + \gamma \rightarrow ^{8}_{4}\text{Be} + ^{1}_{0}\text{n}$$

can be regarded as a photoneutron reaction.

In the deuteron reaction with $^{27}_{13}\text{Al}$ the following products have been observed:

$$^{27}_{13}\text{Al} + ^{2}_{1}\text{H} \rightarrow ^{29}_{14}\text{Si}^{*} \rightarrow ^{25}_{12}\text{Mg} + ^{4}_{2}\text{He}(\alpha),$$
$$^{27}_{13}\text{Al} + ^{2}_{1}\text{H} \rightarrow ^{29}_{14}\text{Si}^{*} \rightarrow ^{28}_{13}\text{Al} + ^{1}_{1}\text{H (p)},$$
$$^{27}_{13}\text{Al} + ^{2}_{1}\text{H} \rightarrow ^{29}_{14}\text{Si}^{*} \rightarrow ^{24}_{11}\text{Na} + ^{4}_{2}\text{He} + ^{1}_{1}\text{H (p)},$$
$$^{27}_{13}\text{Al} + ^{2}_{1}\text{H} \rightarrow ^{29}_{14}\text{Si}^{*} \rightarrow ^{28}_{14}\text{Si} + ^{1}_{0}\text{n}.$$

This emphasizes the fact that the products obtained from the disintegration of a compound nucleus are somehow dependent on the physical conditions of the nucleus at the exact moment of disintegration. All energetically possible reactions occur in proportions determined by selection rules.

An interesting reaction is

$$^{6}_{3}\text{Li} + ^{2}_{1}\text{H (d)} \rightarrow 2\,^{4}_{2}\text{He}(\alpha) + 22\cdot2 \text{ MeV}$$

which gives one of the highest non-fission energies recorded for light atoms.

16.6 Formation of Tritium

An important isotope of hydrogen tritium $^{3}_{1}\text{H}$, is formed in deuteron–deuteron bombardment.

$$^{2}_{1}\text{H} + ^{2}_{1}\text{H} \rightarrow ^{4}_{2}\text{He}^{*} \rightarrow ^{3}_{1}\text{H} + ^{1}_{1}\text{H},$$

where

$$^{2}_{1}\text{H} + ^{2}_{1}\text{H} \rightarrow ^{4}_{2}\text{He}^{*} \rightarrow ^{3}_{2}\text{He} + ^{1}_{0}\text{n}$$

is an alternative.

Tritium is a β^{-}-emitter according to

$$^{3}_{1}\text{H} \rightarrow ^{3}_{2}\text{He} + ^{0}_{-1}\text{e}\,(\beta^{-}) + \tilde{\nu}$$

with a half-life of 12·5 a. Here $\tilde{\nu}$ is an antineutrino.

The β^{-}-energy is very low, 0·018 MeV, and so the use of $^{3}_{1}\text{H}$ in counting experiments demands very sensitive apparatus, which accounts for the fact that tritium was not discovered as a reaction product until 1939.

Tritium is now produced commercially by bombarding a lithium target with reactor neutrons, giving
$$^6_3\text{Li} + ^1_0\text{n} \rightarrow ^3_1\text{H} + ^4_2\text{He}(\alpha).$$
Tritium is the lightest radionuclide there is and is used extensively in tracer work. It also draws attention to the stable ^3_2He nuclide, which is the only nuclide with more protons than neutrons in its nucleus, excepting the ^1_1H nucleus. This shows that there must be a strong attractive force within the ^3_2He nucleus to counteract the strong interproton Coulomb repulsion and so make the nucleus stable.

Problems

(Those problems marked with an asterisk are solved in full at the end of the section.)

16.1* Given the following isotope masses:
$$^7_3\text{Li} = 7.016\,004, \quad ^6_3\text{Li} = 6.015\,125 \quad \text{and} \quad ^1_0\text{n} = 1.008\,665$$
calculate the binding energy of a neutron in the ^7_3Li nucleus. Express the result in u, MeV and joules. (0.007 786 u, 7.3 MeV, 1.18×10^{-12} J)

16.2 Calculate the binding energy in MeV of beryllium of mass 8.005 309 if $^1_0\text{n} = 1.008\,665$ and $^1_1\text{H} = 1.007\,825$. (56.5 MeV)

16.3 From the reaction
$$^{14}_7\text{N} + ^1_0\text{n} \rightarrow ^{15}_7\text{N}^* \rightarrow ^{14}_6\text{C} + ^1_1\text{H} + 0.55 \text{ MeV}$$
calculate the mass of $^{14}_6\text{C}$ if $^{14}_7\text{N} = 14.003\,074$, $^1_0\text{n} = 1.008\,665$ and $^1_1\text{H} = 1.007\,825$. (14.003 323)

16.4 When lithium is bombarded with protons the following reactions can occur:
$$^7_3\text{Li} + ^1_1\text{H} \rightarrow ^8_4\text{Be}^* \rightarrow ^8_4\text{Be} + \gamma + 14.4 \text{ or } 17.3 \text{ MeV}$$
or
$$^7_3\text{Li} + ^1_1\text{H} \rightarrow ^8_4\text{Be}^* \rightarrow ^4_2\text{He} + ^4_2\text{He} + 2.9 \text{ MeV}$$
From these data deduce a simple energy level diagram of the nuclide, ^8_4Be.

16.5 When $^{27}_{13}\text{Al}$ is bombarded with α-particles, protons, deuterons or neutrons about twelve different nuclear reactions may occur. Write down some of these reactions and find which are exoergic and which endoergic.

16.6* In the following deuterium reactions the reaction energy is as stated:
$$^{14}_7\text{N}(d, p)\,^{15}_7\text{N}, \quad Q = 8.61 \text{ MeV},$$
$$^{15}_7\text{N}(d, \alpha)\,^{13}_6\text{C}, \quad Q = 7.68 \text{ MeV};$$
$$^{13}_6\text{C}(d, \alpha)\,^{11}_5\text{B}, \quad Q = 5.16 \text{ MeV};$$
$$^{11}_5\text{B}(\alpha, n)\,^{14}_7\text{N}, \quad Q = ?$$

If $^4_2\text{He} = 4.002\,603$, $^2_1\text{H} = 2.014\,102$, $^1_1\text{H} = 1.007\,825$ and $^1_0\text{n} = 1.008\,665$, what is the Q value of the fourth reaction? (0.15 MeV)

16.7 A light particle a collides with a heavy particle A at rest and a light particle b is emitted along with a heavy partible B. With the usual notation show that the Q value of this reaction is given by
$$Q = E_a\left[\frac{m_a}{M_B} - 1\right] + E_b\left[\frac{m_b}{M_B} + 1\right] - \frac{2}{M_B}(E_a E_b m_a m_b)^{1/2}$$
when b is emitted in the same direction as the path of a. Apply this to the

Rutherford $^{14}_{7}\text{N}(\alpha, p)^{17}_{8}\text{O}$ reaction and compare your result with the known value of Q.

16.8 The Q value of the original Chadwick reaction $^{9}_{4}\text{Be}(\alpha, n)^{12}_{6}\text{C}$ was 5·7 MeV. If the energy of the incident α-particles was 5·3 MeV, show that the energies of those neutrons observed perpendicular to the α-particle direction was about 8·5 MeV.

16.9 What is the Q value of the reactions $^{7}_{3}\text{Li}(p, n)^{7}_{4}\text{Be}$ and $^{7}_{3}\text{Li}(p, \alpha)^{4}_{2}\text{He}$? Comment on their difference.

16.10 A nuclear reaction in which a particle x collides with a stationary nucleus X gives a product particle y and a nucleus Y. The particle y is ejected at right angles to the original xX direction. If the Q value of the reaction is defined as
$$Q = \text{final kinetic energy} - \text{initial kinetic energy},$$
show that the Q value of the $X(x, y)Y$ reaction is
$$Q = E_y(1 + m_y/M_Y^*) - E_x(1 - m_x/M_Y),$$
where the collision is treated non-relativistically and E refers to energy and m, M to mass.

[N]

Solutions to Problems

16.1 The binding energy of the neutron in u is
$$M = (^{6}_{3}\text{Li} + ^{1}_{0}n) - M(^{7}_{3}\text{Li}) = 6·015\,125 + 1·008\,665 - 7·016\,004$$
$$= 0·007\,786 \text{ u}.$$
In MeV this becomes
$$0·007\,786 \times 931 \text{ MeV} = 7·35 \text{ MeV}.$$
Or in joules we get
$$7·35 \times 1·6 \times 10^{-13} = 1·18 \times 10^{-12} \text{ J}.$$

16.6 Treating the four reactions algebraically gives
$$3d \rightarrow p + \alpha + n = 21·45 + Q \text{ MeV}$$
and, using the given masses, we get
$$3 \times 2·014\,102 = 6·042\,306 \text{ u}$$
$$p + \alpha + n = 1·007\,825 + 4·002\,603 + 1·008\,665$$
$$= 6·019\,093 \text{ u}.$$
Hence
$$\Delta M = 0·023\,213 \text{ u}$$
$$= 21·60 \text{ MeV}$$
giving
$$Q = 0·15 \text{ MeV}.$$

Chapter 17

The Measurement and Detection of Charged Particles

17.1 The Wilson Cloud Chamber

The Wilson cloud chamber was devised early this century and it is still one of the most important methods of observing charged particles in modern physics research. As shown diagrammatically in Fig. 17.1 the apparatus consists of a cylinder A which contains air or nitrogen, saturated with water vapour, above a piston B which can be rapidly moved. If the piston is suddenly pulled out, the gas expands and cools rapidly. The water vapour becomes supersaturated and

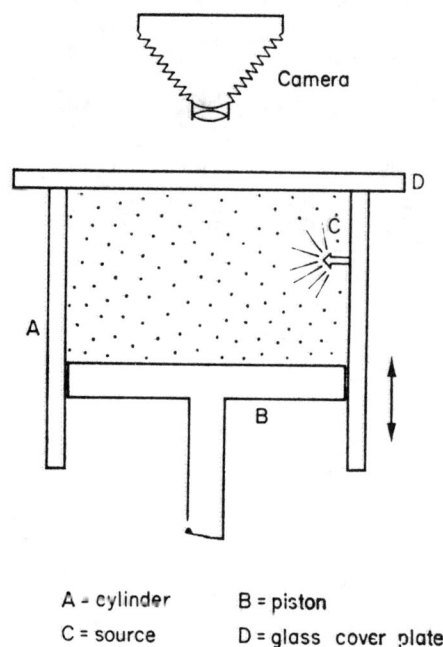

A = cylinder   B = piston
C = source   D = glass cover plate

Fig. 17.1 Showing principle of Wilson cloud chamber.

condenses on any suitable particles present. Charged particles are particularly useful as the ions collect water vapour to form individual droplets which can be seen and photographed. This is done usually through the end plate D with side illumination.

When a small piece of radioactive material, say a thorium salt, C, is put inside the cloud chamber, the paths of the α-particles can be made 'visible' by suddenly expanding the gas, as shown in the photograph in Fig. 15.2 (p. 214).

As the α-particles pass through the water vapour they leave a trail of ions (N_2^+, O_2^+, etc.) and these ions provide centres for condensation droplets. These are large enough to show up the α-tracks — rather like the vapour trails in the wake of a high-flying aircraft — which can be measured for range, and bent in a magnetic field for energy, particle-sign and momentum determinations. Collision processes show up as forked tracks, as in Fig. 16.3 (p. 233). Much of the early qualitative work on α-particles was done with the Wilson cloud chamber in the hands of Rutherford's early research group.

The switching on of the illumination and the photograph takes place immediately after the expansion before the droplets forming the tracks have dispersed. In modern physics the events are photographed stereographically since there is no guarantee that the tracks are parallel to the plane of the chamber. Figure 17.2 shows stereographic pairs of cloud chamber photographs showing the ejection of protons from nitrogen by α-particle bombardment.

In an attempt to use gases at higher pressures, the diffusion cloud chamber was designed in which a stationary layer of supersaturated gas (usually hydrogen) was obtained by allowing vapour to diffuse downward from a warm to a cold surface. The diffusion chamber was therefore horizontal and the sensitive gas layer was not very thick.

17.2 The Bubble Chamber

These methods have now been largely superseded by the bubble-chamber method using the bubbles formed by suitable superheated liquids. One of the disadvantages of the Wilson chamber is the limitation of the track range to a few centimetres by the size of the cloud chamber, and for very long-range particles large chambers are impracticable for technical reasons. Long tracks are due to high-energy particles passing readily through the air which has a low absorption, but if the expansion could be done in a high-density medium much more information would be available. This has been carried out in the bubble chamber, which uses liquids of low surface tension. Pressure is applied and the liquid is heated to a temperature just less than the boiling-point at that particular pressure. At the required signal the pressure is suddenly reduced and the liquid boils. This boiling takes place initially along ion paths and if the chamber is photographed at the right moment tracks of bubbles are seen as in the Wilson chamber. Liquid hydrogen at 27 K is often used and much useful information is obtained owing to the increased density of the medium making collision more probable. Figure 17.3 shows some of the detail found on a bubble-chamber photograph.

Liquid hydrogen, helium or neon may be used at low temperatures whereas organic gases such as propane and freon can be used at room temperatures. The

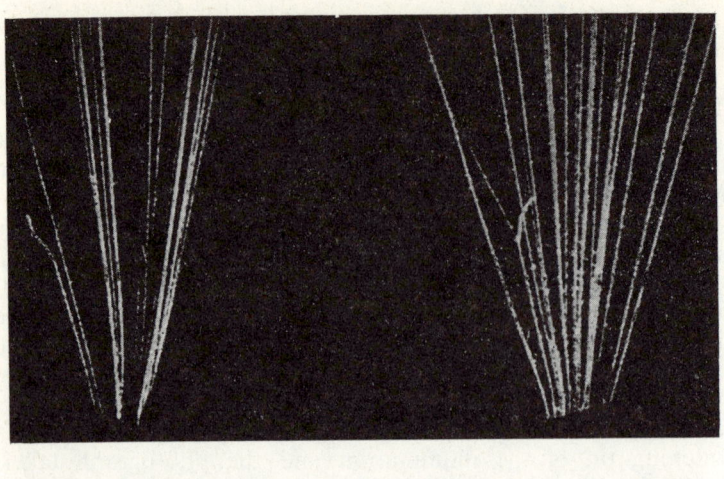

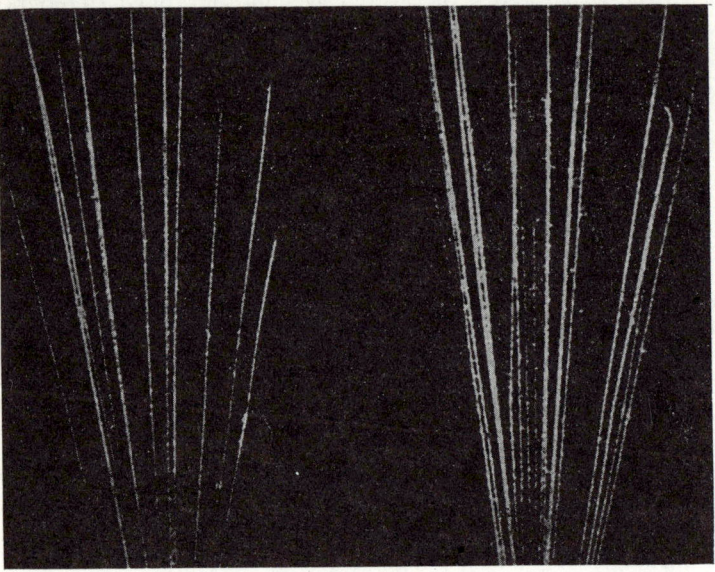

Fig. 17.2 Stereographic pairs of cloud chamber photographs showing ejection of protons from nitrogen atoms when bombarded by α-particles. (Taken from *Radiations from Radioactive Substances* by Rutherford, Chadwick, and Ellis, C.U.P., 1930.)

pressures involved are of the order of 5 atm at low temperatures and about 25 atm at room temperature.

Most of the early work was done with hydrogen because the target nucleus is then always known, viz. the proton. For the organic gases the advantage of having many more protons per nucleus is offset by collisions with the carbon nuclei. Used in conjunction with the magnetic field, the bubble chamber gives

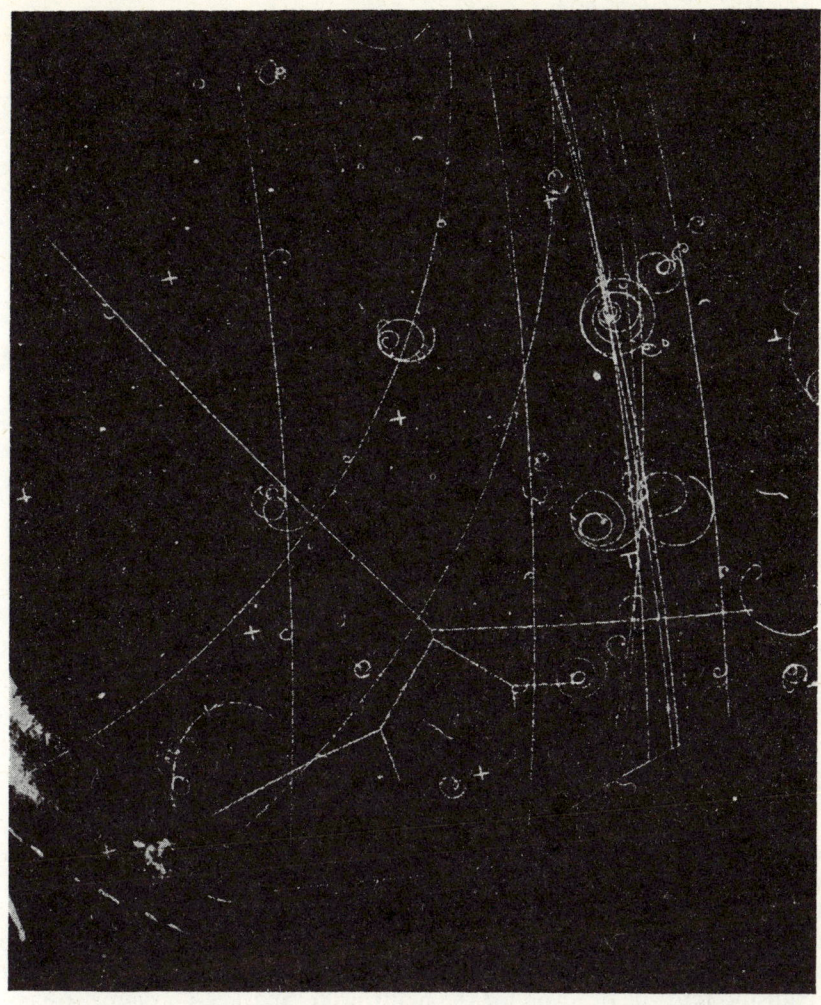

Fig. 17.3(a) Photograph of proton–proton collisions in the liquid hydrogen bubble chamber at the Lawrence Radiation Laboratory. (Photograph by permission of the Lawrence Radiation Laboratory, University of California.)

photographs showing a profusion of events for analysis in terms of identity, energy and momentum.

Since the ions quickly recombine with free electrons no clearing field is required as with the Wilson cloud chamber. The bubble chamber can then be automated to give several photographs per second, but the usual rate is one per second. These can then be stored and examined at leisure. The total weight of a bubble chamber may be several hundred tonnes, involving a field of some 2 T for a chamber of length up to 3 m.

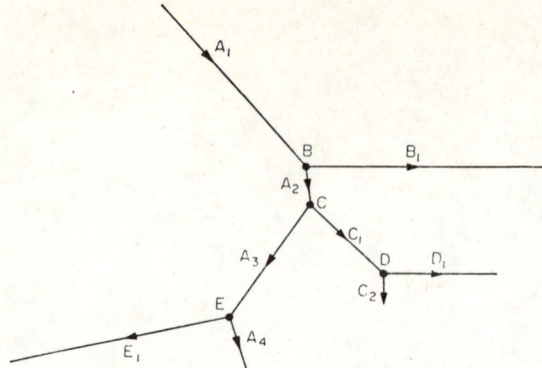

Fig. 17.3(b) Proton collisions of Fig. 17.3(a) shown diagrammatically. The first proton A_1 enters the chamber at the top left hand corner of the picture and collides with the stationary proton B which recoils along B_1. This process is repeated at C, D and E.

Since liquid hydrogen is relatively dense, the bubble chamber cannot be used for low-energy studies or for cosmic ray studies in which the arrival of the ionizing particle is random. Also note that, since recombination takes place so readily, the expansion, the passage of the particles, and the photograph must all take place within a millisecond. The most important role of the bubble chamber is in revealing the events caused by the collisions of particles from the high-flux beams of particle accelerators.

The large bubble chamber at CERN has been called Gargamelle. It weighs 1000 tonnes and contains 10 tonnes of freon, giving a long path length to increase the probability of neutrino collision. Protons from the proton-synchrotron are fired into a 22-m-thick iron shield in front of the bubble chamber. Of the collision products only the neutrinos penetrate this shield on account of their very low collision cross-section. The neutrinos pass into the freon but give no track, of course. However, on collision with a nucleus a profusion of charged particles emerge, giving their tracks for analysis. These are comparatively rare events as the collision must be head-on for the neutrino to interact. Gargamelle has been invaluable in solving the problem of the weak interaction force. See Chapter 28.

17.3 Ionization Chambers

Much of the early radiation detection, particularly in X-ray measurements, was done by means of the ionization chamber. Essentially two electrode plates with an electric field between them are used and the space filled with a gas or vapour to produce ions as required as shown in Fig. 17.4. The potential difference may be of the order of 100 V with a gradient of about $1 \, \text{kV m}^{-1}$. If an ionizing particle passes through the gas, it leaves a trail of ions and produces free electrons. The electrons then drift towards the anode and the positive ions drift very much more slowly to the cathode. The net effect then is a current signal which can be amplified if necessary or displayed electronically so that the output signal is proportional to the intensity of the ionization, which in turn is related to the intense or activity of

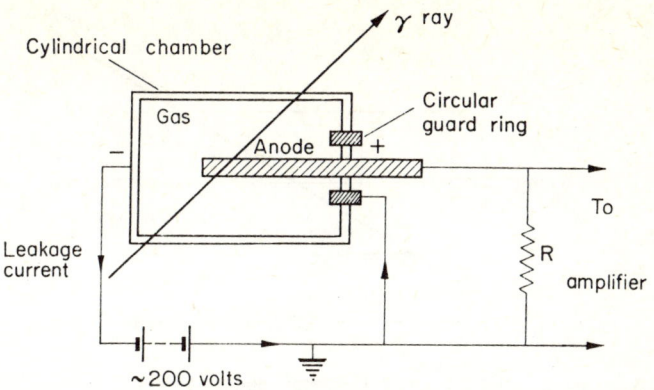

Fig. 17.4 Ionization chamber with guard ring. Note that the leakage current does not pass through the load R.

the source. The pulse size of the signal is therefore proportional to the primary ionization produced by the radiation or the intensity of the X-ray beam.

The pulse size obtained is usually a few tenths of a millivolt, corresponding to direct currents in the range of 100 pA to 10 fA or even less. In order to get full registration of these small effects it is necessary to use electronic amplification. Amplifiers with linear gains of the order of 10^4 or more are required and these are fairly difficult to design. Another difficulty arises from leakage currents, and the ionization chamber must be designed with a guard ring connected to earth, as in Fig. 17.4. It is for these reasons that ionization chambers are never used if reliable results can be obtained with the proportional counter, described in the next section.

The gases used in the ionization chamber are usually air or hydrogen at atmospheric pressure, or at greater pressures for γ-ray detection. If thermal neutrons are to be detected, boron is introduced in the form of boron trifluoride and the potential differences used are of the order of 100 V. In all these cases each ionizing event is registered singly, i.e. all ions and electrons so produced travel to the cathode and anode respectively without further ionization taking place on the way due to collisions with gas molecules. The time constants are so arranged that the result is a continuous small current which is then linearly related to the degree of ionization in the chamber and so to the activity of the source. It is also independent of applied voltage over small ranges when the saturation current has been reached. In order to preserve this relationship all the ancillary electronic apparatus must then give a linear response over the whole range of currents to be measured. The currents are so small that for single-particle counting the ionization chamber is often replaced by the proportional counter, in which this linearity is still preserved, but with an increased current.

17.4 The Proportional Counter

The proportional counter as now used consists of a cylindrical gas-filled tube with a very thin central wire anode, as shown in Fig. 17.5. In the case of the simple ionization chamber the pulse height generated by an event is proportional to the

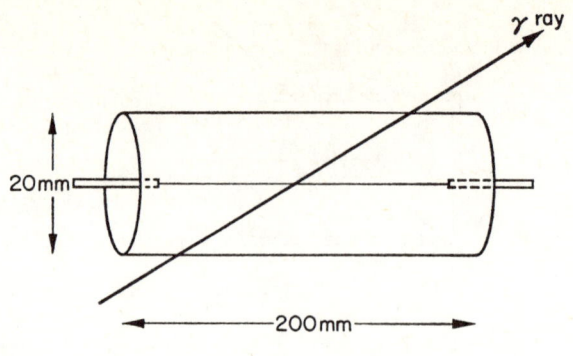

Central wire diameter about 10 μm

Fig. 17.5 Proportional counter (diagrammatic).

intensity of the beam, but because of the comparatively low applied voltages the current produced is always very small and complicated electronic amplifiers are required to measure it with accuracy. If we increase the applied potential in an attempt to produce more ions per millimetre, it is found that the pulse size is no longer independent of voltage over a small range as with the ionization chamber. This voltage range is called the proportional counter region, and differs from the ionization chamber region by virtue of the multiple collisions which take place between electrons and the gas molecules. Since the central wire is very thin (see Fig. 17.5) and the potential difference fairly large, the electric field $E = dV/dr$ is very high, causing the electron velocity to be correspondingly high. Although the pulse size increases with increasing applied voltage in the proportional counter region, it is still proportional to the initial number of ion-pairs produced in the gas at constant voltage.

In Fig. 17.5, where the radius of the wire is a and that of the counter is b, the radial field E at any point distance a from the centre will be $E = k/a$, where k is a constant, and the expression for the potential difference across the tube is given by

$$V = 2 \cdot 3k \log_{10} b/a.$$

For $b = 0 \cdot 01$ m and $V = 1$ kV we have $1000 = 2 \cdot 3k \log_{10}(0 \cdot 01/a)$ giving k in terms of a. We can therefore work out E, the potential gradient, from $E = k/a$ for various values of a, the inner wire radius. This is shown in Table 17.1 in which the increase of field is apparent for the thinner wire. Hence the energy Ve imparted to each electron in this strong field is sufficient to cause further ionization by collision. As more ions are produced by multiple collisions a gas amplification of about 10^3 is achieved. Thus in the proportional region while the pulse height is still linearly related to the intensity of the radiation received the single pulses involved are now much larger, being of the order of a few millivolts. The counter can then be operated with an amplifier having a lower gain than that required for the ionization chamber.

The complete voltage–pulse characteristics of this sort of tube are shown in

TABLE 17.1
Proportional Counter

Wire radius $= a$ m
Tube radius $= b = 0{\cdot}01$ m
Applied voltage $V = 1$ kV

a (m)	k from $V = 2{\cdot}3k\log_{10} b/a$	$E = k/a$ ($V\,m^{-1} \times 10^{-5}$)
10^{-3}	435	4·35
10^{-4}	217	21·7
10^{-5}	145	145
10^{-6}	109	1090
10^{-7}	87	8700

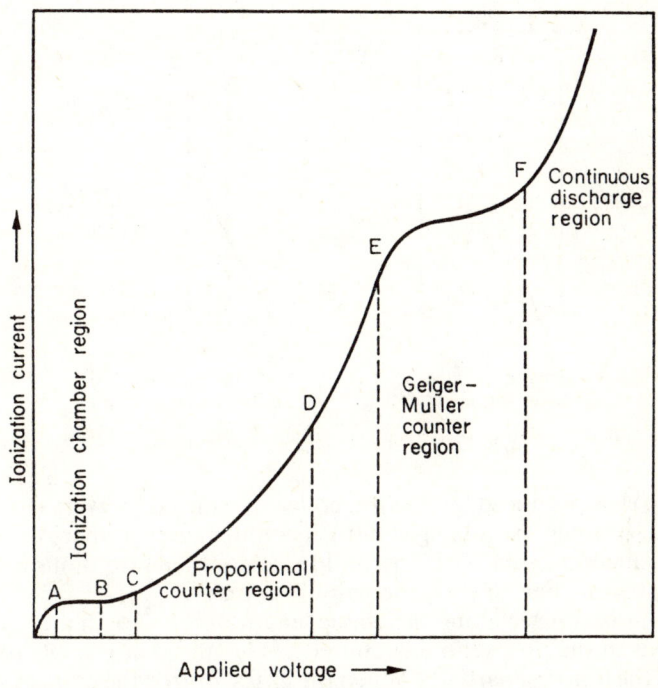

Fig. 17.6 Voltage–current (pulse) characteristic showing main counting regions.

Fig. 17.6. The main regions used for measurements are:

(1) the ionization chamber region AB
(2) the proportional counter region CD
(3) the Geiger–Muller region EF (see next section).

After the point F the tube becomes a simple discharge tube in which the current is produced even after the ionization event has ceased. The tube will only give a measure of ionization intensity if it is used in the region AB or CD. Like the ionization chamber the proportional counter gives single pulses of height proportional to the ionizing power of the radiation.

Argon or methane are common gases used in proportional counters, at a pressure of about 1 atm or a little above.

17.5 The Geiger–Muller Counter

Some typical Geiger–Muller tubes are shown in Fig. 17.7 in which the general construction is seen to be similar to that of the proportional counter. As the potential difference across the tube is increased the gas amplification factor also increases and the pulse becomes very much greater. In contrast with the

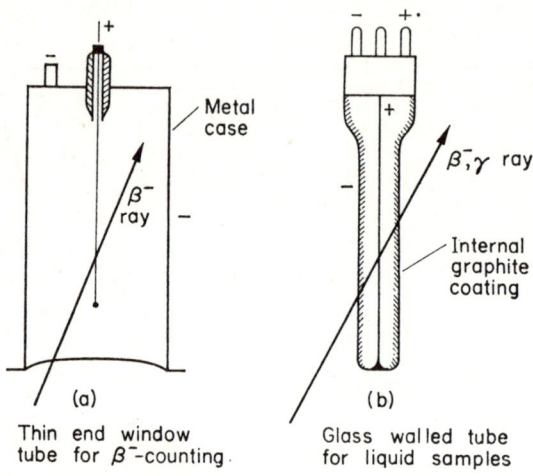

Fig. 17.7 Typical Geiger–Muller tubes.

avalanche of ions produced by electron collisions started by every electron in the proportional counter, in the Geiger–Muller counter every primary *and* secondary electron produces a cascade of ions, and there are therefore thousands of times more ions present than in the proportional counter.

Ionization takes place along the whole length of the wire in a Geiger–Muller tube, whereas in the proportional counter it is localized at a single point by the direction of the ionizing particles. When the gas is ionized the electrons produced in the cascade are immediately drawn to the positive central wire and are counted as a single negative pulse. This whole operation takes less than a microsecond. Being very light compared with the positive ions, the electrons have greater mobility and, after the pulse has been counted, the wire is still sheathed by positive ions which take several hundred microseconds to move away to the outer cathode. During this time the field around the wire is too low to give sufficient energy to further electrons so that the tube remains insensitive until the positive ions have moved away and allowed the pulse to develop in the external circuit. This is the dead-time period of the Geiger–Muller tube, and any ionizing event occurring during this time is not recorded. Thus a Geiger counter cannot resolve events closer than 10^4 per second.

The voltage characteristics of a Geiger–Muller tube are shown in Fig. 17.8. It is to be remembered that the characteristic pulse of the Geiger tube is independent of the size or intensity of the ionizing event as a multiple avalanche can be

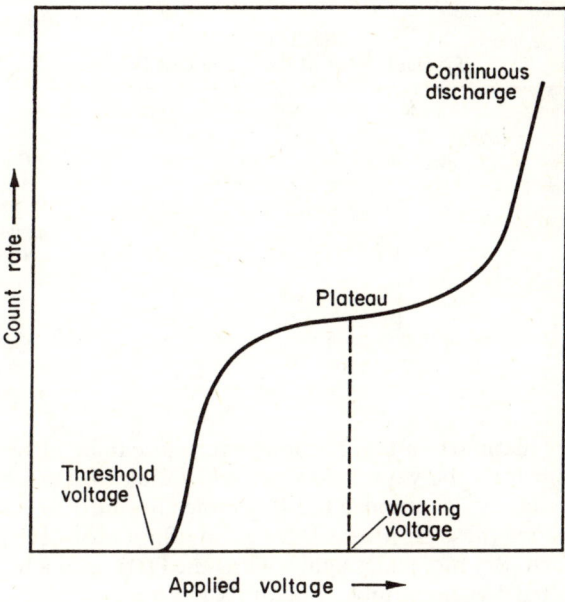

Fig. 17.8 Geiger–Muller tube characteristics.

produced by one or many events and also by primary and secondary electrons. It is seen from this diagram that there is a threshold below which the tube does not work. This can be of the order of a few hundred volts. As the applied potential is increased the counting begins and rises rapidly to a flat portion of the curve called the plateau. This is the Geiger tube region for which the count rate is independent of small changes in potential difference across the tube. Beyond the plateau the applied electric field is so high that a continuous discharge takes place in the tube, as shown in Fig. 17.8, and the count rate increases very rapidly. It does not require any ionizing event for this to happen so that the tube must not be used in this region.

We have seen that the positive ions owing to their mobility being lower than that of the electrons take much longer to reach the cathode. When they do arrive they liberate secondary electrons from the metal of the cathode, which are then drawn inward to start the discharge pattern again. This is an unwanted spurious discharge since the total time of the discharge from a single event may then become several milliseconds. To overcome this, modern Geiger–Muller tubes contain a 'quenching' vapour or gas, which can be an organic vapour such as ethyl alcohol or a halogen such as bromine. As the ions move toward the cathode, they transfer their charge to the vapour molecules which have a lower ionization potential than the argon gas present for ionization (Table 17.2). The ionized vapour molecules reach the wall of the tube and liberate electrons which are quickly used up in decomposing the molecules of the vapour. The organic molecules are decomposed slowly and irreversibly so that a limited tube life of the order of 10^9 pulses is associated with organic quenched Geiger tubes.

TABLE 17.2
Typical Geiger–Muller Tube Gas Fillings

	Fillings	Ionization potential (V)	Pressure used
Ionizing gas	Ne	21·7	Various mixtures
	Ar	15·7	between 10 and
	Kr	14·0	500 mm Hg
Quenching vapours	Cl_2	12·8	1 mm Hg
	Br_2	13·2	1 mm Hg
	Ethyl alcohol	11·3	10 mm Hg
	Ethyl formate	–	10 mm Hg

To extend the lifetime of a tube, bromine vapour can be introduced instead of the alcohol. As before, the vapour ions travel to the cathode where the released electrons immediately decompose the Br_2 molecules into Br atoms by electron collisions and since these Br atoms have great affinity for each other they soon recombine to form Br_2 molecules again. Thus the lifetime of a halogen-quenched tube is not limited by the number of counts registered.

The quenching vapour therefore decreases the probability of positive ions reaching the cathode and producing spurious pulses and it absorbs the photon energy from atoms excited by collision and it readily absorbs energy in self-dissociation. Table 17.2 shows some characteristic gas fillings of Geiger counters.

The general time signal curves are shown in Fig. 17.9 in which the duration of the dead time, paralysis time and the recovery time are clearly shown. The important time is the paralysis time t, which may be as much as a millisecond. No two pulses can be counted separately within this time. However, if t is known, it is possible to get the true count rate N from an observed count rate N_{obs} since the counter is inoperative for $N_{obs}t$ seconds in every second. If the particles come in at the rate of N pwr second (actually) the count does not count $(N_{obs}t)N$ of these. Thus
$$N - N_{obs} = (N_{obs}t)N.$$
Therefore
$$N(1 - N_{obs}t) = N_{obs} \quad \text{or} \quad N = \frac{N_{obs}}{1 - N_{obs}t}.$$

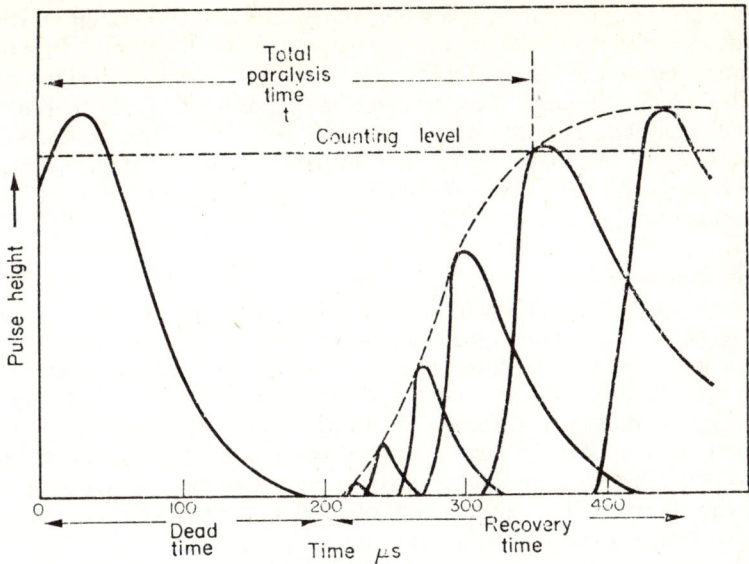

Fig. 17.9 Geiger–Muller counter time characteristics.

This equation is valid for rates of counting less than about $1/10t$ per second and can be used in table form if t is known. It is usual to put in the circuit an electronic device known as the quench probe unit which injects into the counting instrument a known value of t (usually $400~\mu s \pm 5\%$), so that N can be obtained directly from any value of N_{obs}. The actual counting equipment used to display the count is known as a scaler. A scaler may be used with a proportional, scintillation or Geiger counter.

17.6 Scintillation Counters and Semi-conductor Counters

Many radioactive counts take place within a few microseconds so that it is necessary to have counting equipment with a resolution time of this order. A Geiger tube has too long a dead time and therefore a scintillation counter is used for fast counting. Briefly this consists of a scintillation crystal or liquid which absorbs the incident radiation and gives out a pulse of electromagnetic radiation, which need not be in the visible spectral region. This crystal is backed by the photosensitive surface of a photomultiplier so that a very weak pulse is amplified some 10^6–10^8 times before passing to the electronic counting equipment. The resolution time of such crystals is of the order of 10 ns. The crystals themselves are either anthracene or sodium iodide with thallium impurities. Scintillation counters, like proportional counters, give single pulses of height proportional to the radiation energy. The linear proportionality originates in the crystal itself and in order to preserve this linearity the electronic equipment required is a little more complicated than that used in the Geiger–Muller counter.

Semiconductor counters are now available and are the subject of much research. The barrier region in a silicon n–p junction (n = negative electron carrier, p = positive hole carrier) is particularly sensitive to ionizing particles. If these particles can penetrate to the barrier region with the correct potential across the junction, the system acts as a solid ionization chamber. This is a very simple idea which has led to the reproduction of many of the classical scattering experiments without having complicated gas counters. These semiconductor counters are useful in detecting all heavily ionizing radiations such as α-particles, protons, heavy ions and fission fragments but are no good for γ-rays as their specific ionization is too small.

17.7 The Spark Chamber

This has found increasing use in the field of high-energy physics. It consists of a series of large parallel metal plates, a few square metres in area, set in a chamber filled with neon gas at atmospheric pressure. All the plates are isolated from each other but alternate ones are grounded and the others are connected together to a high voltage d.c. pulse generator (~ 15 kV) which gives them a high potential in short bursts of the order of a microsecond each. This is just enough to cause sparks to occur between the plates in such regions as are ionized by a particle entering the chamber. This gives a trail of sparks along the path of the particle which can then be photographed from the side.

Usually the plate separation is a few millimetres and a hundred or so of these plates can make up a volume of a few cubic metres.

One of the main advantages of the spark chamber over the bubble chamber is that triggering and removal of ions by the clearing field are comparatively simple, but, of course, the origin of an event can only be found within an accuracy of one plate thickness. Faster timing is also possible.

With suitable counters at the end of the spark chamber the particular event expected can be made to trigger the high-voltage generator, and so record itself. Thus a selection of events to be studied may be made.

The spark chamber is shown schematically in Fig. 17.10 and a photograph of spark chamber tracks and their interpretation in Fig. 17.11.

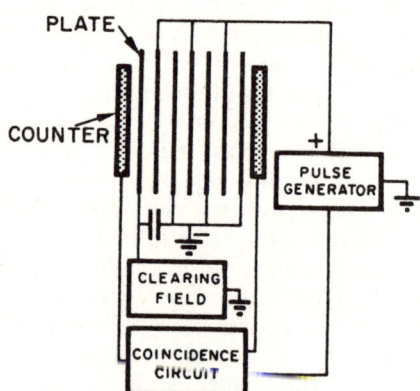

Fig. 17.10 Schematic representation of spark-chamber circuit. (From S. Glasstone, *Sourcebook on Atomic Energy*, 3rd ed.; copyright (c) 1967, D. Van Nostrand Company, Inc., Princeton, N.J.)

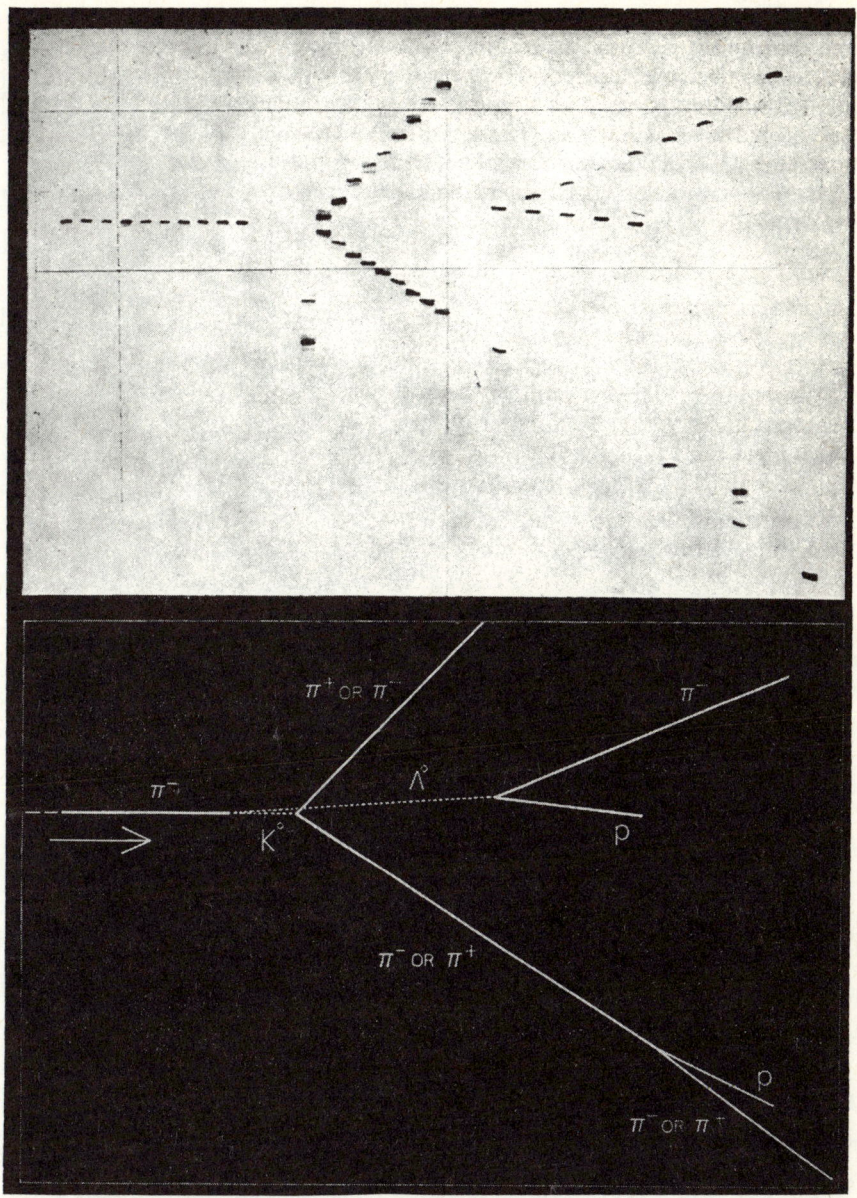

Fig. 17.11 Spark-chamber tracks (above) and their interpretation (below). (Photograph by courtesy of James L. Cronin, University of Princeton; diagram from *The Spark Chamber* by G. K. O'Neill, copyright (c) 1962 by Scientific American, Inc. All rights reserved.)

17.8 The Cerenkov Counter

A phenomenon observed by Cerenkov in 1934 gave another method of measuring the energies of high-energy particles. He found that β^--rays incident on a transparent dielectric medium produced optical radiation whenever the speed of the electrons in the medium was greater than the speed of light in the medium. This condition does not contravene the theory of relativity since the speed of light in, say, glass is of the order of 2×10^8 m s^{-1}. The visible radiation can be amplified by a photomultiplier, as in the scintillation counter.

Theory shows that the light is scattered from the direct electron beam with a maximum intensity at an angle θ given by

$$\cos\theta = \frac{c_0}{nv}$$

where c_0 is the speed of light in air (vacuum), v is the speed of the electrons in the medium and n is the refractive index of the medium.

In Fig. 17.12 the time interval between AC and AB is the same, t, so that

$$AB = c_1 t, \quad AC = vt$$

where c_1 is the speed of light in the medium. One can imagine the electrons being continually accelerated and decelerated by collisions along the path AC, so that electromagnetic radiation is emitted in all directions from AC. These radiations can be regarded as Huygens' secondary wavelets and the maximum intensity will be at the angle θ when the waves are in phase. The wave-front is then BC and

$$\cos\theta = \frac{AB}{AC} = \frac{c_1}{v} = \frac{c_0}{nv},$$

where $v > c_1$. The radiation produced is analogous to the supersonic shock wave of acoustics.

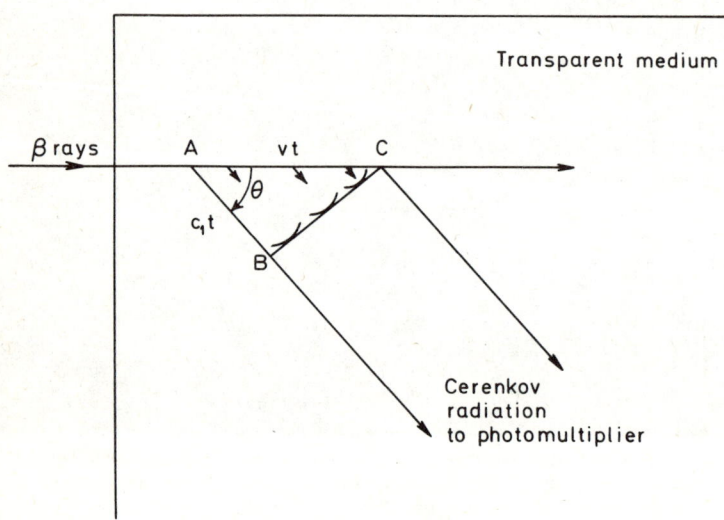

Fig. 17.12 Origin of Cerenkov radiation.

The medium may be solid, liquid or even gaseous so long as it is transparent, and the ionizing particles may be electrons or ions of almost any energy. The measurement of θ is seen to be sufficient to determine v, so that with a photomultiplier we have a very convenient counter for high-energy measurements. When the medium is a gas, $n \simeq 1$, and v itself can approach c_0, so that very high energies can be measured.

17.9 Neutron Counting

Since neutrons have no charge it is not surprising that they produce no paths of ions as they move through a gas. Hence they cannot be observed either in a cloud chamber or in a Geiger tube. As all such counting depends on ionization, use must be made of any ionizing particles produced by a neutron. For example, if boron is bombarded with neutrons, α-particles are produced:

$$^{10}_{5}B + ^{1}_{0}n \rightarrow ^{4}_{2}He(\alpha) + ^{7}_{3}Li$$

and each neutron produces an α-particle. This α-particle in turn will produce an ionization track which can then be used to identify the neutron. Thus for a counter to detect neutrons it must contain some gas which ionizes after neutron collision with its molecules. This is possible with BF_3 gas in which the boron atoms produce the α-particles which in turn produce ionization which can be detected in the usual manner. Neutron counting chambers are either ionization or proportional counting arrangements.

17.10 The Photographic Plate

Photographic plates are darkened by radiation from radioactive substances and the darkening is due to the production of individual tracks as in a cloud chamber. The photographs shown in Chapter 26 are of various particle tracks revealed by the silver grains. In order to record these successfully specially prepared plates are used with an emulsion thickness of several hundred micrometers. Each type of particle has its own particular track as shown and many cosmic ray events have been analysed from such photographs. If neutrons are to be detected by the photographic method the plate must first be soaked in a boron solution.

A disadvantage of the nuclear plate is that, unlike the cloud chamber tracks, nuclear plate tracks cannot satisfactorily be bent in a magnetic field since the large amount of scattering obscures the curvature of the tracks which, in any case, are very short. However it is also true that the very simplicity and cheapness of nuclear plates recommends them for much work in nuclear physics.

17.11 Summary

Ionizing radiations can be measured by ionization chamber, proportional counter and by Geiger–Muller tube methods according to the nature of the investigation. Most simple nucleonic work is done with a Geiger–Muller tube. Neutrons can be counted in specially 'doped' proportional counter tubes.

For fast counting a scintillation counter is used as the resolution time is so much less.

Nuclear emulsion plates can be prepared which will record individual ionization events and even neutron collisions if the emulsion contains sensitive neutron absorbing atoms such as boron. Spark chambers are also used to record individual collisions.

Problems

(*Those problems marked with an asterisk are solved in full at the end of the section.*)

17.1   A sample of uranium, emitting α-particles of energy 4·18 MeV, is placed near an ionization chamber. Assuming that only 10 particles per second enter the chamber calculate the current produced.
One ion pair requires energy of 35 eV. Electronic charge $= 1·6 \times 10^{-19}$. ($1·92 \times 10^{-13}$ A $= 0·192$ pA)

17.2*   An ionization chamber is connected to an electrometer of capacity 0·5 pF and voltage sensitivity of 4 divisions per volt. A beam of α-particles causes a deflection of 0·8 divisions. Calculate the number of ion pairs required and the energy of the source of the α-particles. Use data of Problem 17.1. ($6·25 \times 10^5$ ion pairs, 2·19 MeV)

17.3   It is required to operate a proportional counter with a maximum radial field of 10 MV m^{-1}. What is the applied voltage required if the radii of the wire and tube are 20 μm and 10 mm respectively? (1·24 kV)

17.4   If the mean free path of the electrons in a proportional counter of gas-multiplication factor of 1024 is 1 μm, calculate the distance from the wire for which this multiplication takes place. (10 μm)

17.5   If the wire in Problem 17.4 has a radius of 10 μm and the tube radius is 10 mm, what is the field at the radius giving a gas multiplication of 1024 if the applied voltage is 1200 V? (17·4 MV m^{-1})

17.6   The paralysis time (sometimes simply called the 'dead time') of a Geiger–Muller is 400 μs. What is the true count rate for measured count rates of 100, 1000, 10 000 and 100 000 counts per minute? Express each answer as a percentage counting error and comment on the results. (100·07: 1007: 10 715: 300 000)

17.7*   An organic-quenched Geiger–Muller tube operates at 1 kV and has a wire diameter of 0·2 mm. The radius of the cathode is 20 mm and the tube has a guaranteed lifetime of 10^9 counts. What is the maximum radial field and how long will the counter last if it is used on the average for 30 hours per week at 3000 counts per minute? (18·9 MV/m^{-1}, 3·7 a)

17.8   The plateau of a Geiger counter working at 1 kV has a slope of $2\frac{1}{2}$% count rate (Fig. 17.8) per 100 V. By how much can the working voltage be allowed to vary if the count rate is to be limited to 0·1%?

17.9   A 1·7 MeV β$^-$-particle from $^{32}_{15}$P gives a cascade of $5·5 \times 10^7$ ion pairs each corresponding to an energy loss of 35 eV, in a Geiger counter tube. What is the gas-multiplication factor?

17.10   A 6·4 MeV α particle passes into an ionization chamber which can detect an ionization current of 1 pA. What is the lowest activity in microcuries of the source of these α-particles which the ionization chamber can detect?
Counting efficiency of ionization chamber $= 10%$; 35 eV $= 1$ ion pair.

Solutions to Problems

17.2 Signal voltage $\Delta V = \Delta Q / C$, i.e.
$$\frac{0 \cdot 8}{4} = \frac{\Delta Q}{0 \cdot 5 \times 10^{-12}},$$
$$\Delta Q = 10^{-13} \text{ C}.$$

This is also Ne, where N is the equivalent number of ionic charges each of value $e = 1 \cdot 6 \times 10^{-19}$ C. Hence
$$N \times e = 10^{-13} \text{ C}$$
and
$$N = \frac{10^{-13}}{1 \cdot 6 \times 10^{-19}}$$
$$= 6 \cdot 25 \times 10^5 \text{ ion pairs required}.$$

If the energy of the α-particles in electron-volts is V, then $N = V/35$ since one ion pair requires 35 eV:
$$V = 35 \times 6 \cdot 25 \times 10^5 \text{ eV}$$
$$V = 2 \cdot 19 \text{ MeV}.$$

17.7 Inserting the given data in the equation $V = 2 \cdot 3 k \log_{10} b/a$ we have $1000 = 2 \cdot 3 k \log_{10} 200$, giving $k = 189$.

The field along a radius is
$$E = \frac{k}{r} = \frac{189}{10^{-5}}$$
at the wire surface, or $E_{max} = 18 \cdot 9$ MV m^{-1}.

If the lifetime of the tube is N years the total number of counts recorded will be
$$N \times 50 \times 30 \times 60 \times 3000 = 2 \cdot 7 \times 10^8 \, N \text{ counts}.$$
Therefore
$$2 \cdot 7 \times 10^8 \, N = 10^9$$
giving
$$N = 3 \cdot 7 \text{ a}.$$

Chapter 18

Accelerating Machines as used in Nuclear Physics

18.1 Introduction

The first bombarding particles to be used in nuclear physics were the α-particles available from natural radioactive elements, and we have seen that the upper energy limit of these is a few MeV. It was realized by the Cambridge school in the 1920s that there was a limit to the transmutations that could be obtained with these and that if other particles could be used as missiles the whole range of information would increase, as different types of nuclear reaction became possible. The only other feasible bombarding particles then known were protons, since electrons do not produce nuclear effects. The first research was directed towards the acceleration of protons to energies of a few MeV. This culminated in the Cockcroft–Walton accelerator which appeared in 1932, and was the forerunner of the machines we have today giving energies up to 500 GeV.

The design of successful accelerating machines depends not only on classical physics, electrical engineering, electronics and vacuum techniques, but also on precise mechanical engineering before accurately collimated beams of charged particles can be made available for nuclear bombardment experiments. Moreover, one must remember that the maximum particle energies which can be produced artificially are far less than those energies found in cosmic ray particles. Although cosmic rays have energies of the order of many millions of GeV, the advantage of the particle accelerators lies in the fact that the intensity of the beam is far greater than the intensity of cosmic rays at sea-level.

18.2 The Cockcroft–Walton Proton Accelerator

The principle which Cockcroft and Walton adopted was that of the voltage doubler arrangement shown diagrammatically in Fig. 18.1. In this diagram two capacitor banks $C_1 C_3$ and $C_2 C_4$ are connected across the transformer giving a peak potential of V_0 volts with rectifiers $R_1 R_2 R_3$ and R_4 acting as switches.

In the following argument we shall assume there are no current losses across any of the components. We consider first the simple circuit formed by the transformer, R_1 and C_1, i.e. OPT in Fig. 18.1. For the first half-cycle assume O goes positive and T negative so that the rectifier R_1 conducts and C_1 is charged to

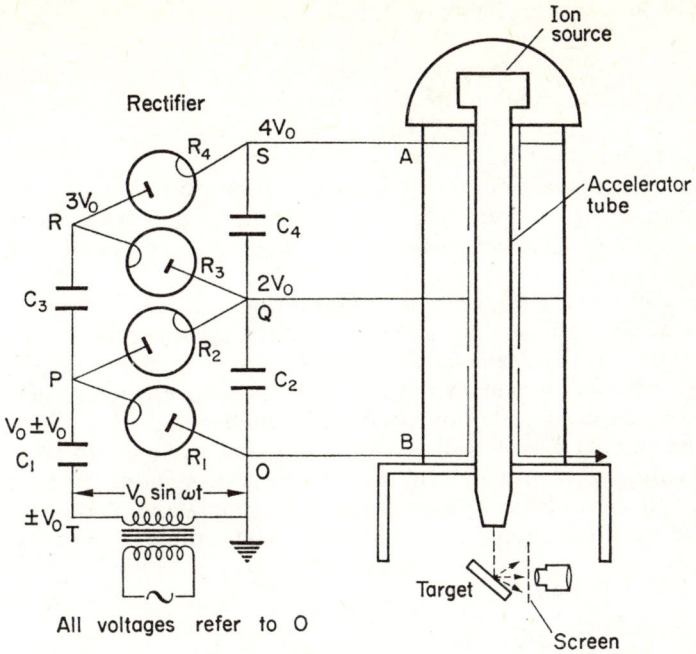

Fig. 18.1 Schematic diagram of Cockcroft–Walton accelerator and discharge tube.

V_0 and the potential of T is $-V_0$. During the second (reverse) half-cycle R_1 no longer conducts, leaving the point P isolated at a potential of V_0 while O drops to $-V_0$, producing a maximum potential difference between P and O of $2V_0$. If now we include R_2, which is conducting during the second half-cycle, the charge accumulated on C_1 is now shared with C_2 and on repeating the first half-cycle C_1 is recharged up to V_0. In this (third) half-cycle C_2 retains its charge, but this is increased by sharing with C_1 again during the fourth half-cycle. After repeating this procedure for a few cycles, C_2 becomes fully charged since it cannot lose charge by current leakage. Eventually an equilibrium is reached in which there is no current through either R_1 or R_2 at any time. The potential of Q is now equal to the *maximum* potential of P with respect to O, i.e. a steady potential difference of $2V_0$ appears across C_2, while the instantaneous potential of P with respect to O consists of a periodic component from the transformer superimposed on the steady V_0 when R_1 is not conducting. Thus an alternating potential difference of peak value V_0 appears across P and Q, with Q always at a steady potential of $2V_0$. Thus if we now add C_3 and C_4 through the rectifiers R_3 and R_4 we can repeat the whole of the above argument and the potential finally appearing at S is $2V_0$ with respect to Q and $4V_0$ with respect to O.

In principle the potential V_0 can be multiplied up to any multiple of V_0 by using the simple voltage doubler in cascade. Cockcroft and Walton reached a final potential of about 0·7 MeV in 1932. This is not very high by modern standards and it is the reason why the early Cockcroft–Walton proton reactions were limited to light elements — lithium, boron, beryllium, etc., as already described (p. 235).

Usually the reaction is a (p, α) reaction. As in α-particle reactions, the direct collisions are rare events but are those of most interest. Thus the reaction
$$^6_3\text{Li} + ^1_1\text{H}(p) \rightarrow ^4_2\text{He}(\alpha) + ^3_2\text{He}$$
is the (p, α) reaction on ^6_3Li.

A survey of (p, α) reactions studied shows that many new isotopes were found by this method which were not possible by the (α, p) type of reaction of Rutherford. The reaction energies involved were of the order of 10 MeV.

18.3 The Van de Graaf Electrostatic Generator

This instrument, although originally conceived as an accelerator for research purposes, is now available in many teaching laboratories as a replacement for the Wimshurst machine. It is shown in Fig. 18.2 and depends for its action on the collection of charge by the hollow conductor which then discharges at the points shown. The endless belt A is driven vertically and picks up a charge at a few thousand volts at B, from a high voltage set. The point C induces a positive charge on to the belt and this is carried up until it is transferred to the sphere by the points E by a corona discharge, and hence to the terminal of the ion source. The

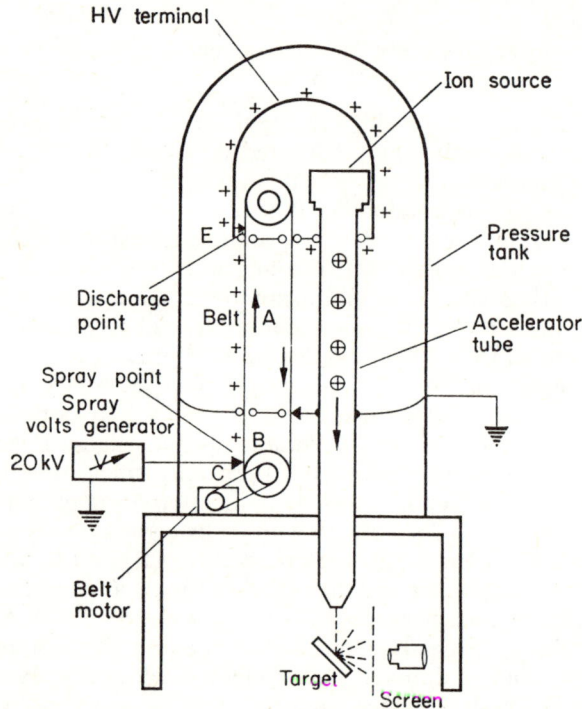

Fig. 18.2 Diagram of Van de Graaff machine and discharge tube.

usual potential is about 6 MeV but the most up-to-date generators can give about 12 MeV and there are now tandem Van de Graaff machines which give up to 20 MeV by electron stripping devices. This technique is to utilize the positive high potential twice, first by accelerating negatively charged particles and subsequently repelling them when they have been stripped of their electrons to become positive ions again. A typical arrangement would be to add electrons to the ions from the ion source, so that the emergent beam consists of a relatively high percentage of *negative* ions which are accelerated down the tube to the positively charged terminal. Here they travel along a stripping tube which removes most of the extra electrons and the resultant positive ion beam is accelerated to earth potential again. The final energy then corresponds to twice the terminal potential, although the ion current is only about 2 μA, compared with the single Van de Graaff current of about 200 μA. The tandem Van de Graaff at Aldermaston is shown in Fig. 18.3.

The Van de Graaff machine can be used to accelerate electrons by reversing the potential of the spray voltage and using a hot filament for thermionic electrons instead of the ion chamber.

18.4 The Linear Accelerator

It is to be noticed that in both the Cockcroft–Walton and the Van de Graaff machines, the high potential is generated by electrostatic devices and applied to the discharge tube containing the ions to be accelerated. In the linear accelerator the energies of the charged particles are increased by a series of linear pulses arranged to give the ions an extra push at the right moment of time, as shown in Fig. 18.4. The accelerator tubes, or drift tubes, are narrow cylinders connected alternately, as shown, to a source of a high-frequency potential. Thus, when cylinders 1, 3, 5 are positive, the cylinders 2, 4, 6, etc., are negative and reversal of potential takes place periodically according to the frequency. The positive ions are generated at S and pass through cylinder C_1 to the gap between C_1 and C_2, where the potential is such that the positive ion is accelerated in the gap into C_2, where it travels with constant velocity to the gap between C_2 and C_3. Here the acceleration process is repeated. The lengths of the cylinders have to be adjusted so that the time taken within the cylinder is just half the period, i.e. the ion always enters the potential field on leaving any one cylinder just as the potential is changing favourably. Since the ions are constantly increasing their velocities the successive cylinders have to be longer and longer. The frequencies required for protons are much higher than for heavy ions and it is now possible to accelerate protons up to about 50 MeV.

The separation between the gaps is governed by the applied high-frequency field and the gap velocity of the ions. It is the distance travelled during one half-cycle and is given by $l = v(T/2) = v/2f$, where v is the instantaneous velocity of the ions and f is the frequency of the applied field. Thus drift tubes of a few centimetres long require oscillating fields with frequencies of the order of hundreds of megahertz. The maximum gain of energy at each gap is Ve, where V is the potential difference across the gap.

The linear accelerator for electrons is different from the proton accelerator. It consists of a tube down which an electromagnetic wave progresses. The tube is

Fig. 18.3 Tandem electrostatic accelerator, Aldermaston, England. This machine and a similar accelerator at Harwell are designed to yield basic information on the behaviour of nuclei. (By courtesy of U.K.A.E.A.)

really a wave-guide since it contains apertures spaced according to the frequency of the travelling wave and the size of the tube. Electrons are injected at about 80 kV in the case of the Stanford University electron linear accelerator, which has an output of 1 GeV (10^9 eV) and is 91 m long.

The most powerful electron linear accelerator in the world is at the Stanford

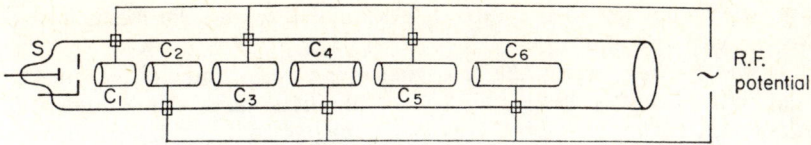

Fig. 18.4 Linear accelerator showing drift tubes of increasing length.

Linear Accelerator Center, U.S.A. It hurls electrons in a straight line over a distance of 3 km (2 miles). Obviously over this distance great precision is required of the engineering components. The maximum electron energy reached is 22 GeV. Such high-energy electrons have been used to investigate the weak electric forces within the nucleus and by firing them into liquid hydrogen and measuring the angular distribution of the scattered electrons the structure of the proton has been studied. We shall discuss the meaning of this work in Chapter 28.

18.5 The Lawrence Cyclotron

It is obvious that in the linear accelerator the length required for really high energies is enormous. Thus, it is possible to see the advantages of bending the charged particles in spirals before finally using them. This was the basis of the famous cyclotron developed in 1930 by E. O. Lawrence and his team in California. Figure 18.5 shows the cyclotron diagrammatically. The source S produces electrons which ionize the gas around S and these ions are then bent in a magnetic field within two hollow conductors, known as 'dees', inside a closed vessel containing hydrogen gas at low pressures. The magnetic field passes across the dees perpendicular to the path of ions. The potential between D_1 and D_2 must change over just as the ions are crossing the gap, as was necessary with the linear accelerator. The magnetic field causes the ions to move in a circular path through

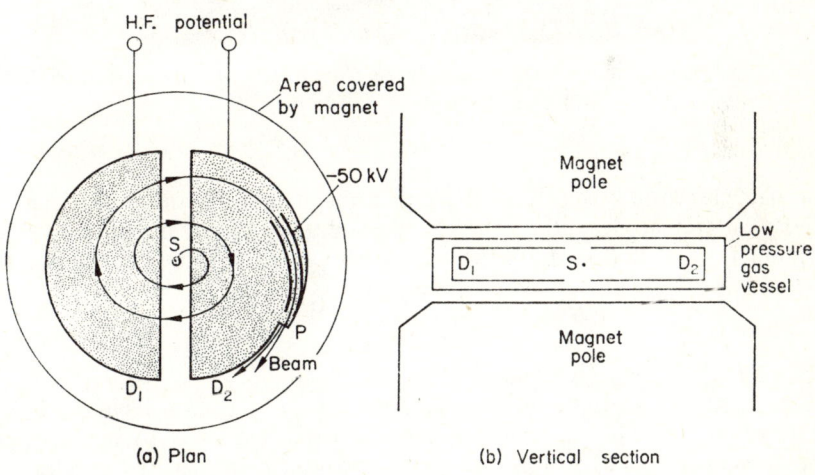

Fig. 18.5 Simplified diagram of cyclotron showing position of dees.

D_1 and when they get to the gap, D_2 goes negative and the ion is accelerated across the gap and so on.

Thus, $mv^2/r = Bev$ and $v = rBe/m$ for an ion of mass m and charge e moving in a circular path of radius r with speed v in a magnetic field of flux density B. The length of path in one dee is $2\pi r/2 = \pi r$ and, if the period is T, the time spent in each dee is $T/2$, where $T = 2\pi r/v = 2\pi m/Be$. The period is therefore independent of speed and radius, and is thus the same for all particles. The ion is always in phase once the potentials on the dees are correctly adjusted so that the energy is increased each time the ion passes a gap. When the ion has reached the maximum radius, it is led out by a channel some 60° long curved to follow the path of the ions with the outer plate at a negative potential to draw the ions away from the magnetic field. They emerge at P. Figure 18.6 shows a photograph of the emergent beam from a cyclotron.

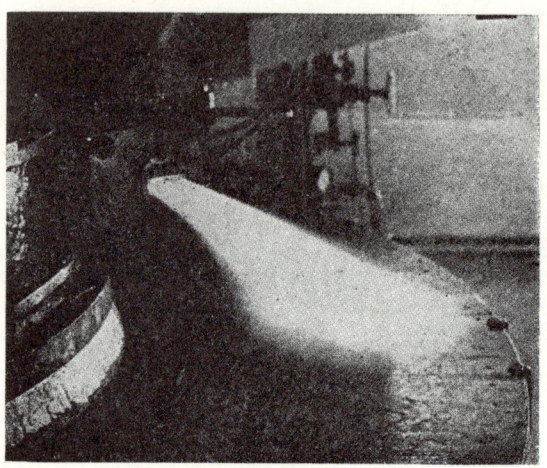

Fig. 18.6 Photograph of an emergent deuteron beam from a cyclotron. (By courtesy of Harvard University Press and A. K. Solomons.)

The particles usually accelerated are protons, deuterons and α-particles, and the energies attained depend on the size of the dees. Since the maximum velocity at circumference $= BeR/m$ (putting $r = R$, where R is the radius of the dees),

$$E = \tfrac{1}{2}mv^2 = \tfrac{1}{2}m\left[\frac{RBe}{m}\right]^2 = \frac{B^2R^2}{2}\frac{e^2}{m}.$$

Therefore $E \propto R^2$ for a given particle.

Thus the size of a cyclotron increases more rapidly than the corresponding increase of energy. The maximum energy of a particle from a fixed-frequency cyclotron is about 40 MeV for α-particles, the limit being set by the relativistic mass increase as well as mechanical engineering difficulties and expense.

18.6 The Synchrocyclotron

It will be noticed in the previous section that the expression used for the kinetic energy was the non-relativistic value $\frac{1}{2}mv^2$. For the early cyclotrons, working at low velocities, this was accurate enough but for higher speeds relativity changes become important. Thus, if $v = 0.8c$, $v^2/c^2 = 0.64$ and $\sqrt{1-\beta^2} = \sqrt{0.36} = 0.6$ and $m = 1.66\ m_0$, where m_0 is the rest mass of the particle and $\beta = v/c$.

Now from $T = 2\pi m/Be$ we see that as m increases so does the period T, and the particle therefore gradually gets out of phase with the high-frequency potential on the dees. The frequency on the dees must therefore be decreased to compensate for the gain in mass. This is carried out by a rotating variable capacitor giving the imposed frequency modulation required. Ions can then be accelerated to very high velocities, and the cyclotron becomes a synchrocyclotron. The famous 4.67 m synchrocyclotron in the Lawrence Radiation Laboratory produced 380 MeV α-particles, and was later redesigned to give 720 MeV protons.

The difference between the cyclotron and the synchrocyclotron is that in the former the output is continuous but in the case of the latter the ions starting out from the centre are subject to a frequency modulation as they approach the periphery and so come out in bursts of a few hundred per second, each burst lasting about 100 μs.

18.7 Electron Accelerating Machines. The Betatron

The possible electron accelerators so far described are the Van de Graaff generator and the linear accelerator. The cyclotron cannot be used for electrons since m_e is so small that the change of mass occurs at low energies. Thus, a 1 MeV electron has a velocity $= 0.9c$, and its moving mass is about three times greater than its rest mass, whereas for a 1 MeV proton the increase of mass factor is only 1.001. An alternative method of accelerating electrons uses an alternating magnetic field rather than an electrostatic one. In the *betatron*, the electrons are contained in a circular tube, referred to as the 'doughnut', placed between the poles of a specially shaped magnet B which are energized by an alternating current in the windings W, see Fig. 18.7. Electrons are produced thermionically and given an initial electrostatic energy of about 50 keV. As the magnetic field builds up during the first half-cycle it induces an e.m.f. inside the doughnut and accelerates the electrons which are already moving in a circular path, by the action of the transverse magnetic field. When the field reaches its first positive maximum it is suddenly stopped and the high-energy electrons leave their circular paths tangentially to strike a target which then emits X-rays. Electrons are always ejected into the target when the magnetic field has just completed its first quarter-cycle and reached its maximum value.

As already explained, the velocities acquired are very high and may approach $0.98c$. If the circumference of the doughnut is ~ 3 m, the frequency is $v/2\pi r = (0.98 \times 3 \times 10^8)/3 = 98$ MHz. If the frequency of the magnetic field is 50 Hz, the time taken for the first quarter-cycle is $\frac{1}{200}$ s and the electrons make $(9.8 \times 10^7)/200 = 4.9 \times 10^5$ journeys per quarter-cycle. If the average energy acquired is 200 eV per cycle, the total energy on ejection is about 100 MeV, the mass now being about 200 m_0. Energies up to 300 MeV are currently available from betatrons

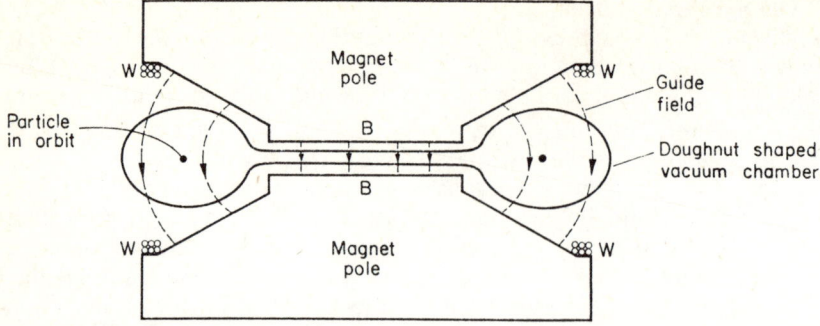

Fig. 18.7 Sectional diagram of betatron showing vacuum chamber doughnut.

which give bursts of X-rays at a repetition rate of one pulse every hundredth of a second.

18.8 Electron Synchrotron

In this machine modulation is provided, as explained for the synchrocyclotron, giving electron energies of the order of 1 GeV, although a 6 GeV electron synchrotron is now operating in the Soviet Union. This is achieved by varying the intensity of the magnetic field used for deflecting the electrons. The arrangement is not unlike the betatron except that the magnet pole pieces are annular and follow the outline of the doughnut (Fig. 18.8). In the central gap some soft iron flux bars serve as the central core of the magnet to start up the machine as a betatron. Part of the interior of the doughnut is coated with copper or silver to give a resonance cavity G, which is attached to a high-frequency oscillator of a few thousand volts. When the oscillator is on, the electron is accelerated each time it crosses through the resonator.

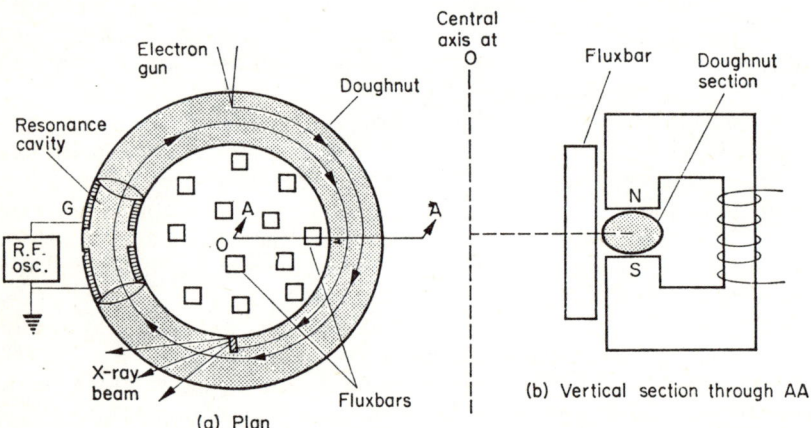

Fig. 18.8 Electron-sycnhrotron. (a) Plan with magnet removed; (b) vertical section through AA showing annular magnet.

Primary electrons are injected into the doughnut at about 100 keV and as the field changes the electrons travel in circular paths and increase their energy as in the betatron. At about 2 MeV the flux bars are magnetically saturated and cannot induce further effects. The betatron action then ceases and the resonant cavity comes into operation. If the potential applied to G operates at the proper frequency the electrons are all kept in phase and receive increments of energy at each revolution. The oscillating potential is switched off when the electrons reach their maximum velocity governed by the maximum intensity of the magnetic field. The electrons then strike the target which gives off short wavelength X-rays or 'Bremsstrahlung'. The rays emerge in pulses as in the betatron.

18.9 Proton Synchrotron

In order to probe farther into the nucleus positive ions of many GeV are necessary and if this is to be done with synchrocyclotron the size and cost would be prodigious. To overcome this the proton synchrotron was devised, based on the electron synchrotron. A ring-shaped magnet is used — much less in mass than the equivalent synchrocyclotron — in which the particle travels with constant radius. There are four quadrants to the magnet covering the annular doughnut, as shown in Fig. 18.9. The protons are injected into the doughnut at low energy from a linear accelerator, or a Van de Graaff machine, and are recovered by magnetic deflection as a pulsed beam after many revolutions. A high-frequency resonator cavity accelerator is used in one of the straight parts with an increasing frequency corresponding to the increased speed of the protons. The field strength of the magnets is also increased to maintain the accelerated protons in a circular path of constant radius. The synchrotron action is applied at the beginning of

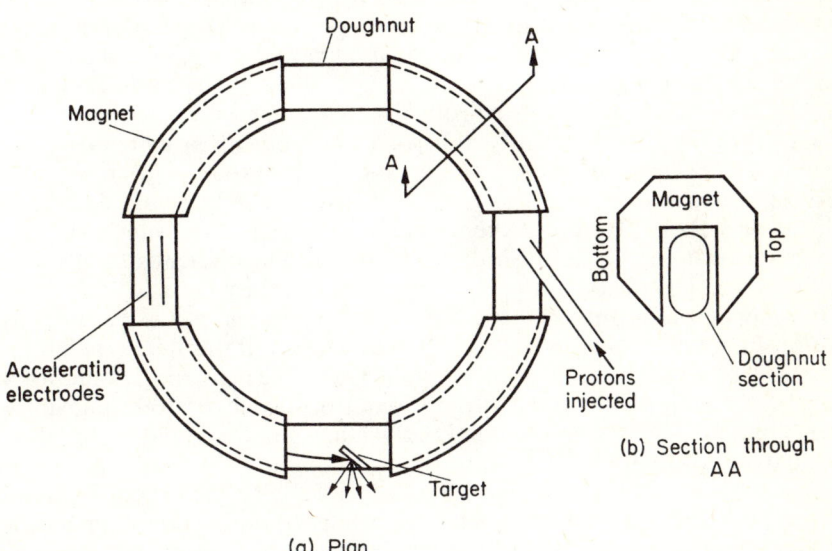

Fig. 18.9 Proton synchrotron. (*a*) Plan; (*b*) section through AA.

each cycle. Energies attained are of the order of 10 GeV. The biggest present-day machines are the Bevatron in the U.S.A., which gives 6·4 GeV protons, and the Synchrophasotron of the U.S.S.R., which operates at 10 GeV.

18.10 The Alternating-Gradient Synchrotron

The upper limit of proton energy in the case of the constant-gradient proton synchrotron is about 10 GeV because to obtain a 50 GeV beam from a machine of this type would require a magnet of well over 100 000 tonnes. In practice it is found that the proton beam deviates appreciably from the circular path when the radial magnetic field gradient is constant and since the whole doughnut is enclosed by the magnetic field in order to confine the beam as much as possible to a circular orbit the size of the magnet becomes an important factor in the design of the big proton synchrotrons.

A method of overcoming this difficulty of beam wandering is to use magnetic fields with alternating gradients to focus the beam, i.e. in successive sections of the radial field the gradient is first towards the centre and then outward from the centre and so on. As the beam travels round its orbit it then passes through sections of the pole pieces of the magnet which cause the beam to be focused vertically and horizontally in rapid succession. At the same time the particles are defocused horizontally and vertically in rapid succession. All this is done by suitably arranging the magnetic field gradients to be inward and outward in the successive sections as described. Thus the field can be regarded as having an alternating gradient and this makes for the strong focusing and defocusing action. In the Brookhaven A.G. proton synchrotron (33 GeV) there are more than 200 sections and because each section is separated by a field free region, the total weight of the magnet is only about 4000 tonnes, the same as that of the Cosmotron C.G. proton synchrotron, also at Brookhaven. The effect of focusing and defocusing the beam in rapid succession is to smooth out the deviations from the ideal circular path required by the constant gradient machine, and to produce a beam which is said to be 'strongly focused'. This is the basis of the CERN alternating gradient proton synchrotron now being used in Switzerland to give 25 GeV protons. More recently a 70 GeV A.G. proton synchrotron has been operating at Serpukhov in the U.S.S.R., and this was the world's most powerful particle accelerator in operation until 1972. Since then there has emerged from CERN the Super Proton Synchrotron (S.P.S.). This is subsidised by the governments of the Western European countries whose scientists collaborate in the research. The maximum proton energy is 400 GeV.

Parallel developments in the U.S.A. were undertaken at the Fermi National Accelerating Laboratory in Batavia, Illinois. The result was the Fermilab Proton Synchrotron which is now operating at 500 GeV. The main ring is some 6·4 km in circumference and the whole project requires a staff of 1300 taken from a consortium of over 50 American universities and is funded by the U.S. Department of Energy.

The latest Fermilab project is a 1000 GeV (1 TeV) Tevatron conceived in 1972. A ring of superconducting magnets, operating at liquid helium temperature of about 4 K, is being installed under the main ring of the present 500 GeV machine. The Tevatron has therefore the same dimensions as the 500 GeV machine but can double the energy of the protons because the superconducting magnets can give fields which are twice as big as the conventional magnets. In order to save power, and all countries are conscious of the need for energy economy, the protons will first be limited to 100 GeV in the main ring before being injected into the Tevatron ring and then accelerated by the increased field to 1000 GeV energy. Colliding beam experiments are being designed in which the equivalent energy on to a target at rest would be about 10^3 TeV. The beam intensity is expected to be 10^{12} protons s^{-1}. The project is an ambitious one and will need massive financial support. It is hoped that the experiments will be ready to start early in the 1980s. Current experiments will be described in Chapter 28.

18.11 Intersecting Beam Accelerators

A new accelerator was conceived in CERN in the early 1970s. This operates on the principle of intersecting storage ring (I.S.R.) collision. This is illustrated in Fig. 18.10. Protons from a small proton linear accelerator (linac) A enter the proton synchrotron at B, where after being accelerated by a small booster, they are accelerated to about 30 GeV. On emerging they are split magnetically at C so that

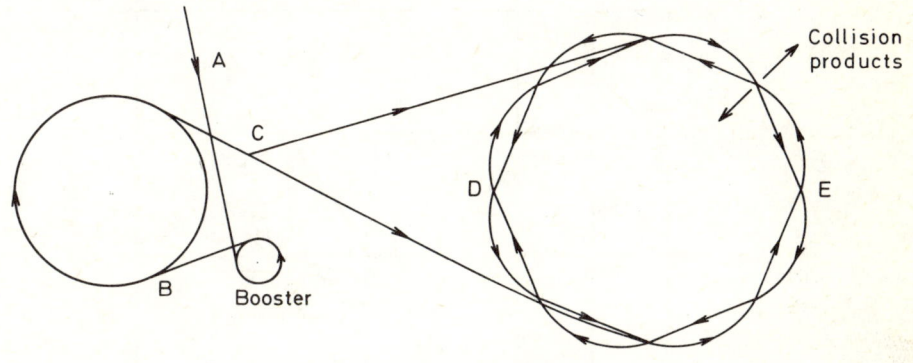

Fig. 18.10 Intersecting beam accelerator.

successive bursts go alternately into clockwise and anticlockwise directions into two large concentric storage rings. The beams travel through separate channels, but at eight intersecting points such as D and E the beams collide head-on at the cross-over points. The total possible energy is very high in terms of a stationary target since there is no loss of energy by recoil. The collision products come out sideways and are recorded by suitable instruments.

18.12 The Growth and Future of Large Accelerating Machines

It has been stressed that the mysteries of the ultimate structure of protons and neutrons, and perhaps other subatomic particles, can only be solved by using intense beams of nuclear particles of very high energies as probes. Since the work of Cockcroft and Walton in 1932 in which the highest potential achieved was something less than 1 MeV, the quest for higher and higher potentials has gone on, together with methods of increasing the ion beam intensity. In this quest the physicists of the U.S.A. have played an outstanding part.

Improvements in the Cockcroft–Walton system were limited to about 1 MeV by the breakdown resistance of the accelerating tube materials. The culminating point in the electrostatic generation of energy is in the tandem Van de Graaff generator working at about 20 MeV.

In order to go beyond the limit of about 10 MeV generated electrostatically, an entirely new method was required. The idea of resonance acceleration was conceived by Lawrence and Livingston, who made the first cyclotron in 1932 giving a proton beam of about 1·2 MeV. In 1940 the betatron was designed giving a beam of 2·3 MeV electrons. The energy limit of this fixed frequency type of accelerating machine is about 25 MeV for protons and 300 MeV for electrons but the linear accelerator is capable of giving about 100 MeV (protons) and about 20 GeV (electrons).

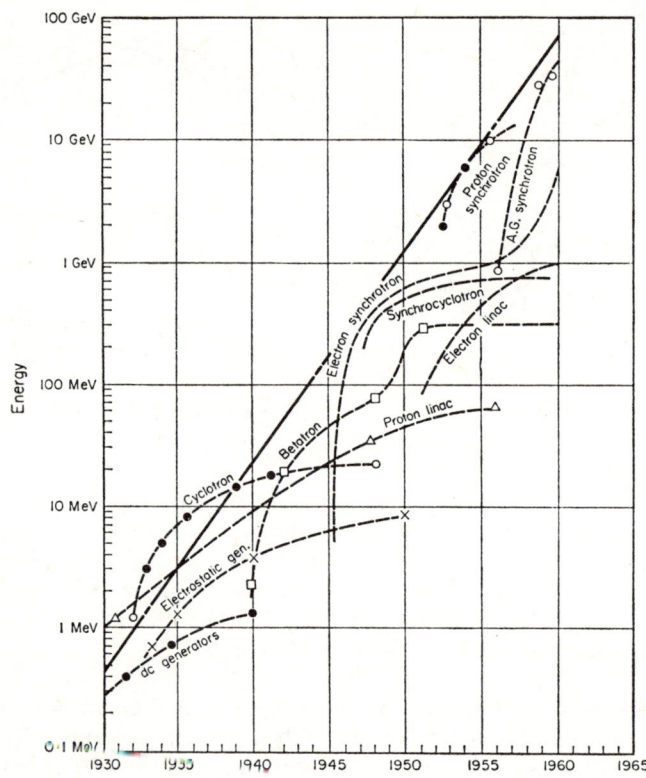

Fig. 18.11 Growth of accelerator energies achieved since 1930. (From *Particle Accelerators*, by Livingston and Blewett, copyright (c) 1962, McGraw-Hill Book Company, Inc. Used by permission.)

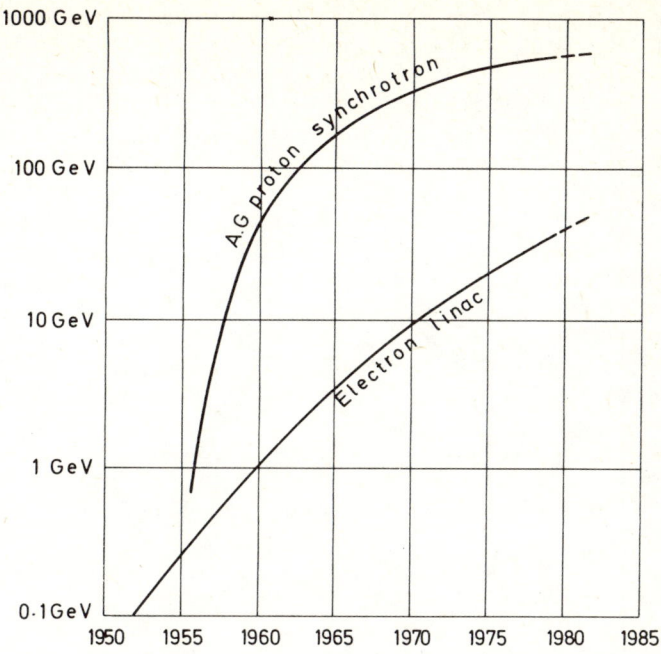

Fig. 18.12 Recent trends in acceleration energy.

The next breakthrough was the application of the principle of phase-stability, i.e. keeping the changing speed of the particle in phase with the high-frequency oscillating potential, giving rise to the synchrotrons for both positive ions and electrons. The electron synchrotron has given energies of the order of 350 MeV while the proton synchrotron can give energies up to about 200 GeV.

Finally, the present designs of collision machines are based on the principle of the alternating gradient method of focussing. This has given energies up to 500 GeV on the Fermilab and CERN machines. Higher energies will be possible when the superconductivity technology is fully developed in the Tevatron. Intersecting particle machines may exceed present upper limit in the near future. Figure 18.11 shows diagrammatically the early growth of various types of accelerator with a very rough linear growth on a logarithmic scale, while Fig. 18.12 shows the more recent advances of two of the more important types of accelerator. The need for new technology at about 1000 GeV is apparent. Table 18.1 shows some details of the various accelerators mentioned.

The cost of one of these big machines is many hundreds of millions of dollars and is so high that in Western Europe and in the U.S.A. national and international co-operation is needed to finance them. Often, then, the important research results are due to international collaboration between scientists, and even the highest energies achieved are only one billionth of the energy of an average cosmic ray particle when 'seen' on a nuclear plate costing a mere dollar! However, it must be remembered that the results which we shall describe in

TABLE 18.1
Details of Some Modern Particle Accelerators

Type and name	Particles	Size	Maximum energy (MeV)	Beam current
D.C. Rectifier				
Cockcroft–Walton	Protons		1·2	
Electrostatic Generator				
Van de Graaff	Protons	9·75 m high	12	20 μA
Tandem Van de Graaff	Protons	1·83 m high	2	250 μA
	Electrons		13·4	2 μA
Magnetic Resonance of Positive Ions				
Oak Ridge Cyclotron		2·18 m dia.	24	1 mA
Birmingham University Cyclotron		1·56 m dia.	20	0·3 mA
Linear Accelerators				
Berkeley	Protons	12·19 m long	32	60 μA
Harwell	Protons	24·38 m long	50	5 mA
Stanford	Electrons	91·44 m long	1000	8 mA
Stanford	Electrons	3 km long	40 000	15 μA
Magnetic Induction				
Betatron	Electrons	2·54 m dia,	315	14 000 rev min^{-1} at 1 m X-rays
Frequency-Modulated Cyclotron				
Synchrocyclotron (CERN, Geneva)	Protons	4·98 m dia.	600	1 μA

Electron Synchrotron			
M.I.T.	2·64 m dia.	300	1000 rev min^{-1} at 1 m X-rays
University of Glasgow	2·74 m dia.	350	
Constant Gradient Proton Synchrotron			
Birmingham University	9·14 m dia.	1000	10^9 protons s^{-1}
Cosmotron (Brookhaven)	21·34 m dia.	3000	10^{11} protons s^{-1}
Betatron (Berkeley)	36·58 m dia.	6400	2×10^{10} protons s^{-1}
Synchrophasotron (Dubra, U.S.S.R.)	60·96 m dia.	10 000	—
Nimrod (Harwell, U.K.)	48·16 m dia.	7000	10^{12} protons s^{-1} at 30 bursts min^{-1}
Alternating Gradient Accelerators			
CERN	Protons 140·21 m dia.	25 000	3×10^{11} protons s^{-1}
Brookhaven	Protons 256·03 m dia.	33 000	3×10^{11} protons s^{-1}
Harvard, U.S.A. (Cambridge)	Electrons 71·93 m dia.	6000	6×10^{12} electrons s^{-1}
Serpukhov (U.S.S.R.)	Protons —	70 000	
CERN	Protons 2·4 km dia.	400 000	
Fermilab (Chicago)	Protons 2 km dia.	500 000	
†Tevatron	Protons 2 km dia.	1 000 000	10^{12} protons s^{-1})

† The Tevatron is projected for the early 1980s.

Chapter 28 could not have been achieved without modern accelerators and complex engineering equipment, giving controlled beams of energy. Specific events can be sought with accelerator beams whereas nuclear plate events are chance encounters.

Problems

(*Those problems marked with an asterisk are solved in full at the end of the section.*)

18.1 A cyclotron with dees of diameter 1·8 m has a magnetic field of 0·8 T. Calculate the energies to which (*a*) protons and (*b*) deuterons are accelerated. ((*a*) 25 MeV, (*b*) 12·5 MeV)

18.2 Briefly compare and contrast the physical principles of machines designed to accelerate electrons and protons to energies of the order of several GeV.

18.3 Describe the development of the wave-guide linear accelerator and explain why such machines are used primarily for the acceleration of electrons.

18.4* Calculate the ratio m/m_0 for electrons, protons, deuterons and α-particles each of energy 1 keV, 1 MeV and 1 GeV. (For 1 MeV protons $m/m_0 = 1\cdot001\,067$ and for 1 MeV electrons, 2·96)

18.5 Show that in the betatron the magnetic flux φ linking the electron orbit is given instantaneously by $\varphi = 2\pi R^2 B$, where R is the radius of the orbit and B is the instantaneous field strength at the orbit.

State carefully the conditions necessary for this equation to be true.

18.6* In a certain betatron the maximum magnetic field at orbit was 0·4 T, operating at 50 Hz with a stable orbit diameter of 1·5 m. Calculate the average energy gained per revolution and the final energy of the electrons. (294 eV, 91 MeV)

18.7* What radius is needed in a proton synchrotron to attain particle energies of 10 GeV, assuming that a guide field of 1·5 T is available? (18 m)

For simplicity put $v = c$. Is this approximation justified?

18.8 In Problem 18.1, what would be the generator frequencies required to accelerate the proton to 25 MeV and the deuterons to 12·8 MeV?

18.9 A synchrocyclotron gives out pulses at the rate of 100 pulses s^{-1}. If the average current in a proton beam is 10 μA m^{-2}, what is the average flux density in the beam?

18.10 Protons are injected into the Brookhaven alternating gradient proton synchrotron with an energy of 50 MeV and accelerated up to 33 GeV. The mean radius of the orbit is 122 m. Find the initial and final frequency of the protons and the strength of the magnetic field.

18.11 Describe the constant gradient proton synchrotron.

Calculate the radius of a proton synchrotron required to accelerate protons to 10 GeV energy in a field of 1·5 T. For simplicity assume that the proton speed is the velocity of light, c. Justify this assumption.

What are the limitations of the proton synchrotron and how have they been overcome? [N]

18.12   In a proton synchrocyclotron the maximum magnetic field is 1·5 T and the limiting radius is 3·0 m.

Show that the maximum kinetic energy of the protons is about 700 MeV.

[N]

Solutions to Problems

18.4   Consider, as an example, 1 MeV protons. From $E = c^2(m - m_0)$ we get

$$E = m_0 c^2 \left(\frac{m}{m_0} - 1\right) \quad \text{or} \quad E = E_0 \left(\frac{m}{m_0} - 1\right).$$

Thus

$$\frac{m}{m_0} = 1 + \frac{E}{E_0},$$

where E is the energy of proton = 1 MeV and E_0 is the rest energy of proton = 938 MeV. Hence

$$\left(\frac{m}{m_0}\right)_p = 1 + \frac{1}{938}$$
$$= 1 + 1{\cdot}001\,067$$
$$= 1{\cdot}001\,067.$$

The rest energy for an electron is 0·51 MeV, so that

$$\left(\frac{m}{m_0}\right)_e = 1 + \frac{1}{0{\cdot}51}$$
$$= 1 + 1{\cdot}96$$
$$= 2{\cdot}96.$$

Hence 1 MeV electron must be treated relativistically but protons of this energy need not be.

18.6   In the betatron the electron velocities are nearly c, so that the total distance travelled in the acceleration time (one quarter-cycle) is $cT/4 = c\pi/2\omega$, and the total number of revolutions is given by

$$N = \frac{c\pi/2\omega}{2\pi R} = \frac{c}{4\omega R},$$

where $\omega = 2\pi \times$ frequency. Thus

$$N = \frac{3 \times 10^8}{8\pi \times 50 \times 0{\cdot}75} = 3{\cdot}1 \times 10^5.$$

Since the electrons must be treated relativistically, we have momentum $= E/c$, where E is the final energy required, i.e.

$$mv = E/c.$$

But

$$\frac{mv^2}{R} = Bev,$$
$$mv = BeR$$

275

or
$$E = BeRc.$$
Hence
$$E = \frac{0.4 \times 1.6 \times 10^{-19} \times 0.75 \times 3 \times 10^8}{1.6 \times 10^{-13}} \text{ MeV}$$
$$= 91 \text{ MeV}.$$

Thus the average energy per revolution is
$$\frac{91 \times 10^6}{3.1 \times 10^5} = 294 \text{ eV}.$$

18.7 The equivalent mass of a 10 GeV proton is 10 GeV + rest mass
$$= 10.938 \text{ GeV}$$
$$= 11.75 \text{ u}$$
$$= 11.75 \times 1.66 \times 10^{-27} \text{ kg}$$
$$= 1.95 \times 10^{-26} \text{ kg}.$$

Now $mv = BeR$ and so
$$R = \frac{mv}{Be} = \frac{1.95 \times 10^{-26} \times 3 \times 10^8}{1.5 \times 1.6 \times 10^{-19}}$$
$$= \frac{5.85}{3.25} \times 10$$
$$= 1.8 \times 10$$
$$= 18 \text{ m}.$$

Chapter 19

Nuclear Models and Magic Numbers

19.1 Introduction
When our knowledge of nuclear structure is compared with that of atomic (i.e. electronic) structure it is evident that the theoretical approach is very difficult. The small size of the nucleus (less than 10 fm) and the fact that the forces concerned do not appear elsewhere, make theoretical nuclear structure more difficult than theoretical atomic structure. Moreover there is no central field of force within the nucleus corresponding to the central field of force for the electrons provided by the positive charge of the nucleus. In nuclear physics the forces of interaction are much larger.

There are many isolated facts which will require explanation when we consider the details of nuclear structure. For example, why does the $^3_2$He nucleus, which contains two protons but only one neutron, remain absolutely stable against the large Coulomb repulsive force between the protons? Why do nuclei emit α-particles and β^--particles when they are known to contain only protons and neutrons? It must be remembered that the β^--particles are not always orbital electrons but do sometimes come out of the nucleus. Why is the binding energy per nucleon almost constant, and why are the $4n$ nuclei particularly stable, as shown in Fig. 14.3? How do we explain the existence of excited states of nuclei and the Geiger–Nuttall rule? There are many other well-established facts associated with nucleonic systematics which must be explained theoretically.

19.2 Neutron Cross-Sections and Nuclear Radii
Just as the first indications of the structure of the atom as a whole came from probe experiments with α-particles, the investigation of the structure of the nucleus also calls for the use of a probe. Since a neutron is very small and electrically neutral it has been used frequently as a bombarding particle in scattering experiments which have provided a great deal of empirical knowledge of the nucleus.

If a beam of neutrons of initial intensity I_0 is attenuated by a sheet of material of thickness x to I, then experimentally, see Fig. 19.1,
$$I = I_0 e^{-\Sigma x},$$
where Σ is the linear absorption coefficient. This is the usual exponential law of absorption. The above equation requires that Σ should have dimensions L^{-1} and

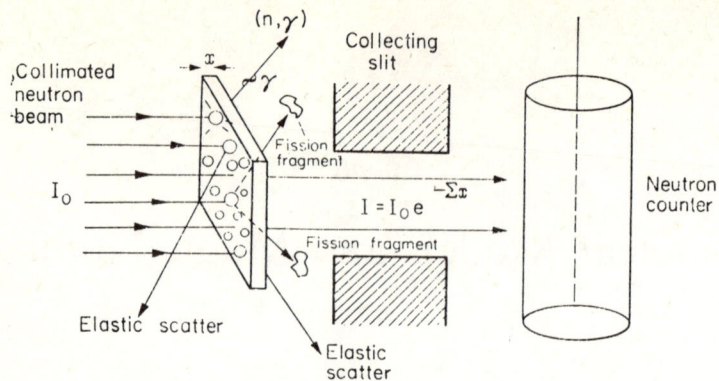

Fig. 19.1 Principle of transmission measurement of total cross-section. Neutrons interacting with nuclei in sample do not reach detector. Correction for small angle scattering is made.

its units are therefore m^{-1}. One would expect Σ to be proportional to the number of nuclei per unit volume in the absorber N_V, so that $\Sigma = \sigma N_V$, where σ is a constant with dimensions of area, since the dimensions of N_V are L^{-3}. We interpret σ as the collision area per nucleus or the neutron cross-section of the absorber nuclei. It is sometimes referred to as the microscopic cross-section of the material. It depends upon the type and the energy of the bombarding particle, and is therefore only constant for the material for a given particle over a narrow energy range.

Now unit volume of material of density ρ contains $N_A \rho / A$ nuclei, where N_A is the Avogadro constant and A is the atomic mass number of the absorber atoms. Thus

$$N_V = \frac{N_A}{A} \rho \quad \text{and} \quad \Sigma = \sigma \frac{N_A}{A} \rho.$$

The unit used for σ is an area of 10^{-28} m^2. This unit is called the 'barn', being always greater than the cross-sectional area of a nucleus.

We have, therefore,

$$\Sigma = \sigma \times \frac{6 \cdot 02 \times 10^{26}}{A} \rho$$

or, approximately,

$$\Sigma = 0 \cdot 06 \frac{\sigma \rho}{A} \text{ m}^{-1},$$

where σ is in barns, and A is the atomic mass number.

Although the unit of Σ is reciprocal length, it is often called the macroscopic cross-section as it corresponds to the total collision cross-section per unit volume of absorber. From the above relation it can be evaluated easily if the microscopic cross-section is known.

We can now write the absorption equation as $I = I_0 e^{-\sigma N_V x}$, which gives $I = I_0 e^{-\sigma N_a}$ by putting $N_V x = N_a$, where N_a is the number of target atoms per unit area of absorber.

Writing $\Sigma = 1/\lambda$, where λ has dimensions of length, we get the equation $I = I_0 e^{-x/\lambda}$ and it can be shown by integration that λ is the mean free path of the neutrons in the target material.

Thus if the intensity of the beam after passing through a thickness x is I and a further transmission of dx produces an alteration dI then the path length of the neutrons giving this change dI is xdI and the total path length of all the neutrons is $\int_{I_0}^{0} xdI$. The average path length per neutron is then

$$\lambda = \frac{\int_{I_0}^{0} xdI}{\int_{I_0}^{0} dI}$$

$$= \frac{\int_{0}^{\infty} xI_0(-\Sigma)e^{-\Sigma x}dx}{I_0}$$

from $I = I_0 e^{-\Sigma x}$ and $\int_{I_0}^{0} dI = I_0$.

So

$$\lambda = -\Sigma \int_{0}^{\infty} xe^{-\Sigma x}dx$$

$$= \frac{1}{\Sigma}$$

(by parts). Hence $\lambda = 1/\Sigma$ is the average path length per neutron or the mean free path. This can then be written

$$\lambda = \frac{A}{\sigma N_A \rho}$$

$$= \frac{A}{0.06 \sigma \rho} \text{ m},$$

where σ is in barns. Thus for beryllium metal $A = 9$, $\sigma = 0.10$ barns, $\rho = 1.847 \times 10^3$ kg m^{-3}, and so

$$\lambda = \frac{9}{0.06 \times 0.1 \times 1.847 \times 10^3}$$
$$= 8.1 \times 10^{-1} \text{ m}$$
$$= 810 \text{ mm}.$$

In Fig. 19.1 we show in principle the measurement of neutron cross-sections. From $I = I_0 e^{-\Sigma x}$ we have $\ln I = \ln I_0 - \Sigma x$ and Σ can then be found by measuring the beam intensity in the detector with and without targets. A plot of

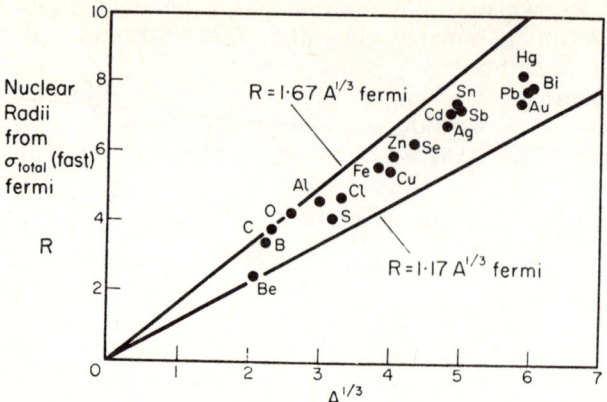

Fig. 19.2 Dependence of nuclear radii on $A^{1/3}$.

ln I against x for several foil thicknesses gives Σ graphically and hence σ. In this experiment the beam energy and the beam geometry have to be carefully controlled.

A collimated beam of neutrons can therefore be used to measure the neutron cross-sections of the elements, and it is found that there is a linear relationship between nuclear 'radius' R, obtained by putting $\sigma = 2\pi R^2$,† and the cube root of A. Thus $R = R_0 A^{1/3}$, as shown in Fig. 19.2, where the unit used for cross-section is the barn. Neutron cross-sections vary a great deal with the energy of the neutron beam as shown in Fig. 19.3. From the graph we find $R_0 = 1 \cdot 3$ to $1 \cdot 4$ fm for neutron cross-sections, the variation of R_0 depending on the manner of measuring σ. The precise value of R_0 is not important to us but the equation $R = R_0 A^{1/3}$ implies that the densities of ALL nuclei are constant, since

$$\rho \propto \frac{A}{V} \propto \frac{A}{R^3} \propto \frac{A}{R_0^3 A} \propto R_0^{-3},$$

i.e. ρ is a constant, independent of A, if R_0 is constant.

As an example we take the case of $^{27}_{13}\text{Al}$ and put $R_0 = 1 \cdot 33$ fm, as a rough average. For the aluminium nucleus, therefore,

$$R = 1 \cdot 33 \times 27^{1/3} \text{ fm}$$
$$= 4 \cdot 0 \text{ fm}.$$

We consider a single nucleus of $^{27}_{13}\text{Al}$ and use 1 u $= 1 \cdot 66 \times 10^{-27}$ kg, we find for the density that

$$\frac{M}{V} = \frac{1 \cdot 66 \times 10^{-27} \times 27}{\frac{4}{3}\pi(4 \cdot 0)^3 \times 10^{-45}} \text{ kg m}^{-3}$$
$$\approx 10^{17} \text{ kg m}^{-3}$$

The nucleus is therefore very dense indeed and all nuclei have this constant value of density. The nuclear physicist uses this fact to describe the nucleus in terms of a model. We have seen how the atomic physicist uses in turn, the Bohr model, the vector model and the wave-mechanical model of the atom to explain

---

† For fast neutrons the experiment measures the total cross-section $\sigma_1 = 2\pi R^2$, equally divided between absorption and scattering, i.e. $\sigma_T = \sigma_A + \sigma_S$.

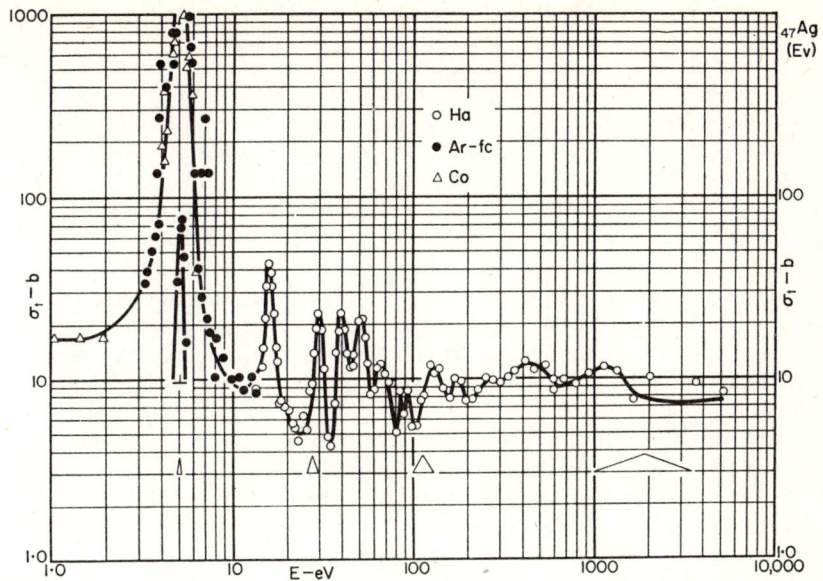

Fig. 19.3 Total neutron cross-section of silver in the low-energy region. (Taken from *Nuclear Physics*, by I. Kaplan, 1963, Addison Wesley, Reading, Mass.)

the various aspects of the electronic properties of the atom. In like manner, the nuclear physicist also uses models to try to explain the behaviour of the nucleus in the nuclear experiments he carries out.

Before we discuss particular models of the nucleus it should be mentioned that there are several other methods of measuring nuclear sizes. An important method is the scattering of fast electrons by nuclei due to the Coulomb interaction between the electrons and the positive nuclear charge, in contrast with the interaction between the neutron and the nucleus. Thus, unlike neutron scattering experiments, electron experiments give information about the distribution of positive charge within the nucleus. 'Electron radii' and 'neutron radii' of nuclei are therefore slightly different, the former being somewhat less than the latter.

19.3 The Liquid-Drop Model

It was Bohr who suggested that since all quantities of a given liquid have the same density under the same conditions, it might be profitable to compare the properties of a nucleus with those of a drop of liquid. There are many similarities such as:

(1) constant density which is independent of size;
(2) latent heat of vaporization, corresponding to constant binding energy per nucleon, see Chapter 14;
(3) evaporation of a drop, corresponding to radioactive properties of nuclei and internal thermal vibrations of drop molecules corresponding to the energy of nuclei;

281

(4) condensation of drops, corresponding to the formation of a compound nucleus and absorption of bombarding particles.

Would it be possible to extend this idea to a more quantitative description of the nucleus? For example, in a drop of liquid the molecules are only influenced by their immediate neighbours — this is the basis of Laplace's molecular theory. This could also be true of the nucleus, where the nucleon–nucleon forces could be short-range forces with a limited sphere of effectiveness. Further — would it be possible to incorporate surface tension effects into the nucleus? The *surface* molecules in a spherical drop of liquid are not so tightly bound as the internal molecules; could this also be true of the surface nucleons of a nucleus? Using these ideas it might be possible to calculate the mass-to-energy ratio of a nucleus from its physical data of A, Z and N. Masses accurate to six figures, see Appendix C, are now available from mass spectrometry, so that empirical formulae can be checked quite easily.

We have already seen that the mass of a nucleus is not exactly equal to the total added masses of its separate nucleons, the difference being equal to the binding energy of the nucleus. Can the binding energy of a nucleus be calculated using the liquid drop analogy just described?

Approximately, the mass of a nucleus is given by
$$^A_Z M = Z M_p + (A - Z) M_n,$$
but more accurately the equation can be written
$$^A_Z M = Z M_p + (A - Z) M_n - B,$$
where B is the binding energy of the nucleus. Hence
$$B = Z M_p + (A - Z) M_n - ^A_Z M,$$
and we have already seen that the binding energy per nucleon, or B/A, is fairly constant after $Z = 10$.

Let us now use the liquid-drop model and consider the factors which contribute to this binding energy.

(1) If there are short-range forces of attraction such that the binding energy per nucleon is constant, and since the density is also constant, we would expect this contribution to the binding energy to be proportional to the total number of nucleons present, that is to the atomic mass number A, since each nucleon contributes the same binding energy. Thus our first contribution is $+a_1 A$, where a_1 is a constant, the positive sign indicating the binding effect of the attractive forces.

(2) Since the Z protons in the nucleus are positively charged and repel one another by Coulomb's law they oppose the binding forces by an electrostatic force of repulsion. Our second factor therefore is a repulsion term depending on the potential energy of the protons in the nucleus, and proportional to $Z^2 e^2 / R$. Since $R = R_0 A^{1/3}$ the second term is $-a_2 Z^2 A^{-1/3}$, where a_2 is a constant.

(3) The third term comes from the liquid-drop model directly. Whereas the argument developed in (1) for the attractive forces assumes that all the nucleons are equally attracted in all directions, this is not so for the surface nucleons. These are weakly held together from inside (equivalent to surface tension in the drop) so that the binding energy represented by the first term

must be decreased by an amount proportional to the surface energy of the nucleus, which in turn depends on the surface area of the spherical nucleus, decrease in binding energy, i.e.
$$\propto R^2 \propto R_0^2 A^{2/3} = -a_3 A^{2/3},$$
where a_3 is the third constant, and the minus sign represents the fact that the first term overestimates the attractive forces of the surface nucleons.

(4) The fourth term was originally due to Fermi and is a distribution term inserted to account for the fact that the stability of nuclei depends on the neutron–proton distribution among the nucleons for a given Z, A. Fermi developed a formula for the effect of this on the binding energy, viz.
$$-a_4 \frac{(Z-A/2)^2}{A}.$$
This is a *weakening* term due to neutron excess over protons in the nucleus, and when $A=2Z$ this term is zero and increases with increasing A, see Fig. 14.1(a).

(5) Finally, to account for the fact that nuclei have various stability characteristics according to their odd–even nucleon properties, there is a small *empirical* correction term δ, which is introduced to allow for the fact that even–even nuclei are more stable than odd–odd nuclei (p. 199); that is to say, an even–even nucleus has a lower energy than an odd–odd nucleus and an even–odd nucleus has an energy intermediate between them. The form of the term δ is

$$\delta = 0 \text{ for } \begin{array}{l} N \text{ even, } Z \text{ odd} \\ N \text{ odd, } Z \text{ even} \end{array} \bigg\} A \text{ odd};$$

$$= +\frac{a_5}{A^{3/4}} \text{ for } N \text{ even, } Z \text{ even, } A \text{ even};$$

$$= -\frac{a_5}{A^{3/4}} \text{ for } N \text{ odd, } Z \text{ odd, } A \text{ even};$$

where a_5 is a constant equal to 33·6 MeV or 0·036 u.

Inserting the five separate constants, the binding energy can now be written as
$$B = a_1 A - a_2 Z^2 A^{-1/3} - a_3 A^{2/3} - a_4 \frac{(Z-\tfrac{1}{2}A)^2}{A} + \delta.$$

The constants are found by fitting this equation to the experimental data, although a_2 can be found from the calculation of the electrostatic repulsion energy of a single proton assuming the nucleus is a uniformly charged sphere. Its calculated value is 0·7103 MeV and the best values of the other constants are $a_1 = 15·753$, $a_3 = 17·80$, $a_4 = 94·77$ and $a_5 = 33·6$, all in MeV. The binding energy can then be written
$$B = 15·753 A - 0·7103 Z^2 A^{-1/3} - 17·80 A^{2/3} - 94·77 \frac{(Z-\tfrac{1}{2}A)^2}{A} + \delta \text{ MeV},$$
where δ is defined above.

The binding energy per nucleon $\bar{B}$ is then
$$\bar{B} = \frac{B}{A} = 15·753 - 0·7103 Z^2 A^{-4/3} - 17·80 A^{-1/3} - 94·77 \frac{(Z-\tfrac{1}{2}A)^2}{A^2} + \delta A^{-1}$$
in MeV, which is the equation of the curve in Fig. 14.3.

The mass of the nuclide $^A_Z M = 0.991\ 75A - 0.000\ 840Z + 0.019\ 114A^{2/3} +$
$0.000\ 762\ 6Z^2 A^{-1/3} +$
$0.101\ 750\ (Z - \tfrac{1}{2}A)^2 A^{-1} - \delta$ u,

where $\delta = \pm 0.036/A^{3/4}$ or 0, as defined above. All these coefficients are based on the ^{12}C scale.

This equation can be tested for agreement with mass spectrometer values of the mass of any nuclide by inserting the appropriate values of Z and A, although a more realistic test is to compare the calculated values of $\bar{B}$ with the experimental values. These agree very closely, as shown in Table 19.1. Notice also that the value of the binding energy per nucleon is nearly constant.

TABLE 19.1
Comparison of Calculated and Measured Values of $\bar{B}$ (MeV)

Nuclide	$\bar{B}$ (calc.) (MeV)	$\bar{B}$ (meas.) (MeV)
$^{17}_{8}$O	8.11	7.75
$^{27}_{13}$Al	8.42	8.33
$^{33}_{16}$S	8.58	8.50
$^{55}_{25}$Mn	8.74	8.75
$^{63}_{29}$Cu	8.75	8.75
$^{98}_{42}$Mo	8.62	8.63
$^{127}_{53}$I	8.37	8.43
$^{195}_{78}$Pt	7.90	7.92
$^{238}_{92}$U	7.56	7.58
$^{245}_{97}$Bk	7.54	7.52

The successes of the liquid-drop model are not to be judged solely on the calculation of atomic masses and binding energies which can be done with a fair degree of accuracy, but also on the prediction of α- and β^--emission properties using the mass equation. Thus it is possible to show why $^{238}_{92}$U is an α-emitter and not a β^--emitter and to calculate the energy of the emission. For β^--emission from $^{238}_{92}$U, we have $^{238}_{92}$U $\rightarrow$ $^{238}_{93}$Np $+\ _{-1}^{0}$e(β^-). Inserting the appropriate values of A, Z in the equation for $^A_Z M$ we find $^{238}_{93}M > ^{238}_{92}M$ by about 0.000 05 u, so that β^--emission is impossible. In the case of emission $^{238}_{92}$U $\rightarrow$ $^{234}_{90}$Th $+\ ^4_2$He (α) we find on substituting values of A, Z in the equation for $^A_Z M$ that $^{238}_{92}M > ^{234}_{90}M + ^4_2$He by about 0.004 u or 3.8 MeV showing that α-emission is possible. This result compares reasonably well with the experimental value (4.18 MeV), and shows the usefulness of the mass equation in predicting radioactive properties of nuclei.

Perhaps the most important success of the liquid-drop model is in the explanation of nuclear fission. As we shall see in Chapter 22 it is possible to predict why $^{235}_{92}$U and not $^{238}_{92}$U is fissile to slow neutrons and also predict the fast fission threshold of $^{238}_{92}$U, as shown in Fig. 22.1.

19.4 Nuclear Shells and Magic Numbers

Modern physics has told us a great deal more about the electrons in an atom than about the nucleus. We have a very successful shell model for the electrons and the frequencies of spectral lines can be calculated very accurately. Resonance and ionization potentials can also be predicted, so that it is reasonable to ask whether or not a shell model can be developed for the nucleus. Can the nucleons exist in well-ordered, quantum-controlled nuclear shells? Is there any evidence for the grouping of nucleons into shells? Can quantum numbers similar to n, l, s, j be applied to the nucleus?

The analogy is so tempting that we make a survey of the literature of nuclear physics and assemble all the measured properties of the nucleus. For instance, if we consider the property of stability, we know that the $4n$ nuclides are relatively stable — the α-particle itself has four nucleons — could this mean the *closing* of a nuclear shell, in the same way as the electron shells of He, Ne, Ar and Kr, at $A=4$?

Is there some numerical rule corresponding to the electronic shell rule of 2, 8, 18, etc., for the closing of nuclear shells? There was little evidence for this in the 1930s but empirical data have since accumulated in favour of a shell structure of the nucleus based upon the fact that nuclei with certain values of N or Z seem particularly stable compared with their immediate neighbours. Some of this evidence is collected below.

(1) The nuclides $^4_2$He ($Z=2$, $N=2$) and $^{16}_8$O ($Z=8$, $N=8$) are particularly stable, as can be seen from the binding energy curve, Fig. 14.3, Numbers, 2, 8 indicate stability.

(2) The biggest group of isotones (N constant) and therefore the most stable, is at $N=82$.
    The next are at $N=50$ and $N=20$. Neutron numbers of 20, 50 and 82 therefore indicate particular stability.

(3) Tin, $_{50}$Sn, has ten stable isotopes, more than any other element, while $_{20}$Ca has six stable isotopes. This indicates that elements with $Z=50$ and $Z=20$ are more than usually stable.

(4) The three main radioactive chains all decay to $_{82}$Pb (see Chapter 4) and $^{208}_{82}$Pb, with $Z=82$ and $N=126$, is the most stable isotope of lead.

(5) It is found that some isotopes produced in β-decay are spontaneous neutron emitters indicating a very low neutron binding energy (see, for example, Fig. 22.7). These are:
$$^{17}_8\text{O}, \quad ^{87}_{36}\text{Kr}, \text{and} \quad ^{137}_{54}\text{Xe}$$
for which $N=9$, 51 and 83, which can be written as
$$8+1, \quad 50+1, \quad \text{and} \quad 82+1.$$

If we interpret this loosely bound neutron as a 'valence' neutron, the neutron numbers 8, 50 and 82 represent greater stability than other neutron numbers. For example the nuclide $^{87}_{36}$Kr with $N=51$ is a neutron emitter because $N=50$ is a stable configuration or a closed nuclear shell. From this and other accumulated evidence the numbers 2, 8, 20, 50, 82 and 126 for either Z or N appear to be associated with high nuclear stability. They are called 'magic numbers', and correspond to closed shells.

There is a great deal of experimental evidence that the numbers 2, 8, 20, 50, 82, 126 are peculiarly favoured when changes of nuclear property with increasing A are studied. There is also much supporting evidence from fast and slow neutron cross-sections, as shown in Fig. 19.4, in which abrupt changes take place at these

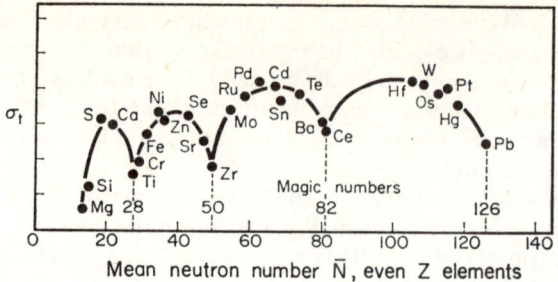

Fig. 19.4 Neutron absorption of even-Z nuclei showing evidence for magic numbers at 28, 50, 82 and 126. (Taken from a paper by H. Rose, *J. Nuclear Energy*, **5**, 4 (1957).)

numbers, showing that these nuclei are particularly stable compared with their immediate neighbours. In general terms, plots of many nuclear properties against Z or N show characteristic peaks, or points of inflection at these peculiar numbers (not unlike some of the evidence for the periodic system of elements).

19.5 The Theory of the Nuclear Shell Model

Magic numbers can be predicted theoretically using some of the ideas discussed in previous chapters. Let us assign a quantum l to the orbital angular momentum of a nucleon about the centre of the nucleus. We put $l = 0, 1, 2, 3, \ldots$, depicting s, p, d, f, ... states of the nucleon, as we did for the electron.

Following the methods applied to the electronic structure of the atom, we write down the appropriate Schrödinger equation and try to solve it. As we have seen, the details of the potential field are entirely unknown, so that we must assume a simple but known nuclear field $V(r)$. Each nucleon is free to move under the action of this field and, whatever stationary energy states are obtained, we must remember that the nucleons are fermions and so much obey the Pauli exclusion principle. The model is based on these facts, together with the assumption that the internucleonic forces (pp, pn and nn) are powerful attractive short-range forces.

For a nucleon with an orbital quantum number l, the wave equation is

$$\frac{d^2 R_{nl}(r)}{dr^2} + \frac{2m}{\hbar^2}\left[E - V(r) - \frac{l(l+1)}{2mr^2}\hbar^2\right]R_{nl}(r) = 0$$

(see p. 144), where R_{nl} is the radial wave function. To solve this we must know $V(r)$ and one of the simplest forms of this potential is that of the harmonic oscillator given by

$$V(r) = \tfrac{1}{2}kr^2 + \text{a constant,}$$

where k is a constant. This gives a parabolic well, as shown on the left-hand side of Fig. 19.5.

The solution of the Schrödinger wave equation for the harmonic oscillator is well known and gives eigenvalues

$$E_N = h\nu(N + \tfrac{3}{2}),$$

where $N = 0, 1, 2, 3, \ldots$ is the oscillator number. Note that when $N = 0$, $E_0 = \tfrac{3}{2}h\nu$, i.e. there is a zero point energy not found in classical physics. In solving the wave equation, some interesting features emerge. For example, it is found that for each

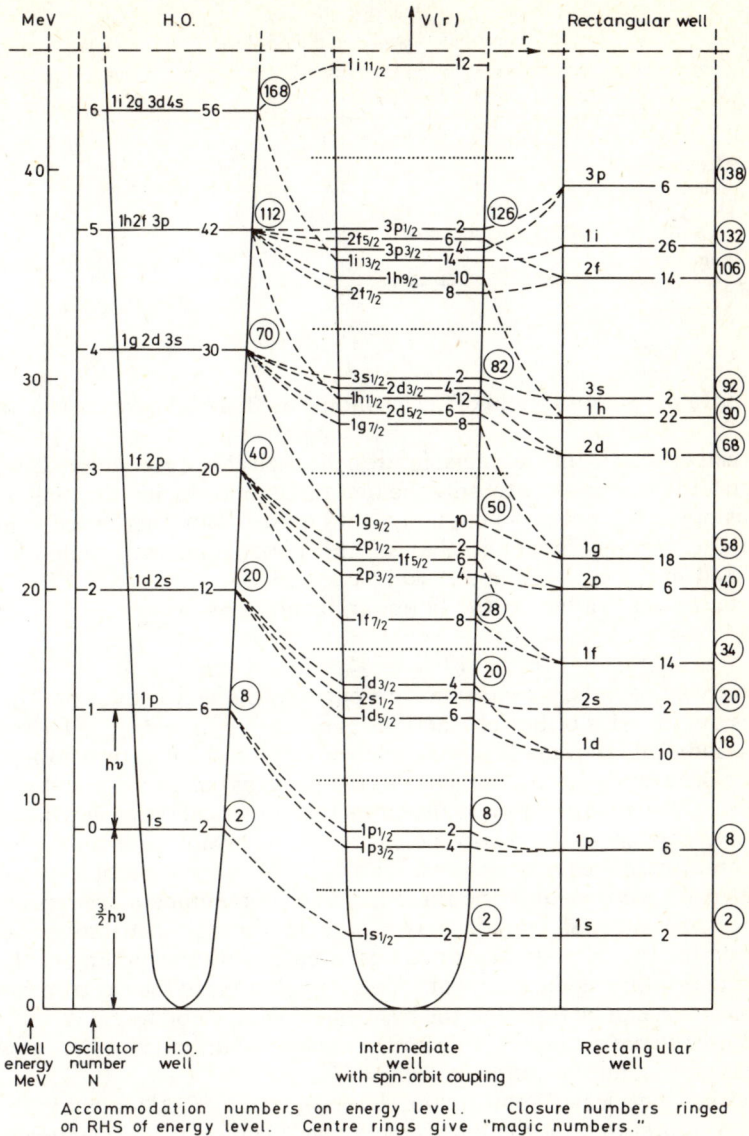

Fig. 19.5 Potential wells for nuclear shell model showing nucleon energy states.

oscillator number N there are degenerate l levels. In all cases $l \leq N$ but in particular l is *odd* when N is *odd*, and l is *even* when N is *even*. Also, the total number of nucleons at each oscillator level N is given by $(N+1)(N+2)$. This is then the occupation number.

With this information we can build up an energy level diagram for the nucleon

TABLE 19.2
Harmonic Oscillator Energy Levels

N	0	1	2	3	4	5	6
l	0	1	0, 2	1, 3	0, 2, 4	1, 3, 5	0, 2, 4, 6
Occupation number $(N+1)(N+2)$	2	6	12	20	30	42	56
Closure number	②	⑧	⑳	㊵	⑺⓪	⑪⑫	⑯⑧

moving in the potential field of the harmonic oscillator. These are shown in Table 19.2.

As can be seen from the ringed figures, the shell closures occur at 2, 8, 20, 40, 70, 112 and 168 nucleons, with only the first three reproducing the magic numbers. This is obviously not good enough, so we must try another expression for $V(r)$.

Another simple model is that of the rectangular well potential already discussed in the case of the electron (see p. 139).

For a nuclear radius R, the boundary conditions are
$$V(r) = 0 \quad \text{for} \quad r > R,$$
$$V(r) = -V_0 \quad \text{for} \quad r < R.$$
When these boundary conditions are put into the wave equation, the solution comes out in terms of Bessel functions of order $(l+\tfrac{1}{2})$, i.e $J_{l+1/2}$. These functions are periodic and the energy levels are given by a radial quantum number $n = 1, 2, 3, \ldots$ whenever $J_{l+1/2}$ is zero (n is *not* the principal quantum number of atomic spectroscopy). Furthermore, n also gives the numerical order in which the states of a given l appear as the energy increases. The eigenvalues depend on both l and n and come out in the order shown in Table 19.3. The spectroscopic notation is n, l.

Again we see that only the three lowest magic numbers are reproduced.

The most successful model has a potential energy distribution somewhere between the two already described and is based on the coupling of the nucleon spin and orbital angular momenta. This is analogous to the electron case and the reasons for it can be traced to the basic nucleon–nucleon force. We are therefore led to use the same coupling rule for the nucleon quantum numbers as we had for the electron quantum numbers, i.e. $j = l \pm \tfrac{1}{2}$.

This has the effect of splitting the l levels into two substates, but in the nucleon case it is found that the state with the higher j value (i.e. $l+\tfrac{1}{2}$) is the more stable and so is found deeper in the potential well. Quantitatively it is found that for the larger N values the higher j values are depressed so much that they merge with the upper states of the lower N values. Furthermore, the calculated energies of the states show large energy gaps immediately after the experimental magic numbers 2, 8, 20, 50, 82, 126, ..., etc. Each single j level may have $2j+1$ neutrons *or* protons and the levels are now designated by the symbol nl_j, where $n = 1, 2, 3, \ldots$ is the numerical order in which a given l appears. Of the $2(2l+1)$ nucleons for each l it is found that $2(l+1)$ are depressed to the lower N group while $2l$ remain with the

TABLE 19.3
Energy Levels for Rectangular Well Potential

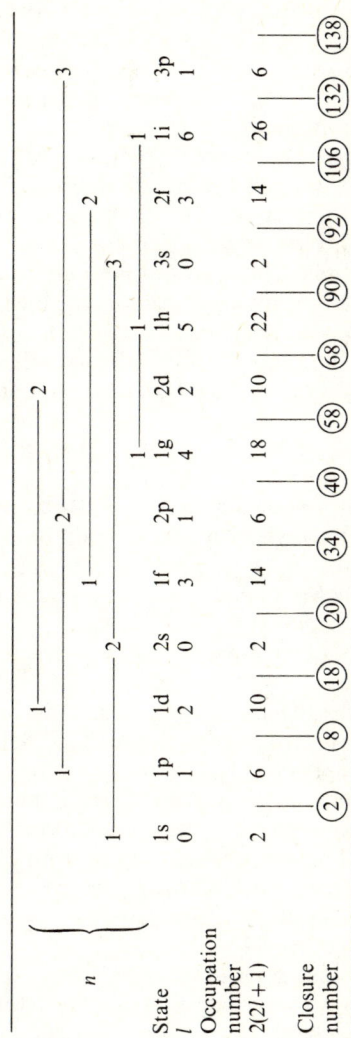

289

original N. In this way the magic numbers are completely reproduced. This is shown in Fig. 19.5, in which the shape of the resultant potential well is somewhere between a rectangular well and an oscillator well. The use of the expressions $2(2l+1)$ and $(2j+1)$ are borrowed from the electron configuration rules and imply that the nucleons obey the Pauli principle.

This single-particle shell model is useful in providing the broad basis for predicting nucleonic spectroscopic states and energy levels, magnetic moments and nuclear spins. However the model is over-simplistic and there are many anomalies which require a more sophisticated model for their description. As mentioned at the beginning of this chapter, in atomic physics the electrons are subject to electromagnetic forces of attraction and repulsion which lead to their binding energies and the operation of the exclusion principle. These forces are small and well understood. In nuclear physics, however, the forces of interaction between the nucleons are much larger and are of the same order of magnitude as the shell level separations. Thus the simple shell model often fails to predict the precise position of the nuclear energy levels. One basic assumption in the model is that the nucleus is spherically symmetrical. But most nuclei are distorted, due to the strong forces of interaction, and are not spherically symmetrical. More advanced models are therefore based on a non-spherical nucleus. One of these is the collective model.

19.6 The Collective Model

In the liquid-drop model the nucleons only affect each other at close range and interact strongly with their immediate neighbours. In the shell model the nucleons do not interact with each other and the nucleons are treated individually, so that the spins and nuclear magnetic moments are due to the last unpaired outer shell nucleons.

In the collective model proposed by Aage Bohr and B. R. Mottleson the potential field in which the nucleons move is perturbed by the usual spin–orbit term, a short-range pairing potential and a long-range deformation potential. The nucleons therefore move in a distorted potential field rather than a spherically symmetrical field. This leads to a new factor due to the angular momentum of the inner, closed shells. A permanent nuclear distortion results which can be described in terms of rotational and vibrational energies within the core of the nucleus. These are small when compared with the single-particle energies of the shell model, but it is now possible to combine these new features of the individual particle model with the strong interaction liquid-drop model to formulate the collective model.

The collective model is apposite when the nucleon numbers deviate a lot from the magic numbers. For odd-A nuclei many of the low-lying energy states can be related to the simple shell model and the last nucleon often gives the nuclear spin correctly, especially for those nuclei near to the magic numbers. On the other hand, for even–even nuclei in which simple shell model pairing abounds, excited states of nuclei require the paired nucleons to be separated (unpaired). This invokes the pairing energy mentioned above and leads to the concept of vibration and rotation states within the core of the nucleus. A detailed description of this very important model of the nucleus is beyond the scope of this book but its importance is reflected by the fact that its originators, Bohr and Mottelson, received the Nobel Prize for Physics in 1975.

Even with the collective model it must not be thought that nuclear energy states can be described in detail. Good agreement with excited states is only found in special cases. However, general trends can be explained and the prediction of nuclear level spectra enhanced. Other more complicated models have been devised for specific cases.

No simple model can give accurate predictions of excited energy states for all nuclei, just as no equation of state can be devised to explain the properties of all gases and vapours. Models in nuclear physics are necessary because the solution of the Schrödinger equations for the many-body problem (i.e. A nucleons in a nucleus), which involves both Coulomb and nuclear forces, is very difficult. In this chapter we have outlined the simpler nuclear models. Each model is right for the nuclear process it describes. Thus the liquid-drop model gives a good interpretation of the energy of nuclear fission (Chapter 22), whereas the shell and collective models aim at evaluating nuclear energy states and nuclear spectra.

One of the most interesting deductions from the nuclear shell model is the possible existence of superheavy elements with double magic numbers.

19.7 Superheavy Elements: Experimental and Theoretical

One of the consequences of the simple shell model is that it is possible to predict those nuclides which are doubly magic, i.e. those for which both N and Z are magic numbers. Magic numbers relate to great nuclear stability, so that nuclides centred around doubly magic numbers are relatively stable. These regions are called 'islands of stability' (cf. islands of isomerism, p. 305). As N and Z increase, it is found that the shell model energy level diagrams for neutrons and protons begin to show differences due to the increasing effect of the Coulomb repulsion potential of the protons. In Fig. 19.6 we show the calculated upper spacings for the double magic number superheavy element $^{342}_{114}?$. The potential field used in the calculations includes the Coulomb repulsion effect of the protons and a spin–orbit coupling term with parameters adjusted to fit the single particle spacings of the actinides at $Z=100$ and extrapolated to higher Z values. To get a better picture of the difference between proton and neutron spacings, Fig. 19.6 should be attached to the top of Fig. 19.5, where it will be seen that the predicted states of the simple shell model are shown much better by neutrons. Magic numbers are shown by gaps. There are noticeable energy gaps after 164, 184 and 228 neutrons, and after 114, 124 (not 126) and 164 protons, as shown. Superheavy nuclei are still limited to certain N/Z values by Coulomb effects so that some combinations of these higher magic numbers give nuclides of high relative atomic mass numbers — superheavy elements more stable than their neighbours.

The possible existence of superheavy elements which retain the trend of the $A:Z$ stability ratio of Fig. 14.1 has stimulated much research, both theoretical and experimental.

Experimentally the existence of superheavy nuclei was first suggested by Gentry in 1970 to explain the giant pleochroic haloes observed by Rutherford in transparent mica. The haloes were due to very small monazite inclusions trapped in the mica. The inclusions contain small quantities of radioactive material, the α-

291

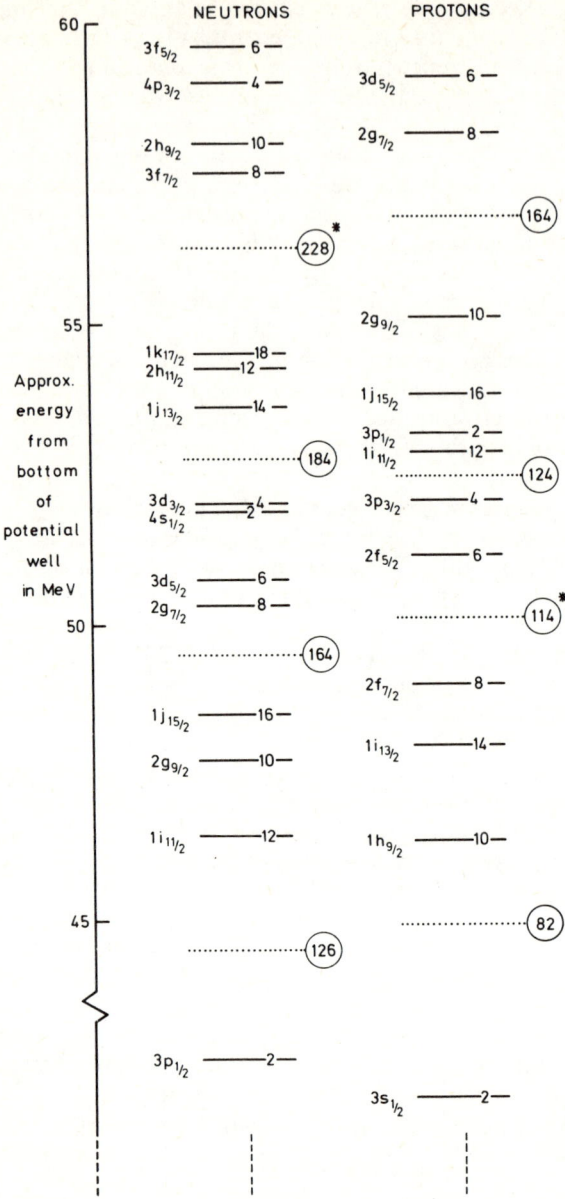

Fig. 19.6 Upper spacings of superheavy elements showing possible magic numbers. Calculated for hypothetical super heavy element with $N=228$ and $Z=114$ i.e. $^{342}_{114}$?

particles from which cause the discolouration of the mica by ionization damage as they pass through it. The circles of discolouration are easily recognized and measured under the microscope. They are, in fact, pictorial representations of the α-particle ionization curve shown in Fig. 15.1. The radius of the halo is a measure of the α-particle energy and in the case of the giant haloes this was about 14 MeV — twice that of any known natural α-energy. Gentry suggested that these very energetic α-particles were emitted by superheavy elements, the remnants of which were left along with the rare earths formed in the monazite. To test this idea the monazite inclusions were irradiated with protons and the X-ray line spectrum analysed (cf. Moseley's experiment, p. 102). The X-ray line wavelengths indicated the presence of elements from two islands of stability. These results have not been confirmed elsewhere.

An alternative explanation is that the haloes are caused by α-particle–proton collisions, the protons being found in water molecules trapped in the mica. The α-particles are contained in small quantities of uranium or thorium salts in the monazite, and the knock-on protons cause the halo discolouration. It is also possible that giant haloes are caused by radioactive α-emitting daughters of long-dead superheavy elements. On the theoretical side the stability of superheavy elements against spontaneous fission or radioactive decay is calculated from a modified liquid-drop model called the 'droplet' model. In this model account is taken of non-spherical shapes of the liquid drop and this extends the parameters of the mass equation from five for the liquid-drop model to nine for the 'droplet' model.

The additional terms due to these nuclear deformations give a complicated 'droplet' semi-empirical mass equation which is a smooth fit for nuclear masses up to $A=240$ except for small local deviations. Using this equation it is possible to test the stability of a decay system by calculating the masses of the parent and daughter nuclides.

Local deviations from the smooth curve are mainly due to shell effects, and the most important shell effect is that of the magic numbers the calculated energies are considerably reduced. When the calculated shell effect term is added to the semi-empirical mass formula it is possible to predict the lifetimes of spontaneous fission and radioactive decay of doubly magic superheavy elements. If a superheavy element is to exist in nature today it must have a lifetime of at least 10^9 a and hitherto the calculations for superheavy elements give lifetimes less than 20 a. If superheavy elements are ever found with certainty it will be necessary to make major changes in the nuclear models if lifetimes of the order of 10^9 a are to be predicted. At present the calculations are very tentative and subject to uncertainties due to the extrapolations and 'best-fit' parameters used, but it seems unlikely that lifetimes of 10^9 a will be predicted by the present model.

This is a good illustration of the difficulties facing theoretical nuclear physicists and the necessity of invoking a nuclear model.

19.8 Latest Developments

The subject of superheavy elements (S.H.E.'s) is of such theoretical and practical interest that an International Symposium was held on S.H.E.'s in March 1978 in Texas. In it many theoretical and experimental studies were reported.

Calculations showed that S.H.E.'s are unstable to spontaneous fission and it was possible to estimate some of the expected half-lives of the S.H.E.'s. It was also predicated that the average number of neutrons per spontaneous fission would be about 10. One of the most fruitful areas of study that of meteorites, which could possibly retain some superheavy material. The average number of neutrons per fission found in meteorites was between 4 and 10, which is much greater than the 2–3 neutrons per fission found in normal uranium spontaneous fission. Also the number of neutrons per spontaneous fission increases with mass suggesting that the high number (4–10) is associated with a high mass number.

Other experiments have involved heavy ion accelerators. A systematic search for superheavy elements based on their possible synthesis by the bombardment of neutron-rich $^{248}_{96}$Cm targets with neutron-rich $^{48}_{20}$Ca ions, has been conducted.

This reaction is judged to be the best for producing superheavy nuclides near the centre of the $Z = 114$ centre of stability. To date the formation and decay of superheavy elements has not been detected, but the experiments are proceeding with other bombarding reactions.

Although superheavy elements have yet to be discovered and confirmed, the possibility of their existence is so strong as to generate definite advances in techniques of theory and experiment. We give one example from geophysics.

19.9 The Melting of the Moon

As an example of a theory based on the existence of superheavy elements we consider the early history of the moon. The returned samples from the Apollo and Luna missions showed that the moon is of the same age as the earth, viz. 4·5 Ga (4 500 000 000 a), but during the first 100 Ma of its life it was completely melted. Thus 4 Ga ago it formed a differential crust and a molten iron core such as exists now in the earth. Theory shows that for an object the size of the moon the gravitational energy released on its formation is insufficient to provide the heat of melting, so that the energy must have come from comparatively short-lived radioactive elements. Calculations show that radioactive potassium, aluminium and uranium could have melted the moon, but on the wrong time scale. If superheavy elements existed in the moon at its formation, they may have been sufficient to provide the radioactive heat of melting in its first few hundred million years.

The effect of the molten core was to provide an internal magnetic field and the palaeomagnetic study of lunar rocks shows that this disappeared about 3·2 Ga ago, either because the molten core solidified or the radioactive heat sources decayed and became inadequate to sustain the core dynamo, action which gives rise to the lunar magnetic field.

Theoretical studies show that superheavy elements could have *melted* the early moon and some of the elements *could* have been soluble in iron in order to drive the core dynamo.

From their positions in the periodic table it is estimated that elements having the atomic numbers 114, 115 and 116 should be soluble in iron. If so it is calculated that their half-lives must be of the order of 100–1000 Ma to account for the disappearance of the magnetic field after 1300 Ma.

All this is very conjectural, but if superheavy elements are ever found with certainty the early history of the moon and the whole solar system may have to be reconsidered; and it all stems from the elementary predictions of the shell model of the nucleus.

Problems

(*The problem marked with an asterisk is solved in full at the end of the section.*)

19.1   Use the semi-empirical mass formula to calculate the binding energies per nucleon of $^{15}_{7}N$, $^{16}_{8}O$, and $^{18}_{9}F$. Account for any differences found. (7·65, 7·95 and 7·80 MeV)

19.2   By calculating the binding energies of the last neutron in the nuclides $^{207}_{82}Pb$, $^{208}_{82}Pb$, and $^{209}_{82}Pb$, discuss the use of this as evidence for the magic numbers 82 and 126.

19.3   A nuclide with $Z = 84$, $A = 219$ is radioactive. Determine whether it is an α-emitter or a β^--emitter. Repeat for the nuclide (72, 170) and check your result from the tables.

19.4   Which of the following experimental characteristics of nuclei can be explained (*a*) by the liquid-drop model and (*b*) by the shell model of the nucleus?

(i)   Approximately constant density of nuclei.
(ii) Discontinuities in nuclear binding energy curves.
(ii) Distribution characteristics of stable isotopes.
(iv) Approximate constancy of the binding energy per nucleon as A is increased.

19.5   What are the main sources of evidence for the existence of nucleons in discrete shells? Contrast the orbital nature of nucleons with that of electrons in atoms.

19.6   What is the importance of a study of the so-called 'magic numbers' in nuclear physics? How far have these features of nuclear systematics been justified (*a*) experimentally and (*b*) theoretically?

19.7*   All odd-A nuclides have a nuclear spin angular momentum given by
$$I = \frac{2n+1}{2} \frac{h}{2\pi}$$
where $n = 0, 1, 2, \ldots$, etc., and the electron, proton and neutron each have a spin angular momentum of $\frac{1}{2}h/2\pi$.

Prove that it is impossible for the electrons of the atom to exist as particles within the nucleus.

19.8   From the shell model prediction of nuclear energy levels (Fig. 19.5) find the ground states of the nuclides $^{15}_{7}N$, $^{17}_{8}O$ and $^{51}_{23}V$. ($f_{1/2}$, $d_{5/2}$ and $f_{7/2}$)

19.9   Calculate the atomic masses of $^{35}_{17}Cl$ and $^{37}_{17}Cl$ from the semi-empirical mass formula. Compare with the accepted values on the ^{12}C scale.

19.10   Show that the magic number after 126 is 184.

19.11   The semi-empirical mass equation of a nuclide Z, A is in atomic mass units:
$$^{A}_{Z}M = 0.992A - 0.000\,84Z + 0.0191A^{2/3}$$
$$+ 0.000\,76Z^2A^{-1/3} + 0.102(Z - \tfrac{1}{2}A)^2 A^{-1} - \delta,$$
where $\delta = \pm 0.036 A^{-3/4}$ or 0 according to the odd–even properties of Z, A.

Using this equation, find the atomic number of the most stable nuclide of the set of isobars with $A=64$ and calculate the mass of this nuclide.

The copper isotope $^{64}_{29}$Cu decays by either β^--emission or β^+-emission. Without working out any exact masses, use the mass equation to explain how this is possible. [N]

19.12 Describe an experiment by which neutron-scattering cross-sections may be measured.

Show how nuclear radii may be obtained from such measurements. How does this information contribute to the concept of the liquid drop model of the nucleus?

The neutron cross-section of the nuclide $^{27}_{13}$Al is 1 barn. What is the radius of the $^{216}_{84}$Po nuclide? [N]

19.13 Give an account of the liquid drop model of the nucleus explaining why it cannot predict nuclear energy states, spins and nuclear quantum systematics in general.

Show how the liquid drop model can be used to derive a semi-empirical mass equation for a nuclide Z, A in terms of Z, A and constant coefficients only. How may these constant coefficients be evaluated? [N]

19.14 Give an account of the shell model of the atomic nucleus and show how the postulate of strong spin–orbit coupling of the angular momenta leads to a satisfactory explanation of the 'magic numbers' found in nuclear systematics, namely:

$$2, \ 8, \ 20, \ 28, \ 50, \ 82, \ 126.$$

[N]

14.15 Discuss the reasons for the use of models of the nucleus in nuclear physics.

Give a brief account of the shell and liquid drop models of the nucleus showing where each is particularly useful. What is the essential difference between the two models? [N]

19.16 Explain qualitatively how the nuclear shell model has been successful in predicting the magic numbers found in the systematic survey of nuclear properties.

What conclusions can be drawn from the fact that the excited nuclide $^{87}_{36}$Kr is a spontaneous neutron emitter?

Solution to Problem

19.7 If there are electrons in the nucleus instead of neutrons, we must have A protons to give mass A, and $(A-Z)$ electrons, so that the net positive charge in nucleus is

$$A-(A-Z)=Z$$

as required. Thus total particles inside nucleus is

$$A+(A-Z)=2A-Z.$$

For odd-A nuclides we require $Z=$odd or $Z=$even.

Case (a) Z odd, then $2A-Z$ is also odd so that the spin of A will be an odd multiple of $\frac{1}{2}h/2\pi$, as required.

Case (b) Z even, then $2A-Z$ is now even and the spin of A should be 0 or an even multiple of $\frac{1}{2}h/2\pi$, i.e. $I=0, 1, 2, \ldots$ units, which is not found for odd A nuclides. Hence electrons and protons cannot be nuclear particles together.

For the nuclear model containing Z protons and $(A-Z)$ neutrons the total number of particles is always A so that the nuclear spin is always odd for odd A and even for even A in terms of $\frac{1}{2}h/2\pi$.

Chapter 20

Artificial Radioactivity

20.1 The Discovery of the Positron

The year 1932 was notable not only for the discovery of the neutron by Chadwick and for the first use of artificially produced nuclear missiles by Cockcroft and Walton, but also for the discovery of the positron by Anderson in America. Anderson was one of R. A. Millikan's cosmic ray workers who used the Wilson cloud chamber method of detection in which the sign of an ionizing particle can easily be determined by the direction of its track curvature in the magnetic field. In cosmic ray work many cloud chamber photographs must be taken and analysed carefully for particles and for collision events. The energies of the particles are measured in terms of absorption in lead sheets placed above the chamber so as to slow down any particle passing through. Measurement of the characteristics of the track (grain density, linearity, etc.) often made it possible to deduce the mass, charge and energy of the particle. The cloud chamber photograph which led Anderson to announce the existence of the positron in 1933 is shown in Fig. 20.1. That the positron was positively charged was established by comparing the curvature of its track with that of a negative electron. The details of the tracks of the two particles were closely similar, suggesting that the masses of the positron and the electron were the same. The mean lifetime of the positive particle was estimated to be about one tenth of a microsecond, so that it could not be a proton, which is stable.

Subsequently, photographs were taken in which there were two tracks of opposite curvature starting at the same point. At the point A in Fig. 20.2 the incident cosmic ray started a collision reaction in which both positive and negative electrons were born simultaneously. This is an example of the creation of the positron–electron pair at the same moment of time. Electrons have now come to be regarded as positrons or negatrons, but the traditional name of 'electron' will be retained for the negatron throughout these descriptions.

Pair creation, an important phenomenon in nuclear physics, is commonly brought about by irradiating a metal foil with high energy nuclear γ-rays. In effect this amounts to the creation of particles from electromagnetic radiation. In order to conserve the spin angular momentum, two electron-like particles must be created simultaneously. The positron is unstable and will quickly lose its identity by recombination with one of the free electrons in the metal. The γ-photon energy must be $\geqslant 2m_0c^2$, where m_0 is the rest mass of the electron and c is the velocity of light, since two electron masses are created. When the positron and another

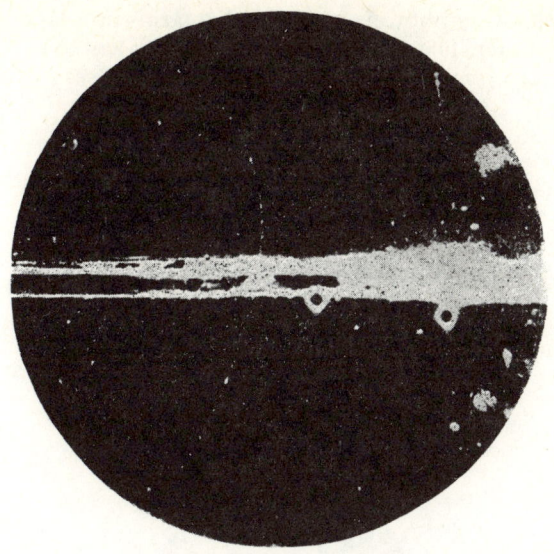

Fig. 20.1 Anderson's original positron cloud chamber photograph. (Taken from Rochester and Wilson, *Cloud Chamber Photographs of the Cosmic Radiation*, Pergamon Press, 1952.)

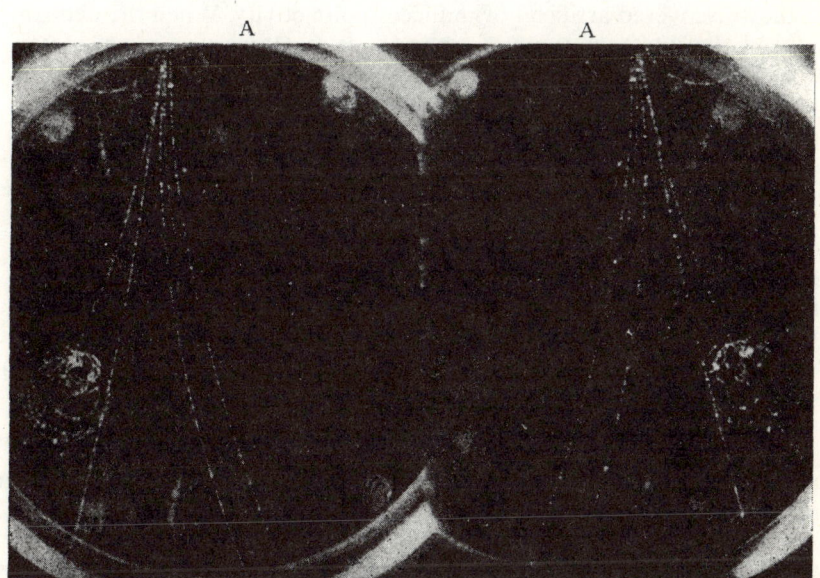

Fig. 20.2 Cloud chamber photograph showing pair production. (Taken from Rochester and Wilson, *Cloud Chamber Photographs of the Cosmic Radiation*, Pergamon Press, 1952.)

electron coalesce, two γ-photons are formed and this process corresponds to the annihilation of matter. The energy E needed is equivalent to two electron masses, so that using $E/c^2 = 2m_0$
$$E = 2 \times 0.511 \text{ MeV}$$
since $m_0 \equiv (1/1836) \times 931.5$ MeV and thus
$$E = 1.022 \text{ MeV per pair}$$
$$= 1.63 \times 10^{-13} \text{ J per pair}.$$

To find the wavelength of the electromagnetic radiation involved in the annihilation of the electron pair we have, for one γ-ray,
$$E = h\nu = \frac{hc}{\lambda} = 1.63 \times 10^{-13} \text{ J}$$
from above, and therefore
$$\lambda = \frac{6.6 \times 10^{-34} \times 3 \times 10^8}{1.63 \times 10^{-13}} \text{ m}$$
$$= \frac{19.8}{1.63} \times 10^{-13} \text{ m}$$
$$= 1.24 \times 10^{-12} \text{ m}$$
$$= 1.24 \text{ pm},$$
which is the wavelength of a hard γ-ray. This wavelength also represents the threshold energy for the creation of a pair from γ-radiation and the above argument illustrates the conditions required for the interconversion of matter and radiation.

Apart from their production by cosmic rays, positrons are also involved in many nuclear reactions, as the researches of I. Curie and F. Joliot (1934) revealed when they discovered artificial or induced radioactivity. When the neutron was discovered the interpretations of some (α, p) reactions was reconsidered and possible alternatives put forward, which included the emission of a positron. The above authors were actually investigating the reaction
$$^{27}_{13}\text{Al}(\alpha, p)^{30}_{14}\text{Si},$$
from which they were measuring the emitted protons. From this reaction they also observed the emission of neutrons and positrons. For neutrons, we have the equation
$$^{27}_{13}\text{Al} + ^{4}_{2}\text{He}(\alpha) \rightarrow ^{1}_{0}\text{n} + ^{30}_{15}\text{P}.$$
When the source of α-particles was removed, the emission of the protons and neutrons ceased, as expected, but the emission of positrons from the isolated aluminium target continued for a long time afterwards. Since $^{30}_{15}$P does not occur in nature, it was presumed to be unstable, emitting positrons according to the equation
$$^{30}_{15}\text{P} \rightarrow ^{30}_{14}\text{Si} + ^{0}_{+1}\text{e}(\beta^+) + \nu.$$

This was confirmed by extracting the phosphorus chemically and showing it to be a true positron-emitter. This radioactive form of phosphorus differed from normal phosphorus only in its mode of preparation, its atomic mass number, and the fact that it was radioactive. Such artificially prepared elements are nearly always radioactive and are called radioisotopes. They can be prepared in several ways and have characteristic half-lives, like those of the naturally occurring radioactive isotopes.

20.2 K-Electron Capture

There are also some cases in which the transmutation taking place has all the characteristics of positron emission yet no positrons can be identified. For example, during the decay of $^{40}_{19}$K it can be shown by radiochemical analysis that the direct products of the decay process are calcium and argon suggesting that both positron and negatron emission occur:

$$^{40}_{19}\text{K} \nearrow {}^{40}_{18}\text{Ar} + {}^{0}_{+1}\text{e}(\beta^+) + \nu \quad \text{not observed,}$$
$$\searrow {}^{40}_{20}\text{Ca} + {}^{0}_{-1}\text{e}(\beta^-) + \tilde{\nu} \quad \text{observed.}$$

Although no positron emission is observed, some equivalent decay process must take place to give the argon detected. Positron emission is equivalent to negatron absorption and this provides the key for the solution of this problem. Since negatron absorption gives the same final result as positron emission a third electron transition becomes possible in which an electron is captured from among the orbital electrons. This is often called K-electron capture.

Occasionally a radionuclide can show all three possible modes of decay. The particular mode of decay followed by a single atom must be governed by chance, since collectively all three modes of decays are observed, as for example in the nuclide $^{64}_{29}$Cu, which decays as follows:

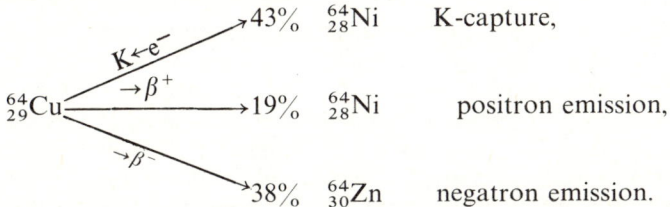

43%	$^{64}_{28}$Ni	K-capture,
19%	$^{64}_{28}$Ni	positron emission,
38%	$^{64}_{30}$Zn	negatron emission.

It is possible to measure the percentage of each product and so deduce the probability of each process taking place. These are the branching ratios shown as percentages.

Proof of the existence of K-capture lies in the fact that the new nucleus of $^{64}_{28}$Ni will be formed with one orbital electron missing whereas the $^{64}_{28}$Ni formed by the positron emission will have a full complement of electrons. The vacancy of the K-electron shell can be filled by an orbital electron transition from one or other of the L, M, N shells. This gives rise to the complete X-ray spectrum of $^{64}_{28}$Ni and its observation provides a clear proof of K-electron capture by the $^{64}_{29}$Cu atom. The wavelength of the Ni K_α line is a known constant, $\lambda = 165.6$ pm, the agreement of the wavelength of the measured K line with this value identifies the $^{64}_{28}$Ni atom as the daughter product with certainty. It should be noted that this cannot be due to the neutral $^{64}_{28}$Ni atom formed by the positron emission.

When the daughter nuclide is stable a simple electron exchange occurs. Sometimes, however, the daughter product is excited and γ-radiation is subsequently emitted. If this photon energy, generated within the nucleus, ejects an electron from the K shell, photoelectrically, there will be an additional emission of the X-ray spectrum which will be identical with that produced by the K-electron capture described above. Since this exchange of energy is not limited to the K electrons it is possible to observe the discrete energies of all the electron shells of the daughter nucleus by bending the emitted electrons in a magnetic field and analysing these discrete β^--energies in the magnetic spectrograph. This is an example of 'internal conversion' in which a line spectrum of discrete electron energies can be produced by the conversion of some fo the γ-ray energy from a radioactive element. These β^--radiations must not be confused with the nuclear β^--radiations which give a continuous spectrum.

20.3 The Origin of Electrons and Positrons within the Nucleus

Since neither positrons nor electrons exist as free particles within the nucleus, we must examine the possible exchanges in the nucleus which would explain the emission of these particles. In all cases the mass number A remains constant but the atomic number Z changes by one unit so that the element itself is also changed. The total number of nucleons remains constant but the total number of protons is altered by one. The explanation lies in a proton–neutron exchange within the nucleus, as follows:

$Z \to Z-1$	$p^+ \to n^0$	$+e^+$	positron emission $+\nu$
$Z \to Z+1$	$n^0 \to p^+$	$+e^-$	negatron emission $+\bar{\nu}$
$Z \to Z-1$	$n^0 \leftarrow p^+$	$+e^-$	K-capture with no neutrino emission
	Inside nucleus	Outside nucleus	

In order to understand why K-capture takes place in preference to positron emission, we must remember that in the latter case neutrons are formed. This requires the parent nuclide to be deficient in neutrons so that the daughter nucleus, on acquiring a further neutron, can still be stable. As we have seen, the nuclei of an element may decay either by position emission or by K-capture. In both cases the parent nucleus is below the line of stability shown in Fig. 14.1(b) and by decreasing its atomic number it can rise towards this line and become more stable. In the case of positron emission the daughter nucleus is m_e lighter than the parent by the loss of an orbital electron in the process of decreasing the atomic number by one unit, and is also a further m_e lighter due to the emission of the positron itself, making a deficit of $2m_e$ in all. Thus for favourable positron emission in radioactive decay the mass of the daughter nucleus must be less than that of the parent by at least $2m_e$, which is equivalent to 1·022 MeV, i.e.

$$M_{Z\,\text{parent}} \to M_{Z-1\,\text{daughter}} + 1\cdot022 \text{ MeV},$$

and the condition for positron emission is $(M_Z - M_{Z-1}) \geqslant 1\cdot022$ MeV. As we have seen previously, the positron is always accompanied by a neutrino, and there is always a continuous spectrum of positron energy. The net energy to be shared between the positron and the neutrino is

$$\{M_Z - M_{Z-1} - 1\cdot022 \text{ MeV}\}.$$

In some cases, however, it is found that $(M_Z - M_{Z-1}) < 1\cdot022$ MeV and the nucleus cannot become more stable by positron emission. The transition from Z to $Z-1$ is still possible by orbital electron capture since there is no energy barrier for this and the small mass difference appears as γ-radiation energy. Hence the only condition for K-capture is that $M_Z > M_{Z-1}$, however small the difference. If then the mass energy difference $(M_Z - M_{Z-1})$ in a beta disintegration is much greater than critical energy 1·022 MeV, the nuclei should decay by copious positron and neutrino emission and there will be relatively little probability of K-capture. However, for a lower mass energy difference, yet still $>1 \cdot 022$ MeV, the relative probability of K-capture increases, while for an energy difference $<1 \cdot 022$ MeV there is no possibility of positron emission.

If the parent nuclide is heavy enough as in the case of $^{64}_{29}$Cu, all three types of electron decay are possible and each mode decays with the same half-life. The probability of each decay is given by the branching ratios, and whether the positrons are impeded by the Coulomb barrier.

20.4 Nuclear Isomerism

A careful study of the half-lives of radionuclides reveals the existence of some nuclides with the same Z and A values but decaying by β^--emission quite differently. This should not be confused with the case of the two excited states of the same nuclide connected by a γ-ray transition, or with the cases of electron transitions discussed in the previous section. When there are two modes of decay having different half-lives apparently from the same nuclide, the phenomenon is called nuclear isomerism and the two nuclide states are nuclear isomers. The isomers may decay independently, as shown in Fig. 20.3(a), or the first transition may be a γ-ray transition to the ground state of the parent nuclide followed by a transition from the ground state to the daughter product. In the latter case the isomers are called genetically related isomers and are shown in Fig. 20.3(b). The γ-radiation releases orbital electrons with discrete energies and the usual X-ray spectra associated with internal conversion results as described in Section 20.2.

The classical example of isomerism comes from a study of natural bromine bombarded with neutrons. The bromine isotopes are ^{79}Br, ^{80}Br, ^{81}Br and ^{82}Br and when a bromide target is bombarded with neutrons radioactive nuclides showing γ-emission as well as β^--decay, with three separate half-lives of 36 h,

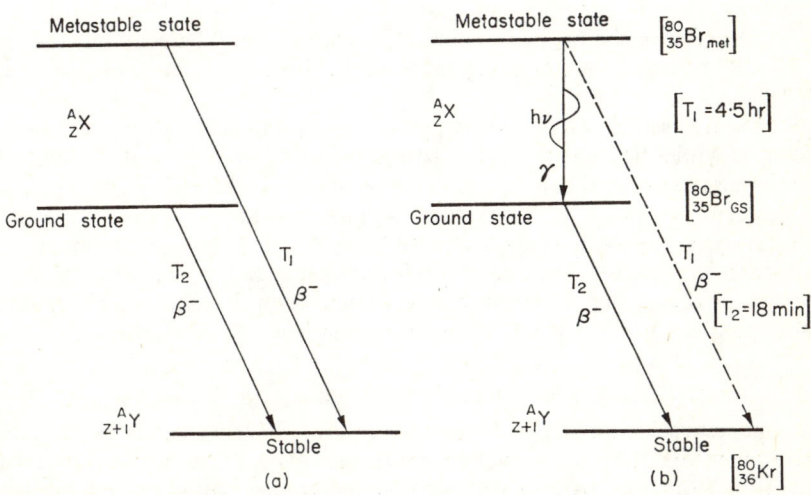

Fig. 20.3 Isomeric decay shows (a) independent decay and (b) genetically related decay as for $^{80}_{35}$Br.

4·5 h and 18 min are formed. Since *natural* bromine only contains two isotopes, ^{79}Br and ^{81}Br, one would expect two half-lives thus:

$$^{79}_{35}\text{Br} + ^{1}_{0}\text{n} \rightarrow ^{80}_{35}\text{Br} + \gamma$$

followed by

$$^{80}_{35}\text{Br} \rightarrow ^{80}_{36}\text{Kr} + _{-1}^{0}\text{e}\,(\beta^-) + \tilde{\nu}\,(T_1)_{1/2}$$

and

$$^{81}_{35}\text{Br} + ^{1}_{0}\text{n} \rightarrow ^{82}_{35}\text{Br} + \gamma$$

followed by

$$^{82}_{35}\text{Br} \rightarrow ^{82}_{36}\text{Kr} + _{-1}^{0}\text{e}\,(\beta^-) + \tilde{\nu}\,(T_2)_{1/2}.$$

When this same bromide target is bombarded with fast deuterons from a cyclotron (d, p) reactions take place as follows:

$$^{79}_{35}\text{Br} + ^{2}_{1}\text{H} \rightarrow ^{1}_{1}\text{H}\,(\text{p}) + ^{80}_{35}Br$$

and

$$^{81}_{35}\text{Br} + ^{2}_{1}\text{H} \rightarrow ^{1}_{1}\text{H}\,(\text{p}) + ^{82}_{35}\text{Br}.$$

Again the same three half-lives are observed from the subsequent decay of the ^{80}Br and ^{82}Br isotopes, showing conclusively that somehow ^{80}Br and ^{82}Br together have three different half-lives, implying that one of them has two separate decay characteristics.

To establish which isotope has the two half-lives the bromide target can be irradiated by γ-rays to give the (γ, n) reaction thus:

$$^{79}_{35}\text{Br} + \gamma \rightarrow ^{1}_{0}\text{n} + ^{78}_{35}\text{Br}$$

and

$$^{81}_{35}\text{Br} + \gamma \rightarrow ^{1}_{0}\text{n} + ^{80}_{35}\text{Br}.$$

These two products *also* have three half-lives, viz. 6·4 min, 18 min, and 4·5 h, the last two being also characteristic of the (n, γ) reactions on bromine. The bromine isotope common to *both* these experiments is ^{80}Br and it is concluded that ^{80}Br had two decay periods, 18 min and 4·5 h, so that ^{80}Br is an example of a nuclear isomer. The metastable state has a half-life of 4·5 h for γ-decay to the ground state which decays by β^--emission as shown in Fig. 20.3(*b*). When radioactive equilibrium between the bromine isomers is reached the β^--decay has the same half-life as the γ-decay from the metastable state. This experiment was first carried out in 1935.

In all similar cases of two β^--decay periods we find that one isomer exists in an excited state while the other is in the ground state. Normally, excited states exist for less than a picosecond before transition to the ground state by γ-ray emission takes place. In some cases, however, the upper energy state is metastable, and can exist for times up to several hours so that it can be regarded as independent of the ground state. This excited state can therefore be regarded as a separate isomer of the nuclide. If it decays by β^--emission, i.e. with its own characteristic half-life, we have independent isomer decay, of which the following are examples:

^{52}Mn $\beta^+ T_1 = 5\cdot5$ days; $\beta^+ T_2 = 21\cdot3$ min;
^{106}Ag   K-capture $T_1 = 43$ days; $\beta^- T_2 = 53$ h.

If the metastable state has a comparatively short lifetime it may be reduced to the ground state of the parent nuclide, emitting γ-rays of half-life T_1, followed by β^--emission from the ground state to the daughter nuclide with half-life T_2. These two isomers are then genetically related. The bromine isotope ^{80}Br is an example of this type in which the isomeric transition of the γ-ray produces instantaneous electrons by internal conversion. The decay of an isomeric state can then be regarded simply as a case of β/γ branching and which radiation is observed depends on the relative decay probabilities.

Nuclear isomerism is a phenomenon which must be explained by any theory or model of the nucleus. It is found that metastable states are favoured if there is a large spin angular momentum difference between the two nuclear states, and the corresponding energy difference is small. This means that the transition probability is low and the lifetime of the upper state long enough to make it independent of the ground state.

When the half-lives of isomers with odd A (with odd Z or odd N) are surveyed there seems to be some confirmation of the magic numbers discussed in Chapter 19. If the observed frequencies of long-lived ($T_{1/2} > 1$ s) isomers are plotted against N or Z they fall into three groups bounded by $N = 50$, 82 and 126, as shown in Fig. 20.4. These groups have been called 'islands of isomerism'. These are the numbers at which 'shells' are closed and greatest stability results, where nuclear spins are paired and there are no resultant nucleons available to provide a

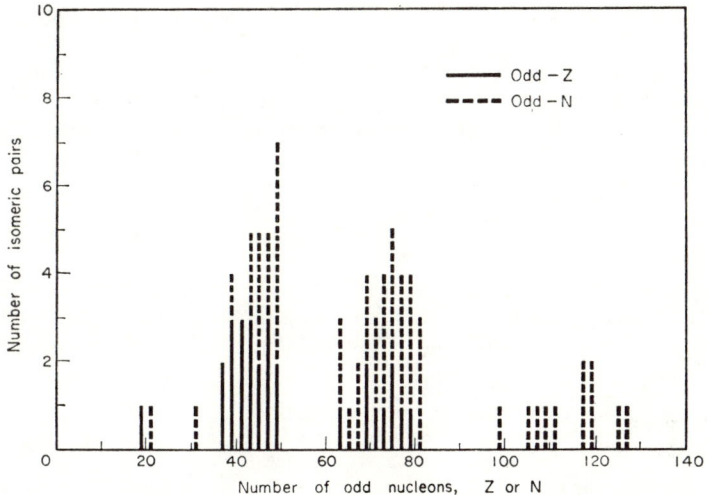

Fig. 20.4 Islands of isomerism. Frequency plot of odd Z and odd N nuclei showing breaks at 50, 82. (From *The Atomic Nucleus*, by R. D. Evans, p. 230, copyright (c) 1955, McGraw-Hill Book Company, Inc. Used by permission.)

large spin angular momentum. Hence the nucleons available to provide a large spin angular momentum. Hence the regions for $N > 50$, $N > 82$ and $N > 126$ contain relatively few isomers as the diagram shows. This is because the lowering of the state of high j in the shell model puts high spin differences just before shell closures (see Fig. 19.5).

20.5 The Production of Radioisotopes

The original artificial radioisotopes were formed by alpha or deuteron bombardment and later by neutron reactions. It is now possible to get plentiful supplies of useful radioactive nuclides by neutron bombardment in nuclear reactors, where the neutron flux may be as high as 10^{16} neutrons m^{-2} s^{-1}. When a substance has to be irradiated, it is placed in a small aluminium cylinder and pushed along a channel into the reactor core for an irradiation time equal to two or three half-lives of the product. The reactions taking place are usually (n, γ) reactions, as, for example, in the production of the useful ^{32}P isotope. The naturally occurring ^{31}P isotope (as phosphate) is irradiated for about two days when the following takes place:

$$^{31}_{15}P + ^{1}_{0}n \rightarrow ^{32}_{15}P + \gamma$$

$$^{32}_{15}P \xrightarrow[14 \cdot 3 \text{ d}]{T_{1/2}} ^{32}_{16}S + ^{0}_{-1}e\,(\beta^-) + \tilde{\nu}.$$

This radioisotope of phosphorus is useful in biology, agriculture, medicine and metallurgy.

An alternative method of preparation is an (n, p) reaction thus:

$$^{32}_{16}S + ^{1}_{0}n \rightarrow ^{32}_{15}P + ^{1}_{1}H(p).$$

This is perhaps more useful since the ^{32}P can be separated from the ^{32}S whereas in the first reaction the two phosphorus isotopes cannot be separated chemically.

Another common stable element, sodium, can be used in a radioactive form. Sodium has only one natural isotope, ^{23}Na, but a useful isotope of atomic mass number 24 can be prepared in the reactor by the action of the neutron flux thus:

$$^{23}_{11}Na + ^{1}_{0}n \rightarrow ^{24}_{11}Na + \gamma$$

followed by

$$^{24}_{11}Na \xrightarrow[15 \text{ h}]{T_{1/2}} ^{24}_{12}Mg + ^{0}_{-1}e\,(\beta^-) + \tilde{\nu}.$$

The half-life of this isotope makes it of little use for long-term investigations. Another radioactive sodium isotope of atomic mass number 22 is obtainable by bombarding $^{24}_{12}$Mg with cyclotron deuterons:

$$^{24}_{12}Mg + ^{2}_{1}H(d) \rightarrow ^{22}_{11}Na + ^{4}_{2}He(\alpha).$$

This isotope is a positron-emitter by the reaction

$$^{22}_{11}Na \xrightarrow[2 \cdot 6 \text{ a}]{T_{1/2}} ^{22}_{10}Ne + ^{0}_{+1}e\,(\beta^+) + \nu.$$

The longer half-life of ^{22}Na makes it useful for long-term investigations.

Radioisotopes are now produced by a variety of methods and the above reactions are only a few of the many possible, even with the same bombarding particle. Thus an alternative to the $^{23}_{11}$Na (n, γ) $^{24}_{11}$Na reaction described above is the (n, α) reaction:

$$^{23}_{11}Na + ^{1}_{0}n \rightarrow ^{20}_{9}F + ^{4}_{2}He(\alpha)$$

followed by

$$^{20}_{9}F \xrightarrow[12 \text{ s}]{T_{1/2}} ^{20}_{10}Ne + ^{0}_{-1}e\,(\beta^-) + \tilde{\nu},$$

when the short half-life of the fluorine nuclide makes it useless in practice.

An alternative source of $^{24}_{11}$Na is the (n, α) reaction on aluminium:
$$^{27}_{13}\text{Al} + ^{1}_{0}\text{n} \rightarrow ^{24}_{11}\text{Na} + ^{4}_{2}\text{He}(\alpha)$$
and the possibility of getting radionuclides of almost any element for particular investigations is now very good.

20.6 Some Uses of Radioisotopes

The most important fact to remember when dealing with radionuclides is that they have the same chemical properties as the stable isotopes of the same element. No one can distinguish ^{24}NaCl from ^{23}NaCl on the dinner table. If the two isotopes are introduced together into some chemical or physical system, they both proceed together and chemical analysis will not differentiate between them. However, since ^{24}Na is a β^--emitter, its progress through the system can be followed by means of a Geiger counter or a scintillation counter. Minute quantities are detectable and when a process has to be followed it is only necessary to mix a 'trace' of the radionuclide with the stable isotope and insert them together. The stable isotope is always accompanied by the active isotope so that the main role of the element can be traced by the presence of its radionuclide. This technique has given rise to the expression 'tracer methods'.

Such methods are widely used in research and industry, in agriculture, in biology and medicine, in metallurgy and engineering. Radioisotopes are used in diagnostic medicine and subsequent treatment, in radiography and in the measurement of thickness and height and in leak detection in underground pipes containing liquids (usually petroleum oils) or gases. A few examples will now be described.

Phosphorus is a necessary element in the complex make-up of any fertilizer, and the phosphorus (phosphate) uptake by growing plants from any type of soil or manure can be studied by 'labelling' the fertilizer with ^{32}P and following its progress through the root system to the foliage by means of a Geiger counter. It has been possible to show that some plants require root feeding whereas others require foliar feeding.

Another type of isotope used is ^{14}C to study the kinetics of plant photosynthesis. By growing plants in an atmosphere containing ^{14}Co$_2$ it has been possible to understand more thoroughly the complicated biochemical reactions involved.

Used as soluble ^{24}NaCl, this radioisotope finds many applications in the study of the transfer of sodium within the human body and provides valuable information concerning the flow dynamics of the body. If radiosodium is injected at one extremity of the body it can be detected within a few seconds at the other extremity. The flow of blood can thus be followed and any constrictions in blood vessels are readily detected. Because of its comparatively short half-life the radiosodium is soon transformed into stable magnesium.

Radioiodine ^{131}I has a half-life of eight days and is useful in medicine because it is known to accumulate in the thyroid gland and in the brain. Being a γ-emitter, radioiodine is useful in locating deep-seated disorders such as brain tumours and malignant thyroid tumours. Ordinary thyroid iodine deficiency can also be treated in a controlled manner using ^{131}I as the tracer nuclide.

^{60}Co emits γ-rays having an energy of about 1 MeV. Such penetrating γ-rays can be used in the radiography of industrial weldings, in which they reveal faults much further inside the metal than would be possible with a 200 kV X-ray set. Moreover, since the source is relatively small, many welds can be inspected simultaneously by placing them in a circle around a γ-source. Medically, the γ-radiation from ^{60}Co can be used therapeutically in the treatment of deep cancerous growths, and this method has largely superseded the older radium methods. Yet another application of this isotope is the gauging of sheet thickness, where, using the feedback principle, the machinery can be made self-adjusting. It can also be used to control the height of filling in packets of commercial powders. Low attenuation of the beam corresponds to an empty packet which can then be rejected automatically.

^{60}Co γ-rays can be used in photodisintegration and irradiation experiments in radiochemistry in the study of photon-induced reactions.

A few of the more modern uses of radioactive isotope are now given. ^{125}I is used in protein iodination since it decays by electron capture and has a very low-energy X-ray. It can be produced with a very high specific activity. ^{35}S is used in animal nutrition and in biochemical studies. ^{14}C and tritium are also used widely as trace elements in biochemical studies of naturally occurring amino acids. ^{51}Cr is used to label red blood corpuscles to study blood changes and blood flow, and the rare gas isotope ^{133}Xe is used in pulmonary circulation tests. Radiogold is used as gold seeds in radiotherapy and low-energy X-emitters are used for mapping out certain organs of the body. The isotope ^{99}Te is widely used for such a scanning technique since it gives a low radiation dose to the patient and has a low half-life of only 6 h. Most organs of the body can be scanned in this way.

The subject of radionuclides and their application is now so vast that the reader must refer to specific books for further information. Millions of pounds are saved annually all over the world by their use and great progress has been made in medical diagnosis and treatment. It is probable that we shall benefit even more in the future by the applications of radioisotopes in factory and hospital.

Problems

(*Those problems marked with an asterisk are solved in full at the end of the section.*)

20.1 Describe the discovery of the positron in cosmic ray cloud chamber photographs. By what reasoning did Anderson reject the possibility of the particle being a proton?

20.2 What are the conditions necessary for (*a*) β^--emission, (*b*) β^+-emission and (*c*) K-capture? When is it possible for an unstable nucleus to decay by all three modes?

20.3 What are 'genetically related' nuclear isomers? Describe experiments to prove that the nucleus $^{80}_{35}$Br has two genetically related isomers.

20.4 The nuclide $^{14}_{8}$O is a positron-emitter decaying to an excited state of $^{14}_{7}$N which decays to its stable state by emitting a γ-ray of energy 2·315 MeV. If the maximum energy of the positrons is 1·835 MeV, calculate the mass of $^{14}_{8}$O. $^{14}_{7}$N = 14·003 074 u and m_e = 0·000 548 u. (14·008 081 u)

20.5 It is required to produce a supply of radiosodium $^{24}_{11}$Na. How could this be carried out:

(a) with a beam of reactor neutrons;
(b) with a beam of cyclotron protons;
(c) with a beam of cyclotron α-particles?

20.6* When a sample of iron is bombarded with cyclotron deuterons to give the (d, p) reaction the half-life of the radionuclide so produced is 46 days. The same radionuclide can be produced by the neutron bombardment of cobalt in which protons are also observed. Identify the radionuclide. ($^{59}_{26}$Fe)

20.7 Write an essay on the use of radioisotopes quantitatively as tracers.

20.8 What radioisotopes would be useful in the study of the surface microstructure of a given steel?

20.9* A dose of 5 mCi of $^{32}_{15}$P is administered intravenously to a patient whose blood volume is 3·5 litres. At the end of 1 h it is assumed that the phosphorus is uniformly distributed. What would be the count rate per millilitre of withdrawn blood if the counter had an efficiency of only 10%; (a) 1 h after injection and (b) 28 days after injection? ((a) $5·3 \times 10^3$ dis s^{-1}, (b) $1·32 \times 10^3$ dis s^{-1})

20.10 The element iodine is readily assimilated by the thyroid gland. A patient is given a dose of 8 mCi of ^{131}I, of which 20% is absorbed by the thyroid after one day. How much energy has been dissipated in the thyroid after 4 days? What will be the disintegration rate per gram of tissue at that time? Mass of thyroid gland = 25 g.

Solution to Problems

20.6 The stable isotopes of Fe are ^{54}Fe, ^{56}Fe, ^{57}Fe and ^{58}Fe. By the (d, p) reaction the possible products are ^{55}Fe, ^{57}Fe, ^{58}Fe and ^{59}Fe, of which ^{57}Fe and ^{58}Fe are stable. Hence the radionuclide produced is ^{55}Fe or ^{59}Fe,

The only stable isotope of Co is ^{59}Co which gives ^{59}Fe by the (n, p) reaction. Thus the radionuclide produced is ^{59}Fe which can be seen from the tables to have a half-life of 46 days.

20.9 Dose per millilitre = 5/3500 mCi. Therefore

$$\text{No. of disintegrations} = \frac{1}{700} \times 3·7 \times 10^7 \text{ s}^{-1} \text{ in blood.}$$

(a) After 1 h, assume no decay. Hence

$$\text{Disintegrations counted} = \frac{1}{10} \times \frac{1}{700} \times 3·7 \times 10^7$$
$$= 5·3 \times 10^3 \text{ dis s}^{-1}.$$

(b) After 28 days, activity = $\frac{1}{4} \times 5$ mCi, since $T_{1/2} = 14$ days. Hence

$$\text{Disintegrations counted} = \frac{1}{10} \times \frac{1}{4} \times \frac{1}{700} \times 3·7 \times 10^7$$
$$= 1·32 \times 10^3 \text{ dis s}^{-1}.$$

Chapter 21

Neutron Physics

21.1 Introduction

We have already discussed the neutron as a nucleon and also as a bombarding particle. Generally speaking the source of a beam of neutrons must be an (α, n) reaction, so that it is possible to have present in the beam other particles together with γ-radiation from the α source. A very common neutron source is the Ra/Be reaction

$$^4_2\text{He}(\alpha) + ^9_4\text{Be} \rightarrow ^{12}_6\text{C} + ^1_0\text{n}.$$
$$\hookrightarrow \text{(from Ra)}$$

Since both neutrons and γ-rays can penetrate deeply, such a source of neutrons must be carefully handled and shielded. Other sources are based on polonium or plutonium as α-emitters using beryllium as the target atom as above. For experiments requiring a high neutron flux density, reactor neutrons are used.

A common neutron source is the ^{2}H(d, n)^{3}He reaction using a small H.T. generator and accelerating tube. There is a satisfactory yield of neutrons at 200 keV, approximately equal to the yield of protons by the ^{2}H(d, p)^{3}He reaction. A photoneutron source using antimony and beryllium is also commercially available.

21.2 Properties of the Neutron

The Mass of the Neutron

Although we have discussed the neutron fairly often we have yet to describe its properties in detail. It is the only elementary particle which is radioactive and reacts with nuclei. It has a definite half-life of about 13 min and its importance in nuclear physics lies in the fact that, due to its electrical neutrality, it can be used as a bombarding particle at all energies from very low to very high values. The mass of the neutron was first determined by Chadwick's early method (p. 43). The photodisintegration of deuterium, later used by Chadwick, provided another method of measuring the mass of the neutron. This reaction is

$$^2_1\text{H} + \gamma(h\nu) \rightarrow ^1_1\text{H(p)} + ^1_0\text{n},$$

where the energy of the γ-ray is known from the radioactive source. Now the masses of all the particles except $^1_0$n were known in this equation, the only unknowns being the energies of the proton and neutron. The proton energy was measured by an ionization method and found to be almost 1·05 MeV. Assuming that the proton and neutrons are ejected with equal energy the total kinetic energy is 2·1 MeV = 0·0023 u. By balancing the equation in the usual way, Chadwick and Goldhaber then calculated

$$M_n = 1·0087 \pm 0·0003 \text{ u}$$

The most precise determination of the mass of the neutron was made using the reaction

$$^1_0n + ^1_1H \rightarrow ^2_1H(d) + \gamma(h\nu)$$

for which

$$E_\gamma = 2·230 \pm 0·007 \text{ MeV}$$
$$= 0·002\ 395 \text{ u}.$$

Thus

$$M_n = ^2_1H - ^1_1H + \gamma$$
$$= 2·014\ 102 - 1·007\ 825 + 0·002\ 395$$
$$= 1·008\ 672 \text{ u}.$$

The present accepted value is $M_n = 1·008\ 665$ u on the ^{12}C scale.

The Half-life of the Free Neutron

The neutron as a free particle is radioactive. It is a β^--emitter with a half-life of 12·8 min decaying according to

$$^1_0n \rightarrow ^1_1H(p) + ^{\ 0}_{-1}e(\beta^-) + \tilde{\nu}.$$

This decay is accompanied by an energy of reaction of about 0·78 MeV as measured in the proton and β^--ray spectrometer, so that the mass difference ($M_n - M_H$) should appear as the decay energy of the reaction.

Now $0·78 \text{ MeV} \equiv 0·000\ 84$ u and

$$M_n - M_H = 1·008\ 665 - 1·007\ 825$$
$$= 0·000\ 840 \text{ u},$$

showing that the mass difference of the particles is indeed the decay energy of the neutron.

Neutron Energies

Since neutrons are neutral it is possible to use them at almost any energy in nuclear reactions. They are arbitrarily classified as follows:

Thermal (reactors only)	$E = 0·025$ eV
Slow	$E = 1$ eV $- 1$ keV
Intermediate	$E = 1$ keV $- 0·5$ MeV
Fast	$E = 0·5$ MeV upwards.

The boundaries of this classification are by no means well defined, so that the ranges are not fixed. It is important to remember that all energies can be used in some nuclear reaction or other.

Neutrons as Waves

Earlier in this book, in Chapter 9, we discussed the wave nature of the electron and mentioned that the argument could equally well be applied to any free particle, the de Broglie wavelength being given by

$$\lambda = \frac{h}{mv}.$$

This is true for any particle having momentum equal to mv.

For electrons,
$$m = 9 \cdot 1 \times 10^{-31} \text{ kg},$$
$$e = 1 \cdot 6 \times 10^{-19} \text{ C},$$
$$h = 6 \cdot 6 \times 10^{-34} \text{ J s}.$$

When an electron is accelerated through V volts the energy equation is $Ve = \tfrac{1}{2}mv^2$ (if V is of the order of a few kilovolts only), so that $mv = \{2Vem\}^{1/2}$ and the wavelength then becomes, in metres,

$$\lambda = \frac{h}{(2Vem)^{1/2}} = \frac{6 \cdot 6 \times 10^{-34}}{(2 \times V \times 1 \cdot 6 \times 10^{-19} \times 9 \cdot 1 \times 10^{-31})^{1/2}}$$
$$= \frac{6 \cdot 6}{(29 \times V)^{1/2}} \times 10^{-9} \text{ m}$$
$$= \sqrt{\frac{1 \cdot 5}{V}} \text{ nm},$$

which is a convenient expression for λ with V in volts, for electrons only.

In the case of a neutron beam

$$\lambda = \frac{h}{mv} \quad \text{becomes} \quad \frac{h}{\sqrt{2mE}}$$

and thus

$$\lambda = \frac{6 \cdot 6 \times 10^{-34}}{(2 \times E \times 1 \cdot 6 \times 10^{-19} \times 1 \cdot 66 \times 10^{-27})^{1/2}} \text{ m},$$

where E is converted to electron volts, giving $\lambda = 28 \cdot 6/\sqrt{E}$ pm, for neutrons. Thus for thermal neutrons, $E = 0 \cdot 025$ eV, and the expression gives $\lambda = 0 \cdot 182$ nm, which is roughly an atomic diameter, and for fast neutrons, $E = 2 \cdot 0$ MeV, $\lambda = 20$ fm, which is approximately the diameter of a nucleus.

We see therefore that the wavelengths of thermal neutrons are of the same order as X-rays and so one would expect the same sort of diffraction effects with crystals as used in X-ray spectrometers. This affords a ready method of measuring neutron wavelengths and confirming the truth of the de Broglie law for neutrons. Fast neutrons have a wavelength of the same order as nuclear radii and can be used for nuclear size determinations and finally very fast neutrons, say 10 GeV, have $\lambda \sim 0 \cdot 1$ fm, so that we now foresee the possibility of a nucleon probe, i.e. the possible investigation of the nuclear structure.

In a beam of reactor neutrons there is always a velocity distribution. The average velocity of reactor thermal neutrons is $v = 2200$ m s^{-1} but there is a spread either side of this. If such a beam passes through a suitable crystal assembly (say graphite) we have the possible application of the Bragg law of diffraction $\lambda(E) = 2d \sin \theta$ so that the crystal lattice picks out its own $\lambda(E)$, from the distribution, to satisfy this equation. The maximum value of λ for graphite will be $\lambda_{max} = 2 \times 0 \cdot 5 \times 1$ since $d = 0 \cdot 5$ nm and thus $\lambda_{max} = 1 \cdot 0$ nm, approximately. This corresponds to a minimum energy given by $0 \cdot 0286/\sqrt{E_{min}} = 1 \cdot 0$ or $E_{min} = 0 \cdot 0008$ eV, so that neutron energies above this will be diffracted and neutrons of energy lower than this will be transmitted. The beam emerging from the graphite column is therefore deprived of all those energies which correspond to reflections of neutrons from sets of planes within the crystal. These energies are mostly the higher energies of the distribution and the emergent neutrons then have energies ~0·001 eV, well below the average energy of thermal neutrons. These neutrons are therefore called 'cold neutrons', and are important for the investigation of the cross-section properties of various reactor materials.

Neutron diffraction is now a tool for research as important as X-ray diffraction and electron diffraction. It has helped considerably in the analysis of crystal structures containing light atoms (H, C, N, O, etc.) which do not scatter X-rays copiously.

An important property of the neutron is that it has a spin angular momentum of $\frac{1}{2}\hbar$ like the electron. For this reason neutron scattering experiments can give information of the spin structure of a material and have enabled the correct spin vectors to be placed on the various lattice sites of magnetic material.

Because of their neutrality and low mass, neutrons are excellent particles for collision studies and they have been used to elucidate some of the properties of nuclear forces in simple neutron–proton scattering experiments and in the location of defects in materials.

None of these topics can be successfully studied by X-ray or electron diffraction because X-ray wavelengths are too long for nuclear scattering and electrons are charged particles and consequently there is the added difficulty of a Coulomb interaction.

21.3 Neutron Bombardment Reactions

The neutrons produced in an (α, n) reaction are never slow neutrons. Their energies are always of the order of 1 MeV so that slow neutrons must be artificially produced by the attenuation of fast neutrons in a slowing-down medium. This is the basis of the moderator action as used in thermal reactors, and one can imagine that the neutrons eventually emerge from the medium with the energy of the thermal motion of the moderating lattice, i.e. 0·025 eV at room temperature.

Slow neutrons must be used with low atomic weight elements if an (n, α) reaction is to result, e.g.

$$^{6}_{3}\text{Li} + ^{1}_{0}\text{n} \rightarrow ^{3}_{1}\text{H} + ^{4}_{2}\text{He}\,(\alpha)$$
$$^{10}_{5}\text{B} + ^{1}_{0}\text{n} \rightarrow ^{7}_{3}\text{Li} + ^{4}_{2}\text{He}\,(\alpha).$$

In both these reactions the α-particle can be regarded as being indicative of the presence of a neutron. As explained before, reactions such as these are used in neutron detectors, either in counters or in the cloud chamber. Slow neutrons are also capable of providing fission reactions, which will be described later, as well as (n, γ) reactions.

The reactions of fast neutrons with light elements are straightforward as, for example, with nitrogen:

$$^{14}_{7}\text{N} + ^{1}_{0}\text{n} \rightarrow ^{14}_{6}\text{C} + ^{1}_{1}\text{H}\,(\text{p})$$

or, less probable,

$$^{14}_{7}\text{N} + ^{1}_{0}\text{n} \rightarrow ^{7}_{3}\text{Li} + ^{4}_{2}\text{He}\,(\alpha) + ^{4}_{2}\text{He}\,(\alpha).$$

An example of a reaction starting and finishing with the same nuclide is

$$^{32}_{16}\text{S} + ^{1}_{0}\text{n} \rightarrow ^{32}_{15}\text{P} + ^{1}_{1}\text{H}\,(\text{p}) + Q$$

followed by
$$^{32}_{15}P \rightarrow ^{32}_{16}S + _{-1}^{0}e(\beta^-) + \tilde{\nu} + E_{\beta^-\text{max}}$$
which is virtually equivalent to writing
$$n^0 = p^+ + e^-$$
since the mass of ^{32}S is nearly equal to that of ^{32}P. ($^{32}S = 31\cdot972$; $^{32}P = 31\cdot974$.) In all such reactions the energy available will be 0·78 MeV, which is equivalent to $(M_n - M_H)$, and $Q + E_{\beta^-\text{max}} = 0\cdot78$ MeV. If the maximum energy of the β^--spectrum of such a β^--emitter is less than 0·78 MeV the whole reaction can be started by slow or even cold neutrons since Q is positive. On the other hand, if the $E_{\beta^-\text{max}} > 0\cdot78$ MeV and Q is negative, it is necessary to use fast neutrons of energy given by $(E_{\beta^-\text{max}} - 0\cdot78)$ MeV numerically. In the case of ^{32}P the β^--energy is 1·70 MeV, so that neutron energies of approximately 1 MeV are needed for this reaction.

21.4 Archaeological Dating by the ^{14}C Method

Carbon has three important isotopes, 12, 13 and 14 having terrestrial abundances of 98·89%, 1·11% and zero, respectively. Of these, ^{14}C is unstable and decays according to the equation
$$^{14}_{6}C \rightarrow ^{14}_{7}N + _{-1}^{0}e(\beta^-) + \tilde{\nu}$$
with $T_{1/2} = 5730$ a and $E_{\beta^-} = 0\cdot158$ MeV.

There is a trace of ^{14}C in the atmosphere due to cosmic neutron bombardment of ^{14}N, thus
$$^{14}_{7}N + ^{1}_{0}n \rightarrow ^{14}_{6}C + ^{1}_{1}H(p).$$
If these two equations are taken together it seems reasonable to assume that over, say, ten half-lives of $^{14}_{6}C$ decay the quantity of $^{14}CO_2$ present in the atmosphere has always been constant. In the atmosphere, therefore, the formation and decay of ^{14}C are in equilibrium. By a similar argument we can assume that the concentration of ^{14}C in all living vegetable tissue is the same, due to the fact that the carbon dioxide taken in by the plants from the atmosphere contains this constant quantity of $^{14}CO_2$. However, when the plant dies and no longer takes in CO_2 from the atmosphere, the ^{14}C equilibrium quantity attained during the life of the plant now begins to decay by β^--emission with the half-life of 5730 ± 30 a.

Suppose a sample of this dead matter (wood, charcoal, book-binding, peat, rope, etc.) is now measured, say t years later. With the usual notation
$$N = N_0 e^{-\lambda t}$$
or
$$\ln N = \ln N_0 - \lambda t.$$
Since
$$\lambda = \frac{0\cdot6931}{T_{1/2}},$$
we get
$$\log N = \log N_0 - 0\cdot3010\, t/T_{1/2},$$
where N_0 is the activity at death and N is the activity at present, referring to the β^--emission of the sample with $T_{1/2} = 5730$ years. Thus if N_0 is the original count rate, it is also the present count rate of the β^--emission from living matter, i.e. the equilibrium activity previously mentioned. This value is about $16\cdot1 + 0\cdot3$

counts per minute per gram of carbon. In order to measure the value of N with any degree of accuracy one must know the efficiency of the counter and the background count with an accuracy of about 2%. When this is done, and using the value of $T_{1/2}$ previously stated, the value of t or the age of the specimen can be found.

Two interesting ages will be mentioned here, one obtained from the charcoal from Stonehenge, England, and the other from the binding of the Isaiah scroll in the Dead Sea scrolls. By the ^{14}C method Stonehenge is found to be 3798 ± 275 years old and the Dead Sea scroll 1917 ± 200 years old. The errors quoted reflect the difficulties associated with ^{14}C dating.

The whole method rests on the assumptions that the ^{14}C content of the atmosphere has been constant over the last 50 000 years, i.e. the cosmic ray intensity over this period has not varied. The ^{14}C method also takes for granted a constancy of ^{14}N over the same period. Finally it must be assumed that there has been no secondary interference during the decay period and the sample has remained the same since its 'death', except for the β^--decay mechanism.

21.5 Tree-Ring Calibration of ^{14}C Dates

The validity of radiocarbon dates depends essentially on the constant concentration of ^{14}C over the last 50 000 years. There are several reasons why this is improbable. The geomagnetic field is not constant with time, and indeed goes through polarity reversals lasting a few thousand years. There have been two such reversals in the past 50 000 years. These geomagnetic changes affect the cosmic ray flux in the atmosphere. Other factors affecting the equilibrium between production and decay of ^{14}C are solar activity and extreme changes in climate, as well as modern nuclear weapons testing. All these factors tend to affect the overall ^{14}C concentration and so introduce errors into the radiocarbon dates.

Recently there has come a more direct method of finding true calendar dates in the shape of tree-ring dating, or dendrochronology. By counting the annual growth rings back from the present in a tree cross-section an accurate estimate of its age can be obtained. By a suitable comparison of tree-ring patterns of young trees with those of older (dead) trees, it is possible to overlap the patterns until the dates of trees which died in antiquity can be found by the tree-ring counting method and also by the radiocarbon method. Such a method depends on a careful survey of the rings and picking out on overlapping patterns the common climatic features, e.g. narrow rings for dry summers and broad ones for wet summers.

The age of the dead tree is then determined by the radiocarbon method and its calendar age by the 'reaching back' principle of comparing its tree-ring pattern with successively younger tree patterns until the present time is reached. By this means it has been possible to calibrate radiocarbon dates in terms of calendar dates. Such a calibration is shown in Fig. 21.1, in which the *smoothed out* calibration curve converts radiocarbon dates to calendar dates. Note that there are deviations in both directions and that the radiocarbon date was just about correct at the A.D./B.C. cross-over point. The calibration curve actually has many kinks and wriggles on it, all of which are of climatalogical importance, but these are not shown.

The main deviations of the radiocarbon scale from the absolute scale were first

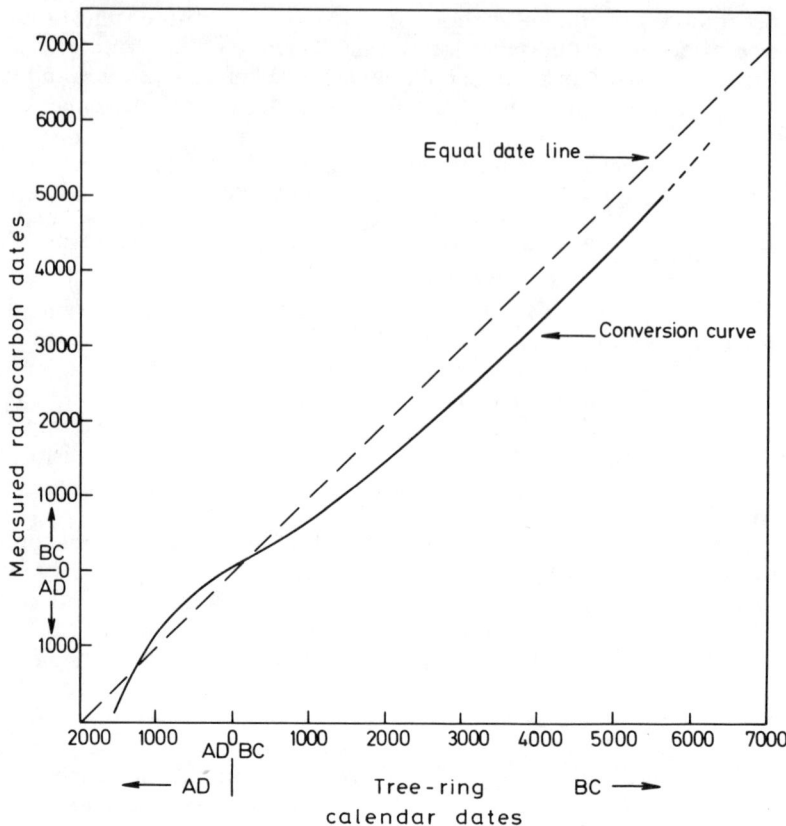

Fig. 21.1 Tree-ring calibration of radiocarbon dates.

noted in 1968 in the ancient bristle-cone pine trees of California. There derives from this the tree-ring calibration curve of Fig. 21.1. Because of this the early archaeological dates given by the radiocarbon method have been revised. Thus Stonehenge appears to be older and the Dead Sea Scrolls younger than at first thought.

Errors are still quite high. Thus a corrected radiocarbon date of 1000 A.D. can be in error by ± 100 a (twice the standard deviation, 95% certainty) and a date of 2430 B.C. (4400 before present) can be in error by ± 240 a. In general an error of about ± 150 a can be assumed.

There is therefore a danger in accepting corrected radiocarbon dates too rigidly — especially when dealing with single dates. When converting from radiocarbon dates to calendar dates it is essential to use the *detailed* calibration curve which contains all the short-term fluctuations.

The calibration does not extend back beyond 10 000 a. Before this time the count rate is very low and massive shielding is required to minimize the background count. Nevertheless the radiocarbon method is still a possible method for dating the last 500 000 a and much information about the dates of recent ice ages and warm periods has been obtained. It still remains one of the most important examples of the use of neutron bombardment reactions.

Problems

(Those problems marked with an asterisk are solved in full at the end of the section.)

21.1* A beam of 'cold' neutrons has a maximum energy of 0·001 eV. Calculate the wavelength of these neutrons and estimate the minimum lattice parameter of the crystal used to produce them. (0·9 nm, 0·45 nm)

21.2 When a beam of neutrons of energy E pass through a scattering medium, the maximum fractional reduction in energy per collision is given by the expression

$$\frac{(\Delta E)_{max}}{E} = 1 - \left(\frac{A-1}{A+1}\right)^2,$$

where A is the atomic mass number of the scattering nucleus. Show that this expression can be obtained from the simple principles of dynamics. Calculate $(\Delta E)_{max}$ for a neutron striking a proton and comment on the result.

21.3 The equation for the cosmic production of $^{14}_{6}C$ is

$$^{14}_{7}N + ^{1}_{0}n \rightarrow ^{14}_{6}C + ^{1}_{1}H + 0.55 \text{ MeV}.$$

Calculate the mass of $^{14}_{6}C$ from this equation.
$^{14}_{7}N = 14.003\,074$ $^{1}_{1}H = 1.007\,825$ $^{1}_{0}n = 1.008\,665$.
(14·003 323 u)

21.4* In the photodisintegration equation for the deuteron the threshold energy is 2.227 ± 0.003 MeV, viz.

$$\gamma(h\nu) + ^{2}_{1}H \rightarrow ^{1}_{1}H + ^{1}_{0}n - 2.227 \text{ MeV}.$$

From the mass spectrometer the mass difference between the doublet $2(^{1}_{1}H)$ and $D(^{2}_{1}H)$ is $(1.5380 \pm 0.0021) \times 10^{-3}$ u. If the mass of the hydrogen atom is $1.007\,825 \pm 0.000\,003$ u, calculate the mass of the neutron. ($1.008\,679 \pm 0.000\,008$ u)

21.5 Repeat Problem 21.4 using the doublet separation of $(1.5495 \pm 0.0024) \times 10^{-3}$ u, which is the nuclear reaction value. Comment on the difference it makes to the calculated mass of the neutron. ($1.008\,623 \pm 0.000\,008$ u)

21.6* It is known that in carbon of living wood there is a total of (16.1 ± 0.3) radioactive disintegrations per minute per gram of carbon. The counter used for measurements on an archaeological specimen of wood was only $(5.40 \pm 0.14)\%$ efficient and registered (9.5 ± 0.1) counts per minute on 8 g of carbon taken from the wood. Without the carbon the counter registered a background rate of (5.0 ± 0.1) counts per minute. If the half-life of radioactive carbon is (5730 ± 30) a, calculate the age of the 'find'. (3600 ± 750 a)

21.7 Calculate the temperature associated with the neutrons of Problem 21.1.

21.8 A copper slab of thickness 5·3 mm will reduce the intensity of a beam of fast neutrons by a factor of 0·5. (Half-value thickness. Compare with half-life in radioactive decay equation.) Find the macroscopic absorption coefficient, the fast neutron cross-section and the radius of the copper nucleus.

21.9 What is the true age of the 'find' in Problem 21.6?

Solutions to Problems

21.1 From the formula $\lambda_{min} = 28.6/\sqrt{E}$ pm we have
$$\lambda_{min} = \frac{28.6}{\sqrt{0.001}} \text{ pm}$$
$$= 0.9 \text{ nm},$$
which is the minimum wavelength transmitted. In the Bragg equation $\lambda = 2d \sin \theta$ the maximum value of $\sin \theta$ is 1, so that the minimum value of d corresponding to $\lambda = 0.9$ nm is 0.45 nm.

21.4 The mass of the neutron is given by
$$M(^2_1H) - M(^1_1H) + \frac{2.227}{931} \text{ u} = M(^2_1H) - 2M(^1_1H) + M(^1_1H) + 0.002\,392 \text{ u}$$
$$= -0.001\,538 + 1.007\,825 + 0.002\,392$$
$$= 1.008\,679 \text{ u}.$$

Errors are:
(1) Energy 2.227 ± 0.003 MeV equivalent to an error of $\pm 0.000\,003$ u
(2) Doublet error $\pm 0.000\,002\,1$ u
(3) Hydrogen atom $\pm 0.000\,003$ u
giving a total maximum error of $\pm 0.000\,008$ u. Hence the mass of the neutron is $1.008\,679 \pm 0.000\,008$ u.

21.6 Count rate for specimen alone
$$= (9.5 \pm 0.1) - (5.0 \pm 0.1)$$
$$= 4.5 \pm 0.2 \text{ counts min}^{-1};$$
therefore
$$\text{No. of disintegrations} = (4.5 \pm 0.2) \times \frac{100}{(5.4 \pm 0.14)}$$
$$= (4.5 \pm 4\tfrac{1}{2}\%) \times \frac{100}{(5.4 \pm 2\tfrac{1}{2}\%)}$$
$$= 83 \pm 7\% \text{ counts per min per 8 g}$$
$$= 10.4 \pm 7\% \text{ counts min}^{-1} \text{ g}^{-1}.$$

Using $A = A_0 e^{-\lambda t}$ we have
$$2.303 \log_{10}\left(\frac{A_0}{A}\right) = \lambda t = \frac{0.693}{T_{1/2}} \times t$$
where $A_0 = 16.1 \pm 0.3 = 16.1 \pm 2\%$, $A = 10.4 \pm 7\%$, $T_{1/2} = 5730 \pm 30 = 5730 \pm \tfrac{1}{2}\%$ a and t is the age of the 'find'. Therefore
$$\frac{A_0}{A} = \frac{16.1 \pm 2\%}{10.4 \pm 7\%} = 1.55 \pm 9\%$$
and
$$\log_{10}(1.55 \pm 9\%) = 0.19 \pm 2.303 \times 9\% = 0.19 \pm 21\%.$$
Thus
$$t = \frac{2.303 \log_{10}(A_0/A) \times (5730 \pm \tfrac{1}{2}\%)}{0.693}$$
$$= \frac{2.303 \times (0.19 \pm 21\%) \times (5730 \pm \tfrac{1}{2}\%)}{0.693}$$
$$= 3600 \pm 750 \text{ a}.$$

Therefore
$$\text{Age of 'find'} = 3600 \pm 750 \text{ a}.$$
Notice the large error involved.

Chapter 22

Nuclear Fission and its Implications

22.1 Introduction

When the neutron and its properties were discovered in 1932, the possibility of new types of nuclear reactions became apparent. The fact that neutrons are extremely small and have no charge makes them ideal nuclear missiles over a large range of energies. We have seen how this led to the production of radioisotopes among light elements and when Fermi in 1934 irradiated the heavier elements, notably uranium, with slow neutrons, many of the products were β^--active, as had been experienced earlier with lighter elements. These were thought to be transuranic elements due to reactions like

$$^{238}_{92}\text{U} + ^{1}_{0}\text{n} \rightarrow ^{239}_{92}\text{U} \rightarrow ^{239}_{93}? \rightarrow ^{239}_{94}?$$
$$\searrow \beta^- \quad \searrow \beta^-$$

When some of the products of such a neutron irradiation experiment were analysed by radiochemical methods, one particular product had a half-life of 3·5 h and the chemical properties of radium or of a radium-like element. When this product was precipitated from the mixed irradiation products by barium chloride, all attempts to separate the radium-like element from the barium failed. A long series of chemical tests finally convinced Hahn and Strassmann in 1938 that the 'radium' compound was actually a barium compound. Another product of the neutron bombardment of uranium was the element lanthanum, which is produced by β^--emission from barium. No simple nuclear transformation equation would account for this production of elements such as barium ($Z=56$) or lanthanum ($Z=57$), so far removed from the parent uranium ($Z=92$).

Assuming that there must be another element or elements of atomic number 36 to make up the original uranium with the barium, Frisch and Meitner in 1939 used the word 'fission' to describe the process which takes place when a heavy nucleus is caused to break down or disintegrate into two (or sometimes more) roughly equal parts known as fission fragments, rather than into one heavy product and one light particle (as in the Rutherford reaction). This was an entirely new type of reaction and as soon as it was discovered many of the world's nuclear research laboratories gave it their immediate attention. By 1940 the following facts had been established:

(1) Natural uranium (0·7% ^{235}U and 99·3% ^{238}U) could be 'fissioned' by either slow neutrons or by fast neutrons, but ^{238}U always required fast neutrons. Eventually it was ascertained that that ^{235}U was fissile to slow neutrons.
(2) The elements thorium ($Z=90$) and protactinium ($Z=91$) could also be fissioned with fast neutrons.
(3) In all cases very large disintegration energies were released, equal to about ten times the order of energies previously experienced.
(4) In all cases fast neutrons were emitted.
(5) Fission fragments were all radioactive and decayed to stable nuclides by a series of β^--emissions.
(6) The atomic mass numbers of the fission products ranged from about 70 to 160, although, of course, one parent uranium nucleus could only produce two fragments. These were all eventually identified by radiochemical methods so proving that intermediate elements were produced by the fission process.

Thus at the beginning of World War II, the process of neutron-induced binary fission was well established.

22.2 The Theory of Nuclear Fission

Although the type of fission referred to in the last section had to be initiated by neutrons, later research showed that fission could also be induced by deuterons and by α-particles from accelerators, and even 'photofission' was possible using incident γ-rays. Further work showed that lighter elements could also be fissioned by high-energy particles, as for example in the case of copper:

$$_{29}^{63}Cu + {}_{1}^{1}H(p) \begin{array}{c} \nearrow {}_{11}^{24}Na \\ \searrow {}_{19}^{39}K \end{array} + {}_{0}^{1}n.$$

Thus the general fission process is by no means limited to the 'classical' case of uranium, and indeed spontaneous fission, in which no bombarding particle is required, was discovered during World War II. The probabilities of such fission processes taking place vary widely. For example the cross-section (or probability) of thermal (0·025 eV) neutron fission of ^{235}U is 580 barns, whereas the fission cross-section for 2 MeV neutrons acting on ^{238}U is less than 1 barn, see Fig. 22.1. The spontaneous fission half-life of ^{235}U is 3×10^{17} a, corresponding to an average of about one spontaneous fission per hour per gram of ^{235}U.

One of the most important facts which emerged from this early work was that ^{235}U was fissionable by neutrons of low energies (e.g. 0·025 eV, thermal neutrons) whereas there seemed to be a threshold energy of about 1·0 MeV before ^{238}U could be fissioned, Fig. 22.1. Thus ^{235}U is fissionable by both slow and fast neutrons, but ^{238}U is fissionable by fast neutrons only. Although this could not easily be explained in simple physical terms, an explanation was eventually given by Bohr and Wheeler using the liquid-drop model (Chapter 19). They obtained a semi-quantitative expression for the neutron energy required to initiate fission in a given nucleus, which agreed well with the experimental value in the case of ^{238}U.

Assuming the drop is held in its spherical state by forces of an internal molecular origin any disturbance of this state will require an external force, which

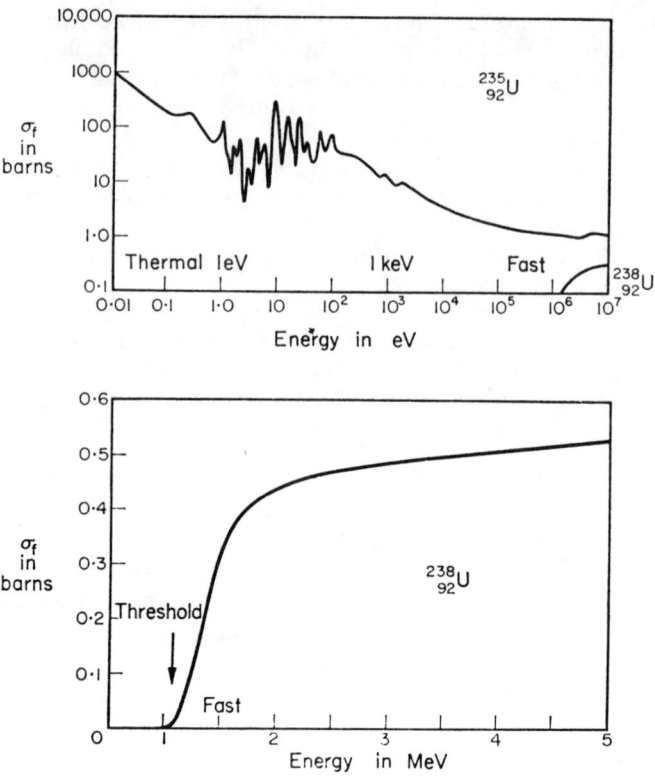

Fig.22.1 Energy cross-section curves of $^{235}_{92}$U and $^{238}_{92}$U.

distorts the sphere into an ellipsoid. If the force is large enough the ellipsoid narrows into a 'dumb-bell' shape and finally breaks at the neck into two major portions with some additional small drops, as in Plateau's spherule when liquid drops break away from a tap under gravity. This process is best understood diagrammatically in Fig. 22.2.

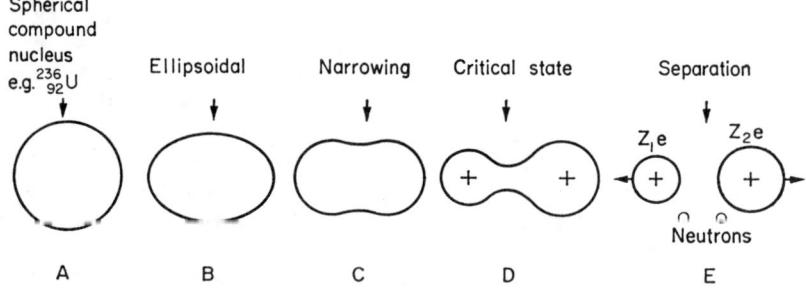

Fig. 22.2 Mechanism of fission in liquid-drop model of nucleus.

When a nucleus undergoes fission, the incident neutron combines with it to form a compound nucleus which is highly energetic. Its extra energy is partly the kinetic energy of the neutron but largely the added binding energy of the incident neutron. This energy appears to initiate a series of rapid oscillations in the drop which at times assumes the shape B in Fig. 22.2. The restoring force of the nucleus arises from the short range internucleon forces. If the oscillations become so violent that stage D is reached, and as each 'half' is now positively charged, the final fission into stage E is inevitable. Thus there is a threshold energy or a critical energy required to produce stage D after which the nucleus cannot return to A, because of the Coulomb repulsion of the two parts.

The critical energy, which must be supplied with the neutron, is best shown in Fig. 22.3, which is a potential energy diagram. In this diagram we see how the energy E_{crit} must be added to the system to enable the energy of the nucleus to become greater than the stability-barrier energy E_b. Once the maximum barrier

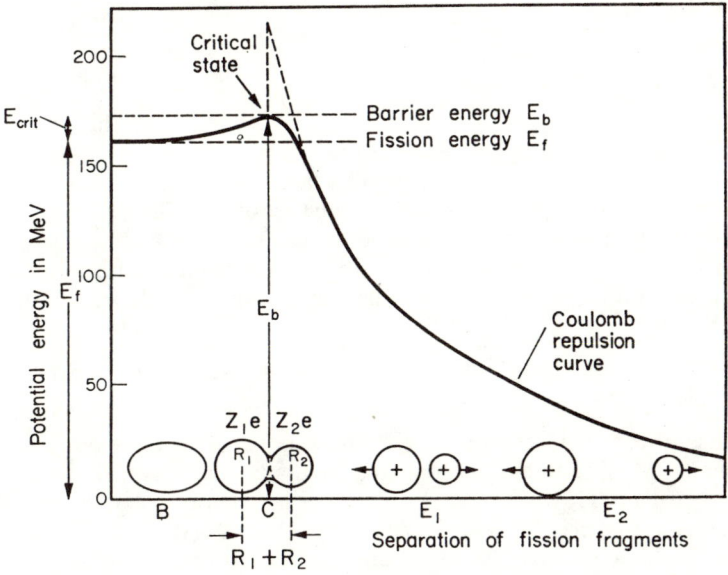

Fig. 22.3 Potential energy curve for fission.

height has been overcome the system 'descends' to the state of lowest potential energy and the fragments separate. When the mass of the compound nucleus is greater than the masses of the total fission fragments, fission is possible and the mass difference is released as energy according to the Einstein relation:
$$\Delta E = c^2 \Delta m$$
The value of the critical deformation energy E_{crit} was first calculated by Bohr and Wheeler on the liquid-drop model (see p. 281). They found

$$E_{crit} = 0.89\, A^{2/3} - 0.02\, \frac{Z(Z-1)}{A^{1/3}}\ \text{MeV},$$

where A is the atomic mass number of the compound nucleus and Z is its atomic number. This formula can be used for uranium, for when $A=236$, $Z=92$ for $^{235}_{92}\text{U}$ fission we have

$$E_{\text{crit}}(^{236}\text{U}) = 0\cdot89 \times (236)^{2/3} - \frac{0\cdot02 \times 92 \times 91}{236^{1/3}}$$

$$= 0\cdot89 \times 38\cdot19 - \frac{0\cdot02 \times 92 \times 91}{6\cdot18}$$

$$= 34\cdot00 - 27\cdot10$$

$$= 6\cdot9 \text{ MeV}$$

for compound nucleus 236, whereas

$$E_{\text{crit}}(^{239}\text{U}) = 0\cdot89 \times (239)^{2/3} - \frac{0\cdot02 \times 92 \times 91}{239^{1/3}}$$

$$= 0\cdot89 \times 38\cdot51 - \frac{0\cdot02 \times 92 \times 91}{6\cdot206}$$

$$= 34\cdot28 - 26\cdot98$$

$$= 7\cdot3 \text{ MeV}$$

for compound nucleus 239.

Thus the ^{238}U nucleus requires rather more total energy than the ^{235}U nucleus to initiate fission. Now we have seen that this energy is added as the kinetic and binding energy of the incident neutron. The latter is calculated from the semi-empirical mass equation (p. 283) by calculating the total binding energy of the 236 compound nucleus and subtracting from it the binding energy of the original 235 nucleus, in the case of ^{235}U fission. One can repeat this calculation for the ^{238}U fission. These subtractions give the binding energies of the added neutron in each case, as follows:

(a) for ^{235}U fission the binding energy of the added neutron $= 6\cdot8$ MeV.
(b) for ^{238}U fission the binding energy of the added neutron $= 5\cdot9$ MeV.

In the case of the ^{235}U fission the binding energy of the neutron supplies 6·8 MeV of the 6·9 MeV required, so that this particular nucleus is fissionable with low energy neutrons. However, for ^{238}U the binding energy is $7\cdot3 - 5\cdot9 = 1\cdot4$ MeV *less* than the required critical energy and so, by this calculation, the ^{238}U nucleus should only be fissionable by neutrons of energy greater than 1·4 MeV. Experimentally this threshold energy is found to be 1·1 MeV but one must remember that the figures used in the above argument are approximate, depending on the choice of constants in the mass equation. The difference in fission properties between ^{235}U and ^{238}U is shown clearly. Compare them again in Fig. 22.1.

In physical terms the difference between the two uranium isotopes is due to the fact that ^{235}U has 92 protons and 143 neutrons and is an even–odd nucleus, whereas ^{238}U has 92 protons and 146 neutrons and is an even–even nucleus. Theory shows that the liberation of fission energy is easier in the case of the odd neutron nuclei than for the even neutron nuclei for a given Z. The reason for the thermal fission of odd-A nuclei is that the resulting even–even nucleus is more tightly bound in the ground state than the emitting even–odd nucleus. Neutron capture then provides more excitation energy and therefore promotes fission. One would expect all even proton, odd neutron nuclei to be fissionable with thermal neutrons whereas the even–even nuclei should require fast neutrons.

Table 22.1 shows some of these facts and we see that in general the difference in fission properties of nuclei is that even nuclei are more stable than odd neutron nuclei, and therefore require incident neutrons of higher energy to cause fission.

TABLE 22.1

Original nucleus	Z	N	Neutrons required for fission
$^{233}_{92}$U	E	O	Slow
$^{235}_{92}$U	E	O	Slow
$^{238}_{92}$U	E	E	Fast
$^{232}_{90}$Th	E	E	Fast
$^{239}_{94}$Pu	E	O	Slow
$^{237}_{93}$Np	O	E	Fast
$^{232}_{91}$Pa	O	O	Slow
$^{236}_{93}$Np	O	O	Slow

22.3 The Energy of Nuclear Fission

Most nuclear reactions, other than fission, have Q values of ~ 10 MeV. The largest known value before 1939 was 22·2 MeV by the deuteron process
$$^6_3\text{Li} + ^2_1\text{H} \rightarrow 2^4_2\text{He}(\alpha) + 22 \cdot 2 \text{ MeV}.$$

Early measurements of the new fission process showed a reaction energy of about 200 MeV, i.e. at least ten times greater than reaction energies normally encountered. This is due to a relatively high mass decrease during fission.

Preliminary calculations of fission energy release can be made using the binding energy curve (p. 207). If we assume probable values of the atomic masses of the fragments of about 95 and 140 we can see from the curve that $\bar{B}$ for $A = 95$ and $A = 140$ is about 8·5 MeV whereas for $A = 236$ it is only 7·6 MeV. The fission energy released is the difference between these two multiplied by the total number of nucleons. Thus:

$$E_{\text{fission}} = 236(8 \cdot 5 - 7 \cdot 6) \text{ MeV}$$
$$= 236 \times 0 \cdot 9 \text{ MeV}$$
$$= 212 \text{ MeV}.$$

Alternatively, if we regard this fission process purely as Coulomb repulsion energy, once the critical stage has been reached we have two spherical nuclei about to repel each other as in Fig. 22.3. The distances apart can be calculated from the radius formula, viz.
$$R = R_0 A^{1/3}$$
$$= 1 \cdot 37 \times 10^{-15} A^{1/3} \text{ m}$$
$$= 1 \cdot 37 \, A^{1/3} \text{ fm}.$$

Thus for $A = 140$, $A^{1/3} = 5 \cdot 19$ and $R_{140} = 6 \cdot 95$ fm. Similarly, for $A = 95$, $A^{1/3} = 4 \cdot 56$ and $R_{95} = 6 \cdot 25$ fm.

Taking probable values of the atomic numbers, the Coulomb repulsion energy is given by

$$E_{\text{Coul}} = \frac{Z(140)Z(95)e^2}{4\pi\varepsilon_0(R_{140}+R_{95})}$$
$$= \frac{52 \times 40 \times (1\cdot6 \times 10^{-19})^2}{(9 \times 10^9)^{-1} \times 13\cdot20 \times 10^{-15}} \text{J}$$
$$= 3\cdot63 \times 10^{-11} \text{ J}$$

and since 1 MeV $= 1\cdot60 \times 10^{-13}$ J, we get

$$E_{\text{Coul}} = \frac{3\cdot63 \times 10^{-11}}{1\cdot6 \times 10^{-13}} \text{ MeV}$$
$$= 227 \text{ MeV}.$$

This result shows that the calculated fission energy is of the order of 200 MeV.

The most reliable calculation of fission energy uses exact mass differences as we have done previously for non-fission reactions. The fission reactions we are considering here can be written in the general form:

$$U + n \to X + Y + \nu n + Q \text{ MeV},$$

where X and Y are the primary fission fragments and ν is the number of fast fission neutrons produced. On the average, over all the uranium atoms in a piece of uranium, $\nu = 2\cdot 5$. Both X and Y are β^--unstable and decay thus:

$$^A_Z X \longrightarrow {}^A_{Z+1} C \longrightarrow {}^A_{Z+2} D \longrightarrow {}^A_{Z+3} E, \text{ etc.}$$

The value of Q is calculated from the exact mass difference of the two sides of the fission equation. To do this we must consider a specific reaction where two stable end points of β^--chains are quoted,

$$^{235}_{92}U + {}^1_0n \to {}^{98}_{42}Mo \text{ (stable)} + {}^{136}_{54}Xe \text{ (stable)} + 2{}^1_0n + Q$$

On the left-hand side of the equation, we have $^{235}_{92}U = 235\cdot044$ u and $^1n = 1\cdot009$ u, with a resulting total of 236·053 u for the total mass of the compound nucleus, assuming zero kinetic energy of the neutrons.

On the right-hand side, we have $^{98}_{42}Mo = 97\cdot905$ u, $^{136}_{54}Xe = 135\cdot917$ u, $2^1n = 2\cdot018$ u, with a resulting sum of 235·840 u for the total mass of the fission products, so that by subtraction

$$\Delta m = 0\cdot213 \text{ u}$$
$$= 0\cdot213 \times 931\cdot5 \text{ MeV}$$
$$= 207 \text{ MeV},$$

which is comparable with the value calculated by the binding energy method above. This is the Q-value.

This figure is typical of all fission energies so that we can always use the approximate figure of 200 MeV in all our fission calculations. This energy is distributed roughly as given in Table 22.2.

This may seem to be very great compared with that of non-fission reactions. In joules it is merely $200 \times 1\cdot6 \times 10^{-13} = 3\cdot2 \times 10^{-11}$ J, which is very small indeed. However, we must remember that this is the energy from each fissioning uranium nucleus. If we could persuade 1 g of ^{235}U to fission completely, the associated energy would be:

$$\frac{N_A}{235} \times 3\cdot2 \times 10^{-11} \text{ J},$$

TABLE 22.2

K.E. of fission fragments	168 MeV
K.E. of fast fission neutrons	5 MeV
β^--decay energy from fission products	4·8 MeV
Neutrino energy from β^--decays	10·0 MeV
Immediate γ-ray energy	4·6 MeV
Fission produce γ-ray energy	6·9 MeV
Total	199·3 MeV per fission

where $N_A = 6\cdot02 \times 10^{23}$ mol^{-1}. Thus,

$$\text{Energy per gram} = \frac{6\cdot02 \times 10^{23}}{235} \times 3\cdot2 \times 10^{-11} \text{ J}$$
$$= 8\cdot2 \times 10^{10} \text{ J}$$
$$= 82 \text{ GJ, or } 5\cdot1 \times 10^{29} \text{ MeV,}$$

which would be sufficient to heat 200 Mg of water to the normal boiling point.

This is indeed a very large quantity of energy from only about 50 mm³ of uranium. It is equivalent to the energy liberation in an explosion of 20 tons of T.N.T. If it is all released at once, i.e. all the uranium atoms fission in about 1 μs we then have a nuclear bomb ... if we *control* the release of this energy we have a nuclear reactor. Nuclear fission can be used as either a source of energy for military or peaceful purposes.

The unit of energy used in describing nuclear bombs is the ton of T.N.T. equivalent. The original 1945 atomic bombs contained roughly 1 kg of fissile material. This is equivalent to $5\cdot1 \times 10^{26}$ MeV or about

20 000 tons T.N.T.

or

$8\cdot2 \times 10^{13}$ J

or

$2\cdot3 \times 10^7$ kW h

or

7000 tons of coal burnt.

This is the energy equivalent of a 20 kiloton bomb. A 20 megaton bomb would have an energy equivalent to 7 million tons of coal burnt in a few microseconds so that the whole of the annual coal output of the U.K. is equivalent to a mere twenty hydrogen bombs of 20 megaton equivalents each.

22.4 The Distribution of Fission Products

We have seen that a fissionable nucleus gives only *two* fission fragments which thereafter decay by β^--emission to a stable end product. What particular fragment nuclides are produced by the given nucleus is a matter of chance and the range of gross fission products is roughly from bromine to barium in the periodic table. The concentration of fission nuclides depends on the atomic mass, and the distribution curve has a curious saddleback shape as shown in Fig. 22.4, which is

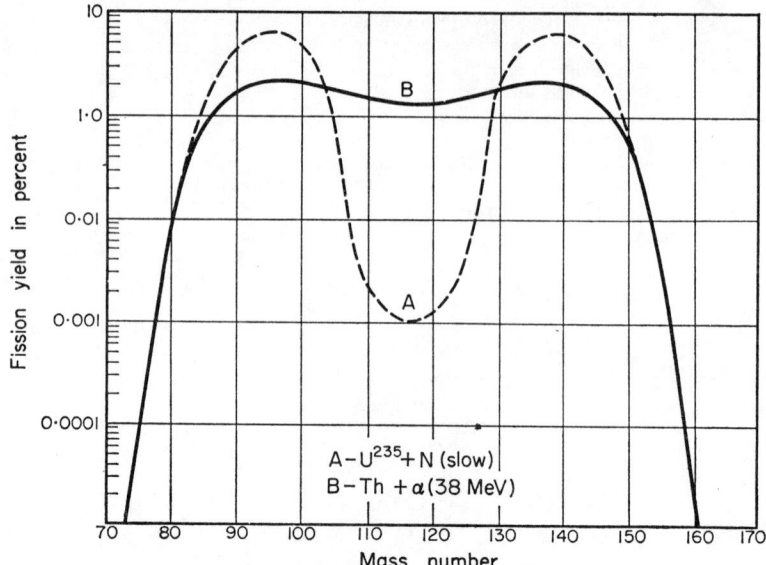

Fig. 22.4 Fission produce yield curves. Curve A $^{235}_{92}$U with slow neutrons; curve B Th with α-particles. (Taken from *Radioactivity and Nuclear Physics*, J. M. Cork, Van Nostrand, 1957.)

the well-known diagram of the fission yield curve from ^{235}U. There are two well-defined maxima, at $A=95$ and $A=140$ roughly. The total yield is 200% since there are *two* fragments per fission. Note that the ordinate on a logarithmic scale and that the concentration of the most probable nuclides is only 6%. The total number of identified fission nuclides is about 300, including nearly 200 different β^--emitters. The assymmetric fission yield curve shown in Fig. 22.4 is shown by all nuclei which can be fissioned by thermal neutrons, but with fast neutrons and other particles the 'trough' in the curve tends to fill up.

The energy distribution of fission products can be obtained by assuming that the two fragments are ejected with equal and opposite moments so that
$$M_1 V_1 = M_2 V_2$$
and thus
$$\frac{E_2}{E_1} = \frac{\tfrac{1}{2}M_1(M_2 V_2)^2}{\tfrac{1}{2}M_2(M_1 V_1)^2} = \frac{M_1}{M_2} = \frac{95}{140},$$
giving $E_2/E_1 = 2/3$ roughly, for the two peak positions. This is shown in Fig. 22.5 which is the energy curve corresponding to the nuclide distribution of Fig. 22.4.

22.5 Characteristics of Fission Neutrons

The harnessing of nuclear energy either in a reactor or in a bomb depends essentially on the production of fast fission neutrons. The distribution of fission neutron energy is shown in Fig. 22.6 in which the average energy is 2·0 MeV and the most probable energy about 0·7 MeV. This fission energy spectrum is most important when considering the neutron cycle within the moderator of a thermal

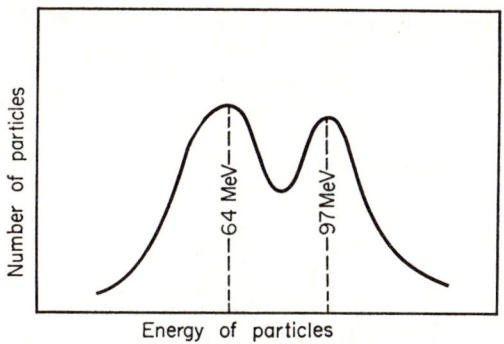

Fig. 22.5 Energy distribution of fission fragments. (Taken from *Radioactivity and Nuclear Physics*, J. M. Cork, Van Nostrand, 1957.)

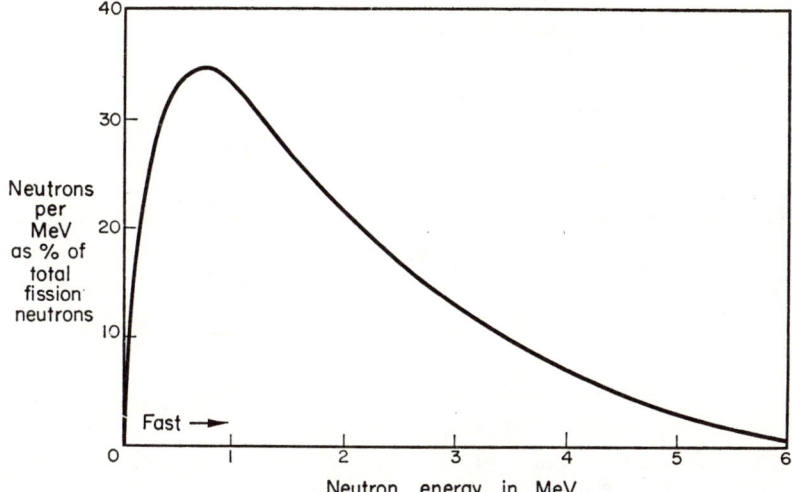

Fig. 22.6 Fission neutron energy spectrum.

reactor. The number of neutrons born per individual fission is always an integer but the gross average is about 2·5 neutrons per fission, for thermal fission.

In addition to the neutrons just mentioned, which are born at the moment of fission and therefore may be called prompt neutrons, there are a few cases where neutrons are produced in the middle of one of the β^--active decay chains. These amount to about 0·75% of all the neutrons and are called delayed neutrons because at some part of the chain a neutron and a β^--particle are ejected simultaneously. The best-known example is $^{87}_{35}\text{Br} \rightarrow {}^{87}_{36}\text{Kr}$ decay scheme for β^- radiation, with $T_{1/2} = 55\cdot 6$ s, shown in Fig. 22.7. The $^{87}_{36}\text{Kr}$ is formed in highly excited states, one of which instantaneously emits a neutron to form stable $^{86}_{36}\text{Kr}$. The energy appears as the neutron energy of 0·25 MeV and, since the excited $^{87}_{36}\text{Kr}$

decays immediately, the half-life of the neutron formation is the same as that of the parent $^{87}_{35}$Br, i.e. 55·6 s. Not all the possible excited states are shown in Fig. 22.7 and the γ-emissions from them as the ground states are formed are also emitted.

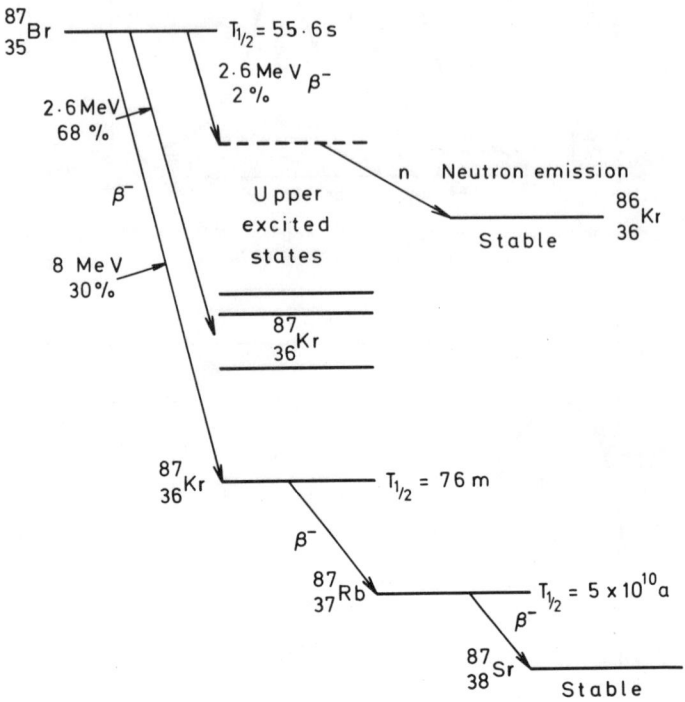

Fig. 22.7 Delayed neutrons from the $^{87}_{35}$Br precursor.

It must be mentioned here that the decay time of 55·6 s is very much greater than fission neutron lifetimes, so that the presence of these *delayed* neutrons increases the mean lifetime of all the reactor neutrons. It is for this reason that reactor control is possible, as explained in Section 22.7. The successful running of a thermal reactor requires a moderator between the pieces of uranium to slow down the neutrons of 2·0 MeV average energy to the thermal energy of the lattice vibrations of the moderator, which is about 0·025 eV and equivalent to about 300 K. As can be seen by trying to fit a neutron of 0·025 eV energy into the neutron spectrum curve, Fig. 22.6, this slowing down process represents the moderation of nearly all the fission neutrons, since the average energy is about 2·0 MeV.

22.6 The β^--Decay Chains of Fission

From Table 22.2 we see that the fission product β^--radiation energy is about 15 MeV including the neutrino energy. From a complete radiochemical analysis of gross fission products it has been possible to allocate fission nuclides to their respective β^--decay chains and the details of many of these chains are now fully understood. Some are long and some are short and since the concentrations of the individual nuclides are not constant, the gross decay law is not exponential. One long chain is the following:

$$^{143}_{54}\text{Xe} \xrightarrow{1\text{ s}} {}^{143}_{55}\text{Cs} \xrightarrow{\text{v.v. short}} {}^{143}_{56}\text{Ba} \xrightarrow{\text{short}} {}^{143}_{57}\text{La}$$

$$19 \text{ min}$$

$$(\text{stable}) \; {}^{143}_{60}\text{Nd} \xleftarrow{13\cdot 7 \text{ d}} {}^{143}_{59}\text{Pr} \xleftarrow{32 \text{ h}} {}^{143}_{58}\text{Ce}$$

and a similarly placed short chain is

$$^{140}_{54}\text{Xe} \xrightarrow{16\text{ s}} {}^{140}_{55}\text{Cs} \xrightarrow{66\text{ s}} {}^{140}_{56}\text{Ba} \xrightarrow{12\cdot 8\text{ d}} {}^{140}_{57}\text{La} \xrightarrow{40\text{ h}} {}^{140}_{58}\text{Ce (stable)}.$$

Note that these are isobaric and not isotopic chains.

Two β^--chains of great importance in reactor control are the following:

$$^{135}_{52}\text{Te} \xrightarrow{2\text{ m}} {}^{135}_{53}\text{I} \xrightarrow{67\text{ h}} {}^{135}_{54}\text{Xe} \xrightarrow{9\cdot 2\text{ h}} {}^{135}_{55}\text{Cs} \xrightarrow{20\,000\text{ a}} {}^{135}_{56}\text{Ba (stable)}$$

and

$$^{149}_{60}\text{Nd} \xrightarrow{1\cdot 7\text{ h}} {}^{149}_{61}\text{Pm} \xrightarrow{50\text{ h}} {}^{149}_{62}\text{Sm (stable)}$$

in which the isotopes of xenon and samarium have extremely high cross-sections to thermal neutrons and are therefore regarded as reactor poisons. Table 22.3 (p. 332) shows Xe and Sm cross-sections compared with those of other reactor materials. In the β^--chains we see that some nuclides are short-lived and some are long-lived, but all decay individually according to the exponential law. Most of the β^--chain nuclides are formed in excited states and therefore emit γ-rays. From Table 22.1 we see that this γ-ray energy is 6·9 MeV per fission.

It is the γ- and β^--radiations of these fission products that constitute the long-term biological hazard for reactor operators and also the fall-out hazard in nuclear bomb bursts. The gross fission product decay law is

$$A_t = A_1 t^{-1\cdot 2}$$

for *any* unit of time, where A_t is the activity at t units after fission (burst) and A_1 is the activity after 1 unit of time. This is largely an empirical law, and is roughly equivalent to an attenuation factor of 0·1 for a time ratio increase of seven. This law is the net result of the exponential decay of many fission products in the concentration of Fig. 22.4, each with its own individual half-lives.

22.7 Controlled Fission—Nuclear Reactors

The nuclear fission chain reactions in which successive generations of neutrons are used to cause further fissions and release further energy in uranium nuclei can take place almost instantaneously (bomb) or over an extended period (reactor, or pile). The conditions for each must be carefully chosen. Not all neutrons are used in producing fission as some are inevitably lost in non-fission nuclear processes and others as geometrical leakage beyond the range of the uranium. On average, if a chain reaction is to be sustained in a lump of uranium, at least *one* of the 2·5 neutrons born per fission must be preserved for further fission. We can define a constant k (sometimes called the neutron multiplication factor) equal to the number of net effective fission neutrons born per generation. If $k > 1$ the fission neutron population increases and fission is sustained, but if $k < 1$ the neutron population decreases and a chain reaction fails to develop. The equation for the neutron behaviour is

$$N = N_0 \exp\left(\frac{k-1}{k} \cdot \frac{t}{\tau}\right),$$

TABLE 22.3
Thermal Neutron Cross-Section for Materials Important in Reactor Design

Material	Absorption cross-sections σ_a	Scattering cross-sections σ_s	Fission cross-sections σ_f	Remarks
H_2O	0·66	~62		Moderator
D_2O	0·00046	~14		Moderator
Be	0·10	7		Moderator
BeO	0·10	11·2		Moderator
C (graphite)	0·0045	4·8		Moderator
B	769	~3·8		Control rods
Cd	2550	65		Control rods
^{135}Xe	$2·7 \times 10^6$			Neutron poison
^{149}Sm	$5·0 \times 10^4$			Neutron poison
^{235}U	694		582	Fissile
^{238}U	2·8			Fissile (fast neutron)
Natural U	7·7		4·2	Fissile
^{239}Pu	1025		738	Fissile. Produced in the reactor

All cross-sections are in barns

where N is the number of neutrons present at a given time t, N_0 is the number of neutrons present at a given time $t=0$, k is the neutron multiplication factor defined above, τ is the mean neutron lifetime between fissions within the uranium itself (this is about 1 ns).

The expression $(k-1)/k$ is sometimes called the reactivity ρ so that the equation

$$N = N_0 e^{\rho t/\tau}$$

expresses variation of the neutron population with time. If $k>1$ then the reactor neutron population increases, but if the reactivity is negative, or $k<1$, the number of neutrons present decreases exponentially.

Since the loss of potentially fissioning neutrons from the surface of a roughly spherical piece of uranium will be proportional to the surface area $4\pi R^2$ and the production of fission neutrons will be proportional to the volume $\frac{4}{3}\pi R^3$ the ratio of loss to production will be inversely proportional to R. From this we see that the larger the piece of uranium the smaller will be the neutron loss and the greater will be the probability of having $k>1$. Thus, there should be a critical size for a lump of uranium for which $k=1$. When $k<1$ a fission chain reaction cannot take place but for $k>1$ the reaction will be sustained.

The original type of 'atomic bomb' consists of two subcritical pieces of fissile uranium or plutonium separated by a sufficient distance to render them harmless. When they are suddenly impelled together so that the total mass becomes supercritical and, in the presence of a neutron source, fission commences and continues until all the uranium is fissioned, mainly by fast neutrons. For example, total mass of uranium used in the early atomic bombs was roughly 1 kg and for $k=1\cdot1$, we have $\rho=0\cdot1$ and

$$N = N_0 e^{0\cdot1t/10^{-9}}.$$

If $t=1$ μs,

$$N = N_0 e^{100} \quad \text{or} \quad N/N_0 = e^{100} = 10^{43}.$$

This equation means that in 1 μs the neutrons would have multiplied many millions of times over, sufficient to cause fission in all the 10^{24} atoms of uranium present, and that the $8\cdot2 \times 10^{13}$ J associated with the 1 kg of uranium will have been released in less than 1 μs, which constitutes a large explosion.

To control this release of energy use is made of the fact that natural uranium has the *two* isotopes 235 and 238 in the ratio 1:137·8 and that *only* 235 is fissionable by slow neutrons. Hence, if the fast fission neutrons from a piece of uranium can be slowed down *before* reaching the next piece of uranium the time factor involved would be increased and the fission would take place in the 235 isotope only, although some thermal neutrons are lost in the 238 isotope of uranium by absorption. The uranium is distributed in a regular way inside a slowing down medium or moderator in such a way that before a fission neutron from one piece reaches another it has slowed down and commenced to diffuse towards the next piece with the lattice energy of the graphite, i.e. 0·025 eV. All the British civil reactors use graphite as the moderator. On entering the second piece of uranium the neutron causes fission in the 235 isotope and the whole cycle is repeated. The average lifetime of fission neutrons in the graphite reactor is increased to about 1 ms, and when the delayed neutrons are also included, the overall neutron lifetime is increased further to 0·1 s. Thus, taking $\rho=0\cdot1$ for comparison with the previous calculation, we get $N = N_0 e^{0.1t/0.1} = N_0 e^t$ and for $t = 1$ μs, $N = N_0 e^{10^{-6}}$ and $N \approx N_0$.

The neutron growth in a reactor is therefore very much slower than in a bomb. In fact a million millionfold growth in a bomb takes t seconds, given by $e^{0.1 \times t/10^{-9}} = 10^{12} = e^{28}$, so that $t = 28/10^8 = 3 \times 10^{-7}$ s, approximately.

For the moderated neutrons in a reactor the corresponding time t is given by $e^{0.1 \times t/10^{-1}} = 10^{12} = e^{28}$, so that $t = 28$ s for the reactor.

Such calculations show that when the neutron lifetime is increased from 1 ns to 0·1 s control of the reactor becomes feasible and is achieved by using rods of cadmium/boron steel inside the reactor core. These rods can be moved in and out mechanically and since both cadmium and boron have high thermal neutron cross-sections, as shown in Table 22.3, the neutron flux can be absorbed and controlled at will.

There are many features of nuclear reactors we cannot discuss here, and to trace the life history of a neutron from birth as a fast fission neutron until it finally causes further fission as a slow neutron in $^{235}_{92}U$ is very complicated. Nuclear reactors depend for their working on a precise knowledge of the life history of a reactor neutron. Reactors can be heterogeneous or homogeneous depending on the moderator system used. They may use fast or thermal neutrons, natural uranium or enriched (with $^{235}_{92}U$) uranium, the moderator may be water, heavy water, graphite, organic liquid or beryllium. Reactors may be used to produce power, or for research purposes such as in the testing of materials at high temperatures in high neutron fluxes. They may also be used to produce plutonium for enriching other fissile fuel. The production of plutonium is discussed in the next chapter.

22.8 Nuclear Power Reactors

Of the many types of nuclear reactor mentioned above, the designs suitable for nuclear power generation are comparatively few. Nevertheless there are now more than 200 power stations in the world producing electricity. This shows that nuclear power is regarded as of prime importance for industrial development. The strategic power of a nation — and therefore its political power — is strongly dependent on its available electrical power. Hence the annual global consumption of electricity is increasing rapidly. In the U.K. it is estimated that by the year 2000 the installed electrical capacity will be 123 GW(E) of which 48 GW(E) will be nuclear. Compare this with the 1980 figures of 88 GW(E) total 11 GW(E) nuclear, respectively. The projected figures anticipate the gradual run-down of North Sea oil so that by 2030 the figures will be about 256 GW(E) total of which 196 GW(E) will be nuclear. At present (1979) the U.K. has ten nuclear power stations operating at 5 GW(E) or about 12% of the total power requirements. Nuclear-produced electricity is the cheapest — coal electricity costs 1·07p, oil 1·27p and nuclear 0·69p per kW h. These are 1978 prices. On this basis alone it would seem sensible to continue the production of nuclear electricity by improved designs of reactors. The original Calder Hall type of power reactor is the Magnox reactor, a thermal reactor using natural uranium metal in a magnesium aluminium alloy can as fuel, having graphite as moderator and being gas-cooled by carbon dioxide. At present the U.K. has nine of these Magnox power reactors operating commercially. Based on experience with these the Advanced Gas-Cooled Reactor (A.G.R.) has been designed, one of which is now operating successfully. The A.G.R. is more efficient than the Magnox reactors, since its fuel is ceramic uranium oxide which can operate at a much higher temperature than natural uranium metal. It is still graphite-moderated and cooled by carbon dioxide. Four other A.G.R. reactors will be commercially operating by 1982.

Since the mid-1960s much research at Dounreay has concentrated on the fast breeder reactor. This is cooled by liquid sodium and requires no moderator. It uses a plutonium oxide/uranium oxide mixture as a fuel with a blanket of depleted uranium (i.e. nearly pure ^{238}U) in which plutonium can be formed by the fast neutron reaction

$$^{1}_{0}n + ^{238}_{92}U \rightarrow ^{239}_{92}U \xrightarrow{\beta^{-}} ^{239}_{93}Np \xrightarrow{\beta^{-}} ^{239}_{94}Pu$$

It can be arranged so that the amount of plutonium produced is equal to or greater than that consumed in the fuel. This extra plutonium can then be used for the fuel of further reactors, hence the term 'breeder'. Initially the plutonium must be obtained from the spent thermal reactor fuel rods. The original fast breeder research reactor at Dounreay has been replaced by a bigger version in the Prototype Fast Reactor, which was commissioned in 1974 and now operates at a thermal power level of 600 MW(Th) and an electrical power output of 250 MW(E). In view of the finite life of all fossil fuels, including natural uranium, it is considered that the real future of nuclear fission power lies in the commercial development of the fast breeder reactor producing the nuclear fuel plutonium artificially. In addition to this programme the U.K. Government has given an option to the C.E.G.B. to order two pressurized water reactors to be started in 1982. These are based on the Westinghouse P.W.R. so successfully operating in the U.S.A. and if accepted could be commercial reactors by about 1987.

Another factor in favour of developing the fast breeder reactor is that the lifetime of a Magnox reactor is only 20 a and that of an A.G.R. is 25 a.

The oldest Magnox reactors were commissioned in 1962, so that the run-down will start in the mid-1980s. By 1990 the Magnox reactors will be at the end of their life, the A.G.R.s at about the middle of theirs' and the P.W.R.s will have just started operating.

22.9 Nuclear Power Prospects

The inevitable conclusion drawn from the previous section is that the future of nuclear fission power depends on the fast breeder reactor. The fast reactor has about eight times the enrichment of fossil material in of a thermal reactor so that its rating, i.e. the heat released per unit mass of fuel, is correspondingly eight times as great. The fuel core of 20% PuO_2 and 80% UO_2 is surrounded by a neutron-absorbing blanket of depleted uranium which produces plutonium by the fast neutron reaction. Some plutonium is produced in the core also, so that the plutonium balance can be written:

Pu in core + Pu in blanket − Pu burned up in core.

This balance can be manipulated to be +ve, −ve, or even zero. This is quite different from the thermal reactors, in which it can only be +ve, i.e. in the core. Usually there is a breeding gain of plutonium of about 0·2, i.e. the breeding ratio is 1·2:1, while extracting about 50 times more energy than a thermal reactor for the same fuel consumption.

The next stage is the development of a commercial fast reactor in Britain, although no firm decision has been taken on such a reactor. If this goes through the reactor will be the CFR1 and it is expected to have an output of 1300 MW(E) and to be generating about 1990. Figure 22.8 shows the design of the prototype fast reactor on which the design of the CFR1 will be based. The molten sodium is pumped round the core, from which it extracts the heat. This is transferred to an intermediate heat exchange to a secondary circuit of molten sodium which is used in turn to raise steam in the heat exchanger. This steam is used to generate electricity in a conventional turbine electrical plant.

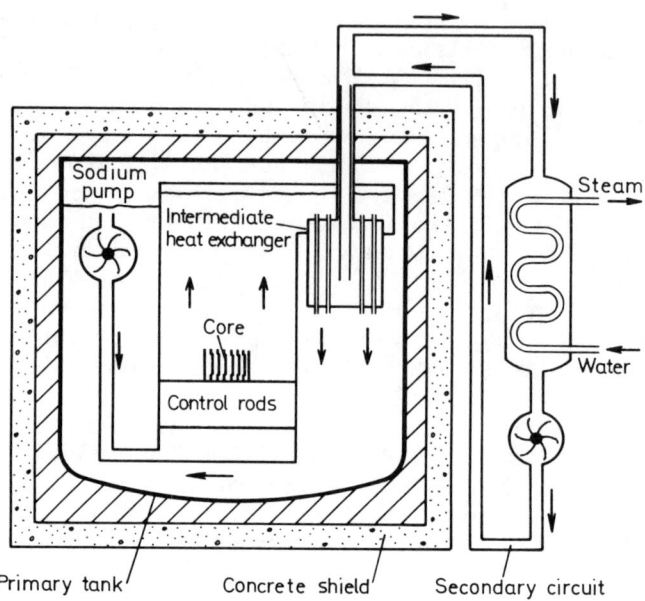

Fig. 22.8 Prototype fast reactor showing circulation of liquid sodium.

It can be seen that the CFR1 will have many technological problems to solve before it safe and efficient. No mention has been made here of the safety of nuclear reactors, the fact that all of them produce a military weapon fuel — plutonium — and the fact that the spent thermal reactor fuel rods, having had their plutonium extracted, are left in a highly radioactive state and this represents a disposal problem which is deplored by all those who seek to preserve the environment.

These are the short-term proposals. What is required in the long term is a safe nuclear system offering no environmental problems with a limitless supply of fuel, and a system which will not produce a military weapon fuel as a by-product.

Perhaps within the next hundred years this goal may be achieved in the shape of the thermonuclear, or fusion, reactor which will be discussed in Chapter 24.

Problems

(*Those problems marked with an asterisk are solved in full at the end of the section.*)

22.1 What are the features of a nuclear fission reaction which make it different from any other type of nuclear reaction?

22.2 Explain in physical terms why more than one fission neutron is produced from a single primary neutron.

22.3 Describe in physical terms the concept of 'cross-section' as used in neutron physics. Draw the diagrams of the fission cross-sections of $^{235}_{92}U$ and $^{238}_{92}U$ and their variation with energy. Comment on the differences between them.

22.4 Show how far the liquid-drop model is successful in explaining why $^{238}_{92}U$ is not fissile to slow neutrons.

22.5* Calculate the electrostatic potential energy between two equal nuclei produced in the fission of $^{235}_{92}U$, at the moment of their separation. Use the formula $R = R_0 A^{1/3}$. (240 MeV)

22.6 To what extent would you expect (a) the liquid-drop model and (b) the shell model of the nucleus to be able to explain the shape of the fission product distribution curve?

22.7 Write down the β^--decay fission chains which produce $^{135}_{54}Xe$ and $^{149}_{62}Sm$. Why are these two nuclides called 'fission poisons'?

22.8 In some of the β^--decay chains the β^--emitting nuclides are in excited states. This means that some such nuclei can decay by neutron emission as well as β^--emission, with the same half-life. An example of this is given on p. 330.

The percentage of delayed neutrons in the total fission yield is about 0·75% and this increases the *average* neutron lifetime to about 0·1 s.

Show how this makes possible the control of the fission chain reaction.

22.9 Write an essay on the types and uses of nuclear fission reactors.

22.10* The usefulness of cadmium in a nuclear reactor depends on the high thermal absorption cross-section of the 113 isotope, viz. σ_a (^{113}Cd) = 21 000 barns. If the density of cadmium is $8 \cdot 7 \times 10^3$ kg m^{-3}, calculate the macroscopic cross-section of ^{113}Cd and hence the thickness required to attenuate a neutron beam to 0·01% of its original intensity. (97 000 m^{-1}, $x = 0 \cdot 95$ μm)

22.11 The average energy released in the fission process is 200 MeV per fission. If a reactor is working at a power level of 6 MW calculate the number of fissions per second required to produce this power. ($1 \cdot 88 \times 10^{17}$ fissions s^{-1})

22.12 In the neutron-induced binary fission of $^{235}_{92}U$ (235·044) two stable end products $^{98}_{42}Mo$ (97·905) and $^{136}_{54}Xe$ (135·917) are often found. Assuming that these isotopes have come from the original fission process, find (a) what elementary particles are released, (b) the mass defect of the reaction and (c) the equivalent energy released. (2n, $4\beta^-$, 0·2215 u, about 206 MeV)

22.13 How many fissions per second are induced in a natural uranium rod (density 19×10^3 kg m^{-3}) by a thermal neutron flux of 10^{18} neutrons m^{-2} s^{-1}.

22.14 Assuming that the power level of a reactor is proportional to the total number of neutrons in the reactor, calculate the reactivity required to increase the

337

power level by a factor $e = 2.718$ in 10 s with $\tau = 1$ ms for the reactor. Ignore delayed neutrons.

22.15 Recalculate the power-level increase of the previous problem by taking into account the delayed neutrons. Comment on the difference.

22.16 Show that the power density of a reactor is given by the expression $3.2\Sigma_f nv \times 10^{-11}$ W m^{-3}, where Σ_f is the macroscopic fission cross-section, n is the average neutron density, v is the average neutron velocity, nv is the average neutron flux.

22.17 Describe briefly the working of a thermal neutron graphite moderated nuclear reactor.

A sample of natural uranium ($^{238}_{92}$U = 99.3%) is subjected to a thermal neutron flux of 10^{16} neutrons m^{-2} s^{-1}. Determine the thermal energy produced in a 1 cm diameter rod of natural uranium of length 30 cm by $^{235}_{92}$U fission. σ_f $^{235}_{92}$U = 590 barns, energy per fission is 200 MeV, density of uranium is 19×10^3 kg m^{-3}.
[N]

22.18 What is meant by the term 'total neutron cross-section' as used in nuclear physics?

Draw rough sketches of the fission cross-section–energy curves for neutron-induced fission in $^{235}_{92}$U and $^{238}_{92}$U. Comment on their shapes.

From the following data show whether or not $^{234}_{92}$U is fissile to thermal neutrons. The critical energy for fission is 6 MeV. Mass of neutron is 1.0087 u, mass of $^{234}_{92}$U is 234.0409 u, mass of $^{235}_{92}$U is 235.0439 u.
[N]

Solutions to Problems

22.5 If R is the radius of either nucleus at moment of fission then
$$R = 1.3 \times \left(\frac{236}{2}\right)^{1/3} \times 10^{-15} \text{ m} = 1.3 \times 4.88 \times 10^{-15} \text{ m},$$
taking $R_0 = 1.3 \times 10^{-15}$ m. Then
$$E = \frac{(Z/2)^2 e^2}{4\pi\varepsilon_0 \times 2 \times R}$$
$$= \frac{(46)^2 \times (1.6 \times 10^{-19})^2}{4\pi \times 8.85 \times 10^{-12} \times 2 \times 1.3 \times 4.88 \times 10^{-15}} \text{ J}$$
$$= \frac{2116 \times 2.56 \times 10^{-1}}{12.56 \times 8.85 \times 2.6 \times 4.88} \times \frac{1}{1.6 \times 10^{-13}} \text{ MeV}$$
$$= 240 \text{ MeV}.$$

22.10 Macroscopic cross-section Σ is given by
$$\Sigma = \frac{N_A \rho \sigma_a}{A}$$
$$= \frac{6.02 \times 10^{26} \times 8.7 \times 10^3 \times 21\,000 \times 10^{-28}}{113}$$
$$\Sigma = 97\,000 \text{ m}^{-1}.$$
From $I = I_0 e^{-\Sigma x}$ we have
$$\Sigma x = \log I_0/I$$
$$= 2.3 \log_{10} 10\,000$$
$$= 2.3 \times 4 = 9.2,$$

giving
$$x = \frac{9 \cdot 2}{97\,000} \text{ m}$$
$$= 95 \times 10^{-6} \text{ m}$$
$$x = 95 \ \mu\text{m}.$$

Thus a thickness of only 0·1 mm is required to reduce the neutron flux to one ten-thousandth of its original value.

Chapter 23

The Transuranic Elements

23.1 Neptunium ($Z=93$) to ?

A glance at any pre-1939 textbook of physics or chemistry will show that the list of the elements and their relative atomic masses, as well as the periodic table, ended with the element uranium ($Z=92$). This had been the case for the previous forty years. Since then several new elements have been added to the list, mainly by United States scientists. All these are man-made and radioactive and some are useful because they are fissionable. The new elements are:

$Z=93$	94	95	96	97	98
Np	Pu	Am	Cm	Bk	Cf
Neptunium	Plutonium	Americium	Curium	Berkelium	Californium

$Z=99$	100	101	102	103	104
Es	Fm	Md	No	Lw	Ku
Einsteinium	Fermium	Mendelevium	Nobelium	Lawrencium	Kurchatovium

$Z=105$
Ha
Hahnium

These are the so-called transuranic elements and are the direct result of the impetus given to neutron bombardment experiments by the discovery of the fission process. It is possible that there may be traces of neptunium and plutonium in the earth's crust, but the quantities involved are so small that these elements may be regarded as truly artificial.

From the Russian laboratory at Dubna has come the report of the discovery of elements 106 and 107, as yet unnamed. These have not been confirmed and other scientists are being cautious. Before a new element can be accepted it should satisfy the following criteria:

(1) the atomic number must be established by chemical analysis if possible;
(2) there must be some proof of a genuine decay process to an established end product;
(3) identification of the characteristic α-rays, by half-life and energy, of this decay process must be possible.

There is no doubt that the new transuranic elements will satisfy these three points, although the quantities available make even microchemical analysis very difficult.

23.2 Formation of Transuranic Elements

The fission reaction in uranium produces medium mass elements which are β^--active. There is, however, a probability that fission does not take place in some of the ^{238}U nuclei present and an alternative reaction produces an isotope ^{239}U of uranium. This itself may be β^--active (to account for the obvious β^--activity of the residue) according to:
$$^{239}_{92}U \rightarrow {}^{0}_{-1}e\,(\beta^-) + {}^{239}_{93}? + \tilde{\nu},$$
the question mark representing some unknown element of atomic number 93. This was the reaction pursued by the early workers (1934) in their research, and after the fission reaction work had abated many of them returned to work on these alternative transuranic β^--emitters. The identification of the new materials proceeded along the familiar lines of half-life determination and radiochemical analysis wherever possible. A good deal of this early work was done in Berkeley, California, which was the centre of much nuclear research based on the accelerating machines built there.

The half-life of the 239 isotope of uranium was found to be 23·5 min and a second β^--decay with half-life of 2·3 d was also found. After separation, the 2·3 d β^--emitter was shown to have properties similar to those of the rare earths, although it could not be a rare earth since its atomic mass was too high. Chemical tests pointed to the new material having an atomic number 93, so that this first transuranic element was identified and named neptunium, since Neptune is the next planet beyond Uranus. Many new isotopes of neptunium were subsequently discovered, mainly by deuteron and alpha bombardment by the big machines at Berkeley. An early experiment with cyclotron deuterons showed that the following reaction is possible:
$$^{238}_{92}U + {}^{2}_{1}H \rightarrow {}^{238}_{93}Np + 2{}^{1}_{0}n.$$
This isotope is also β^--active, having a half-life of two days.

One can see that in general an (n, γ) reaction followed by successive β^--emissions will lead to a new series of elements in the same way that isobars are formed in the fission β^--chains. Thus:
$$^{238}_{92}U + {}^{1}_{0}n \rightarrow {}^{239}_{92}U + \gamma,$$
followed by
$$^{239}_{92}U \rightarrow {}^{239}_{93}? \rightarrow {}^{239}_{94}? \rightarrow {}^{239}_{95}?$$
$$\searrow \beta^- \quad \searrow \beta^- \quad \searrow \beta^-$$
etc. On paper there is no limit to this method of producing transuranic elements, but one must realize that other competing modes of decay may also be possible. The full range of isotopes of a new element can only be realized after extensive experiments with high-energy particles other than neutrons.

23.3 Neptunium, Np ($Z = 93$)

Many of the β^--emitting isotopes of the transuranic elements have such short half-lives that it is very difficult to carry out chemical analyses on them. The production of $^{237}_{93}$Np, which is an α-emitter of half-life of the order of one million years was, therefore, a useful step forward in the research into the chemical properties of the element.

This isotope was discovered in 1942 by the action of *fast* neutrons on $^{238}_{92}$U, the

result being the formation of $^{237}_{92}U$ according to
$$^{238}_{92}U + ^{1}_{0}n \rightarrow ^{237}_{92}U + 2^{1}_{0}n$$
followed by
$$^{237}_{92}U \xrightarrow[6.7 \text{ d}]{T_{1/2}} {}^{237}_{93}Np + _{-1}^{0}e(\beta^-) + \tilde{\nu}$$
and
$$^{237}_{93}Np \xrightarrow[2.2 \times 10^6 \text{ a}]{T_{1/2}} {}^{233}_{91}Pa + {}^{4}_{2}He(\alpha)$$

This long-lived isotope is the precursor of the $(4n+1)$ radioactive series first mentioned in Chapter 4. Because this isotope is nearly stable the use of neptunium chemical salts is now almost as common as the use of uranium salts. Among the more common neptunium isotopes so far identified are:
231, 232, 233, 234, 235, 236, 237, 238, 239, 240 and 241
of which all but 231, 233 and 237 and β^--emitters.

23.4 Plutonium, Pu (Z=94)

The element plutonium is perhaps the most important of the transuranic elements on account of its 239 isotope, which is comparable with $^{235}_{92}U$, being fissile in a similar manner. It is thus to be regarded as a reactor and a bomb fuel.

Following the general pattern the β^--decay of neptunium must lead to the formation of plutonium thus:
$$^{239}_{93}Np \xrightarrow[2.33 \text{ d}]{T_{1/2}} {}^{239}_{94}Pu + _{-1}^{0}e(\beta^-) + \tilde{\nu}$$
followed by
$$^{239}_{94}Pu \xrightarrow[25\,000 \text{ a}]{T_{1/2}} {}^{235}_{92}U + {}^{4}_{2}He(\alpha)$$

These reactions show how $^{235}_{92}U$ is produced by the plutonium decay. This isotope is important in reactor technology because it is the isotope required for the production of fission energy. The β^--chain of activity virtually stops at $^{239}_{94}Pu$. However, following the (d, 2n) reaction on $^{238}_{92}U$ the neptunium product, viz. $^{238}_{93}Np$, gives another plutonium isotope $^{238}_{94}Pu$ by β^--emission which is again an α-emitter. Many new isotopes of plutonium are formed by cyclotron α-bombardment as well as by the β^--decay of the corresponding neptunium isobars. Known isotopes have all mass numbers from 232 to 246.

The modern production of plutonium is the result of the growth of the 239 isotope in reactors. All the British Atomic Energy Authority and the Civil Power Reactors so far in use are natural uranium reactors in which, as has been shown, the production of $^{239}_{94}Pu$ is a natural byproduct. In fact, the original reactors were designed to produce plutonium and *not* electrical power. When the uranium rods are removed from the reactor core they have to be sent back to the Atomic Energy Authority for plutonium extraction. This is carried out by a complicated chemical process and the purified plutonium is stored away in carefully designed subcritical arrays for future use in enriched reactors or bombs.

23.5 Americium, Am ($Z=95$), and Curium, Cm ($Z=96$)

The bombardment of $^{238}_{92}U$ with cyclotron α-particles leads to the formation of plutonium isotopes according to

$$^{238}_{92}U(\alpha, n)^{241}_{94}Pu \quad \text{or} \quad ^{238}_{92}U(\alpha, 2n)^{240}_{94}Pu.$$

The $^{241}_{94}Pu$ isotope was found to be a fairly long-lived β^--emitter whose product was an α-decay isotope analogous to $^{239}_{94}Pu$. Thus:

$$^{241}_{94}Pu \xrightarrow[13\ a]{T_{1/2}} {}^{241}_{95}Am + {}^{0}_{-1}e(\beta^-) + \tilde{\nu}$$

forming americium, and then

$$^{241}_{95}Am \xrightarrow[500\ a]{T_{1/2}} {}^{237}_{93}Np + {}^{4}_{2}He(\alpha).$$

This isotope has been used extensively in the study of the chemistry of americium. Isotopes in the range 237–246 have been formed by a series of reactions with reactor neutrons, and using high-energy particles from accelerators.

Both americium ($Z=95$) and curium ($Z=96$) were discovered in 1944, the latter being named after Marie Curie. As the first transuranic elements became available in small quantities they were each subjected to the usual bombarding experimental techniques now so well established. One of the early results was the production of $^{242}_{96}Cm$ from the reaction

$$^{239}_{94}Pu(\alpha, n)^{242}_{96}Cm,$$

where $^{242}_{96}Cm$ is an α-emitter of half-life 162·5 d. The chemistry of curium is made difficult owing to the lack of a really long-lived isotope. There are thirteen isotopes known, ranging from $^{238}_{96}Cm$ to $^{250}_{96}Cm$. One of the longest-lived seems to be $^{246}_{96}Cm$ with a half-life of 4000 a which, although an α-emitter, can be used in further studies of the chemistry of curium.

23.6 Berkelium, Bk ($Z=97$), and Californium, Cf ($Z=98$)

The names of these elements betray their origin. Berkelium was discovered in 1949 with the cyclotron reaction

$$^{241}_{95}Am(\alpha, 2n)^{243}_{97}Bk,$$

where $^{243}_{97}Bk$ has a half-life of 4·5 h. The longest lived isotope within the range (243–250) is $^{247}_{97}Bk$ whose half-life is 1000 a. This is an α-emitter, and forms the basis of a close study of the chemistry of berkelium compounds.

At present there are eight isotopes of berkelium available, but only in relatively small quantities. Their masses range from 243 to 250.

By cyclotron α-bombardment of $^{242}_{96}Cm$, the new element californium was found in 1950:

$$^{242}_{96}Cm(\alpha, n)^{245}_{98}Cf$$

with a half-life of 44 min. Eleven californium isotopes in the mass range 244–254 have been found. An interesting feature of the production of Cf is the use of cyclotron ions heavier than α-particles. Thus $^{12}_{6}C$ and $^{14}_{7}N$ ions have been used, as follows:

$$^{238}_{92}U(^{12}_{6}C, 6n)^{244}_{98}Cf \quad \text{and} \quad ^{238}_{92}U(^{14}_{7}N, p\ 3n)^{248}_{98}Cf.$$

The chemistry of californium is not quite so well established since there is comparatively little material available, and experiments have to be carried out at tracer level.

23.7 Einsteinium, Es ($Z=99$), and Fermium, Fm ($Z=100$)

These two elements were first found in the debris samples of the so-called 'Mike' thermonuclear explosion tests in the Pacific in 1953, but have since been found in nuclear reactors and also by the heavy-ion bombardment of $^{238}_{92}U$, viz.
$$^{238}_{92}U\,(^{14}_{7}N,\,5n)\,^{247}_{99}Es.$$
Mass numbers between 245 and 256 have been reported, one long-lived isotope being $^{254}_{99}Es$ with $T_{1/2} = 500$ d, used in the further study of einsteinium.

Similarly the production of Fm by
$$^{238}_{92}U\,(^{16}_{8}O,\,4n)\,^{250}_{100}Fm \quad \text{and} \quad ^{240}_{94}Pu\,(^{12}_{6}C,\,4n)\,^{248}_{100}Fm$$
have been observed. Fermium has isotopes within the range 248–257.

The names of Einstein and Fermi are perpetuated in these elements.

23.8 Mendelevium, Md ($Z=101$), and Nobelium ($Z=102$)

Great names are also honoured in the next two elements to be described.

The discovery of mendelevium was announced in 1955 after cyclotron α-particle bombardment of $^{253}_{99}Es$. The quantity of einsteinium used was infinitesimal and the number of Md atoms produced was only a few atoms at a time, being separated from the residue and successfully identified radiochemically as $^{255}_{101}Md$. This decays by orbital electron capture to $^{255}_{100}Fm$ with a half-life of 30 min. $^{256}_{101}Md$ also exists, with a half-life of 90 min.

An element of atomic number 102, nobelium, was first sought in 1957 by a team of British, Swedish and United States scientists working on the Swedish cyclotron. In this experiment $^{244}_{96}Cm$ was bombarded with $^{13}_{6}C$ ions in the hope of producing $^{253}_{102}No$ or $^{251}_{102}No$ by
$$^{244}_{96}Cm(^{13}_{6}C,\,4n)^{253}_{102}No \quad \text{or} \quad ^{244}_{96}Cm(^{13}_{6}C,\,6n)^{251}_{102}No$$
where the $^{13}_{6}C$ ion is used so that the product nucleus has an odd mass number. This was for technical reasons connected with the Swedish accelerator programme. This work was not fully confirmed in America, but the existence of nobelium was finally proved by ion bombardment in the later experiments at Berkeley, California (1958). Using $^{12}_{6}C$ ions instead of $^{13}_{6}C$ ions on $^{246}_{96}Cm$ the following reactions were observed
$$^{246}_{96}Cm(^{12}_{6}C,\,4n)^{254}_{102}No$$
followed by
$$^{254}_{102}No \rightarrow ^{250}_{100}Fm + ^{4}_{2}He(\alpha)$$
and
$$^{250}_{100}Fm \rightarrow ^{246}_{98}Cf + ^{4}_{2}He(\alpha).$$
There are now confirmed isotopes of nobelium with atomic masses 253, 254, 255 and 256.

23.9 Lawrencium, Lw ($Z=103$)

Lawrencium was synthesized in 1961 in the Lawrence Radiation Laboratory in California and named after E. O. Lawrence. An isotope $^{257}_{103}Lw$ has been reported having a half-life of 8 ± 2 s. It emits an α-particle of energy 8·6 MeV and is formed by bombarding californium (mixture of 249, 250, 251 and 252 isotopes) with a beam of $^{10}_{5}B$ and $^{11}_{5}B$ ions. The isotope $^{256}_{103}Lw$ was produced in the U.S.S.R. by the reaction $^{243}_{95}Am(^{18}_{8}O,\,5n)^{256}_{103}Lw$. It has a half-life of 45 s. Known isotopes of Lawrencium have masses of 255, 256 and 257.

23.10 Elements with $Z = 104$, 105, 106 and 107

In 1967 it was reported that Soviet scientists had synthesized an element of atomic number 104 and named it kurchatovium after the Soviet nuclear physicist Kurchatov. The work was probably carried out in 1965/66 but was not made public until 1967.

The method used is basically the cyclotron–heavy ion method using, in this case, $^{22}_{11}\text{Ne}$ ions on $^{249}_{94}\text{Pu}$. The plutonium target was actually encased in the cyclotron chamber and the reaction reported is

$$^{242}_{94}\text{Pu}(^{22}_{10}\text{Ne}, 4n)^{260}_{104}\text{Ku}.$$

The half-life of the new element, which was collected almost one atom at a time, was given as 0·3 s, decaying by spontaneous fission.

Evidence that it was really an element with $Z = 104$ was adduced from its similarity to hafnium. Hafnium is the first element immediately after the lanthanide series, and an element with $Z = 104$ would be expected to be the first element after the actinide series (see next section). It was shown that the chloride of kurchatovium, like the chlorides of hafnium and zirconium, was more easily volatilized than the chlorides of the rare earth elements. So small was the yield of kurchatovium that each experiment was carried out continuously for 50–60 h. The mass range is 257–260.

The same Russian laboratory subsequently synthesized an element of atomic number 105 by bombarding $^{243}_{95}\text{Am}$ with $^{22}_{10}\text{Ne}$ ions to give $^{261}_{105}?$ or $^{262}_{105}?$. The half-life is said to be 0·1 s. This is now called hahnium.

More recently, in 1977, the production of elements with $Z = 106$ and 107 has been reported from Dubna, U.S.S.R. These new elements have not yet been confirmed elsewhere (see Section 23.1).

23.11 The Actinide Series

Transuranic elements are unstable and their longest half-life is of the order of 10^6 a, which means that none of them has survived the age of the earth, 4.55×10^9 a, with the possible exception of Np and Pu. Even these are not primeval elements but have been formed by neutron bombardment. Unknown elements with atomic numbers greater than about 110 decayed by spontaneous fission as soon as they were formed, since it can be shown that for $Z > 110$, spontaneous fission is the major mode of decay. The average lifetimes of these spontaneous fission decay processes is extremely small, i.e. less than 1 μs. Other heavy elements in the range $Z = 90$–100 can also decay by spontaneous fission, but this is not the major mode of decay.

The transuranic elements as a whole are interesting in that they form a series beginning at actinium analogous to the rare earth (lanthanide) series starting at lanthanum. They are sometimes referred to as members of the actinide series. Thus we have the rare earth series and the actinide series compared in Table 23.1. Note that the actinide series finishes at Lw ($Z = 103$), so that Ku ($Z = 104$) should have chemical properties like the tetravalent column of atoms in the periodic table, viz. C, Si, Ti, Zr and Hf (see Fig. 12.7). Indeed the chemical similarity of kurchatovium to hafnium has now been shown.

The elements of the lanthanide series are characterized by incomplete 4f shells

TABLE 23.1
Rare Earth (Lanthanide Series) 4f

Atomic number	57	58	59	60	61	62	63	64	65	66	67	68	69	70	71
Element	La	Ce	Pr	Nd	Pm	Sm	Eu	Gd	Tb	Dy	Ho	Er	Tm	Yb	Lu

Actinide Series 5f

Atomic number	89	90	91	92	93	94	95	96	97	98	99	100	101	102	103
Element	Ac	Th	Pa	U	Np	Pu	Am	Cm	Bk	Cf	Es	Fm	Md	No	Lw

The elements Lu(71) and Lw(103) each start a new block of transition elements.

TABLE 23.2
Suggested Outer Electron Configurations for Transuranic Elements

Ac	$6d7s^2$	Np	$(5f^46d7s^2)^c$	Bk	$(5f^86d7s^2)^d$	Md	$5f^{13}7s^2$
Th	$(6d^27s^2)^a$	Pu	$5f^67s^2$	Cf	$5f^{10}7s^2$	No	$5f^{14}7s^2$
Pa	$(5f^26d7s^2)^b$	Am	$5f^77s^2$	Es	$5f^{11}7s^2$	Lw	$5f^{14}6d7s^2$
U	$5f^36d7s^2$	Cm	$5f^76d7s^2$	Fm	$5f^{12}7s^2$	Ku	$5f^{14}6d^27s^2$
						Ha	$5f^{14}6d^37s^2$

Alternatively: a $5f6d7s^2$, b $5f6d^27s^2$, c $5f^57s^2$, d $5f^97s^2$.

while those of the actinide series have incomplete 5f shells, the outer s, p shells being full in each case.

There is much evidence to support this comparison with the rare earth series, particularly magnetic susceptibility data and spectroscopic data which lead to the suggested electron configurations shown in Table 23.2. Note that the $7s^2$ is always saturated, as is the $6s^2$ state in the rare earth series, the inner shells being partly filled. (See Table 12.6.)

There is much further evidence to show the similarity of these two series, both chemical and crystallographic. The chemical oxidation states are comparable and ionic radii, as deduced crystallographically, show a smooth increase when going up either series, indicating the gradual filling of inner electron shells.

Problems

(*The problem marked with an asterisk is solved in full at the end of the section.*)

23.1 Write down the equations for the production of $^{239}_{94}Pu$ and explain why this nuclide is important.

23.2* When $^{239}_{94}Pu$ is bombarded with α-particles two neutrons are liberated. What is the final product? What is the product if only one neutron is liberated? How could these two possibilities be identified?

23.3 How does the element neptunium fit in to the fourth radioactive series (i.e. the $4n+1$ series)?

23.4 Write down the equations in which $^{243}_{96}Cm$ is produced from $^{241}_{95}Am$ by α-particle bombardment.

23.5 Write an essay on the production of new elements at about $Z=100$, pointing out whether or not you consider that the methods you describe could produce new elements indefinitely.

23.6 Make out a list of transuranic elements and their isotopes which have the necessary nuclear properties for fission. Which do you consider feasible?

23.7 Suggest an electron configuration for kurchatovium ($Z=104$). What elements would you expect it to resemble?

Solution to Problem

23.2 The required equations are:
$$^{239}_{94}\text{Pu} + ^{4}_{2}\text{He} \rightarrow 2^{1}_{0}\text{n} + ^{241}_{96}\text{Cm}$$
and
$$^{239}_{94}\text{Pu} + ^{4}_{2}\text{He} \rightarrow ^{1}_{0}\text{n} + ^{244}_{96}\text{Cm}.$$

Both curium isotopes are α-emitters with almost identical energies but with different half-lives.

Chapter 24

Thermonuclear Reactions and Nuclear Fusion

24.1 Introduction

Power from nuclear fission is now a reality both on land and sea, and in those countries where the coal measures are rapidly being worked out one can envisage their future economy depending on nuclear fuels, especially when oil is regarded as a costly import. On a long-term basis an infinite supply of uranium and plutonium is required, but in time the uranium-bearing minerals will disappear and we shall be dependent on the breeder reactor for our plutonium.

An alternative to the fission reaction as a source of energy is the fusion reaction. We have seen in Chapter 14 and Fig. 14.4 that when low atomic weight elements are joined together they move to states of numerically higher binding energy per nucleon and so release energy. This is the basis of all fusion reactions and it is also the source of stellar energy and the power of the hydrogen fusion bomb. Fusion depends for its action on the collision of two very energetic nuclei, a subsequent rearrangement of the nucleons and the release of energy in the form of the kinetic energy of product particles and their excitation energy. Since the primary nuclei are positively charged and repel each other electrostatically, the initial kinetic energy must be high enough to overcome this repulsion effect. A large kinetic energy implies a high temperature in order that the fusion energy produced should be sufficient to provide enough secondary particle energy to make the whole reaction self-sustaining. Experimentally it is found that the necessary primary kinetic energy increases rapidly with atomic number, so that the most promising experiments have been carried out with the three hydrogen isotopes, particularly deuterium. Now since the deuterium–hydrogen ratio in water is 1:7000 the possible terrestrial supplies of thermonuclear deuterium fuel are very large indeed, and in time will permanently solve the problem of the depletion of our chemical, mineral and fossil fuels. Although thermonuclear reactors are not yet feasible, their inherent safety (a fission reactor could possibly 'run away'), and particularly the absence of radioactive byproducts, makes them more attractive in the long run than currently designed fission reactors.

24.2 The Source of Stellar Energy

It has been calculated that the sun (our nearest star) discharges energy at the rate of about 10^{26} J s^{-1}. This can be taken as typical of hot stars whose interiors are at a temperature of about 20×10^6 K. The age of the sun is at least 5×10^9 a, so that the total loss of energy in this time is exceedingly high. How can the sun have maintained this energy output for so long — what is the source of all stellar energy?

H. A. Bethe in the United States suggested in 1939 that the production of stellar energy is by thermonuclear reactions in which protons are continuously transformed into helium nuclei. For comparatively low stellar temperatures he proposed the following cycle:
$$^1_1H + {}^1_1H \to {}^2_1D + {}^0_{+1}e(\beta^+) + v,$$
$$^2_1D + {}^1_1H \to {}^3_2He,$$
$$^3_2He + {}^1_1H \to {}^4_2He + {}^0_{+1}e(\beta^+) + v;$$
so that by addition we have
$$4{}^1_1H \to {}^4_2He + 2{}^0_{+1}e(\beta^+) + 2v + 27 \text{ MeV},$$
with an energy release of about 27 MeV.

This is the so-called proton–proton cycle and is an important source of energy in the sun. It predominates in stars of comparatively low temperatures.

In 5×10^9 a it is obvious that a great deal of hydrogen has been converted to helium, so that we may expect to find that the older stars are richer in helium compared with the younger ones.

For the main sequence stars (the sun is only a small star) Bethe suggested an alternative to the proton–proton cycle — the carbon–nitrogen cycle:
$$^{12}_6C + {}^1_1H \to {}^{13}_7N,$$
$$^{13}_7N \to {}^{13}_6C + {}^0_{+1}e(\beta^+) + v,$$
$$^{13}_6C + {}^1_1H \to {}^{14}_7N,$$
$$^{14}_7N + {}^1_1H \to {}^{15}_8O,$$
$$^{15}_8O \to {}^{15}_7N + {}^0_{+1}e(\beta^+) + v,$$
$$^{15}_7N + {}^1_1H \to {}^{12}_6C + {}^4_2He.$$
So that, on addition, we again have
$$4{}^1_1H \to {}^4_2He + 2{}^0_{+1}e + 2v + 27 \text{ MeV}$$
This conversion of hydrogen to helium is a mass exchange reaction which will continue until the whole of the star's supply of protons is used up. In the case of the sun both the above cycles take place with roughly equal probability and it is estimated that it will be about 3×10^{10} a before the protons have all been converted to helium, so that the sun is still in its youth.

In the above summary of the source of stellar energy we have seen that the fusion of four protons to form helium is only possible because of the high initial temperature and the fact that the carbon and nitrogen atoms act as true catalysts. If we try to reproduce these reactions in the laboratory in order to obtain a source of heat power, the reaction rates of both the above cycles would be far too low, so that the possibility of fusion of light nuclei has to be considered. This requires an initial temperature of many millions of degrees in which the state of a gas is known as the plasma.

24.3 The Plasma

Because of the initial high temperatures required for sustained fusion, estimated at about 5×10^7 K, the atoms are fully ionized and these ions and the free electrons are moving about very rapidly. It is possible that the separation of positive nuclei and free electrons is never very large because of their electrostatic attraction but they do move much more independently of each other than at ordinary temperatures. The mixture is still electrically neutral, of course, and the whole state is called the 'plasma' state, a sort of second gaseous state. The matter contained in stars and galaxies is largely in the plasma state but the setting up of a plasma in the laboratory requires artificial conditions such as the passage of a heavy electrical discharge through the gas — approaching 1 MA, at which current the Joule heating is enough to give the particles sufficient kinetic energy to cause fusion to take place. Another method is to inject high-energy neutral particles into the plasma.

Plasma physics and plasma engineering are now very important subjects for research and are occupying the time of a large proportion of the nuclear physicists of the world. Plasma physics laws are still not widely understood. The study of plasma physics becomes the study of magnetohydrodynamics (M.H.D.) because of the similarity of the plasma and its containment to a continuous conducting fluid in a magnetic field.

24.4 Nuclear Fusion Reactions in the Plasma

There are four feasible hydrogen reactions, all of which probably take place in a hydrogen plasma. These are:

$$^2_1H + {}^2_1H \rightarrow {}^3_2He + {}^1_0n + 3\cdot25 \text{ MeV}, \qquad (1)$$

$$^2_1H + {}^2_1H \rightarrow {}^3_1H + {}^1_1H + 4\cdot0 \text{ MeV}, \qquad (2)$$

$$^3_1H + {}^2_1H \rightarrow {}^4_2He + {}^1_0n + 17\cdot6 \text{ MeV}, \qquad (3)$$

$$^2_1H + {}^3_2He \rightarrow {}^4_2He + {}^1_1H + 18\cdot3 \text{ MeV}. \qquad (4)$$

We notice here that the simplest fusion reaction, $^2_1H + {}^2_1H \rightarrow {}^4_2He + Q$, does not proceed owing to the non-conservation of linear and angular momentum between the three particles. However, the $^4_2He^*$ nucleus can be regarded as the compound nucleus for the first two reactions quoted.

Thus the four reactions are simply nuclear rearrangements and possibly the word 'fusion', in its narrowest sense, is a misnomer. The reactions (3) and (4) of the above are really between the reaction products of (1) and (2) so that in fact we could imagine the overall conversion of six deuterons as follows:

$$6{}^2_1H \rightarrow 2{}^4_2He + 2{}^1_1H + 2{}^1_0n + 43 \text{ MeV}.$$

This is equivalent to the production of about 10^5 kW h per gram of deuterium as compared with about 10^4 kW h per gram of $^{235}_{92}U$ in fission, a useful increase.

The reaction cross-sections for the above reactions have been studied carefully and vary between about 10^{-4} barn and 1 barn, as shown in Fig. 24.1. These curves show the finite possibility of fusion occurring even at very low energies. This is a quantum-mechanical effect which cannot be explained classically. Gamow has shown theoretically that the very sensitive energy dependence is such that the predicted cross-section changes from 3×10^{-18} barn at 1 keV to $1\cdot5 \times 10^{-5}$ barns at 10 keV, an increase of 13 decades per decade energy increase.

In the above reactions when there are only two product particles the lighter particles carry away the majority of the energy, so that in the first reaction the neutron takes with it three quarters of the reaction energy, i.e. about 2·4 MeV, and so could be detected as a fast neutron.

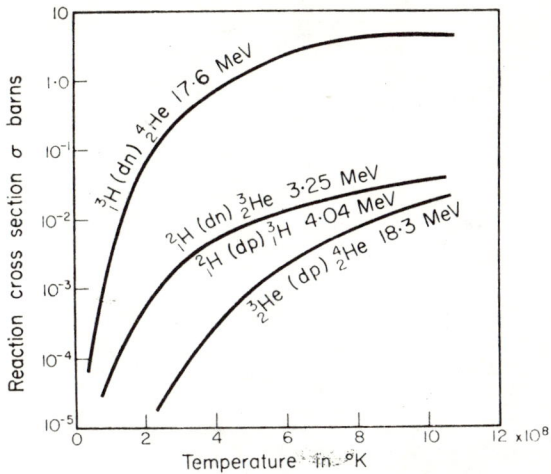

Fig. 24.1 Reaction cross-sections for some feasible fusion reactions.

The most likely reaction for controlled fusion power generation is the d, t reaction (3) above. This produces high energy, 17·6 MeV; it has the largest collision cross-section and the lowest threshhold temperature of all the feasible hydrogen fusion reactions. The neutrons take 14·0 MeV of this energy and can be used to replace the used tritium from the lithium breeder reactions

$$^{6}_{3}\text{Li} + ^{1}_{0}\text{n} \rightarrow ^{4}_{2}\text{He} + ^{3}_{1}\text{H} + 4\cdot8 \text{ MeV}$$

and

$$^{7}_{3}\text{Li} + ^{1}_{0}\text{n} \rightarrow ^{4}_{2}\text{He} + ^{3}_{1}\text{H} + ^{1}_{0}\text{n} - 2\cdot5 \text{ MeV}.$$

These equations are essential to the future designs of fusion power reactors.

An important parameter for the actual realization of fusion energy is the reaction rate. The factors influencing the reaction per unit volume are particle density n, mutual reaction cross-section σ and velocity v. For two different product nuclei the reaction rate is given by

$$R_{12} = n_1 n_2 \langle \sigma v \rangle \text{ reactions m}^{-3} \text{ s}^{-1},$$

where $\langle \sigma v \rangle$ is the average value of the cross-section and velocity produced so that the reaction rate depends on the second power of the particle density, which is of the order of 10^{20} particles m^{-3}. From the particle distribution of velocities the average value $\langle \sigma v \rangle$ can be found as a function of energy by numerical integration. Figure 24.2 shows the plot of $\langle \sigma v \rangle$ against temperature measured as kinetic energy in keV, where 1 keV $\equiv 1\cdot16 \times 10^7$ K. Again we notice the rapid fall-off at very low temperatures, although the value of $\langle \sigma v \rangle$ is always finite. At the lower energy range it is only the faster particles in the distribution which contribute.

24.5 Conditions for a Maintained Fusion Reaction

Ultimately we require the thermonuclear reaction to be self-sustaining and energy producing. This can only be possible if the rate of generation of energy exceeds the rate of loss at all times. When these are just balanced the plasma is in a critical state, or at a critical temperature which must be exceeded for the reaction

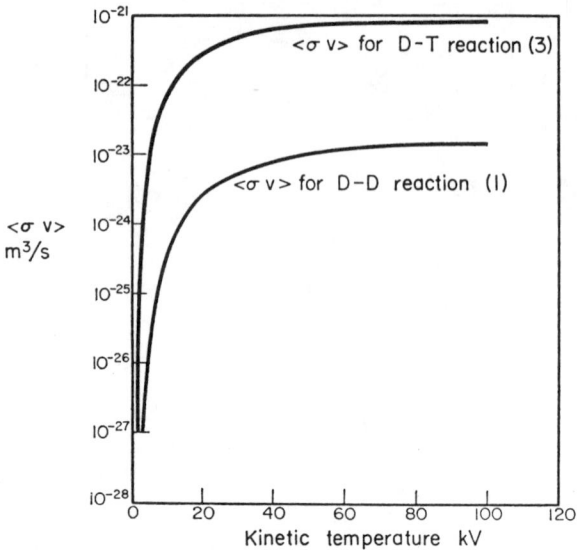

Fig. 24.2 Variation of $\langle \sigma v \rangle$ with kinetic temperature for reactions (1) and (3) (see text).

to proceed. Energy is lost from the plasma largely by means of X-rays and Bremsstrahlung and these unavoidable losses set the minimum critical temperature, which for a hydrogen plasma is about 5×10^7 K.

Thus, if we can heat the hydrogen in an evacuated chamber to this temperature and prevent it reaching the chamber walls, where it would lose further energy by conduction, it might be possible to create a self-sustaining nuclear fusion chain reaction. This is by no means easy and some of the problems associated with this will now be discussed.

(i) *Containment.* This is the problem of holding the plasma away from the vessel walls for long enough for fusion to occur. In the case of stellar thermonuclear energy discussed in Section 24.2, the contraction under gravity when the reaction slows down is sufficient to raise the temperature and pressure again to speed up the reaction. A star is therefore a self-controlled system.

The only method of containment which is feasible at present in the laboratory, is the 'magnetic bottle' method, where the movement of the plasma, which is of course electrically conducting, is controlled by magnetic fields. This may be produced by the passage of the heavy heating current, of the order of a million amperes, down the tube. This produces a circular magnetic field which then reacts with the plasma to 'pinch' it down to a thin filament. This is analogous to two parallel wires with currents in the same direction being drawn together as a result of the left-hand motor rule and is called the 'pinch' effect. See Fig. 24.3.

Another method is to maintain a high-frequency alternating magnetic field inside a cavity containing the plasma. The plasma is ionized and the reaction with the field causes the plasma to contract under the influence of the 'magnetic pressure'.

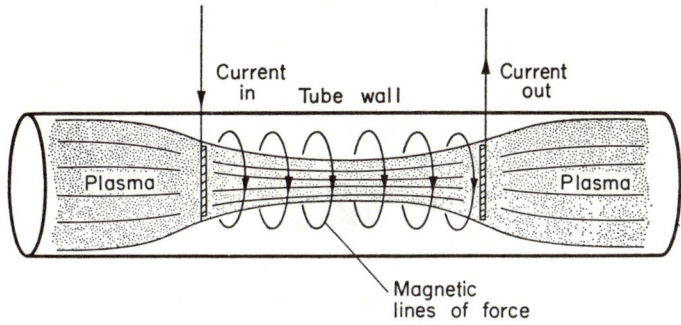

Fig. 24.3   Principle of pinch effect in hot plasma. In practice the electric current is induced in the conducting plasma.

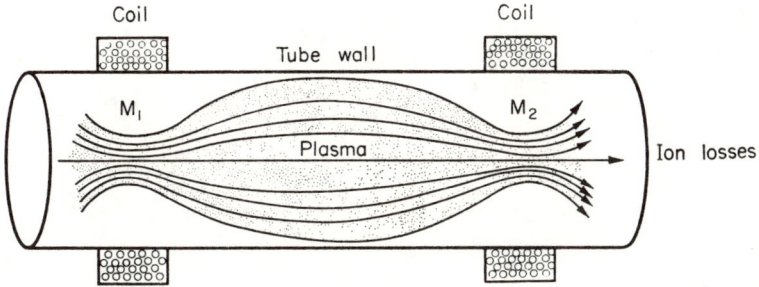

Fig. 24.4   'Mirror'-machine with ions trapped between two reflecting regions of converging magnetic fields.

A third method, shown in Fig. 24.4, is to use magnetic coils to provide a reflecting region for the moving ions. With the arrangement shown, the lines of force of the field are inhomogeneous and most of the charged particles on reaching M_1 or M_2 will be reflected. Axial particles will be lost so that the plasma would gradually die away. Figure 24.4 is the basis of the so-called 'mirror'-machine and another feasible design in which the coils are in opposition, is the cusp-machine, shown diagrammatically in Fig. 24.5, in which the plasma is held between opposing magnetic fields.

(ii) *Instability of plasma.* Ideally the contained plasma is in a thin continuous line filament. In practice, due to magnetic and electrostatic leakages, the plasma filament is very distorted, giving rise to the well-known 'wriggle' shown in Fig. 24.6. This snake-like effect can only be reduced by a series of correctly placed magnetic fields. Generally the 'wriggle' can be straightened out by the use of an axial magnetic field, as shown in Fig. 24.7, and Fig. 24.8 shows an actual plasma filament. These are called M.H.D. or gross instabilities.

(iii) *Reaction times.* Having contained the plasma and heated it to the right temperature it must now be maintained long enough in these conditions for the nuclei to react. The holding time required depends on the density of the plasma,

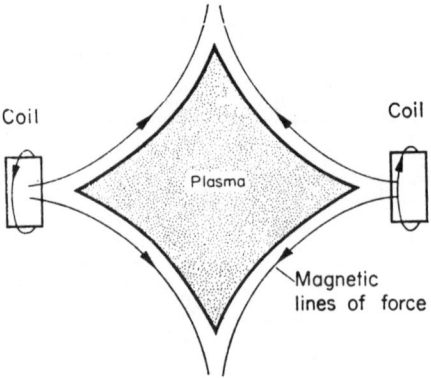

Fig. 24.5 Cusp machine showing plasma trapped by magnetic fields of oppositely directed currents.

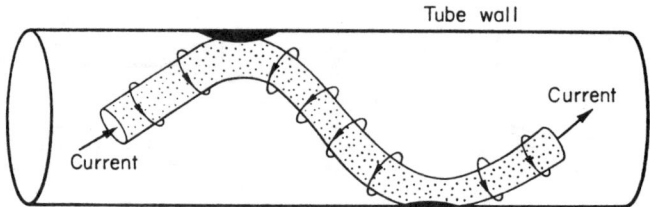

Fig. 24.6 Plasma touching tube walls showing 'wriggle'.

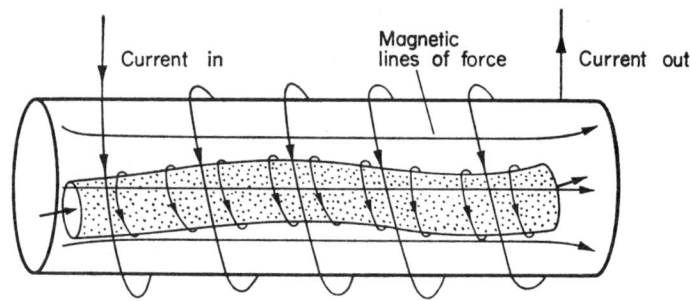

Fig. 24.7 'Wriggle' straightened out by stabilizing magnetic field along tube axis.

but the time taken for the field to rise to its final value must be much smaller. Some relevant figures are about 1 T for a holding time of 0·1–1·0 s or 10 T for 1–10 s, where the field 'rise time' must only be about 100 μs. This involves many difficult engineering problems of storing, switching and transmitting electrical energy of many megajoules during these short pulses. Thus in Zeta (the U.K.A.E.A. fusion device at Harwell) the capacitor bank was about 1600 μF with working potentials up to 25 kV, giving a stored energy of 0·5 MJ discharged at 50 kA for 3 ms.

Fig. 24.8 Photograph of pinched discharge. (By courtest of the Los Alamos Scientific Laboratory.)

24.6 The Possibility of a Fusion Reactor

Some of the problems of plasma physics and engineering have been indicated in the previous section, but there are more general requirements to be met before a successful thermonuclear reactor is feasible. Some of these are as follows:

(i) *Power density*. A plot of power density against particle density for two possible reactions is shown in Fig. 24.9. The upper limit of power density is set by engineering facilities at about 100 MW m^{-3}, so that from the graphs the particle densities required are only of the order of 10^{-4}–10^{-5} of ordinary atmospheric gas densities. This shows that the actual plasma would be very tenuous and a mean reaction time of about 1 s sets the target for the time of containment. The plasma internal energy (heat) content U is given by

$$U = \frac{3}{2} nkT,$$

where n is the particle density, k is the Boltzmann constant and T is the temperature in kelvins. Thus

$$U = \frac{3}{2} \times 10^{20} \times 1 \cdot 38 \times 10^{-23} \times 10^8$$
$$= 2 \cdot 1 \times 10^5 \text{ J m}^{-3},$$

so that the actual energy flux for a suitable plasma is not very large, and there is really no danger of the container walls being evaporated.

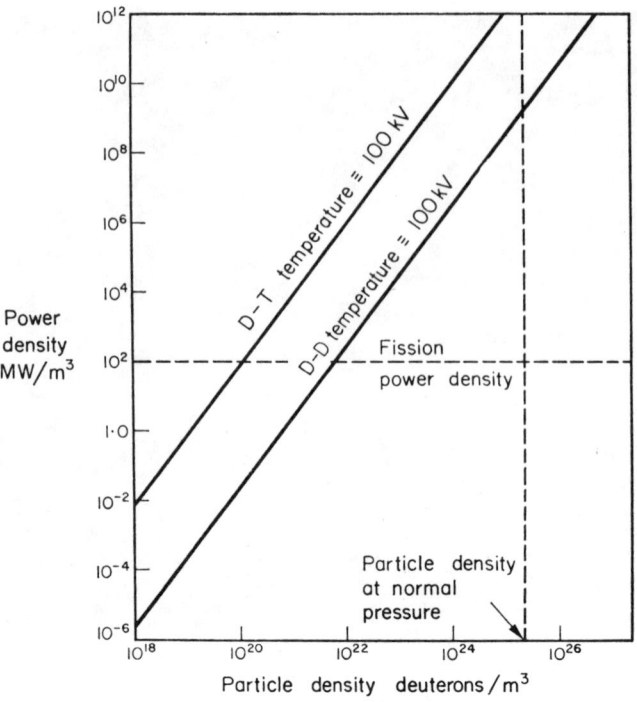

Fig. 24.9 Power density for D–D and D–T reactions.

(ii) *Balance of power.* There can only be *available* power when the electric power generated by the nuclear reactions is greater than *all* the power required to keep the plasma above its critical condition. For example, the generation of the magnetic field requires a large amount of power to overcome the resistance of the coil windings and for fields about 10 T the calculations indicate that tubes of the order of several metres in diameter and many times longer will be necessary.

In the realization of fusion power the speed of the colliding nuclei must be due to thermal kinetic effects rather than electrical accelerations, and in order to be certain that true thermonuclear reactions occur it is necessary to show experimentally that the measured neutrons are not all from accelerated deuterons colliding with stationary atoms.

(iii) *Lawson's criterion.* Calculations show that a sustained fusion reaction will take place if the plasma can be maintained at 10^8 K for 1 s or more. A high degree of thermal containment is therefore necessary. This is expressed in terms of a factor τ_e, the energy confinement time, equivalent to the ratio of the total heat energy of the plasma divided by the total rate of energy loss to the surroundings. If the particle density is n m^{-3} the product $n\tau_e$ then represents the confinement time per cubic metre of plasma. By combining the data on the physics of the fusion process together with estimates of the energy output and various engineering factors, J. D. Lawson concluded that for a feasible sustained fusion reaction the product $n\tau_e$ must exceed 10^{20} s m^{-3}. This is known as Lawson's criterion, and the

ultimate aim of fusion research is to get a system with
$$n\tau_e \sim 3 \times 10^{20} \text{ s m}^{-3}$$
The required *minimum* values of the individual reactor plasma parameters are: $T = 1-2$ MK; $n = 2-3 \times 10^{20}$ m^{-3}; $\tau_e = \frac{1}{2} - 1$ s.

In Table 24.1 we have gathered together some data showing the improvement in the Lawson factor since 1955.

We see from the above that modern research into fusion systems has to solve three main problems:

(i) to find a way of heating a plasma to 100 MK,

(ii) to devise methods of controlling and isolating the hot plasma for an adequate time, and

(iii) to design a practical fusion reactor to produce electricity directly and economically.

Many experimental arrangements have overcome some of the above difficulties on a small scale. The names of Zeta in England and Perhapsatron and Stellarator in the U.S.A. were early projects associated with fusion. At present there is no way of extracting the fusion energy usefully and it will be many years before the fusion power reactor will be a reality. But it will come, and research in controlled thermonuclear reactions continues. In the 1970's there was much technological progress, largely due to a better understanding of plasma dynamics and the development of the so-called Tokamak (toroidal magnetic chamber) torus system first proposed by the U.S.S.R. physicists in 1969.

24.7 Tokomak Fusion Systems

Many geometrical arrangements have been tried in order to fulfil the above conditions for a maintained fusion system. The most favoured system is the Tokomak configuration pioneered by the U.S.S.R. It is shown diagrammatically in Fig. 24.10. Experiment showed that the best stability of the plasma was achieved by a combination of two magnetic fields, the toroidal field B_ϕ provided by a series of circumferential coils giving a magnetic field concentric with the circular plasma current, and the poloidal field B_θ, at right angles to B_ϕ and provided by a transformer linked with the conducting plasma, acting as the secondary winding. The circular poloidal field is therefore at right angles to the current direction and serves to confine the plasma while the poloidal current is parallel to the plasma flow and heats it by Joule heating. Fig. 24.10 shows the relevant geometry.

In the original Zeta experiment a reversed-field pinch was used. The two fields, B_ϕ and B_θ, were of about the same magnitude but the toroidal field was reversed in the outer regions of the plasma. This meant that, for a given field, a plasma of higher pressure could be confined. A larger machine with a reversed-field pinch is now being built in the U.K. at Culham, which will have currents of ~ 400 kA with $\tau_e = $ a few ms with a toroidal vacuum chamber of 0·8 m major radius and 0·25 m minor radius. Design studies are being pursued based on this experiment. It is hoped to confirm the good confinement properties of the reversed-field pinch in more reactor-like plasmas. The plasma current is expected to be 1 MA and the torus 1·8 m × 0·6 m.

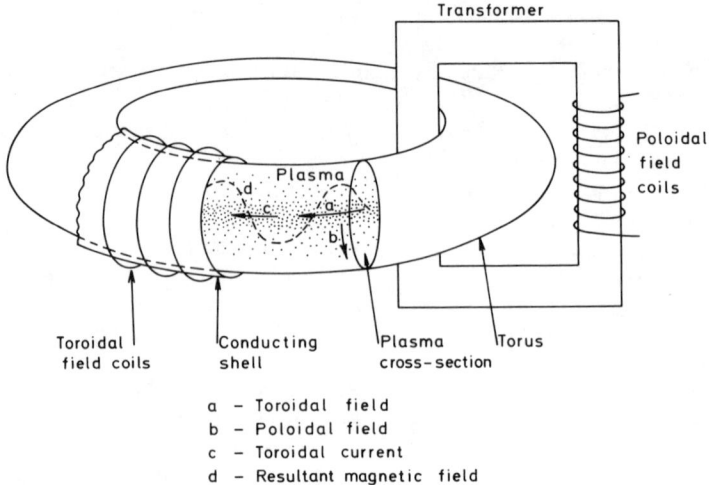

a - Toroidal field
b - Poloidal field
c - Toroidal current
d - Resultant magnetic field

Fig. 24.10 Schematic drawing of a Tokomak fusion reactor.

For true Tokomaks the theory shows that the merit factor is the product

$$\frac{r}{R}\cdot\frac{B_\phi}{B_\theta}=q$$

where r is the minor radius of the torus and R the major radius. Maximum plasma stability is obtained when $q \geqslant 3$ and this implies that for a workable size of torus with $r:R$ about 1:3 the field B_ϕ must be much greater than B_θ. This is the basis of all Tokomak configurations. The high ratio of $B_\phi:B_\theta$ limits the twist of the resultant magnetic field (magnetic shear) to less than one helical turn per revolution, giving a much higher degree of stability than when the two fields are nearly equal. The twisting net field is showing in Fig. 24.10. The two fields combine to pinch the plasma. Expansion to the outside walls can also be reduced by an external conducting magnetic shell. The post-1970 results for the Lawson criterion in Table 24.1 are all based on the Tokomak configuration.

We see, then, that it is desirable to make the following factors as large as possible: the energy confinement time τ_e, the toroidal field B_ϕ; the major radius R; the temperature T; and the plasma current I_p. To maximize all these factors simultaneously presents great technical difficulties before a fusion reactor can be made. There may be upper limits to some of these parameters. Even if the Lawson criterion can be reached, there is still the technical problem of extracting the heat and designing the exchanger. The tritium used up in the plasma must be continuously replaced. Designs are based on the use of a breeder blanket of lithium salts surrounding the torus in which the reactions as follows will take place:

and

$$_0^1n + {}_3^6\text{Li} \rightarrow {}_2^4\text{He} + {}_1^3\text{H(t)} + 4{\cdot}8 \text{ MeV}$$

$$_0^1n + {}_3^7\text{Li} \rightarrow {}_2^4\text{He} + {}_1^3\text{H(t)} + {}_0^1n - 2{\cdot}5 \text{ MeV}.$$

Breeder reactions.

TABLE 24.1
Growth of Fusion Systems

Year	τ_e (s)	T (K)	Lawson's criterion $n\tau_e$ (s m^{-3})
1955	10^{-5}	10^5	10^{15}
1960	10^{-4}	10^6	10^{16}
1965	2×10^{-3}	10^6	10^{17}
1970	10^{-2}	5×10^6	5×10^{18}
1975 D.I.T.E. Culham, U.K.	5×10^{-2}	10^7	10^{19}
1978 P.L.T., U.S.A.	0·1	7.5×10^7	3×10^{19}
J.E.T. design 1982	0·5–1	5×10^7	10^{20}
Target 2010	>1	10^8	3×10^{20}

A temperature of 10^8 K corresponds to an ion temperature T_i of 10 keV in energy units.

The fusion reaction is $^2_1H(d) + ^3_1H(t) \rightarrow ^4_2He + ^1_0n + 17.6$ MeV. The energy is taken out of the system by both radiation and particle energy, and the neutron kinetic energy is converted to heat in the lithium blanket, which is then injected into the heat exchanger for steam raising. The tritium produced is fed back into the plasma and the only waste product is the helium gas.

The main advances in fusion research with a view to producing a fusion reactor are all based on the Tokomak principle. There are now about 100 Tokomaks in operation, the largest being the Russian T.10, the Princeton Large Torus P.L.T. of the U.S.A. and the Culham (U.K.) Torus D.I.T.E. The design parameters of these are shown in Table 24.2. The gap between the reproducible attainment of these systems and the target of the Lawson criterion is narrowing, as Table 24.1 shows.

In Europe the Joint European Torus (J.E.T.) is being built on a site adjacent to the Culham Laboratory. The design of this fusion reactor requires particle densities approaching 10^{20} m^{-3} at temperatures greater than 50 MK. J.E.T. is a proposed major nuclear fusion experiment. It will be ten times bigger than the current Tokomaks with a corresponding increase in the plasma current. The object of J.E.T. is to study plasma conditions with parameters approaching the reactor range. On the completion of the experimental programme the results from J.E.T. will be sufficient to establish dimensions, parameters and the plasma

TABLE 24.2
Design Parameters of Some Existing Tokomaks

Tokomak	T(MK)	I_p(MA)	R(m)	r(m)	B_ϕ(T)
D.I.T.E. (U.K.)	10	0·3	1·1	0·23	2·8
T.10 (U.S.S.R.)	20	0·8	1·5	0·35	5·0
P.L.T. (U.S.A.)	60	1·4	1·3	0·45	5·0

TABLE 24.3
Tokomak Parameters

	Culham torus D.I.T.E.	Projected J.E.T. Tokomak	Conceptual superconducting power reactor
In operation	1975	1982	2010
Plasma current, I_p	300 KA	2·6–4·8 MA	20 MA
Toroidal field, B_ϕ	2·8 T	2·8 T	10 T
Poloidal field, B_θ	0·3 T	0·3 T	1 T
Minor radius, r	0·3 m	Elliptical 1·25 m horizontal 2·1 m vertical	D-shaped 2·1 m horizontal 3·7 m vertical
Major radius, R	1·0 m	3 m	7·4 m
Volume of torus, V	2 m^3	~100 m^3	~1000 m^3
Total cost of plant	£1 M	£100 M	£1 G
Lawson criterion	10^{19} sm^{-3}	10^{20} sm^{-3}	$>10^{20}$ sm^{-3}

behaviour to be expected of a future power reactor. Such a reactor would require superconducting magnets operating in liquid helium at 4 K to maintain magnetic fields of the order of 10 T.

A comparison of the J.E.T. design figures with those of a possible power reactor conceptual design is shown in Table 24.3. This is the Culham design. There are other design studies taking place in the U.S.A., U.S.S.R., W. Germany and Japan. The question of a development timetable depends on the willingness of governments to use public funds. However, the U.S. has laid down a tentative timetable proposing an experimental power reactor for 1997 costing $1G leading to a fusion power station producing electricity in the year 2015.

An artist's impression of the conceptual Culham power reactor is shown in Fig. 24.11. The overall diameter of such a reactor will probably be of the order of 50 m and the volume of the torus greater than 1000 m^3.

24.8 Energy in the Future

The success of developing fusion power Tokomaks will depend very much on financing the research, the enormous cost of which is why nations are joined together in a common effort such as J.E.T. Another example of this sort of cooperation is in the development of the big particle accelerators. In addition, the problem of getting a current of 20 MA to flow for periods greater than 1 s might well require the output of a small power station.

It should be mentioned that other methods of inducing fusion reactions are being studied. One depends on the high concentration of energy in a laser beam. A small pellet of a deuterium–tritium mixture is bombarded simultaneously by several laser beams, which cause the pellet to be compressed until its centre reaches about 100 MK giving a d, t thermonuclear reaction. The pellet

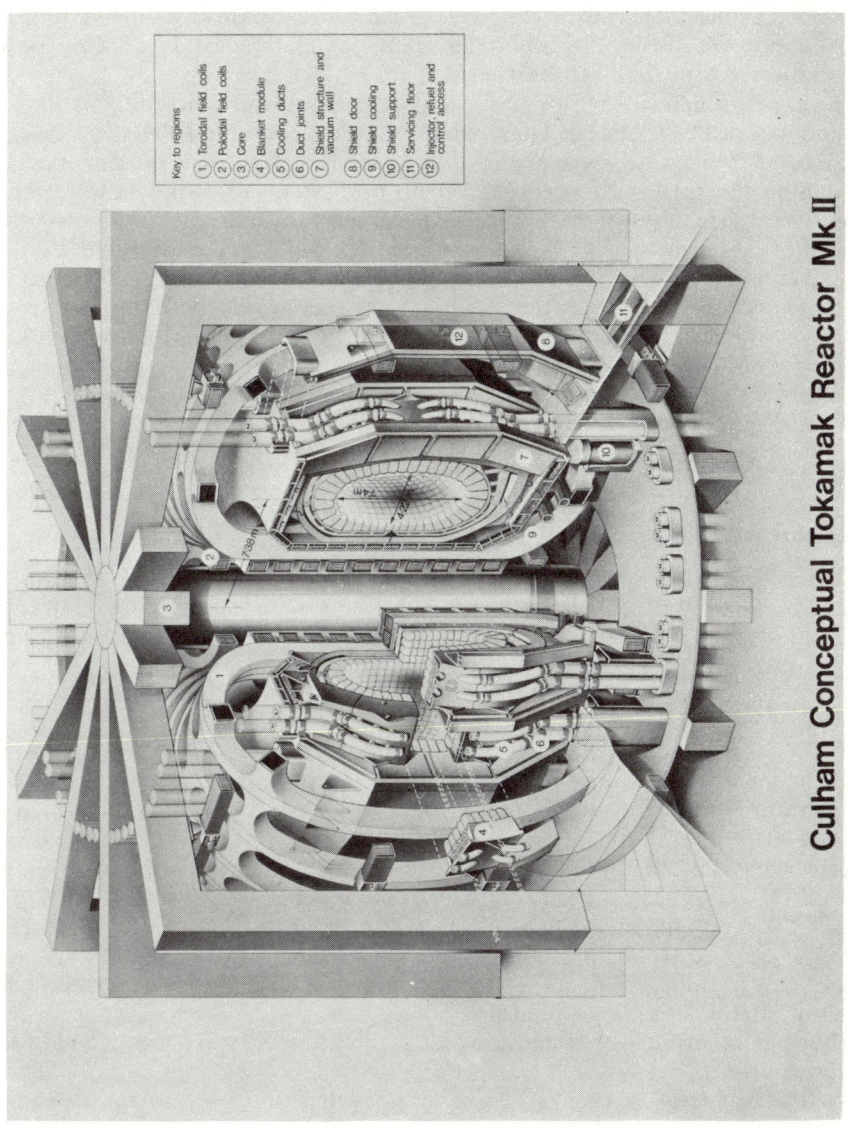

Fig. 24.11 Possible 21st century fusion reactor showing the D-shaped toroidal field coils, with the poloidal field coils around the iron core. (By courtesy of the U.K.A.E.A. Culham Laboratory, Abingdon, England.)

consequently expands rapidly releasing the extractable kinetic energy of the 14 MeV reaction neutrons. In another scheme high velocity (25 GeV) heavy ions from an accelerator are fired into a d, t fusion pellet. Heavy ions such as Cs^+ and Xe^+ are feasible. Generally, these ideas are at present part of academic research and cannot yet be looked upon as main contenders for large-scale power production within the next 100 years.

In all these schemes Lawson's criterion will still hold and a breeding blanket for tritium will still be required. These are the features common to all these designs. In all cases the cost will be very high. However, the world is at present spending something like £1000 M per year on fusion systems showing how seriously research into the physics of the reactor plasma is taken. It is only some 40 years since the first Fermi 'pile' was made, yet fission power is well established. Perhaps this gives us a time scale for the realisation of fusion power and it seems probable that because of the high cost of these devices they would be built only by the richest countries or by groups of countries. The first operational fusion reactor giving economically produced electricity may possibly be built and financed by an international consortium from Europe, U.S.A., U.S.S.R. and Japan.

Two review articles worth reading are *New Scientist*, **82**, 24 May 1979, and *Physics Today*, **32**, May 1979.

Problems

(*Those problems marked with an asterisk are solved in full at the end of the section.*)

24.1* Estimate the volume of water in the Great Lakes and calculate the amount of energy released if all the deuterium atoms in the water are used up in fusion. ($1 \cdot 5 \times 10^{39}$ MeV)

24.2 Write an account of the Bethe explanation of the apparent infinite supply of heat and energy in the hot stars.

24.3* A thermonuclear device consists of a torus of diameter 3 m with a tube of diameter 1 m. It contains deuterium gas at 10^{-2} mmHg pressure and at room temperature (20°C). A bank of capacitors of 1200 μF is discharged through the tube at 40 kV. If only 10% of the electrical energy is transformed to plasma kinetic energy, what is the maximum temperature attained? Assume the energy is equally shared between the deuterons and electrons in the plasma. ($4 \cdot 75 \times 10^5$ K)

24.4 Discuss the meaning of 'electron temperatures' and 'deuteron temperatures' as applied to hot plasmas.

24.5 Write an account of the main difficulties in the design of a thermonuclear reactor capable of giving useful power.

By what experimental evidence would a thermonuclear reaction be judged to have occurred?

24.6 Calculate the rate of loss of mass of the sun as a result of its radiant energy emission.

24.7 How much motor fuel (petrol, gasoline) would provide the same energy as that obtained from 1 litre of water used in a thermonuclear reactor?

Energy value of petrol 10 kW h kg^{-1}.

Assuming a car is used for 16 000 km per year, consumption of 10 km l^{-1}, how long would this amount of petrol last?

24.8   What is the temperature equivalent of 40 keV deuterons used in a thermonuclear reaction?

24.9   In problem 24.3, if the deuteron density is 10^{22} m^{-3}, and $\langle \sigma v \rangle$ is 10^{-23} m^3 s^{-1}, calculate the power density due to the primary reactions.

Solutions to Problems

24.1   The area of the Great Lakes is about 256 000 km². The average depth will be about 80 m, say, i.e. roughly 0·08 km, making the volume about 20 500 km³.

Thus mass of water is $20\,500 \times (10^3) \times 10^3$ kg
$$= 2 \cdot 05 \times 10^{16} \text{ kg}.$$

Relative molecular mass of water = 18, so that the number of molecules of water in $2 \cdot 05 \times 10^{16}$ kg is
$$\frac{2 \cdot 05 \times 10^{16} \times 6 \cdot 02 \times 10^{26}}{18} = 6 \cdot 86 \times 10^{41} \text{ molecules of water}.$$

The abundance of deuterium is 0·0156% so that the total number of deuterium atoms is
$$6 \cdot 86 \times 10^{41} \times 2 \times 0 \cdot 0156 \times 10^{-2}$$
$= 2 \cdot 14 \times 10^{38}$ deuterium atoms in the Great Lakes water.

Now from the text the fusion of six deuterium atoms gives an energy release of 43 MeV = 7·17 MeV per atom. Thus
$$\text{Total energy release} = 2 \cdot 14 \times 10^{38} \times 7 \cdot 17 = 1 \cdot 5 \times 10^{39} \text{ MeV}.$$

This is equivalent to 3×10^9 hydrogen bombs of 20 megatons T.N.T. equivalent each.

24.3   Cross-sectional area of torus $= \frac{1}{4}\pi 1^2$
$$= \frac{\pi}{4} \text{ m}^2,$$
$$\text{Circumference} = 3\pi \text{ m},$$
$$\text{Volume of torus} = \frac{3}{4}\pi^2$$
$$= 7 \cdot 4 \text{ m}^3,$$
$$\text{Pressure} = 10^{-5} \text{ mHg}$$
$$= 10^{-5} \times 13 \cdot 6 \times 10^3 \times 9 \cdot 81$$
$$= 1 \cdot 34 \text{ Pa}.$$

From the equation $PV = NkT$, where N is the total number of deuterium molecules present, we have $1 \cdot 34 \times 7 \cdot 4 = Nk \times 293$ or $Nk = 0 \cdot 0338$ for the deuterium gas.

Now $\bar{E} = \frac{3}{2}NkT_k$ for the average kinetic energy (is this valid?), where T_k is the kinetic temperature of the plasma particles. From the discharge,
$$\bar{E} = \frac{1}{2}CV^2$$
$$= \frac{1}{2} \times 1200 \times 10^{-6} \times (40\,000)^2$$
$$= 96 \times 10^4 \text{ J}$$
and energy used is $\frac{1}{10}\bar{E} = 9 \cdot 6 \times 10^4$ J. Therefore
$$4 \times \tfrac{3}{2}NkT_k = 96 \times 10^3$$
since each deuterium molecule produces two ions and two electrons.

This gives $T_k = 4 \cdot 75 \times 10^5$ K.

Chapter 25

Cosmic Rays

25.1 Discovery

As long ago as 1900, C. T. R. Wilson and others found that the charge on an electroscope always 'leaked' away in time, and this could never be prevented, no matter how good the insulation. When the properties of radioactive radiations were better known Rutherford showed that the rate of leakage was considerably reduced by shielding the electroscope with thick slabs of lead, but there was always a residual leakage of charge which could not be eliminated. It was thought therefore that the initial conduction in the enclosed gas was probably due to ionizing radiations from radioactive minerals in the ground. When it was shown that over the sea, where mineral radioactive effects are negligible, the rate of leakage was still pronounced, and was only partially diminished by shielding, it was concluded that the ionizing radiations were descending as well as ascending. The famous experiment of Hess in 1912 in which he sent up an ionization chamber in a balloon and found that the intensity of ionization actually increased up to a height of 5000 m and then decreased again, showed beyond doubt that these ionizing radiations travel down to earth through the air. A further observation showed that the intensities were the same for night or day, indicating that the origin of these radiations was not solar. Hess suggested therefore that these rays were of cosmic origin, and they were finally called 'cosmic rays' by Millikan in 1925.

Millikan and others conducted some early researches on cosmic rays and found that there were two components, soft and hard, and that the hard, or very penetrating component, was not fully absorbed by many feet of lead or even at the bottom of lakes as deep as 500 m. This showed that the energy of cosmic rays was many times that of any other natural or artificial radiation known at that time.

In 1927 Clay found that the intensity of cosmic rays depended upon latitude, being a minimum at the equator and a maximum at the poles. This is a geomagnetic effect supporting the suggestion that cosmic rays are charged particles entering the earth's magnetic field from a great distance. At this stage the really intensive study of the properties of cosmic rays and their uses in nuclear physics had begun.

25.2 Nature of Cosmic Rays

Primary cosmic rays have their origin somewhere out in space. They travel with speeds almost as great as the speed of light and can be deflected by planetary or intergalactic magnetic fields. They are unique in that a single particle can have an energy as high as 10^{20} eV but the collective energy is only about 10 μW m^{-2} for cosmic rays entering the atmosphere, which is roughly equal to the energy of starlight. In starlight the energy of a single photon is only a few electron volts, compared with the average for cosmic rays of 6 GeV per particle.

The composition of cosmic rays entering the earth's atmosphere is fairly well known from balloon experiments, and it is found that these primary cosmic rays consist mainly of fast protons. There are very few positrons, electrons or photons, and the 'particle' composition is mainly 92% protons, 7% α-particles and 1% 'heavy' nuclei, carbon, nitrogen, oxygen, neon, magnesium, silicon, iron, cobalt and nickel stripped of their electrons. The average energy of the cosmic ray flux is 6 GeV, with a maximum of about 10^{20} GeV. (Compare this with 10^3 GeV, the maximum energy of the artificially accelerated particles.) The radiation reaching the earth is almost completely isotropic.

As soon as the primary rays enter the earth's atmosphere multiple collisions readily take place with atmospheric atoms, producing a large number of secondary particles in showers. Thus when a primary proton strikes an oxygen or nitrogen nucleus a nuclear cascade results. These secondary atmospheric radiations contain many new particles, neutral and ionized, as well as penetrating photons, but little if any of the primary radiation survives at sea-level. Secondary cosmic rays at sea-level consist of about 75% muons and about 25% electrons and positrons, although some α-particles, γ-photons and neutrons may be present in negligible quantities. The muon will be described later.

The collision cross-sections for the primary component of cosmic rays are of the order of 10^{-1} barns and the mean free path for a collision process at the top of the atmosphere may be as high as several kilometres. The new particles produced after primary collisions give in their turn more secondary radiations by further collisions until a cascade of particles has developed, increasing in intensity towards the earth. This is shown diagrammatically in Fig. 25.1 and an actual photograph of a cascade shower deliberately produced in lead is shown in Fig. 25.2.

The energy spectrum of the primary cosmic rays ranges from 10^9 eV to about 10^{20} eV and can be written $dN/dE = K(E + m_0 c^2)^{-\gamma}$, where N is the number of nuclei with a kinetic energy per nucleon $> E$ (in GeV), $m_0 c^2$ is the nucleonic rest energy and K and γ are constants for a given cosmic ray component. This is represented in Table 25.1. Above 1·5 GeV per nucleon the cosmic ray intensity is fairly steady with time, with flux values in space of about

Protons	1500 nuclei m^{-2} s^{-1} sr^{-1},
α-particles	90 nuclei m^{-2} s^{-1} sr^{-1},
'Heavies'	10 nuclei m^{-2} s^{-1} sr^{-1}.

At lower energies the cosmic ray intensity is not constant with time but depends on the activity of the sun. It is found that during periods of high sunspot activity the cosmic ray intensity is low, presumably due to the trapping of the charged primaries high above the earth by the increased magnetic field of the sun at these times. Corresponding to the 11-year cycle of maximum sunspot activity there is therefore a cycle of minimum cosmic ray intensity.

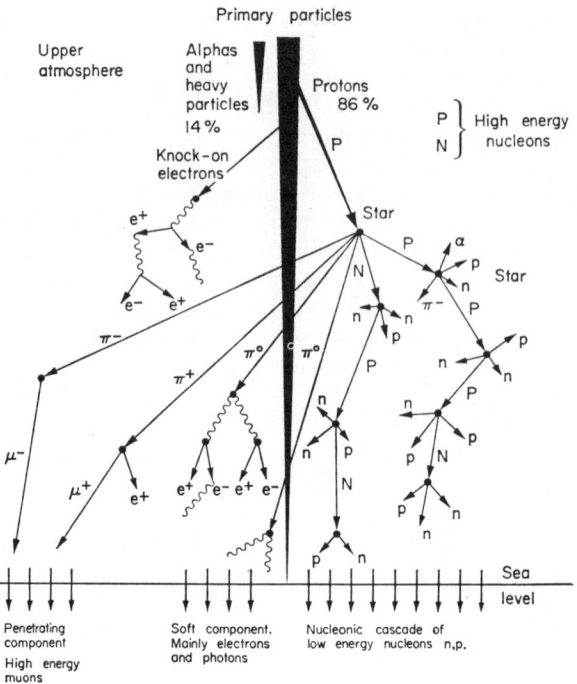

Fig. 25.1 Secondary products from a primary cosmic ray particle collision in the atmosphere.

25.3 The Origin of Cosmic Rays

An early observation on cosmic ray intensities showed that the sun itself must actually be the source of at least some of the low-energy primaries, since at times of solar flares the cosmic ray intensity increased. However, this can only account for a small fraction of the total, and since cosmic rays are nearly isotropic around the earth their origin in such a 'point source' as the sun is precluded and we must look much further into the depths of space.

An interesting feature of the composition of the primary rays is the existence of heavy nuclides up to relative atomic masses of about 60, and the fact that the distribution of the elements in cosmic rays shows a similar trend to that in the sun, stars, nebulae and in the non-volatile parts of meteorites, although the primary cosmic radiations are significantly richer in heavy nuclei compared with the general matter of the universe. This seems to indicate a cosmic ray origin in which matter is present and where the conditions are of relatively low energy (compared with cosmic ray energies), possibly in supernovae explosions.

Fermi suggested that the cosmic rays have their origin in interstellar space and are accelerated to high energies, as they stream through the arms of a galaxy, by the associated galactic magnetic field which is about 10^{-9} T or 1 nT. The cosmic ray particle is injected into the galactic magnetic field from the surface of a star with an appreciable initial energy and is caused to spiral in this field. It will

Fig. 25.2 Cascade shower produced in lead plates. Cloud chamber photograph. (From Rochester and Wilson, *Cloud Chamber Photographs of the Cosmic Radiation*. Pergamon, 1952.)

eventually 'collide' with another region of high magnetic field which is approaching it with a high velocity. The cosmic ray particle is reflected or repelled with increased energy since the magnetic field is moving towards it. When a cosmic ray particle is trapped between two such fields it gains energy by multiple repulsions and the more energetic particles of the distribution finally escape into space with a high velocity of projection. This model is not unlike the 'mirror'-machine discussed in the previous chapter. The trapping and ejecting mechanism can be repeated until the particle reaches the solar system where it is observed.

TABLE 25.1
Composition of Primary Cosmic Rays Entering Earth's Atmosphere

Nucleus	% Composition	Energy range (GeV nucleon^{-1})	Flux, i.e. no. of particles $m^{-2}\ s^{-1}\ sr^{-1}$
H	92	2–20	4000 $E^{-8/7}$
He	7	1·5–8	460 $E^{-7/4}$
Li Be B	0·18	—	12 $E^{-7/4}$
C N O F	0·36	3–8	24 $E^{-7/4}$
Ne and beyond	0·15	3–8	16 E^{-2}

E is the total energy per nucleon in GeV.

It is concluded, therefore, that cosmic rays acquire their energies in the vicinity of magnetically active stars, especially supernovae. This is supported by the observations on radio stars which show intense radio noise due to very fast electrons moving in magnetic fields, suggesting that cosmic rays may also be associated with stellar events of great violence. Since the cosmic rays are pushed about in all directions by these great belts of stellar magnetic fields, in which they undergo multiple reflections and changes of direction, they surround the earth isotropically so that the earth can be regarded as a simple body in a whole sea of cosmic rays.

25.4 Geomagnetic Effects

Compton and Millikan in 1935 carried out a world-wide survey of cosmic ray intensities and showed that the lines of equal cosmic ray intensity followed closely the earth's geomagnetic latitude indicating that some, at least, of the primaries must be charged particles affected by the variations in the geomagnetic field.

The earth has a magnetic moment of about 10^{23} Am^2 with a magnetic field of flux density 30 μT at the equator. As shown in Fig. 25.3, for the particles that enter the earth's atmosphere 'vertically' and parallel to the geomagnetic lines of force at the poles, there is little interaction between the magnetic field and the charged particles near the poles. However, near the equator the magnetic field is perpendicular to the direction of the cosmic rays and the interaction is therefore much greater so that the less energetic particles are deflected out of their original path. Only those exceeding a critical energy reach the earth's surface. This critical energy is equivalent to a 'cut-off' in the energy spectrum, and depends on the latitude.

The minimum particle momentum, corresponding to the cut-off energy, is given by $P_{min} = 14\cdot85 \cos^4 \lambda$, where λ is the magnetic latitude and the unit of momentum is GeV c^{-1}, where c is the velocity of light.

No particle below this limit can reach the earth at a given latitude λ and the maximum value of P_{min} is 14·85 at the equator and about 0·9 at $\lambda = 60°$. It is probable that some of the low-energy components in the primary radiations are trapped in the earth's field at very high altitudes giving rise to the Van Allen radiation belts discovered in the American satellite experiments in 1958. These are toroid-shaped regions containing circulating particles of low energy but high intensity. The axis of these belts coincides with the geomagnetic axis. See Fig. 25.4.

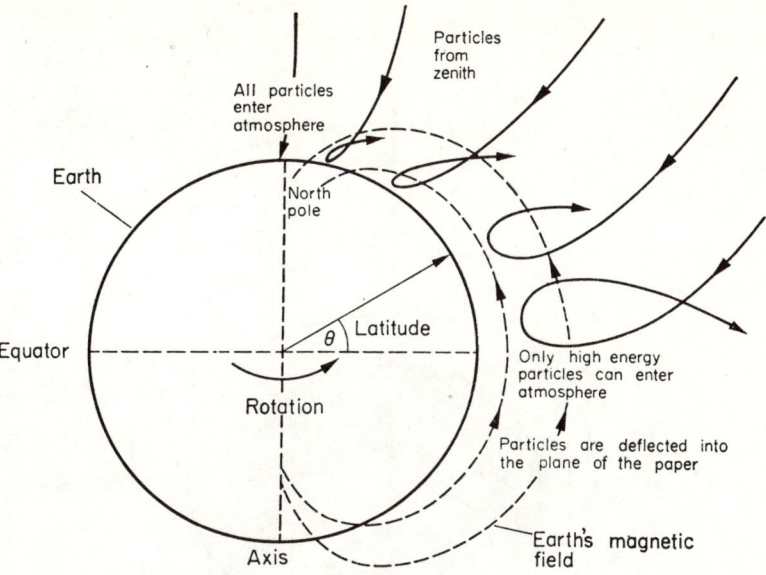

Fig. 25.3 Deflection of cosmic ray particles approaching from zenith showing action of earth's magnetic field.

Since the main geomagnetic field is directed from south to north over the surface of the earth, and assuming the primary particles are positively charged, the moving cosmic ray nuclei are deflected towards the east in accordance with the left hand motor rule. This gives an east–west effect in which the observed intensity of cosmic rays incident from the west is about 20% greater than that incident from the east. Thus slow cosmic ray particles come in more readily from the west than from the east. This asymmetry has been fully demonstrated experimentally, thus supporting the view that primary cosmic rays are positively charged and consist largely of protons.

25.5 Cosmic Rays at Sea-Level

Secondary cosmic rays as measured at sea-level have a different distribution of particles from the primary rays. Very few primary ray protons reach sea-level, where the penetrating or hard cosmic rays consist mainly of charged muons. We shall deal with the properties of these new subnuclear particles in the next chapter. It is sufficient to say here that upper atmospheric cosmic rays contain largely the so-called π-mesons or pions (mass $273m_e$) and μ-mesons or muons (mass $207m_e$) of both signs. There are further secondaries, positrons, electrons and photons, occurring in showers of innumerable particles. These make up the soft component, being absorbed by 100–200 mm of lead. At sea-level muons and electrons of both signs predominate.

When π^0-mesons, created by fast proton collisions with oxygen, nitrogen and other nuclei in the atmosphere (see Fig. 25.1), decay into γ-rays of over 100 MeV energy, the latter produce electron–positron pairs of almost the same energy.

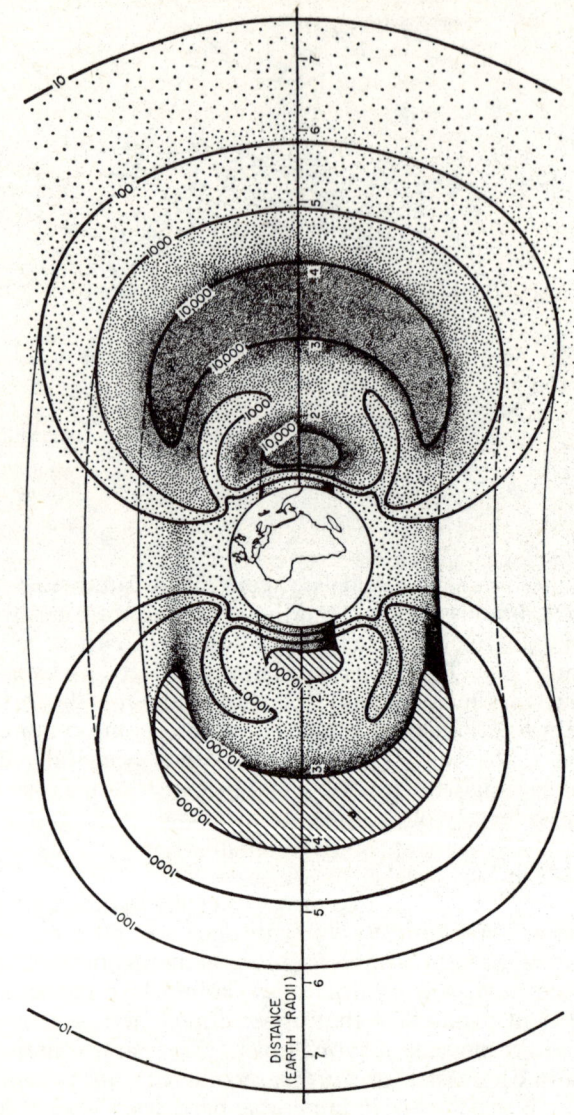

Fig. 25.4 Structure of radiation belts revealed by contours of radiation intensity (black lines) is shown schematically by shading (left); dots (right) suggest distribution of particles in the two belts. Contour numbers give counts per second; horizontal scale shows distance in Earth radii (about 6400 km) from the centre of the Earth. Particles in the inner belt may originate with the radioactive decay of neutrons liberated in the upper atmosphere by cosmic rays; those in the outer belt probably originate in the sun. (From *Scientific American*, J. A. Van Allen, March, 1959, **200**, no. 3, p. 39.)

These then generate new and very energetic photons by Bremsstrahlung, production having a continuous spectrum with a maximum energy given by
$$h\nu_{max} = E - m_e c^2,$$
where E is the energy of the incident electron of mass m_e. The new Bremsstrahlung γ-rays create further electron–positron pairs and they in turn produce Bremsstrahlung and so the process continues until the whole of the initial π^0-decay energy is dissipated. This multiplication process is called an electron–photon 'shower'. According to this theory the number of positrons and electrons in cosmic rays should increase as the earth is approached. This is actually true to within about 15 km of the earth's surface, below which height the intensity decreases again, as originally found by Hess.

These electron–photon cascade shower lengths are short enough in metals to be observed experimentally. In the air the electron shower path length is about 30 km and in lead about 5 mm, so that they can easily be observed in a cloud chamber as shown in Fig. 25.5.

25.6 Extensive Air Showers

In addition to the narrow electron–photon showers just described, there are extensive air showers containing hundreds of millions of particles reaching the earth together and covering many thousands of square metres. These large air showers are due to the ease with which the low energy electrons and photons are deviated from the main path of the shower by multiple collisions with atmospheric nuclei. Since the total energy of a shower should be about equal to the energy of the primary particle (proton) causing it, we can get some idea of the energy of the latter by measuring the total energy of the shower particles. By this means a figure of 10^{20} eV for the maximum energy of the cosmic ray primary component is obtained.

If a 10^{20} eV particle collides with an air particle one can imagine the next generation of particles having sufficient energy to give many further energetic collisions. Many mesons and nucleons are so produced, giving rise to a penetrating shower which we could call a nucleonic cascade in contrast to the electronic cascade described in the previous section. The main components of these nucleonic showers are π-mesons and nucleons which are the penetrating component at sea-level. Most of the components of the nucleonic showers are radioactive.

25.7 The Detection of Cosmic Ray Particles

Most methods described in Chapter 17 can be used in the detection of the charged particles contained in cosmic rays. The oldest method is the Wilson cloud chamber in which much of the early research was done and in which physicists were able to recognize the tracks of α-rays, β^--rays, protons, etc., very readily. Cloud chambers were used extensively until 1947 when the nuclear emulsion method was developed as a complementary technique. Nuclear emulsions are still used on a large scale where cheapness is an important item in a research budget. The bubble chamber cannot be used for cosmic rays because the lifetimes of the events are too short, and the cosmic rays arrive at random. This method of detection is ideal when used in conjunction with the pulsed beams from, say, the Bevatron, which are also of this order of duration. The bubble

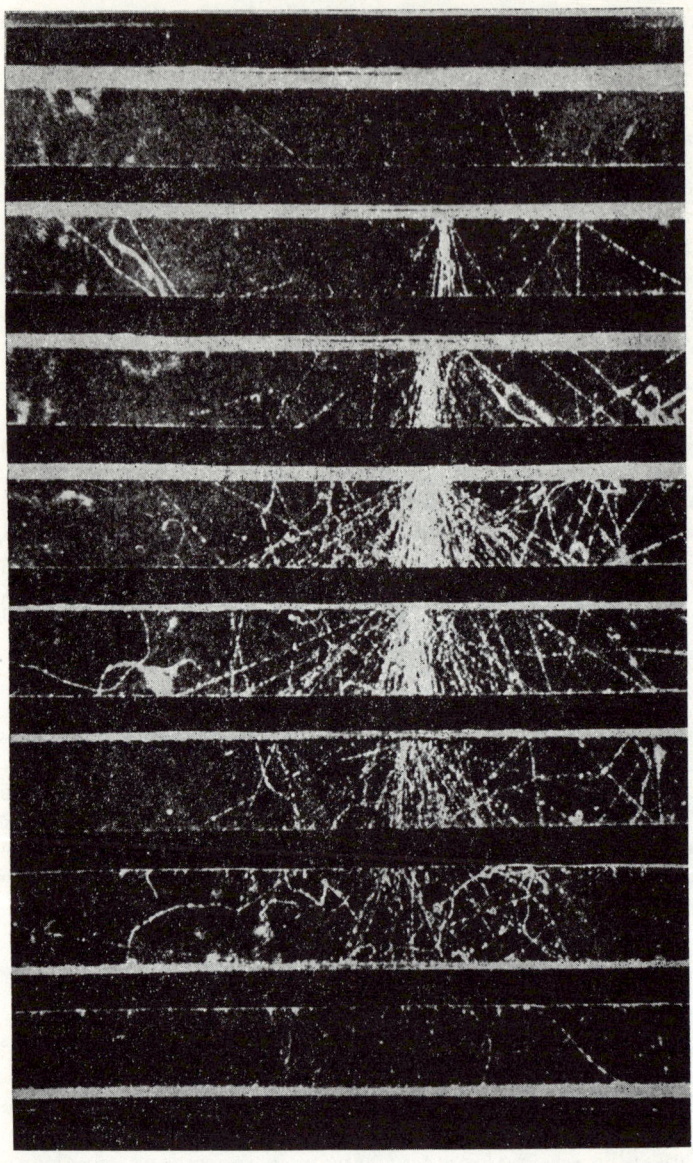

Fig. 25.5 Photon–electron cascade shower passing through lead plates. Cloud chamber photograph. (From Rochester and Wilson, *Cloud Chamber Photographs of Cosmic Radiation*, Pergamon, 1952.)

chamber can therefore be used to investigate artificially produced strange particles rather than those produced in cosmic ray bursts.

The direction of a cosmic ray burst can be determined with the Geiger-counter 'telescope'. Three or more Geiger tubes are arranged parallel to each other like the rungs of a ladder so that when a particle passes down the 'ladder' it discharges the whole set of counters simultaneously. When such a coincidence takes place the electronic amplifiers record a 'count'. Particles incident obliquely to the ladder cannot trigger-off all the tubes and no count is recorded. Thus a direction can be selected and the cosmic ray angular intensity determined by scanning. Unfortunately the identification of individual particles is impossible with this arrangement.

In the case of the nuclear emulsion plate each individual particle leaves a characteristic track which can be identified by the skilled worker. Features which are used for identification are track length, grain density and track 'wobble', and plates are now put together in stacks so that details of the whole event can be followed. A careful measurement of track characteristics gives an estimate of the mass of the particle, but the sign is not so easy to find as in a cloud chamber, which can easily be operated in a deflecting magnetic field.

The use of counters arranged in coincidence, anticoincidence and in delayed coincidence, together with the use of counter-controlled expansion chambers and various emulsion techniques, forms the basis of nearly all cosmic ray measurements of direction, and intensity. This is particularly true for atmospheric and sea-level investigations of very energetic multiparticle events. It is only at great heights, where the unwanted background is low, that the single counter can be used successfully. All events recorded on photographic plates in cosmic ray work or in high-energy particle collision physics can now be computer scanned and analysed.

25.8 The Future of Cosmic Ray Research

As we have seen, a few cosmic rays can be detected with energies somewhat greater than 10^{20} eV. These high energies must be due to particle accelerations taking place within or beyond our own galaxy. Primordial source points must therefore be galactic or extra-galactic. During the acceleration process the particles must travel many interstellar distances and pass through many magnetic fields before reaching the earth. Details of this acceleration process are not known, nor do we know for certain the origins of the source points of cosmic rays. Current theories are that pulsars and supernovae explosions are likely sources of primary cosmic rays, at least up to 10^{17} eV.

Modern research (1978) is broadly in two directions: (*i*) to find a theory and account for the origin of cosmic rays and to evaluate them experimentally; and (*ii*) to use the high-energy particles in the cosmic ray spectrum to investigate nuclear collision products in the search for new particles and the structure of nucleons.

The Origin of Cosmic Rays

In considering the origin of cosmic rays, it has been convenient to divide the spectrum into three bands, the low-energy band of 10^9–10^{10} eV, the intermediate band at about 10^{12}–10^{17} eV and the high-energy band above 10^{18} eV. Most work has been done on the low-energy band where the new subject of γ-ray astronomy has indicated a galactic origin for these cosmic rays. Next, the 10^{12}–10^{17} eV band has been investigated in terms of anisotropies in the intensities and directions of these cosmic rays. The predicted anisotropies are small, but on balance point to a galactic origin for this band. For energies $> 10^{18}$ eV there are difficulties in containing the particles within the galaxy because of their high energies. An extragalactic 'universal' model has therefore been proposed for these high-energy cosmic rays. On this model there should be a sharp energy cut-off in the spectrum at 6×10^{19} eV due to the attenuation of the primary protons by the 2·7 K black body radiation field, a relic of the big bang hypothesis of cosmological theory. This cut-off is not observed experimentally. In fact, several particles with energies *above* 10^{20} eV have been recorded and the shape of the energy spectrum found by plotting the number of particles with a given energy against that energy above 10^{19} eV is quite different from that expected, showing a tendency to flatten out rather than to drop to zero. Furthermore, many particles are observed arriving from directions nearly perpendicular to the galactic plane indicating that some of the highest energy cosmic rays have an extragalactic origin.

The proponents of *galactic* origin above 10^{18} eV suggest that the particles are mainly heavy nuclei, such as iron, for which the galactic trapping mechanism is stronger than for lighter particles. Such measurements on mass as have been made indicate that the particles are mainly protons and the situation is at present unresolved. Many workers favour a compromise with most of the particles above 10^{18} eV or 10^{19} eV coming from 'local' extragalactic sources such as a local explosion galaxy.

Experimentally the problem is very difficult because at these high energies the cosmic ray flux is very small; for example, above 10^{19} eV the flux is only 1 particle per square *kilometre* per *year*! At such a low flux the cosmic ray components are difficult to resolve so that the presence of iron and other heavy components is not easy to detect. For this reason the cosmic ray detectors used in this work cover very large areas. At Haverah Park, near Harrogate, England, the U.K. array is 12 km^2 in area.

The origin of the low-energy (10^9–10^{10} eV) band can be inferred from a study of cosmic ray γ-photons, and it is here that most progress is being made. The γ-rays arise from the interaction of primary particles with interstellar matter. Primary electrons will produce *Bremsstrahlung* radiation while protons will give γ-rays from neutral pions as shown in Fig. 25.6. These γ-rays travel in straight lines and it is found experimentally that there is a concentration of γ-rays in the plane of the

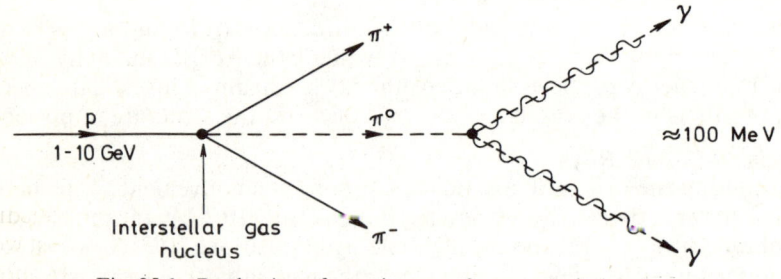

Fig. 25.6 Production of cosmic γ-rays from neutral pion decay.

galaxy to such an extent as to support a galactic rather than a 'universal' model. Some γ-rays also come from the pulsars of the Crab and Vela supernovae remnants. This is known because the γ-ray intensity pulses at the same rate as the radio pulses and several of these γ-ray pulsar sources have now been detected. The very fact that some γ-rays have been coming from pulsars and that γ-rays are themselves part of the cosmic ray flux means that at least we know that some cosmic rays have their origin in pulsars. Whether or not protons are accelerated in these sources as well is not known with certainty, but it seems likely that some, and perhaps most, are.

The mid-energy band of $10^{12} - 10^{17}$ eV is also thought to be of galactic origin although the measurements, based on cosmic ray anisotropies (i.e. the dependence of intensity on direction in space), are not so conclusive as the γ-ray evidence for the low-energy band.

Returning to the high-energy end of the spectrum, an alternative suggestion to explain the flattening of the spectrum above 10^{19} eV has been made. This involved the escape of neutrons from clusters of galaxies. Very energetic nuclei are probably produced in certain galaxies and these interact with the gas and the photons in the clusters. The charged fragments are then trapped by the intergalactic magnetic fields but the neutrons escape and produce fast knock-on protons by striking gas nuclei or decay in flight into protons. Calculations show that at 10^{19} eV a neutron would have a mean free path of the same order as cluster dimensions and thus some neutrons could escape above 10^{19} eV and produce the relatively high flux at this energy.

All modern cosmic ray theories are speculative but experimental data are being collected at such a rate through large-scale laboratory measurements and satellite observations that these theories can be tested much more rapidly than before.

A crucial factor in interpreting cosmic ray data is the mass composition of the rays in the various energy bands, and as the mass spectrum becomes more exactly known the problem of the 6×10^{19} eV cut-off will ultimately be solved — and with it the origin of these cosmic rays.

High-Energy Particle Collisions

We now turn to the second line of research using cosmic rays as high energy bombarding particles. Although cosmic rays have a low flux their high energies compensate for this in single particle collision investigations.

In the never-ending search for the detailed structure of the nucleus and of the nucleon it is essential that the probe particles have energies as high as possible. Accelerating machines can now give energies up to 1000 GeV (10^{12} eV) with a very high flux compared with that of the cosmic rays. Thus nuclear plates used in cosmic ray research require relatively longer exposure times, but often lead to events not produced anywhere else. Many of the sub-nuclear particles were discovered in cosmic rays, starting with the discovery of the positron, the muon and the pion. However, much of the current work on fundamental particles is largely confined to accelerators. The role of cosmic rays is now to be found in examining gross features of interactions at energies well above those available from accelerators and the search for exotic particles such as free quarks (Chapter

28) still continues in cosmic rays. These particles are playing a large part in our search for the details of nucleon structure and form the subject of the next three chapters.

Problems

(*The problem marked with an asterisk is solved in full at the end of the section.*)

25.1   Compile a list of relative abundances of elements from $Z=1$ to $Z=60$ as found in the universe and in cosmic rays (see *Radioactivity and Nuclear Physics* by J. M. Cork, D. Van Nostrand Co., p. 300). What is the importance of this comparison?

25.2*   From the information given in Table 25.1 calculate the intensity of the various primary particles in $\mu W\ m^{-2}\ sr^{-1}$ at 10 GeV per nucleon. How would you expect this to vary with latitude? (Protons, 0·46 $\mu W\ m^{-2}\ sr^{-1}$)

25.3   Discuss the fact that the primary cosmic rays do not contain appreciable numbers of electrons, positrons or photons.

25.4   Write an essay on the origin of cosmic rays.

25.5   What is '*Bremsstrahlung*' radiation and how does it differ from γ-radiation? Describe the production and role of *Bremsstrahlung* radiation in cosmic ray showers.

25.6   Estimate the dose rate at sea level at 45°N from cosmic ray secondary particles.

25.7   In a cloud chamber using a field of 2 T the radius of an electron track in a cosmic ray shower is 20 m. What is the energy of the electron?

25.8   Calculate the meson flux density at sea level if the ionization chamber used is a cylinder of 400 mm diameter and 400 mm long filled with air at 4 atm pressure, and gives a current of 56 fA.

1 meson gives 7 ion pairs per millimetre path length in air at normal pressure. Assume the mesons enter the flat end only of the ionization chamber.

Solution to Problem

25.2   Consider protons only.

From Table 25.1 the flux is given by $4000/E^{8/7}$ particles $m^{-2}\ s^{-1}\ sr^{-1}$, where E is in GeV. Therefore

$$\text{Flux} = \frac{4000}{10^{8/7}}$$
$$= 290 \text{ particles}$$

each with 10 GeV energy. Thus

$$\text{Energy intensity} = 290 \times 10^4 \times 1{\cdot}6 \times 10^{-13} \text{ J m}^{-2}\ s^{-1}\ sr^{-1}$$
$$= 0{\cdot}46\ \mu W\ m^{-2}\ sr^{-1}$$

Chapter 26

Stable and Semi-Stable Particles

26.1 Introduction

In the previous chapter we saw that primary cosmic rays consist largely of protons having energies between 1 GeV and 10^{11} GeV. When such fast protons encounter the nuclei of atoms in the atmosphere, high-energy nucleon–nucleon collisions take place which cannot easily be reproduced in the laboratory. It is not surprising, then, that many new particles were discovered in cosmic ray events, particles with very strange properties compared with the early elementary particles known to physicists. The first new particle to be found was the positron, soon to be followed by the μ-meson, which has a mass lying between that of the electron and the proton. Many other mesons have since been discovered and now we know that when a fast cosmic ray proton strikes a nucleus it reacts strongly with the nucleons and in the ensuing rearrangement a shower of many mesons can be ejected. Figure 26.1 shows this type of event in a nuclear emulsion plate.

The discovery of the first meson as an elementary particle was made in 1937 by Neddermeyer, Anderson, Street and Stevenson in Wilson cloud chamber cosmic ray observations. This particle was then called a mesotron and could be either positive or negative. We now call this particle the μ-meson or muon. Its mass was estimated to be about three hundred times the mass of the electron. The existence of a particle with these peculiar characteristics had actually been predicted by Yukawa in 1935. We have already seen that the strong short-range n–n, n–p and p–p forces of attraction in the nucleus are the binding forces which give the nucleus its stability. In his study of the nuclear force-field Yukawa predicted quantum-mechanically the existence of a system of nucleon exchange forces rather like the electron exchange forces in the H_2^+ ion, where the two protons are held together by the continuous exchange of the single electron between them. Yukawa put forward a theory of nuclear attraction forces which required the existence of a particle, with either positive or negative charge, and of mass equivalent to two or three hundred electronic masses, in order to give the correct distance over which the short range forces act, viz. about 1 fm. He also suggested that this particle might help to explain some of the difficulties in β^--decay theory by suggesting that when a neutron in a nucleus changes to a proton it first emits a negative meson, which immediately decays to an electron and a neutrino. These mesons of Yukawa are the quanta of the nuclear force-field and it was natural to identify the 1937 experimental meson with Yukawa's nuclear photon. This was

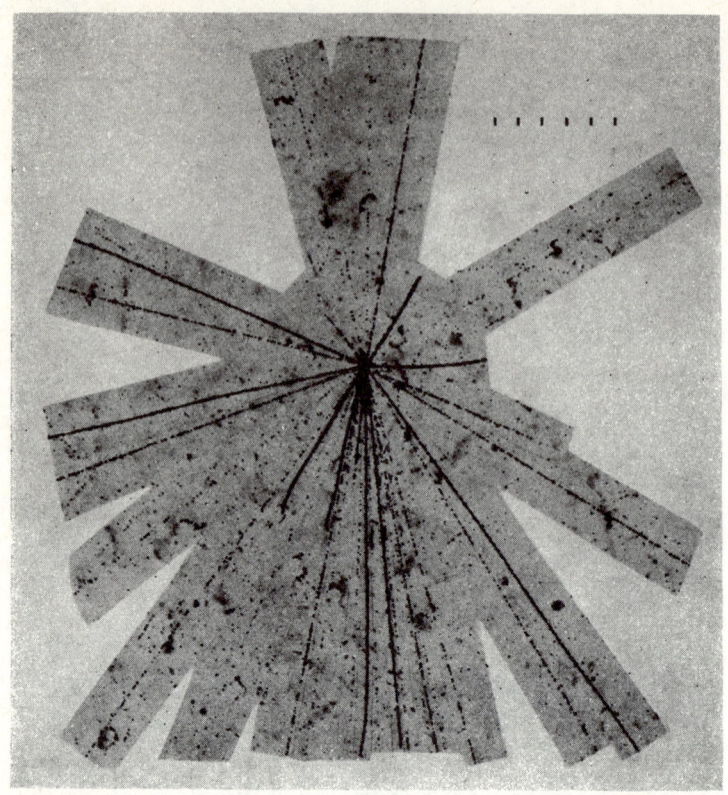

Fig. 26.1 Disintegration of an emulsion nucleus by a high-energy proton. The proton enters the plate top centre and produces a star by collision with a silver or bromine nucleus. The shower particles are emitted largely in the forward direction and several shorter range particles are also visible. (From Powell *et al.*, *The Study of Elementary Particles by the Photographic Method*, Pergamon, 1959.)

then interpreted as the strong force-carrier within the nucleus, shuttling backwards and forwards among the nucleons and so binding them together.

One of the properties of the negative meson, as predicted by the theory, was that when stopped by ordinary solid materials it should be absorbed very rapidly, showing strong nuclear interaction, but in time it was found that μ-mesons have only a weak interaction with nuclei, a result incompatible with Yukawa's requirements. Experimentally it was found that fast μ-mesons could pass through thick plates of lead without being absorbed, showing their weak interactions with nuclei of solid substances. Hence the possibility of identifying the experimental μ-meson with Yukawa's particle was open to doubt, and when in 1947 the first nuclear emulsion plates exposed to cosmic rays at high altitudes were examined by C. F. Powell and his team at Bristol, and the existence of *another* meson of mass about 300 m_e was suspected, this doubt was strengthened, since the Yukawa theory did not postulate two different mesons. In the early nuclear emulsion plates some of the mesons were found to decay at the end of their range. These were later identified as positive π-mesons which decay when stopped in the emulsion to secondary mesons. The new π-mesons were also shown to be produced in disintegration processes and it was found that they reacted very strongly with the nucleons of the emulsion nuclei. Figure 26.2 shows one of the first photographs of the $\pi \to \mu \to \beta^-$-decay scheme in which the characteristics of meson tracks are clearly seen. The successful identification of such new particles depends on a careful examination of their track characteristics with respect to their difference from those of the known elementary particles.

For example electron tracks are very irregular owing to the strong Coulomb scattering, and these can readily be recognized at the end of μ^+-meson tracks. Meson tracks are never quite linear, showing less sharp deviations than electron tracks, and have a grain density which increases towards the end of the track. Negative mesons are often identified by the nuclear disintegrations at the end of their range. Thus the characteristics which are examined are track length or range, sudden termination of the track in the emulsion, grain density, track 'wobble' or scattering. Plates are now put together in piles or stacks so that a whole event can be traced in three dimensions from start to finish, showing all the constituent particles. Figures 26.3 and 26.4 are photographs showing the tracks of electrons and protons as examples of nuclear particle tracks as used in the analysis of collision events.

26.2 The Positron: Particles and Antiparticles

Until 1932 the only elementary particles known in physics were the proton, the electron and the photon. In that year two new elementary particles were discovered, viz. the neutron and the positron, which have already been discussed.

The existence of a positron had been predicted by Dirac in 1930 in his relativistic theory of the free electron. Dirac set up the relativistic wave equation for the electron and showed that solutions were possible for all values of the total energy E_t whenever

$$E_t \geqslant +mc^2 \quad \text{or} \quad E_t \leqslant -mc^2,$$

where m is the actual mass of the moving electron and c is the velocity of light. According to the Dirac theory there exists a set of mathematically possible positive energy states with energies greater than mc^2 and also a set of possible negative energy states with energies less than $-mc^2$. The full quantum-mechanical treatment of this problem is beyond the scope of this book, but the following simple argument may help the reader.

In Appendix A we show that the relativistic equation connecting the momentum and total energy of a moving particle is $E_t^2 = m_0^2 c^4 + p^2 c^2$, where m_0 is the rest mass of the moving particle and p is its momentum. Thus

$$E_t = \pm \sqrt{m_0^2 c^4 + p^2 c^2},$$

and for any given value of the momentum p the energy E_t can be positive or negative. In the case of the electron there are therefore positive energy states corresponding to observable energies, but the negative energy states have no simple physical meaning and can only be interpreted mathematically.

Since electrons in the positive states would make radiative transitions to the

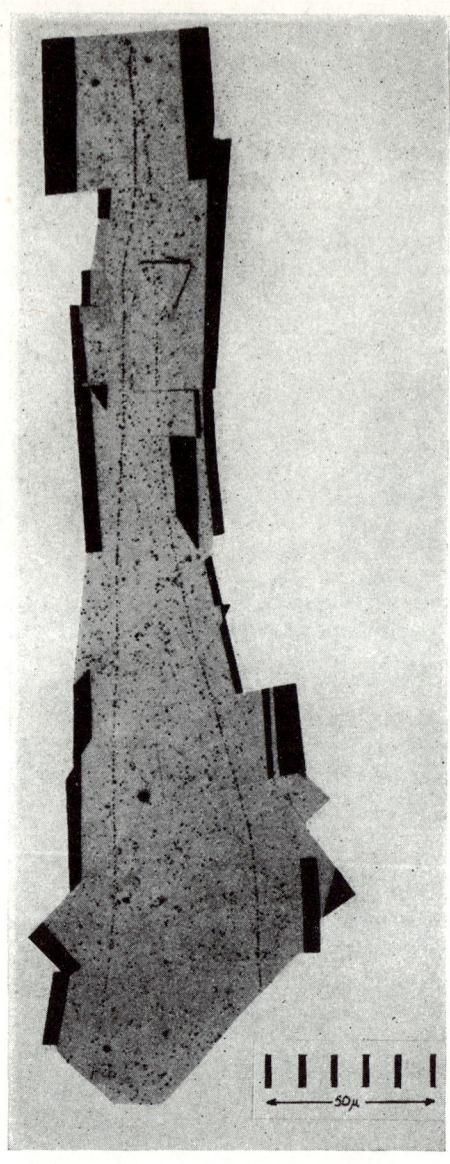

Fig. 26.2 First observation of π-meson decay in a nuclear emulsion. The pion enters the plate at the bottom left-hand corner and reaches the end of its range at the top. A secondary μ-meson is ejected nearly backwards along the line of approach of the pion. Note the increase of grain density of these particles at the end of their respective ranges. (From Powell, *ibid.*)

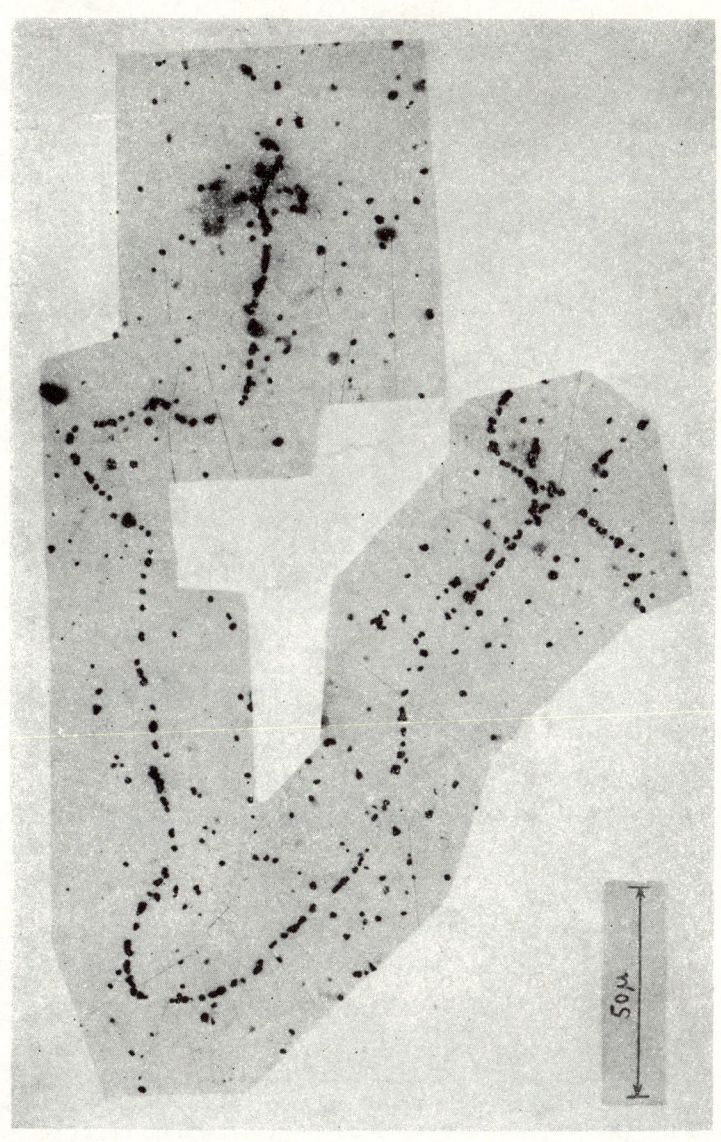

Fig. 26.3   Electron tracks in nuclear emulsion. (From Powell, *ibid*.)

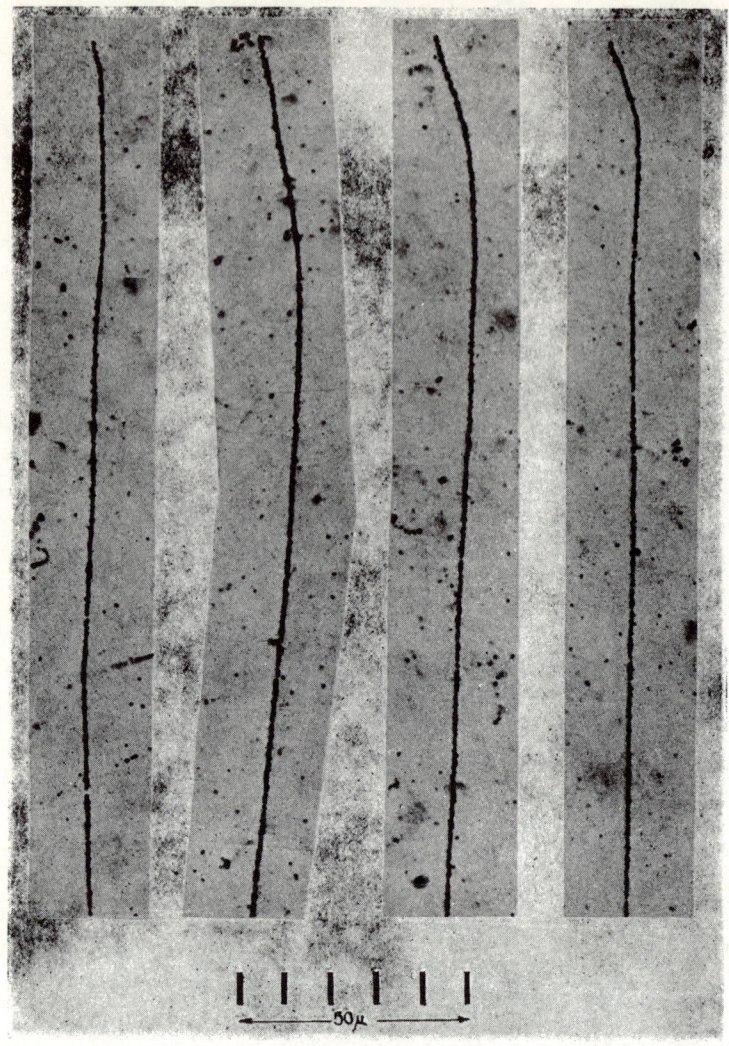

Fig. 26.4 Proton tracks in nuclear emulsion. (From Powell, *ibid*.)

negative states, and since this is not observed, Dirac proposed that all the negative energy states in a perfect vacuum were completely occupied by electrons, whereas all the positive states were normally empty. The negative states are therefore completely filled and are unobservable until a vacancy occurs in one of them by the removal of an electron to a positive, observable, energy state by the interaction of the electron with the electromagnetic field. This leaves a positively-charged hole or vacancy which is manifest as a particle with the same mass as its companion electron but with opposite charge and spin (conservation of angular momentum). The energy required for this upward transition will be $\geqslant 2m_0c^2$, since two particles are created, which means that a positron–electron pair cannot be created by bombarding particles or photons of less than the threshold energy of 1·02 MeV, see Fig. 26.5.

The above is a simplified description of the formation of a positron–electron pair. In free flight most positrons are slowed down to low energies before annihilation and many of these are then captured by electrons to give so-called positronium atoms. A positronium atom is a short-lived positron–electron pair. The spins may be antiparallel, giving a singlet state whose lifetime is about 0·1 ns, or parallel, giving a triplet state with a lifetime of 0·1 μs. The former state decays into two γ-photons, while the latter decays with the emission of three γ-photons.

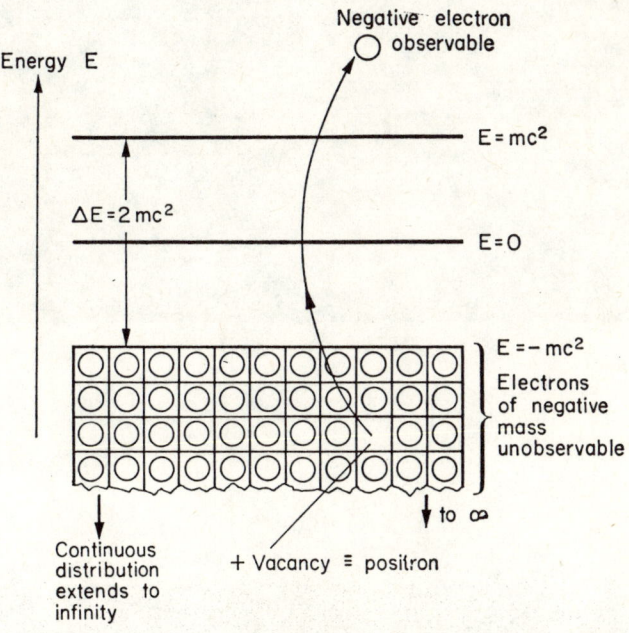

Fig. 26.5 Creation of a positron–electron pair, according to Dirac.

The relativistic equation for E_t holds for all free particles of spin $\tfrac{1}{2}h/2\pi$, so that all such particles have 'antiparticles'. When a particle and its antiparticle meet, great energy is created by their mutual annihilation. As we have seen, the positron–electron pair requires about 1 MeV for its creation so that we can expect the creation of a proton–antiproton pair to require at least 1836×1 MeV, i.e. an energy of nearly 2 GeV relative to the centre of the mass of the pair. Calculation shows that the initial kinetic energy of the bombarding proton in a proton–proton collision producing an antiproton by

$$p^+ + p^+ \to (p^+ + p^+) + (p^+ + \overline{p^+})$$

is $6m_pc^2 = 5\cdot 6$ GeV, when allowance is made for conservation laws and relativistic

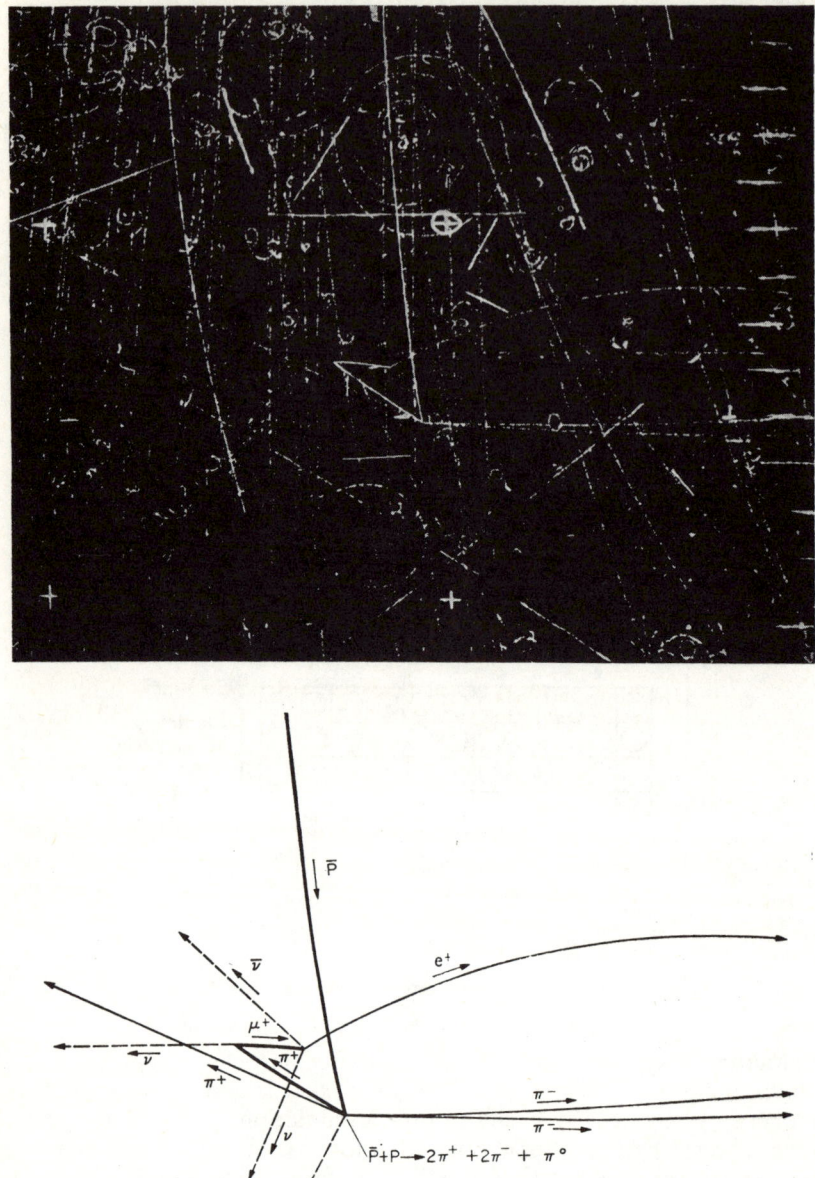

Fig. 26.6 Antiproton annihilation in hydrogen bubble chamber. (From Jay Orear, *Fundamental Physics*, Wiley, 1961, p. 348.)

effects at these high energies. Note that $\overline{p^+}$ is negatively charged.

Antiprotons were first produced at Berkeley, California in 1955 by using protons of kinetic energy about 6 GeV from the Bevatron on metal targets. The collision products were analysed magnetically and antiprotons were found. An American bubble chamber photograph of the annihilation of a proton by an antiproton is shown in Fig. 26.6 and a similar event in a nuclear emulsion is shown in Fig. 26.7. Various collision products are possible. Thus by bombarding a copper target with protons of 6·2 GeV energy from the Bevatron many $p^+:\overline{p^+}$ pairs have been observed.

By allowing a selected beam of such antiprotons to enter liquid hydrogen it was estimated that 0·3% of the collisions were due to $\overline{p^+}+p^+\to n^0+\overline{n^0}$. The antineutron annihilates with a neutron and the resulting energy flash was detected by a scintillation counter.

Positive evidence of the existence of an antineutron was obtained in 1958 when a beam of antiprotons was directed into a propane bubble chamber. There was no track visible at the point of the $(\overline{p^+}+p^+)$ event, but some way from it a carbon atom disintegrated with the formation of a four-pronged star. It was shown that the 'invisible track' was due to an antineutron which annihilated in the star.

Figure 26.8 shows a similar event in hydrogen.

The detection of the antineutron is obviously very difficult, since it can only be deduced from the energy count at the point at which it annihilated as indicated by the broad arrow in Fig. 26.8.

Particles and antiparticles have opposite charges but always have the same mass. They interact strongly in pairs where they create great energy from the annihilation of particle matter. The existence of the antiproton and the antineutron as well as the positron gives strong support to the Dirac theory of free particles.

26.3 Pions, Muons and Kaons

We have already mentioned ahat π-mesons (pions), which have mass about 273 m_e, are created when primary cosmic ray protons collide with atmospheric nuclei and cause energetic nuclear disintegrations. Pions are radioactive and have a very short lifetime $\overline{T}$. They can exist in all three states:

$$\pi^+, \text{ with } \overline{T}=25 \text{ ns}$$
$$\pi^-, \text{ with } \overline{T}=25 \text{ ns}$$
$$\pi^0, \text{ with } \overline{T}=0\cdot1 \text{ fs.}$$

These lifetimes, which refer only to mesons at rest, are so short that only a fraction of cosmic ray pions can reach sea-level. They are attenuated in the atmosphere because of the strong nuclear interaction. The charged pions decay to muons and neutrinos as follows:

$$\pi^+\to\mu^++\nu+33 \text{ MeV}$$
$$\pi^-\to\mu^-+\tilde{\nu}+33 \text{ MeV.}$$

The charged muons are also unstable, emitting electrons:

$$\mu^\pm\to e^\pm+2\nu+105 \text{ MeV.}$$

The radioactive decay of μ-mesons can be followed easily in the nuclear emulsion plate and several examples of the $\pi\to\mu\to e$ scheme are shown in Fig. 26.9.

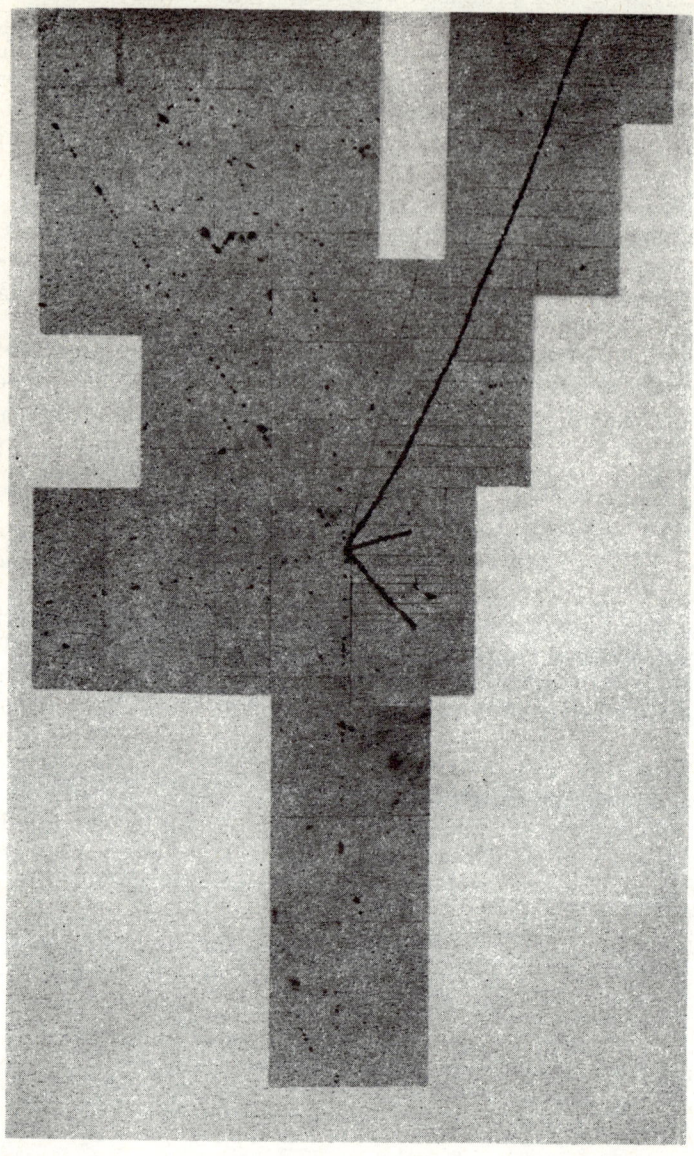

Fig. 26.7 Emulsion photograph showing annihilation of antiproton. The antiproton enters from the top left-hand corner and is annihilated at the end of its range. The annihilation energy is then distributed among the secondary charged particles of which four pions and two protons are shown. (From Powell, *ibid.*)

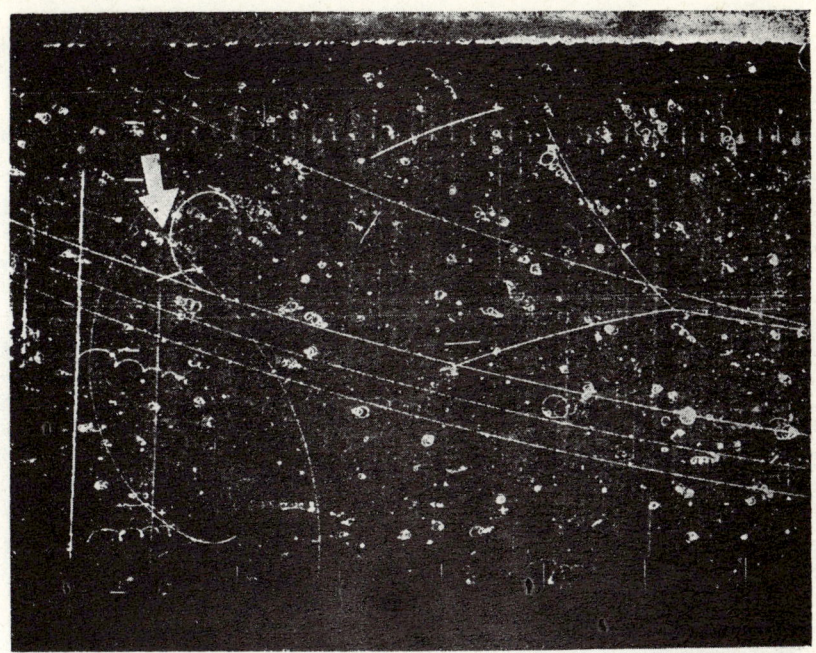

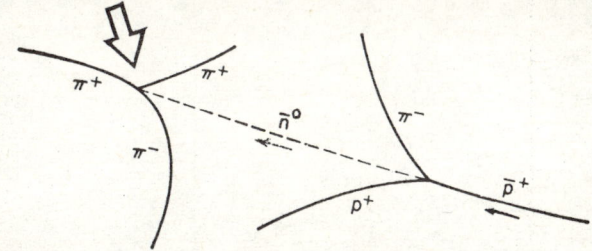

Fig. 26.8 Bubble-chamber photograph of the reaction $p^+ + p^+ \to n^0 + p^+ + \pi^-$. The antineutron then annihilates into three charged pions at the arrow. (From C. K. Hinrichs, B. J. Moyer, J. A. Poirier and P. M. Ogden, *Phys. Rev.* **127**, 617 (1962) by courtesy of the Lawrence Radiation Laboratory, Berkeley, California.)

The above equations are of great importance. Since the muon decays from rest, the conservation of linear momentum requires that two particles must be ejected. The only particle observed is the electron with its continuous energy spectrum. Since the charge is conserved, one expects the second particle to be a neutrino. However, since the spin of the muon is $\frac{1}{2}$ and the sum of the electron and neutrino spins can only be 0 or 1, but never $\frac{1}{2}$, the conservation of angular momentum requires another particle of opposite spin $\frac{1}{2}$ to be ejected at the same time. This must be an antineutrino, so that the muon decay equation should read

$$\mu^\pm \to e^\pm + \nu + \tilde{\nu} + 105 \text{ MeV}.$$

Spins $\qquad\quad \frac{1}{2} \to \frac{1}{2} \ + \ \underbrace{0}$

387

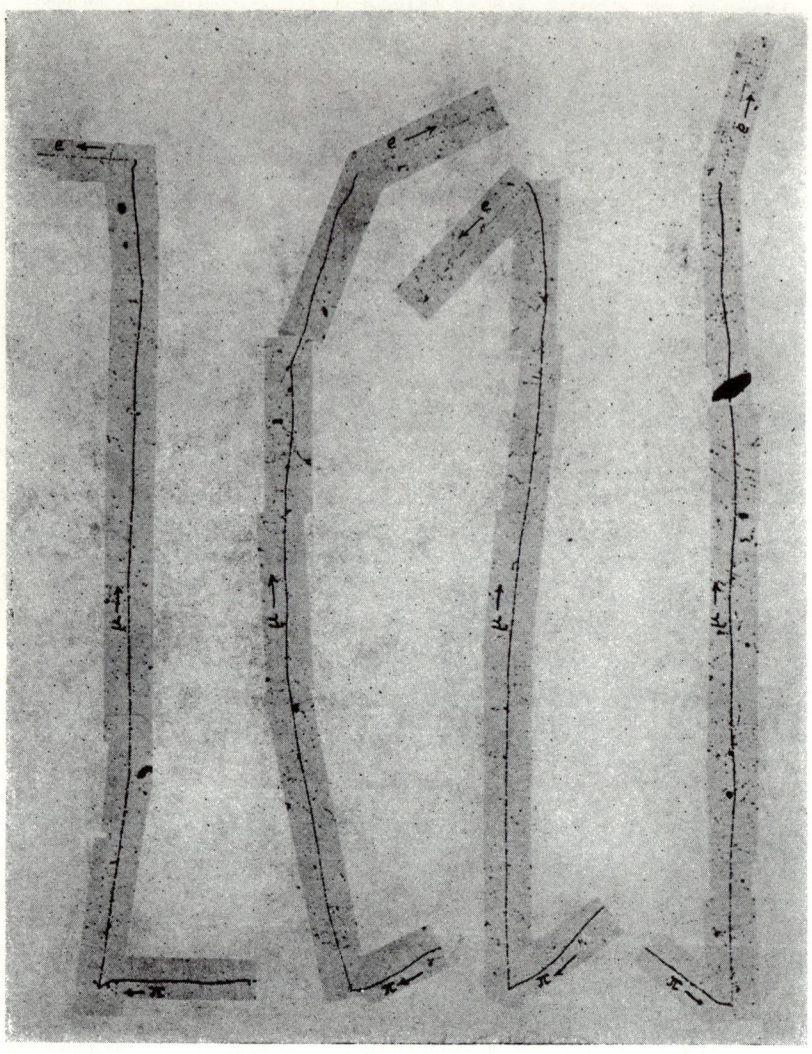

Fig. 26.9 Decay scheme $\pi \to \mu \to e$ in nuclear emulsion. (From Powell, *ibid.*)

This raises another difficulty. If the neutrino and the antineutrino are ejected simultaneously, one would expect that they would occasionally annihilate in flight to produce a γ-ray by $\nu + \bar{\nu} \to \gamma$, i.e. the apparent decay

$$\mu^{\pm} \to e^{\pm} + \gamma,$$

an event which would be recognizable by the γ-ray conversions. This has never positively been found; so we conclude that ν and $\tilde{\nu}$ are *not* antiparticles to each other. As they must be neutrinos, they can only be neutrinos of a different species.

Two types of neutrino are therefore postulated:

(a) that associated with electrons and positrons, which we have previously called simply neutrinos, v and $\tilde{v}$, but which we now write as v_e and $\tilde{v}_e$;

(b) that associated with muon decay, which we now write as v_μ and $\tilde{v}_\mu$.

The muon decay equation should now be written

$$\mu^- \to e^- + \tilde{v}_e + v_\mu + 105 \text{ MeV},$$
$$\mu^+ \to e^+ + v_e + \tilde{v}_\mu + 105 \text{ MeV}.$$

The annihilation γ-rays are not produced, since

$$v_e \neq v_\mu \quad \text{and} \quad \tilde{v}_e \neq \tilde{v}_\mu.$$

The decay equations we have previously discussed are therefore rewritten as:

$$\left. \begin{array}{l} n^0 \to p^+ + e^- + \tilde{v}_e \\ p^+ \to n^0 + e^+ + v_e \end{array} \right\} \begin{array}{l} \text{free decay} \\ \text{within nucleus} \end{array}$$
$$\pi^+ \to \mu^+ + v_\mu$$
$$\pi^- \to \mu^- + \tilde{v}_\mu$$
$$\mu^+ \to e^+ + v_e + \tilde{v}_\mu$$
$$\mu^- \to e^- + \tilde{v}_e + v_\mu.$$

The muon itself (i.e. μ^-, not the antimuon μ^+) is a very interesting particle. In all respects it behaves as a superheavy electron, and most electron decay schemes have a corresponding muon decay scheme. As we shall see later, negative muons can replace electrons in atomic Bohr orbits to form mesic or muonic atoms. Thus, apart from its extra mass, a muon behaves exactly like an electron. The reason for this still remains a mystery.

Neutral pions give energetic γ-rays by $\pi^0 \to \gamma + \gamma + 133$ MeV and this is the mechanism of the cosmic ray electron shower production by photon bombardment, which has been discussed in Chapter 25.

Negative pions react strongly with nuclei (e.g. C, N or O in the emulsion of the nuclear emulsion plate) to give characteristic star patterns; see Fig. 26.10. The basic reaction in the nucleus is $\pi^- + p^+ \to n + Q$, where the energy Q is very large and causes the star. It is this strong interaction with nucleons which provides the nuclear binding forces, whereas the interaction of muons with nucleons is so weak that they can find their way readily down to sea-level in a cosmic ray burst. There they are observed as the penetrating component of secondary cosmic rays. Their weak reaction with protons is by

$$\mu^- + p^+ \to n^0 + v_\mu.$$

All three types of pion can be produced artificially by high-energy protons or photons on metal targets. Muons are then available from the decay of the pions in flight.

Many other unstable particles have been observed in cosmic ray studies and subsequently confirmed in the accelerator experiments. Examples are the K-mesons (mass $\sim 970\, m_e$), and the hyperons which have masses greater than that of the nucleon. K-mesons (kaons) were first discovered in high-altitude cosmic ray experiments with emulsions but now are readily available from the accelerators. Kaons exist as K^+ and its antiparticle K^-, and also as K^0 and its antiparticle $\overline{K^0}$. Their masses are slightly different, as shown in Table 26.1. They are similar to

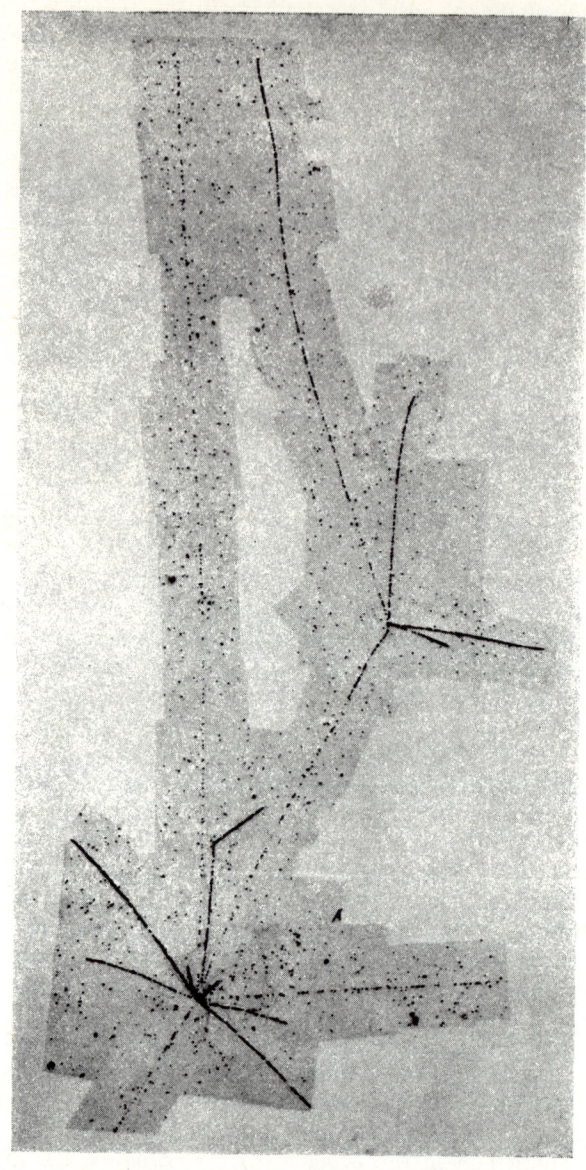

Fig. 26.10 Creation of a π-meson in a nuclear emulsion. The pion is created in the lower disintegration and proceeds to the upper where it reaches the end of its range and is captured by a light element causing a second disintegration. (From Powell, *ibid*.)

TABLE 26.1

Mass–Spin Spectrum of Stable and Semi-Stable Particles and their Antiparticles

Antiparticles are shown by $\bar{x}$ notation and $\overline{x^+} \equiv x^-$

	Rest mass m_e	Spin $\hbar$	Bosons	Fermions		Mean lifetime	Some possible decay schemes
				Leptons	Baryons		
	0	1	Photon γ				Stable
	0	$\tfrac{1}{2}$		Neutrino $\nu_e, \bar\nu_e$; $\nu_\mu, \bar\nu_\mu$			Stable
Leptons	1	$\tfrac{1}{2}$		Electron $e^-;e^+$			Stable
	207	$\tfrac{1}{2}$		Muon $\mu^-;\mu^+$		2·2 μs	$\mu^\pm \to e^\pm + \nu + \bar\nu + 105$ MeV
	264	0	Pion π^0			0·22 fs	$\pi^0 \to 2\gamma + 133$ MeV
Mesons	273	0	Pion $\pi^+;\pi^-$			25 ns	$\pi^\pm \to \mu^\pm + \nu + 33$ MeV
	966	0	Kaon $K^+;K^-$			10 ns	$K^+ \to \mu^+ + \nu$
	975	0	Kaon $K^0;\overline{K^0}$			10 ns	$K^0 \to \pi^+ + \pi^-$
Nucleons	1836	$\tfrac{1}{2}$			Proton $p^+;\bar p^+$		Stable
	1838	$\tfrac{1}{2}$			Neutron $n^0;\overline{n^0}$	1013 s	$n^0 \to p^+ + e^- + \bar\nu$
	2182	$\tfrac{1}{2}$			Lambda $\Lambda^0;\overline{\Lambda^0}$	0·27 ns	$\Lambda^0 \to p^+ + \pi^- + 37$ MeV
Hyperons	2340	$\tfrac{1}{2}$			Sigma $\begin{cases}\Sigma^-;\overline{\Sigma^-}\\ \Sigma^0;\overline{\Sigma^0}\\ \Sigma^+;\overline{\Sigma^+}\end{cases}$	0·16 ns; 0·1 ns; 80 ps	$\Sigma^- \to n^0 + \pi^- + 118$ MeV; $\Sigma^0 \to \Lambda^0 + \gamma + 80$ MeV; $\Sigma^+ \to p^+ + \pi^0 + 116$ MeV
	2582	$\tfrac{1}{2}$			Xi $\begin{cases}\Xi^-;\overline{\Xi^-}\\ \Xi^0;\overline{\Xi^0}\end{cases}$	0·1 ns; 0·1 ns	$\Xi^- \to \Lambda^0 + \pi^- + 66$ MeV; $\Xi^0 \to \Lambda^0 + \pi^0 + 70$ MeV

pions but have many more decay possibilities. Common modes of decay are

$$K^{\pm} \to \pi^{\pm} + \pi^0$$
$$K^+ \to \mu^+ + \nu_\mu$$
$$K^+ \to \pi^+ + \pi^+ + \pi^- \text{ with } \bar{T} = 12\cdot 4 \text{ ns.}$$

K^+- and K^--mesons have masses of about 966 m_e while K^0-mesons have masses of 975 m_e, and decay by

$$K^0 \to \pi^+ + \pi^-.$$

All kaons and pions have zero spin.

26.4 Hyperons

Hyperons are unstable particles, having masses greater than that of the nucleon, and the first were discovered in cosmic rays by Rochester and Butler in 1947. They have lifetimes of the order of 0·1 ns, and three groups are now known:

Λ^0 (lambda) particles, zero charge only;
Σ (sigma) particles, with +, 0, − charges;
Ξ (xi) particles, with 0, − charges only.

The Λ^0-particles were so named from the fork-like tracks produced by the secondary charged particles (Fig. 26.11). Some possible modes of decay are:

$$\Lambda^0 \to p^+ + \pi^- \qquad \Sigma^+ \to p^+ + \pi^0$$
$$\Lambda^0 \to n^0 + \pi^0 \qquad \Xi^- \to \Lambda^0 + \pi^-$$
$$\Sigma^- \to n^0 + \pi^- \qquad \Sigma^+ \to n^0 + \pi^+.$$
$$\Sigma^0 \to \Lambda^0 + \gamma$$

Hyperons are produced in the laboratory by pion or kaon interactions with protons. Energies greater than 1 GeV are required for these reactions. Typical reactions are:

$$\pi^- + p^+ \to \Lambda^0 + K^0,$$
$$\pi^- + p^+ \to \Sigma^- + K^+,$$
$$K^- + p^+ \to \Xi^0 + K^0,$$
$$K^- + p^+ \to \Sigma^+ + \pi^-,$$
$$K^- + p^+ \to \Xi^- + K^+.$$

All hyperons have spin $\tfrac{1}{2}\hbar = \tfrac{1}{2}h/2\pi$.

26.5 Classification of the Elementary Particles

Elementary particles are classified into groups according to their mass and spin properties. These are, referring to their masses:

(1) the *photon* with zero rest mass and spin 1. It is a massless boson.
(2) the *leptons* or light particles. These are the electrons, muons and neutrinos and their antiparticles, all with masses less than the pions and with spin $\tfrac{1}{2}$. For reasons connected with statistical mechanics they are also called *fermions*. Leptons interact weakly with other particles.
(3) the *mesons* or intermediate particles, so called because their masses are between those of the muons and the nucleons. They are the pions and the kaons and have zero or integral spin.
(4) the *baryons*. These are the heavy particles of nucleon mass and above. *Hyperons* have masses *greater* than the nucleons. The baryons are therefore the nucleons and the hyperons.

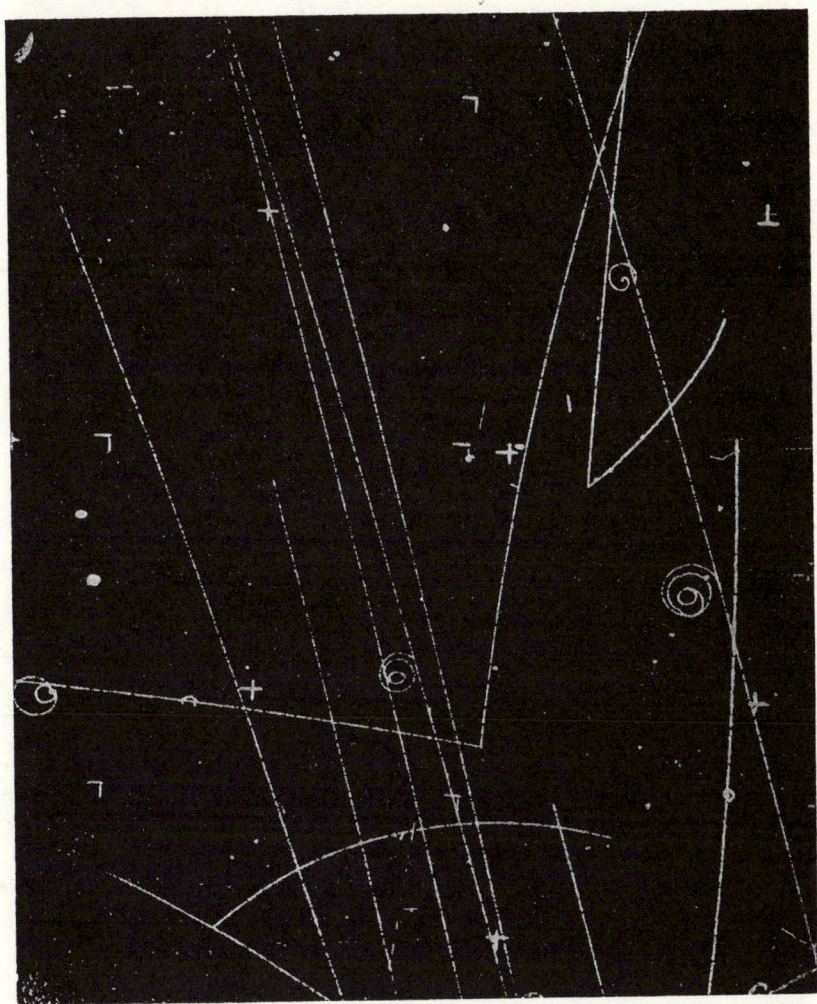

Fig. 26.11(a) The decay of two fundamental particles by a weak-interaction process is illustrated in the bubble chamber photograph above, made by Luis W. Alvarez and his colleagues at the University of California. The events in the photograph are traced in the drawing on the next page. A high-energy negative pion (π^-), produced by the Berkeley Bevatron, enters the chamber at lower right. It strikes a proton in the liquid hydrogen of the bubble chamber, giving rise to a neutral K-meson (K^0) and a lambda particle (Λ^0). Being uncharged, these two particles leave no track. The neutral K-meson decays into a negative pion and a positive pion; the lambda particle into a proton (p) and a negative pion. (Taken from *Scientific American*, March 1959, 'The Weak Interactions', by S. B. Treiman.)

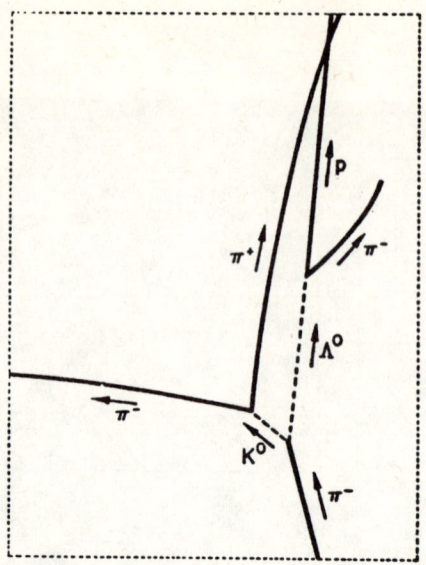

Fig. 26.11(b) Interpretation of Fig. 26.11(a).

All baryons have half-integral spins. Mesons and baryons are strongly reacting particles, and collectively they are called *hadrons*.

An important concept in all nuclear reactions is the conservation of spin angular momentum and, from a study of this applied to individual events, it is possible to assign a quantum number to each particle in terms of the unit $h/2\pi$ ($\hbar$). Baryons and leptons with half integral spins are called fermions while mesons with zero or integral spins are called *bosons*. Thus the muon (μ-meson) is really a lepton with spin $\frac{1}{2}$ and therefore a fermion, whereas the photon is a fundamental boson with spin 1. Based on these definitions it is possible to classify some 32 of these particles according to Table 26.1. Some particles are shown with their antiparticles which are distinguished by a bar over the symbol. This table is reproduced diagrammatically in Fig. 26.12, except that the muon neutrinos are omitted.

26.6 Mesic Atoms : The Muonium Atom

We have seen that a pion can be regarded as a nuclear photon for nuclear structure calculations. Similarly, kaons can be regarded as photons associated with shorter-range forces than pions. This leads to the concept of a mesic cloud in the nucleus analogous to the electron cloud of the atom. These mesic clouds are converted to real particles when the proton is struck by a particle of sufficiently high energy to sweep away the meson cloud and cause a rearrangement of the residuals. This often requires baryon or meson collisions since these particles interact strongly within the range of the kaon and pion forces whilst leptons have only weak interaction.

The interaction of negative muons with matter arises from their relatively long lifetimes (1 μs). During this time they are rapidly slowed down to rest and are able

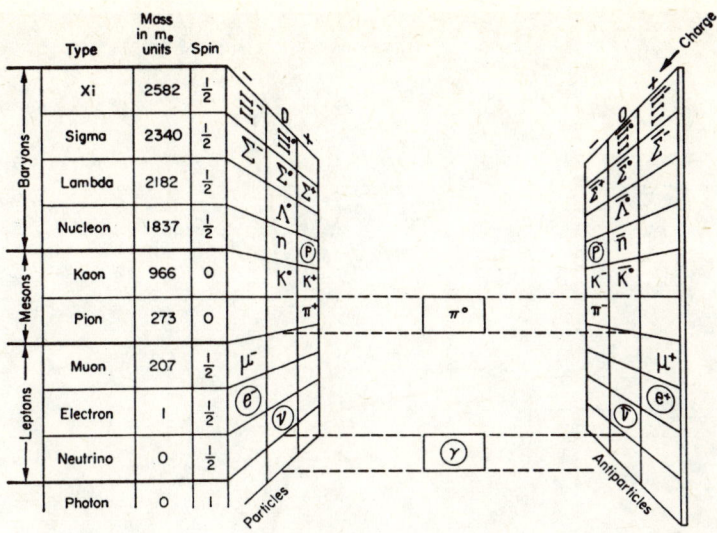

Fig. 26.12 Particles and their antiparticles. (From Orear, *Fundamental Physics*, Wiley, 1961.) Muon neutrinos not shown.

to replace orbital electrons to produce electrically neutral atoms with μ^- orbits. These are mesic atoms, and since $m_\mu > > m_e$ the mesic orbit has a very small radius compared with the electron orbit, as can be seen from the formula for the first Bohr radius

$$r = \frac{n^2 h^2 4\pi\varepsilon_0}{4\pi^2 m e^2}.$$

This mesic orbit is so near to the nucleus that in its lowest state it may spend as much as 50% of its time *within the nucleus*. This is therefore a penetrating orbit which can give information about the nuclear charge distribution. For example, it is possible to deduce the size of the nucleus from mesic transitions corresponding to electron transitions. The size is given by $R = 1\cdot 2 \times A^{1/3}$ fm, slightly less than the 'neutron' size.

The electric field around the nucleus of a mesic atom is then so contracted that it can more readily penetrate the field of a hydrogen atom and form a mesic molecule. This is depicted in Fig. 26.13 which is an American photograph from the hydrogen–deuteron bubble chamber showing the probable formation of a mesic hydrogen molecule HD. The μ^- is captured in a higher electron orbit in the HD molecule, which then becomes a molecular ion consisting of $(p + d + \mu^-)$ in which the μ^- is in orbit.

In this case the p and d particles are held closer together by the comparatively small meson orbit and they eventually overcome their Coulomb repulsion to form ^{3}He by

$$^1_1H + ^2_1H \rightarrow ^3_2He + \gamma + 5\cdot 49 \text{ MeV}.$$

This reaction can only be explained quantum mechanically since classically the

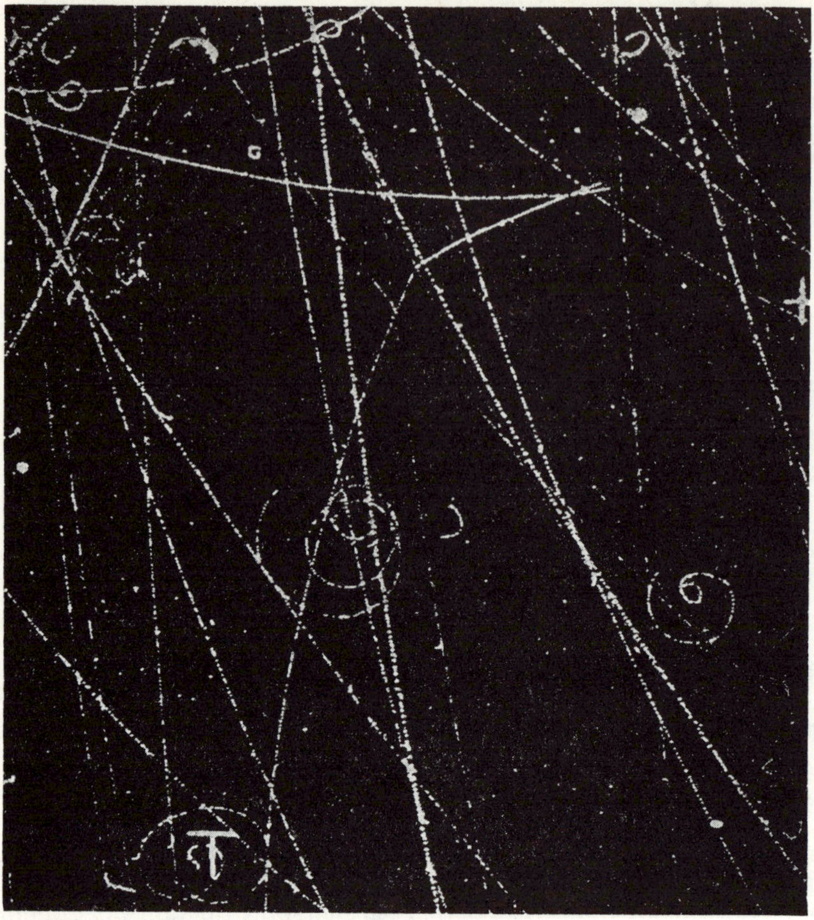

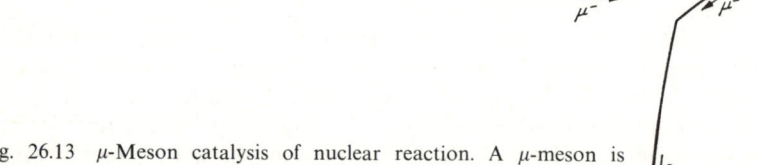

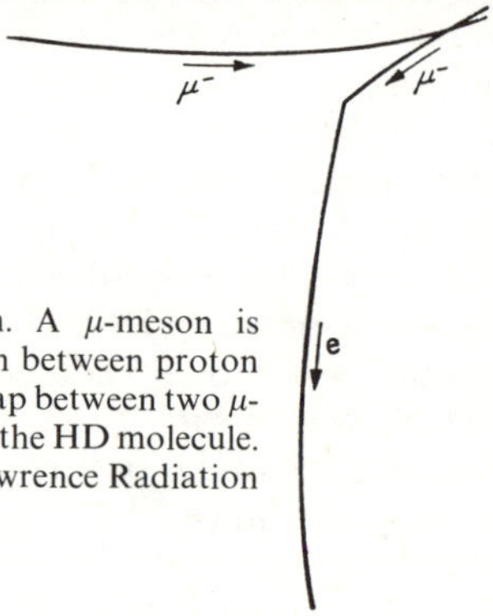

Fig. 26.13 μ-Meson catalysis of nuclear reaction. A μ-meson is absorbed in hydrogen and, in the subsequent reaction between proton and deuteron, carries off the bulk of the energy. Note gap between two μ-meson tracks while it is in a molecular mesonic orbit in the HD molecule. (By courtesy of Professor Luis. W. Alvarez and the Lawrence Radiation Laboratory.)

Coulomb potential barrier is too high for the reaction to proceed from rest particles. The whole reaction can be written

$$(p+d+\mu^-) \rightarrow {}_2^3\text{He} + \mu^- + \gamma + 5\cdot49 \text{ MeV},$$

in which the muon is unchanged and may take the whole of the 5·49 MeV energy. Its role is therefore that of a catalyst. Any particular muon could repeat this reaction to give a catalysed chain reaction. This has only a finite length partly due to the loss of energy to the γ-radiation, but largely due to the muon decay.

The above description refers to muons in orbit. What about the opposite picture, i.e., a muon acting as a nucleus to an electron? The muon would be μ^+, of course, and the electron would revolve around it. This 'atom', the muonium atom, has actually been found.

Muonium is a lighter atom than the hydrogen atom, and the two particles do not annihilate, since e^- and μ^+ are not antiparticles to each other. However, muonium is unstable with the same lifetime as the muon, viz. 2·2 μs. This is the lightest 'atom' we know (the positronium 'atom' has no central core) and it is thought that a study of muonium will lead to a further understanding of the muon–electron problem.

Problems

(*Those problems marked with an asterisk are solved in full at the end of the section.*)

26.1* In the muon decay $\mu^- \rightarrow \beta^- + \nu_\mu + \tilde{\nu}_e$ the electron is ejected with relativistic energy. Why are two neutrinos emitted? Calculate the maximum available energy for the process and the average electron energy. (105 MeV, 35 MeV)

26.2 When an antiproton of energy 72 MeV comes to rest in a nuclear emulsion it is annihilated by a proton to form two pairs of positive and negative pions and a neutral pion. See Fig. 26.6. Calculate the average kinetic energy of each pion assuming they all have the same energy. What are the main assumptions made in this type of calculation? (250 MeV)

26.3 In Problem 26.2, it can be assumed that all the pions decay without proton interaction with other nuclei. Use Table 26.1 to find how many electrons, positrons, neutrinos, antineutrinos and γ-photons may be expected. ($2\beta^-$, $2\beta^+$, 4ν, $2\tilde{\nu}$, 2γ)

26.4 What is the antiparticle to the photon? Why?

26.5 The observed lifetime of a pion is 75 ns as measured in the laboratory. Find the velocity, mass, kinetic energy and momentum of the moving pion. ($\beta = 0\cdot94c$, $m = 810\ m_e$, $E = 270$ MeV, $p = 270/c$ MeV c^{-1})

26.6* Calculate the energy of the neutron produced when a slow negative pion is captured by a proton. Should the neutron be treated relativistically? (Approx. 9·0 MeV)

26.7 Assuming that an atom can be converted to a mesic atom with a μ^--meson in the first Bohr circular orbit, $n=1$, find at what value Z the meson orbit will just penetrate the nucleus. ($Z=45$)

Taking this particular Z value, calculate the value of n for the meson orbit which is just inside the K-electron shell. ($n=15$)

26.8* If the pion decays from rest to give a muon of 4·05 MeV energy, what is the kinetic energy of the accompanying neutrino? What is the mass of the neutrino in this process? (30·1 MeV, 95 m_e)

Solutions to Problems

26.1 Taking $M_\mu = 207\ m_e$, the total available energy of the reaction is 206 m_e which is $206 \times 0\cdot 51$ MeV $= 105$ MeV.

If this energy is equally divided between the three particles the average energy of the electron is 35 MeV.

26.6 From the equation $\pi^- + p^+ \rightarrow n^0 + (hv) + Q$ and putting masses in MeV units, we have
$$M_\pi = 273 m_e = 273 \times 0\cdot 51 = 139 \text{ MeV}.$$
Therefore
$$139 + 938 = 939 + (hv) + Q,$$
$$Q = 138 \text{ MeV} = E_\gamma + E_n.$$

Case (a): For a non-relativistic neutron we have, from rest,
$$m_n v_n = \frac{E_\gamma}{c} \quad \text{(conservation of momentum)},$$
$$\frac{E_n}{E_\gamma} = \frac{1}{2} \frac{m_n v_n^2}{c m_n v_n} = \frac{1}{2} \frac{v_n}{c} = \frac{1}{2} \frac{E_\gamma}{m_n c^2} = \frac{E_\gamma}{1878},$$
$$\frac{E_n}{E_\gamma + E_n} = \frac{E_\gamma}{1878 + E_\gamma}.$$
Therefore
$$\frac{E_n}{138} = \frac{E_\gamma}{1878 + E_\gamma}$$
and
$$\frac{E_n}{138} = \frac{138 - E_n}{1878 + 138 - E_n} = \frac{138 - E_n}{2016 - E_n}.$$
Thus
$$2016 E_n - E_n^2 = 138^2 + 138 E_n,$$
$$E_n^2 - 2154 E_n + 138^2 = 0,$$
$$E_n = \frac{2154 \pm \sqrt{2154^2 - 4 \times 138^2}}{2} = \frac{2154 \pm 2136}{2}$$
and so
$$E_n = 9 \text{ MeV}.$$

Case (b): Relativistically, we have $138 = E_n + E_\gamma$, as before, but with
$$m_n v_n = \frac{m_{0n} v_n}{\sqrt{1-\beta^2}} = \frac{E_\gamma}{c} \qquad (1)$$
and
$$E_n = m_{0n} c^2 \left[\frac{1}{\sqrt{1-\beta^2}} - 1\right]. \qquad (2)$$

From (1),
$$m_{0n}\frac{v_n}{c} = \frac{E_\gamma}{c^2}\sqrt{1-\beta^2};$$

therefore
$$m_{0n}^2 \beta^2 c^4 = E_\gamma^2 (1-\beta^2),$$

so that
$$\beta^2 = \frac{E_\gamma^2}{m_{0n}^2 c^4 + E_\gamma^2}$$

and
$$1 - \beta^2 = \frac{m_{0n}^2 c^4}{m_{0n}^2 c^4 + E_\gamma^2},$$

thus
$$\frac{1}{\sqrt{1-\beta^2}} = \frac{\sqrt{m_{0n}^2 c^4 + E_\gamma^2}}{m_{0n} c^2}.$$

Substituting in (2) we get
$$\frac{E_n}{m_{0n} c^2} + 1 = \frac{\sqrt{m_{0n}^2 c^4 + E_\gamma^2}}{m_{0n} c^2}$$

or
$$E_n = \sqrt{m_{0n}^2 c^4 + E_\gamma^2} - m_{0n} c^2.$$

Therefore
$$(E_n + m_{0n} c^2)^2 = (m_{0n}^2 c^4 + E_\gamma^2),$$
$$E_n^2 + 2 E_n m_{0n} c^2 + m_{0n}^2 c^4 = m_{0n}^2 c^4 + E_\gamma^2,$$

and so
$$E_\gamma^2 - E_n^2 = 2 E_n m_{0n} c^2.$$

$$\frac{138(E_\gamma - E_n)}{E_n} = 2 m_{0n} c^2 = 1880,$$

$$\frac{E_\gamma}{E_n} - 1 = \frac{1880}{138} = 13 \cdot 6,$$

$$\frac{E_\gamma}{E_n} = 14 \cdot 6$$

or
$$\frac{E_n}{E_n + E_\gamma} = \frac{1}{15 \cdot 6}$$

so that
$$E_n = \frac{138}{15 \cdot 6} = 8 \cdot 8 \text{ MeV}.$$

The relativistic calculation gives almost the same result as the non-relativistic.

26.8   From $\pi^{\pm} \to \mu^{\pm} + \nu_{\mu} + Q$ we have
$$Q = M_{\pi} - M_{\mu}\\ = 67\, m_e\\ = 67 \times 0{\cdot}51 \text{ MeV}\\ = 34{\cdot}17 \text{ MeV}$$
and, since $E_{\mu} = 4{\cdot}05$ MeV,
$$E_{\nu} = 30{\cdot}12 \text{ MeV}.$$
The mass of this neutrino is
$$\frac{30{\cdot}12}{0{\cdot}51}\, m_e = 59\, m_e.$$

Chapter 27

Short-Lived Resonance States

27.1 Forces and Fields

In the previous chapter we discussed the formation and properties of some of the older elementary particles and frequently referred to strong and weak interactions. Physical problems are often solved by studying interactions between the components of a system. In the pre-neutron days the components of all atomic and nuclear systems were the electron, the proton and the photon and mutual interaction between these took place through the electromagnetic field. Bodies also attract each other through another fundamental interaction known as the gravitational force. Both these fundamental interactions obey the inverse square law but their magnitudes are widely different. If we consider two protons at a given distance apart a simple calculation shows that the Coulomb force is about 10^{36} as strong as the gravitational force. Thus in dealing with atoms and nuclei the gravitational force is negligible and only becomes significant on a planetary and stellar scale.

Since 1932, the number of fundamental particles has increased enormously, and the description of these new particles and their interactions was soon found to be inadequate in terms of the two fields just discussed. Since the diameter of a nucleus is measured in femtometres (10^{-15} m) while an atomic diameter is about 0·1 nanometre (10^{-10} m) the repulsive force between two nuclear protons will be 10^{10} times larger than the electrostatic force between a nuclear proton and an orbital electron in an atom. But nuclear protons do *not* repel each other (consider the stable nucleus $^3_2$He) and so we conclude that there must be an even stronger attractive force within the nucleus, *between protons*, which overcomes the strong Coulomb repulsive force. This force, which is associated with the production of mesons, is the third field of force and is involved in the so-called strong interactions. It occurs between nucleons and is a short-range force acting at distances appreciably less than a nuclear diameter. Theory shows that the strong interaction is about 137 times as great as the electromagnetic interaction within the nucleus. It is the interaction considered by Yukawa in his original theory of meson production.

The fourth and last type of force, known as the weak interaction, is also a nuclear force which governs the radioactive meson decay processes. It is involved in lepton changes and is only about 10^{-10} times the strength of the electromagnetic field.

Thus there are four basic force fields in physics, each of which has a 'source', such as charge for the electromagnetic field or mass for the gravitational field, and a field particle associated with the energy changes of the system. These are shown in Table 27.1, which includes a rough guide to the relative interaction strengths.

Just as the photon is the quantum of the electromagnetic field the meson is the quantum of the nuclear field. The 'graviton' and the 'intermediate boson' have not yet been found.

TABLE 27.1
Comparison of the Four Basic Interactions

Field	Relative magnitude	Associated particle	Characteristic time
Gravitational	10^{-39}	'Graviton'	'10^{16} s'
Electromagnetic	10^{-3}	Photon	10^{-20} s
Strong interaction	1	Meson	10^{-23} s
Weak interaction	10^{-13}	'Intermediate boson'	10^{-10} s

Associated with each of these fields is a characteristic time. The range of the strong interactions, 10^{-15} m or 1 fm corresponds to about 10^{-23} s, which is the minimum time for a signal to travel across a nucleus of diameter 3 fm.

This is the basic nuclear time for comparison purposes, so that an 'event' taking place in a shorter time interval than this has no meaning. The strength of the electromagnetic field is 10^{-3} of the strong field so that the associated time will be correspondingly greater, viz. $10^3 \times 10^{-23} = 10^{-20}$ s. Most electromagnetic interactions have lifetimes of the order of 10^{-15}–10^{-20} s, which corresponds roughly to the time taken for a photon to pass across an atom, i.e., $\frac{1}{3} \times 10^{-18}$ s.

Table 27.1 also shows that the strength of the weak interaction as 10^{-13} times that of the strong interaction, so that the corresponding weak interaction time will be $10^{13} \times 10^{-23} = 10^{-10}$ s. Most weak decay processes have a mean lifetime of 10^{-8}–10^{-10} s, which is very long compared with the time associated with strong interactions. The word 'stable' is used to describe all particles except the strong interaction particles, i.e. all particles immune to strong decay.

Physical phenomena are ultimately measured in terms of energy changes arising from four basic types of physical force. All atomic and nuclear interactions can be described in terms of electromagnetic, strong and weak interactions or forces. Strong interactions involve particles of high energy whereas lepton decay processes are the result of weak interactions. The electromagnetic interaction is proportional to the charges involved. The name 'hadron' is used for particles that interact with each other through the strong interaction.

27.2 What is an Elementary Particle?

Fifty years ago it was easy to build a system of atoms and nuclei using only protons and electrons and even with the advent of the neutron there was little difficulty in setting up models in terms of three elementary particles as units. With the discovery of the first antiparticle, the positron, and the emergence of the neutrinos and mesons, it became clear that use of the word elementary as referring to the permanent units of an atom was obsolete. The words 'elementary' and 'fundamental' which have often been used to describe the particles discussed in the previous chapter, became meaningless. Of these particles only the electrons, proton and neutrinos are infinitely stable. The others have comparatively short lifetimes, so that it is impossible to recognize them all as fundamental or elementary. However, as these particles have discrete masses it is not impossible to regard them as higher quantum states of a basic state or states. We shall return to this point in our discussion of resonance particles and quarks. None of the particles so far discussed has a lifetime characteristic of strong interactions. Thus the lifetime of 10^{-10} s may be regarded as long compared with the strong interaction characteristic time, and in this chapter all particles with this lifetime are regarded as stable. (The meaning of 'fundamental' is discussed in Chapter 28.)

27.3 Short-Lived or Resonance Particles

Much of the foregoing refers to the particles described in Chapter 26. The shortest-lived of these particles is the neutral pion π^0 — the lightest of all the strongly interacting particles with a mean lifetime of about 10^{-16} s characteristic of electromagnetic decay. During the last few years there has been a profusion of new particles which have increased the number already known to more than 100. These are the new resonance particles which are extremely unstable with lifetimes of about 10^{-23} s showing that they are strongly interacting particles. They are called resonance particles because they are recognized by the resonance peaks in a normal energy spectrum of an event. Thus if protons were collected at various energies in a $\pi^+ + p^+$ collision, the energy distribution curve could be as shown in Fig. 27.1, which is purely schematic.

Peak I is the main peak of the proton beam and peaks II, III and IV are inelastic (high-absorption) scattering peaks coinciding with resonance states between the two particles. Notice that this type of experiment is very similar to the inelastic

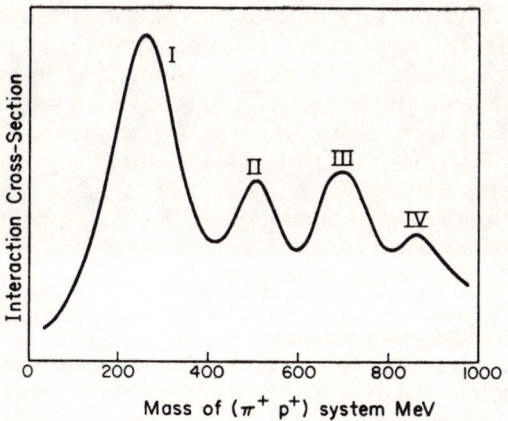

Fig. 27.1 Schematic resonance curve of π^+:p^+ collision.

Fig. 27.2(a) Y* resonance particle and a negative pion 'π^- from 0 in drawing on next page) are produced in this bubble-chamber collision between a negative K meson (K^-) and a proton at 0. The resonance particle disintegrates before it can leave a track into a neutral lambda (Λ^0) particle, which leaves no track (broken line), and a positive pion (π^+). The lambda decays into a proton (p^+) and a negative pion (π^-). The photograph was made by the experimental team under Luis W. Alvarez at the Lawrence Radiation Laboratory. (By couetesy of the Lawrence Radiation Laboratory, Berkeley, California.)

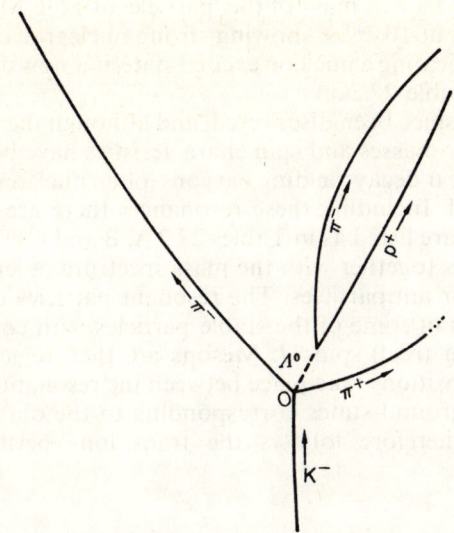

Fig. 27.2(b) Line diagram of Fig. 27.2(a).

electron scattering in mercury vapour carried out by Franck and Hertz. This curve shows that the system, like the mercury atom, can exist in a set of intermediate short-lived excited states. These new enhanced probability, or resonant states, can be assigned mass, charge and spin consistent with the conservation laws. Their independence is momentary, as decay times are only 10^{-7} times the previous shortest lived particle, namely the π^0-meson. Although too short to measure, this time is sufficient for the excess energy to reassemble in the form of mesons and other particles. Resonances can therefore only be inferred by their decay products and this is how such particles have been found.

The first resonant particle to be discovered was the N* particle, in 1951, by Fermi, but it remained unnamed. In 1960 the reaction $K^- + p^+ \rightarrow \Lambda^0 + \pi^+ + \pi^-$ was being studied by Alvarez and his group at the Lawrence Radiation Laboratory and many hundreds of plates like that shown in Fig. 27.2 were analysed by a computer. Some of the results suggested that the conservation of linear momentum law was being violated and *two* resulting particles were indicated rather than three. Possibilities were

$$K^- + p^+ \rightarrow Y^{*+} + \pi^- \quad \text{or} \quad K^- + p^+ \rightarrow Y^{*-} + \pi^+,$$

where Y* is a suggested new resonance particle (or an excited baryon state) showing strong nuclear decay in 10^{-23} s into

$$Y^{*+} \rightarrow \Lambda^0 + \pi^+ \quad \text{or} \quad Y^{*-} \rightarrow \Lambda^0 + \pi^-.$$

The analysis of a large number $K^- + p^+$ events gave a most probable Y* mass of 1385 MeV and a decay time of 10^{-23} s, showing the Y* particle to be a strong interaction particle. This is now designated as an excited Σ state. (See Table 27.2C.)

The Fermi particle of 1951 was eventually named the N* particle. The scattering cross-section in pion–proton collisions gave a resonance peak at about

200 MeV corresponding to a rest mass of the 'particle' of 1236 MeV. Again the estimated lifetime was about 10^{-23} s, showing strong nuclear decay. Originally called a N* resonance, indicating a nucleon excited state, it is now designated as Δ baryon resonance. (See Table 27.2C.)

Other resonances have since been discovered, and although the recognition of such states is difficult their masses and spin characteristics have been measured. They all show strong nuclear decay yielding baryons (often nucleons) and mesons which are easily observed. Including these resonances there are now nearly a hundred 'particles' which are listed as in Tables 27.2 A, B and C. These show the long-lived 'stable' particles together with the mass spectrum of leptons, mesons and baryons without their antiparticles. The resonant particles can be looked upon as the excited states of some of the stable particles with correspondingly greater masses and higher (real) spins J. Mesons are then regarded as mass-energy emission when transitions take place between the resonant particles, and to the (relatively) stable ground states corresponding to the old particles. The production of mesons therefore follows the transitions permitted by the

TABLE 27.2A
Stable Particles with $\bar{T} \gg 10^{-23}$ s

	Class	Particle	Rest mass (MeV)	Spin J	Mean lifetime (s)	Some observed decay modes
	Boson	Photon (γ)	1	0	Stable: infinite	—
Fermions	Leptons	$\nu_e : \nu_\mu$	0	$\frac{1}{2}$	Stable: infinite	—
		e	0.511	$\frac{1}{2}$	Stable: infinite	—
		μ	105.7	$\frac{1}{2}$	2.2×10^{-6}	$e\nu\bar{\nu}$
Hadrons	Bosons / Mesons	$\pi^\pm$	139.6	0	2.6×10^{-8}	$\mu\nu_\mu$
		π^0	135.0	0	8.9×10^{-15} EM	$\gamma\gamma$
		$K^\pm$	493.8	0	1.24×10^{-8}	$\mu\gamma$: $\pi\pi^0$: $\pi\pi^-\pi^+$
		K^0	497.8	0	1.24×10^{-8}	50% K_S^0: 50% K_L^0
		K_S^0	—	0	0.87×10^{-10}	$\pi^+\pi^-$: $\pi^0\pi^0$
		K_L^0	—	0	5.3×10^{-8}	$\pi^0\pi^0\pi^0$: $\pi^+\pi^-\pi^0$
		η^0	548.8	0	3×10^{-14} EM	$\gamma\gamma$: $\pi^0\gamma\gamma$
	Baryons and Fermions / Nucleons	p^+	938.3	$\frac{1}{2}$	Stable: infinite	—
		n^0	939.6	$\frac{1}{2}$	1.01×10^3	$p^+ e^- \bar{\nu}_e$
	Hyperons	Λ^0	1115.5	$\frac{1}{2}$	2.52×10^{-10}	$p^+\pi^-$; $n^0\pi^0$
		Σ^+	1189.5	$\frac{1}{2}$	0.81×10^{-10}	$n^0\pi^+$; $p^+\pi^0$
		Σ^0	1192.5	$\frac{1}{2}$	1×10^{-14} EM	$\Lambda^0\gamma$
		Σ^-	1197.4	$\frac{1}{2}$	1.66×10^{-10}	$n^0\pi^-$; $n^0 e^- \bar{\nu}_e$
		Ξ^0	1314.9	$\frac{1}{2}$	2.9×10^{-10}	$\Lambda^0\pi^0$: $p^+\pi^-$
		Ξ^-	1321.3	$\frac{1}{2}$	1.73×10^{-10}	$\Lambda^0\pi^-$: $n^0\pi^-$
		Ω^-	1672	$\frac{3}{2}$	1.1×10^{-10}	$\Xi^0\pi^-$: $\Lambda^0 K^-$

The above particles are well established and immune to strong decay.

TABLE 27.2B
Established Mesons and Meson Resonances
Lifetimes: Stable 10^{-10} s: Resonances 10^{-23} s

Mass (MeV)	Charge Q −1	0	+1		Particle and mass (MeV)
1700					
1600					
1500		•			η^0 1514
1400	•	⁞	•	RESONANCES	K_V 1419
1300	•	•	•		$\pi(A_2)$ 1305
1200		•			η^0 1260
1100					
1000		•			η^0 1019
900	•	⁞	•		η^0 958
800	•	⁞	•		$K \times$ 893
700					η^0 783
					$\pi \times$ 765
600		•			η^0 549
500	•	⁞	•	STABLE	K 496
400					
300					
200					
100	•	•	•		π 137
0					

Multiplicities: $\eta^0 = 1$, $\pi = 3$, K = 4.

TABLE 27.2C
Established Baryons and Baryon Resonances
Lifetimes: Stable 10^{-10} s: Resonances 10^{-23} s

Mass (MeV)	Charge Q: −1	0	+1	+2		Particle and mass (MeV)
2200		•	•			N 2190
2100		•	•			Λ⁰ 2100
2000	•	•	•			Σ 2030
1900	•	•	•	•	RESONANCES	Δ 1920
1800		••	•			Λ⁰ 1830, Λ⁰ 1815, Σ 1770
1700	ST •	••••	••••	•		N 1710, Λ⁰ 1690, N 1688, N 1680, Ω⁻ 1672 ST, Λ⁰ 1670, Δ 1640
1600	•	•	•			
1500	•	••••	•••			N 1550, Ξ 1530, N 1525, Λ⁰ 1520, N 1470, Λ⁰ 1405
1400	•	••	•			Σ 1385
1300	•	•			Stab.	Ξ 1318
	•	•	•	•	Res.	Δ 1236
1200	•	•	•			Σ 1193
1100		•			STABLE	Λ⁰ 1115
1000						
900		•	•			N { n⁰ 939·6, p⁺ 938·3

Multiplicities: Λ⁰ = 1, N = 2, Ξ = 2, Σ = 3, Δ = 4, Ω = 1. Antibaryons are not shown.

appropriate conservation laws. A simple example is the production of excited pions of spin one from the transitions shown in Fig. 27.3. This is only part of many quantum exchange possibilities between resonance and long-lived states discussed in Chapter 26.

The significance of the quantum number I will be explained later.

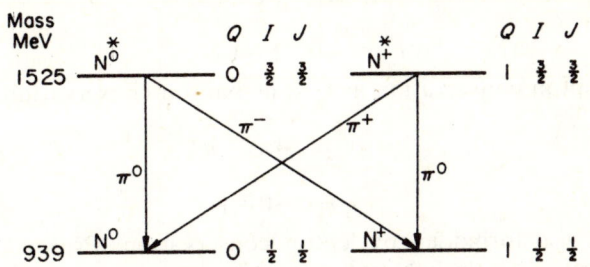

Fig. 27.3 Production of pions by transitions from resonance states.

27.4 Conservation Laws: Baryon and Lepton Conservation

We are already familiar with many conservation laws in atomic and nuclear systems, such as the conservation of charge, mass/energy, linear momentum and angular momentum. In atomic physics we know that the application of these laws leads to selection rules for allowed spectra and in nuclear physics to the prediction of new particles, e.g. neutrinos. In the field of subnuclear physics we are now presented with a whole new list of particles which are observed in collision experiments and in different modes of decay. Some modes of decay are never observed, and it is natural to suppose that these are prevented by some unknown law of conservation. Thus new laws of conservation have been deduced from a study of all possible types of particle reaction and decay, as well as mathematically.

One of the great mysteries of nuclear physics is the stability of the proton. We know that the free neutron is unstable to β^--decay by
$$n^0 \to p^+ + e^- + \tilde{\nu}_e$$
so why not $p^+ \to \pi^0 + \mu^+$ since spins would still be conserved? Some law must prevent this. This is the law of conservation of baryon number in which all baryons are assigned a baryon number $B=1$, all antibaryons have $B=-1$, and all mesons and leptons have $B=0$. Thus for
$$n^0 \to p^+ + e^- + \tilde{\nu}_e$$
we have
$$B = 1 \to 1 + 0 + 0$$
so that this reaction 'goes'; but for
$$p^+ \to \pi^0 + \mu^+$$
we have
$$B = 1 \to 0 + 0.$$
This decay does not occur as the baryon number is not conserved.

Similarly it can be shown that lepton numbers must also be conserved if we assign a lepton number $l=1$ or -1 as follows to the leptons, remembering that $l=0$ for mesons and baryons, and treating muons and electrons differently,

$$l_e = 1 \text{ for } e^-, \nu_e, \quad l_e = -1 \text{ for } e^+, \tilde{\nu}_e,$$
$$l_\mu = 1 \text{ for } \mu^-, \nu_\mu, \quad l_\mu = -1 \text{ for } \mu^+, \tilde{\nu}_\mu.$$

The equation

$$n^0 \to p^+ + e^- + \tilde{\nu}_e$$

then has

$$l_e = 0 \to 0 + 1 - 1$$

so obeying lepton conservation as well as baryon conservation. However, the decay

$$p^+ \to \pi^0 + \mu^+$$

has

$$l_\mu = 0 \to 0 - 1$$

so that it is *also* forbidden by lepton conservation. Proton decay is really forbidden because it is the lightest baryon in the mass spectrum. See Table 27.2C.

The muon decays we discussed in the last chapter, viz.

$$\mu^+ \to e^+ + \nu_e + \tilde{\nu}_\mu$$

has

$$l_e = 0 \to -1 + 1 + 0,$$
$$l_\mu = -1 \to 0 + 0 - 1$$

and

$$\mu^- \to e^- + \tilde{\nu}_e + \nu_\mu$$

has

$$l_e = 0 \to 1 - 1 + 0,$$
$$l_\mu = 1 \to 0 + 0 + 1$$

are seen also to conserve the lepton numbers and therefore 'go'. Since muon and electron decays are all weak interactions, i.e. strong interactions do not produce leptons, it follows that lepton conservation does not apply to decay by strong interactions.

27.5 Multiplet Structure — Isospin and Hypercharges

As far as strong interactions are concerned, the neutron and the proton are the two states of equal mass of a nucleon doublet. A glance at Tables 27.2B, C shows that particles can be grouped in multiplets of equal mass but different charges. Examples are shown below:

				Multiplet number, M	Mass (MeV)
Nucleons	N^+	N^0		2	939
Pions	π^+	π^0	π^-	3	765
Kaons	K^+	$K^0, \overline{K}^0$	K^-	2, 2	496
Xi baryon		Ξ^0	Ξ^-	2	1318

An important outcome of the multiplicity number M of an equal mass group is the concept of *isospin* I. This is not a true mechanical spin but its quantum-mechanical derivation follows similar lines to that of electron spin in spectroscopy and obeys similar rules so that we put $M = 2I+1$ for the charge multiplicity, just as in spectroscopy we put $R = 2S+1$, where R is the multiplicity of the energy levels associated with the electron spin S quantum number of the atom.

In the case of the nucleon doublet it is reasonable to assume that they are two states of the same nuclear particle. They are distinguished only by their charge and thus by the interaction of the proton with the electromagnetic field. The isospin $I = \frac{1}{2}$ is assigned to all nucleons but with the component $I_3 = \frac{1}{2}$ for the proton and $I_3 = -\frac{1}{2}$ for the neutron. Here $I_3 = \pm\frac{1}{2}$ is the z component of I in exactly the same way that we had $s_z = \pm\frac{1}{2}$ for the z component of the electron spin $s = \frac{1}{2}$. Thus the proton and neutron form a doublet with the same isospin. Similarly for the pion triplet we have $M = 3$ and $I = 1$ giving

$$I_3 = \begin{cases} +1 & \text{for } \pi^+, \\ 0 & \text{for } \pi^0, \\ -1 & \text{for } \pi^-. \end{cases}$$

The kaons are grouped into two pairs with $I_3 = \pm\frac{1}{2}$ for each pair. These are

$$K^+, I_3 = +\tfrac{1}{2} \quad \text{and} \quad K^0, I_3 = -\tfrac{1}{2},$$
$$K^-, I_3 = -\tfrac{1}{2} \quad \text{and} \quad \overline{K^0}, I_3 = +\tfrac{1}{2}.$$

Finally, the delta particle Δ has four states, viz.

$$\Delta^{++}, \Delta^+, \Delta^0 \text{ and } \Delta^-$$

for which $I = \frac{3}{2}$ and the I_3 values are $\frac{3}{2}, \frac{1}{2}, -\frac{1}{2}$ and $-\frac{3}{2}$ respectively.

In general there are $2I+1$ isospin states for a particle of given I, just as there were $2l+1$ magnetic states for each value of l in the case of the electron.

Using these data we now obtain isospins as follows:

	I	I_3
Nucleons	$\frac{1}{2}$	$\pm\frac{1}{2}$
Pions	1	$0, \pm 1$
Kaons	$\frac{1}{2}$	$\pm\frac{1}{2}$
Xi baryons	$\frac{1}{2}$	$\pm\frac{1}{2}$
Delta baryons	$\frac{3}{2}$	$\pm\frac{3}{2}, \pm\frac{1}{2}$

Isospin I is conserved only in strong interactions so that it applies only to hadrons and not to leptons. See Table 27.2A.

Continuing our consideration of particles in groups or multiplets the concept of hypercharge Y is now introduced. This is a charge number equal to twice the average charge $\bar{Q}$ of a multiplet, so that for the above examples we have the hypercharge numbers shown below, where

$$\bar{Q} = \frac{1}{M}(Q_1 + Q_2 + \ldots + Q_M):$$

				$\bar{Q}$	$Y = 2\bar{Q}$
Nucleons	N^+	N^0		$1/2$	1
Pions	π^+	π^0	π^-	0	0
Kaons	K^+	$K^0, \overline{K^0}$	K^-	$\frac{1}{2}, -\frac{1}{2}$	$1, -1$
Xi baryons		Ξ^0	Ξ_0^-	$-\frac{1}{2}$	-1
Delta baryons	Δ^{++}, Δ^+	Δ^0	Δ^-	$\frac{1}{2}$	1

It is found that, like isospin I, the hypercharge number Y is conserved in strong interactions. It is also conserved in electromagnetic interactions, but not in weak interactions.

These are strange new ideas of conservation. They can be justified empirically by applying them to established decays and reactions. The strangest quantum number of all is actually called the 'strangeness' number S. (This is not a *spin* quantum number.)

Various relations between the quantum numbers already mentioned can be derived. Thus we can see that

$$Q = I_3 + \tfrac{1}{2}B$$

for baryons.

Proton, $Q = \tfrac{1}{2} + \tfrac{1}{2} = 1$;
Neutron, $Q = -\tfrac{1}{2} + \tfrac{1}{2} = 0$;
Δ^{++} baryon, $Q = \tfrac{3}{2} + \tfrac{1}{2} = 2$, etc.

If we extend this to pions, for example, for which $B = 0$, we have

π^+, $Q = 1 + 0 = 1$;
π^0, $Q = 0 + 0 = 0$;
π^-, $Q = -1 + 0 = -1$.

Now for the baryon Λ^0 for which $I_3 = I_0 = 0$ since it is a singlet (Table 27.2A) we have, from the $Q = I_3 + \tfrac{1}{2}B$ formula,

$$Q = 0 + \tfrac{1}{2}$$

i.e. $Q_{\Lambda^0} = \tfrac{1}{2}$, obviously incorrect. Thus Λ^0 was regarded as a 'strange' particle.

Similarly the neutral kaon K^0 has $B = 0$ and $I_3 = -\tfrac{1}{2}$, from which we get $Q_{K^0} = -\tfrac{1}{2}$, again incorrect. The K^0 particle was another 'strange' particle. Difficulties such as this led to the new concept of strangeness. Nucleons, not strange, were assigned $S = 0$ and putting K^0, $S = +1$ and Λ^0, $S = -1$, where S is the strangeness quantum number, the charge formula was written

$$Q = I_3 + \tfrac{1}{2}B + \tfrac{1}{2}S$$

giving

$$Q_{\Lambda^0} = 0 + \tfrac{1}{2} + -\tfrac{1}{2} = 0$$

and

$$Q_{K^0} = -\tfrac{1}{2} + 0 + \tfrac{1}{2} = 0,$$

as required.

Thus we can write $Q = I_3 + \tfrac{1}{2}Y$, where Y is the hypercharge quantum number given by $Y = S + B$.

Since Y and B are each conserved in strong and electromagnetic interactions S must also be so conserved. As S is a function of the quantum numbers Y and B it becomes redundant if Y and B are used, although it is still frequently used. It must not be confused with the electron spin quantum number S of the atom.

Summary of New Quantum Numbers

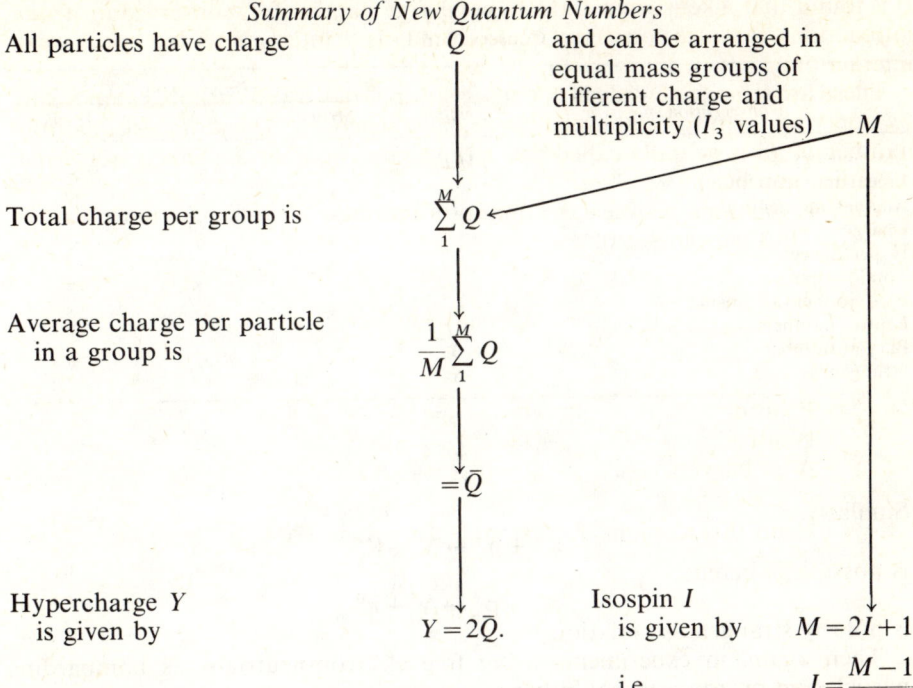

All particles have charge Q and can be arranged in equal mass groups of different charge and multiplicity (I_3 values) M

Total charge per group is $\sum_1^M Q$

Average charge per particle in a group is $\dfrac{1}{M}\sum_1^M Q = \bar{Q}$

Hypercharge Y is given by $Y = 2\bar{Q}$.

Isospin I is given by $M = 2I + 1$ i.e. $I = \dfrac{M-1}{2}$

In addition:

Baryon number B is given by $B = \pm 1$ for all baryons, $= 0$ for all other particles.

Strangeness number S is given by $S = Y - B$.

Lepton number l is given by $l = \pm 1$ for all leptons, $= 0$ for all other particles.

The number of possible conservation laws has increased with the number of new particles discovered. Some apply to all three types of reaction while others apply only to parts. They are tabulated in Table 27.3.

If these conservation laws are applied to high-energy pion–nucleon collision which often give large quantities of kaons, examples of possible equations are:

(i) $\pi^+ + n^0 \rightarrow \Lambda^0 + K^+$
(ii) $\pi^+ + n^0 \rightarrow K^0 + K^+$.

Using the conservation of baryons and strangeness, for

(i) $S = 0 + 0 \rightarrow -1 + 1$ we get $\delta S = 0$
$B = 0 + 1 \rightarrow 1 + 0$ $\delta B = 0$

(ii) $S = 0 + 0 \rightarrow 0 + 0$ we get $\delta S = 0$
$B = 0 + 1 \rightarrow 0 + 0$ $\delta B \neq 0$

So that reaction (ii) violates baryon conservation and therefore cannot occur.

TABLE 27.3
Conservation Laws

		Interactions		
Conservation of	Symbol	Strong	E.M.	Weak
Mass/energy	M.E.	✓	✓	✓
Linear momentum	—	✓	✓	✓
Angular momentum	J	✓	✓	✓
Charge	Q	✓	✓	✓
Hypercharge	Y	✓	✓	×
Total isospin	I	✓	×	×
z Component of isospin	I_3	✓	✓	×
Lepton numbers	l_e, l_μ	—	✓	✓
Baryon number	B	✓	✓	✓
Strangeness	S	✓	✓	×

Similarly,
$$\pi^- + p^+ \to \Lambda^0 + K^0$$
is possible, whereas
$$\pi^- + p^+ \to \Lambda^0 + \pi^0$$
is not, by strangeness violation.

Then again, in experiments using free electron-neutrinos as bombarding particles on protons, we could have

(i) $\tilde{v}_e + p^+ \to n^0 + \mu^+$,
(ii) $\tilde{v}_e + p^+ \to n^0 + e^+$.

Applying lepton number considerations we have for (i)

$$l_e = -1 + 0 \to 0 + 0 \qquad \delta l_e \neq 0$$
$$l_\mu = 0 + 0 \to 0 - 1 \qquad \delta l_\mu \neq 0$$

whereas for (ii)

$$l_e = -1 + 0 \to +-1 \qquad \delta l_e = 0$$
$$l_\mu = 0 + 0 \to 0 + 0 \qquad \delta l_\mu = 0$$

showing that the first reaction is impossible. Consider next the collision event illustrated in Fig. 26.6.
$$\bar{p}^+ + p^+ \to 2\pi^+ + 2\pi^- + \pi^0.$$
Applying various conservation laws, we have

$$Q = -1 + 1 \to 2 - 2 + 0 \qquad \delta Q = 0$$
$$B = -1 + 1 \to 0 + 0 + 0 \qquad \delta B = 0$$
$$Y = -1 + 1 \to 0 + 0 + 0 \qquad \delta Y = 0$$
$$S = 0 + 0 \to 0 + 0 + 0 \qquad \delta S = 0$$
$$I_3 = -\tfrac{1}{2} + \tfrac{1}{2} \to 2 - 2 + 0 \qquad \delta I_3 = 0$$

so that this event is possible. These are but a few of the many examples we could choose to illustrate the new conservation laws.

27.6 Classification of Elementary Particles

By inspection of the list of particles now available and the multiplet structures to which they conform it is possible to regroup the mesons and baryons (all the strongly interacting particles), using only three of the conserved quantities just discussed. These are B, Y and I and we can now refer to these as conserved quantum numbers. This gives only four basic meson groups eta η, pi π, and the two kaon groups according to their B, Y and I values. These are shown in Table 27.4, where M is the charge multiplicity.

TABLE 27.4
Meson Groups $B=0$ from Tables 27.2A, B, C

Meson	Y	I	M	No. of masses
η	0	0	1	6
π	0	1	3	3
K	+1	$\frac{1}{2}$	2	3
Anti-K	−1	$\frac{1}{2}$	2	3

This leads to 27 mesons altogether so that the complete spectrum is as follows:

Meson	M	Masses (MeV)
η	1	549 783 958 1019 1260 1514
π	3	137 765 1305
K^+, K^0	2	496 893 1419
K^-, K^0	2	496 893 1419

Total 8   Each mass has a multiplicity M.

Notice that each named basic meson covers more than one actual meson, e.g. there are *six* singlet η-mesons with different masses and three triplet π-mesons of different masses. These are the resonance mesons.

Similarly the *baryons* can be reduced to *six* basic groups in terms of their common Y, I and M. These are designated by capital Greek letters in Table 27.5

TABLE 27.5
Baryon Groups $B=1$

Baryon		Y	Total I	M	No. of masses
Lambda	Λ	0	0	1	8
Sigma	Σ	0	1	3	4
Nucleon	N	1	$\frac{1}{2}$	2	8
Xi	Ξ	−1	$\frac{1}{2}$	2	2
Omega	Ω	−2	0	1	1
Delta	Δ	1	$\frac{3}{2}$	4	3

which shows how the 53 baryons are arranged in six groups. The complete spectrum is shown below—

Baryon	M	Masses (MeV)							
Λ	1	1115	1405	1520	1670	1690	1815	1830	2100
Σ	3	1193	1385	1770	2030				
N	2	939	1470	1525	1550	1680	1688	1710	2190
Ξ	2	1318	1530						
Ω	1	1672							
Δ	4	1236	1640	1920					

Total 13   Each mass has a multiplicity M.

These are the well-established resonances. There are many more which are not fully confirmed but which would still fit the above patterns. They are being actively researched.

It is evident then that the large number of particles can be reduced to simpler descriptions of families or regularities by the application of the conservation laws. These regularities or symmetries are not fortuitous. Is it possible therefore to devise a physical or mathematical model which would enable us to explain the above properties of all the known particles, and so help us in our search for *new* particles by predicting their properties in much the same way as searching for unknown elements in the periodic table?

27.7 Particle Symmetries

The details of particle symmetries are expressed mathematically by using special algebras. In the special unitary symmetry group known as SU(3), we have a particular method of classifying strongly reacting particles. SU(3) denotes the symbol for the mathematics of the special unitary group of 3×3 arrays of quantum numbers. Sometimes this is called the eightfold way, because it covers relations between eight conserved quantities and it groups particles into 1-, 8-, 10- or 27-member families.

These 8- and 10-member groups are of particular interest. Theory shows that the eight particles of the meson octet with zero (real) spin should be arranged as follows:

$$\left. \begin{array}{l} \text{one triplet with } Y=0 \text{ and total isospin } I=1, \\ \text{one doublet with } Y=1 \text{ and total isospin } I=\tfrac{1}{2} \\ \text{one doublet with } Y=-1 \text{ and total isospin } I=\tfrac{1}{2} \\ \text{one singlet with } Y=0 \text{ and total isospin } I=0 \end{array} \right\} J=0$$

When this was first postulated there were seven ground state mesons known with the following values of Q, Y and the z-component of I:

		Q	Y	I_3
Triplet pions	π^+	1	0	+1
	π^0	0	0	0
	π^-	−1	0	−1
Quartet kaons	K^+	1	+1	$+\tfrac{1}{2}$
	K^0	0	+1	$-\tfrac{1}{2}$
	$\overline{K^0}$	0	−1	$+\tfrac{1}{2}$
	K^-	−1	−1	$-\tfrac{1}{2}$

The pions obviously form a triplet obeying the conditions required by the theory, and if the kaons are taken to represent two doublets, viz.

		Q	Y	I_3
K	K^+	1	+1	$+\tfrac{1}{2}$
	K^0	0	+1	$-\tfrac{1}{2}$
Anti-K	K^-	−1	−1	$-\tfrac{1}{2}$
	$\overline{K^0}$	0	−1	$+\tfrac{1}{2}$

We have exactly the right combinations of quantum numbers for seven out of the eight particles required for the octet. These can now be plotted on Q, Y axes to give a symmetrical hexagonal lattice array of particles, as in Fig. 27.4, in which the I_3 lattice is also shown. This forms a hexagonal lattice in which the points $(-1, 1)$ and $(1, -1)$ shown are not required by the theory. But it is seen that the *singlet* $(0, 0)$ demanded by the theory is missing. The missing meson was expected to have $Q=0$, $Y=0$, $I=I_3=0$ and a rough calculation of the mass gave 567 MeV. The meson was found in 1961 and had a mass of 549 MeV which was a satisfactory vindication of the theory. This is the eta meson η^0 shown at the origin of Fig. 27.4.

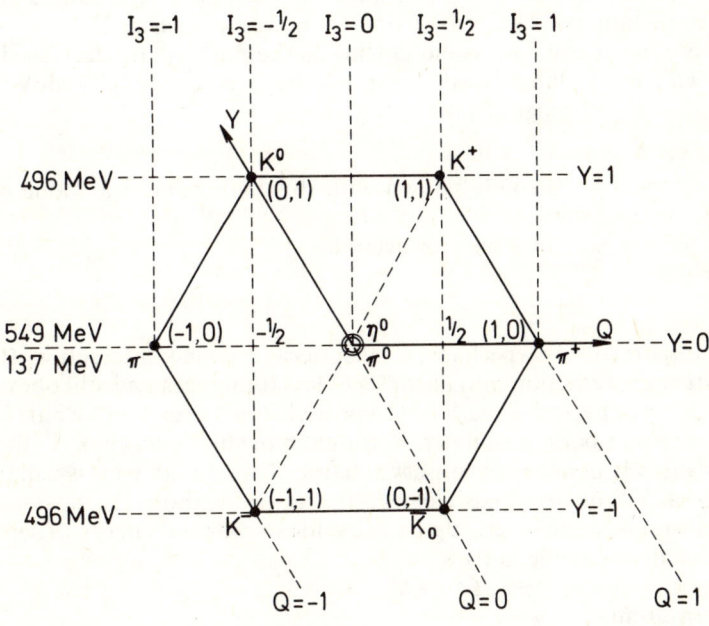

Fig. 27.4 Meson octet for $J=0$ showing position of predicted η^0-meson on a Q, Y, I_3.

A more remarkable prediction was the case of the 10-fold baryon group with real spin $J=\tfrac{3}{2}$. Theory predicted that there should be a group of 10 members as follows:

one quartet with $Y=1$ total $I=\frac{3}{2}$ $M=4$
one triplet with $Y=0$ total $I=1$ $M=3$
one doublet with $Y=-1$ total $I=\frac{1}{2}$ $M=2$ 10 members
one singlet with $Y=-2$ total $I=0$ $M=1$

The known resonances in 1963 with $J=\frac{3}{2}$ were

		Q	Y	I	(Mass (MeV)
Quartet delta	Δ^{++}	2			
	Δ^{+}	1	1	$\frac{3}{2}$	1236
	Δ^{0}	0			
	Δ^{-}	−1			↘149
Triplet sigma	Σ^{+}	1			↗
	Σ^{0}	0	0	1	1385
	Σ^{-}	−1			↘145
Doublet xi	Ξ^{0}	0	−1	$\frac{1}{2}$	1530↗
	Ξ^{-}	−1			
Singlet: unknown	?	−1	−2	0	?

These values are seen to be equidistant from each other and can be displayed on a simple rectangular lattice by using Q, Y axes as in Fig. 27.5.

Theory predicted equal mass increments as the multiplicity decreased so that the unknown particle, later called Ω^{-}, should have a mass of 1677 MeV together with the following characteristics:

$Q = -1$;
$Y = -2$, the only particle known with this high numerical value of Y;
$I = 0$, i.e. a singlet;
$B = 1$ as it is part of a baryon decuplet;
$S = Y - B = -3$;
Spin $J = \frac{3}{2}$

and, most important of all perhaps, it must decay by weak interaction. It cannot decay by strong interaction into particles of less total mass and still obey the rule of conservation of hypercharge. The Ω^{-} particle must decay into a baryon and a meson to conserve baryon number. If, for example, the baryon is Λ^{0} the meson must be negatively charged to conserve charge. The available mass will be 1677−1115 = 562 MeV. The only possible negative meson of about the mass is the K^{-}-meson at about 496 MeV, leaving 66 MeV for the kinetic energy of separation. The possible decay mode is then

$$\Omega^{-} \rightarrow \Lambda^{0} + K^{-}$$

with baryon number
$$B = 1 \rightarrow 1 + 0 \qquad \delta B = 0$$

and isospin
$$I = 0 \rightarrow 0 + \tfrac{1}{2} \qquad \delta I \neq 0,$$

showing that the isospin I is *not* conserved. Similarly hypercharge and strangeness are not conserved. Hence the decay of Ω^{-} is by weak interactions, since these conservations only hold for strong interactions. A few more trials will

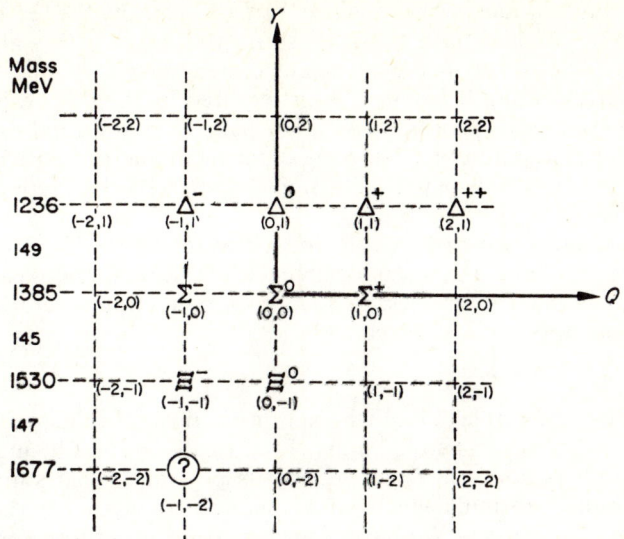

Fig. 27.5 Baryon decuplet for $J=\frac{3}{2}$, $B=1$ on Q, Y plot. Unknown particle is a singlet with Q, Y coordinates $(-1, -2)$ and mass 1677 MeV.

show that plausible decay schemes do not obey isospin conservation and therefore Ω^- decay is by weak interaction which is not subject to this restriction. The decay lifetime was therefore predicted to be that of weak interactions, viz. 10^{-10} s, and the Ω^- particle is therefore classed as a 'stable' hyperon in Table 27.2A and C.

This prediction in 1963 was an almost unique opportunity for those research workers with the appropriate facilities for high-energy collisions, because, once all the properties of the particle were known, it was possible to plan an experiment to find it. The search for the Ω^- particle was undertaken at Brookhaven with the 33 GeV proton synchrotron. First it was necessary to generate a strangeness of -3 for this particle. This gave the following collision process as a possible scheme:

$$K^- + p^+ \rightarrow K^+ + K^0 + \Omega^-$$

and thus

$$S = -1 + 0 \rightarrow 1 + 1 - 3,$$
$$I = \tfrac{1}{2} + \tfrac{1}{2} \rightarrow \tfrac{1}{2} + \tfrac{1}{2} + 0,$$

as required. The relativistic threshold energy was calculated to be 3·2 GeV. The proton collisions on the target produced a beam of K^- mesons which were separated from the other particles by various magnetic devices. The K^- beam of about 5 GeV was then led into the 2-m hydrogen bubble chamber where the K^-, p^+ collisions occurred. Possible decay modes which conserve baryon numbers are

$$\Omega^- \rightarrow \Xi^- + \pi^0,$$
$$\Omega^- \rightarrow \Xi^0 + \pi^-,$$
$$\Omega^- \rightarrow K^- + \Lambda^0,$$

and these decay products were sought in order to identify the Ω^- particle. After many photographs had been taken — some 100 000 in all — all three decay schemes were discovered in 1964. The calculated mass was 1686 MeV and the measured lifetime about 0.7×10^{-10} s, as required by the theory. Note that this time is long enough to produce a visible track. This is a good example of the theory of an investigation preceding the experiment and the experimental result amply confirming the theory. It thoroughly indicated the mathematics of the SU(3) theory.

A photograph of the production and decay of Ω^- is shown in Fig. 27.6. Although the particular algebra associated with these predictions cannot solve all the problems of particle physics, its success in predicting several new particles shows that the approach is correct.

27.8 Quarks

It is obviously desirable to find whether or not the multiplicity of particles can be built up from other simpler units. Empirically we could juggle with the quantum numbers concerned and see if it is possible to get some underlying grouping system or formula which would enable us to predict all the mesons and baryons and their masses. We have a meson spectrum of four members and a baryon spectrum of six members, each member having the same combinations of quantum numbers. These two groups must be obtainable from our formula.

Analysis has shown the possible existence of three basic units which could be the real fundamental particles. They are called quarks, two having $I = \frac{1}{2}, S = 0$, one having $I = 0$ and $S = -1$, their full quantum numbers being:

Quark	I	I_3	S	Q	B	Y
A	$\frac{1}{2}$	$\frac{1}{2}$	0	$\frac{2}{3}$	$\frac{1}{3}$	$\frac{1}{3}$
B	$\frac{1}{2}$	$-\frac{1}{2}$	0	$-\frac{1}{3}$	$\frac{1}{3}$	$\frac{1}{3}$
C	0	0	-1	$-\frac{1}{3}$	$\frac{1}{3}$	$-\frac{2}{3}$

All the possible combinations of these three quarks are set out in Table 27.6 for baryons only, i.e. three quarks are necessary for each baryon to give $B = 1$. Quark quantum numbers are added algebraically. Likewise mesons are generated by a quark and an antiquark as in Table 27.7.

From Tables 27.6 we see that the 13 basic baryons and the eight basic mesons,

TABLE 27.6
Baryons Generated from Three Quarks (i.e. $B = 1$)

Quarks			S	Q	Y	I_3	Particle
A	A	A	0	2	1	$\frac{3}{2}$	Δ^{++}
B	B	B	0	-1	1	$-\frac{3}{2}$	Δ^-
A	A	B	0	1	1	$\frac{1}{2}$	Δ^+, N^+
A	B	B	0	0	1	$-\frac{1}{2}$	Δ^0, N^0
A	A	C	-1	1	0	1	Σ^+
A	B	C	-1	0	0	0	Σ^0, Λ^0
B	B	C	-1	-1	0	-1	Σ^-
A	C	C	-2	0	-1	$\frac{1}{2}$	Ξ^0
B	C	C	-2	-1	-1	$-\frac{1}{2}$	Ξ^-
C	C	C	-3	-1	-2	0	Ω^-

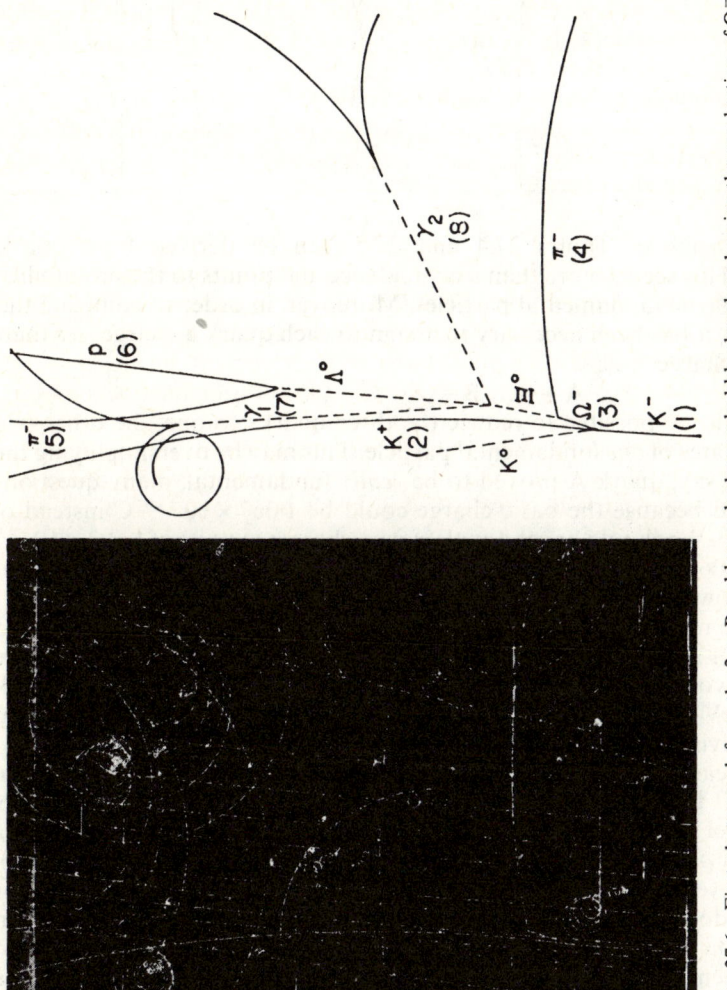

Fig. 27.6 First photograph from the 2-m Brookhaven hydrogen bubble chamber showing the production of Ω^-. Line diagram on right shows sequence of events beginning with the collision of the fast moving K^- particle (1) with a proton and the production of the Ω^- particle (3), a K^+ particle (2) and the invisible K^0 particle. The subsequent decay of Ω^- into π^- and Ξ^0 is clearly shown. This is a possible decay mode as explained in the text. The unnamed curved tracks are positron and electron pairs created from the γ-rays (7) and (8). The identification of unknown particles is accomplished by applying conservation laws to the various events until the properties of particle (3) are deduced. These were as required by the prediction of the Ω^- particle. (By courtesy of Brookhaven National Laboratory.)

TABLE 27.7
Mesons Generated from a Quark and an Antiquark (i.e. $B=0$)

Quarks		S	Q	Y	I_3	Particle
A	$\bar{A}$	0	0	0	0	
B	$\bar{B}$	0	0	0	0	π^0, η^0
C	$\bar{C}$	0	0	0	0	
A	$\bar{B}$	0	1	0	1	π^+
A	$\bar{C}$	1	1	1	$\frac{1}{2}$	K^+
B	$\bar{A}$	0	-1	0	-1	π^-
B	$\bar{C}$	1	0	1	$-\frac{1}{2}$	K^0
C	$\bar{A}$	-1	-1	-1	$-\frac{1}{2}$	K^-
C	$\bar{B}$	-1	0	-1	$\frac{1}{2}$	$\overline{K^0}$

i.e. all the hadrons of Tables 27.4 and 27.5, can be derived from quark combinations. This seems more than a coincidence and points to the probability of quarks being true fundamental particles. Moreover, in order to complete the hadron charges, it has been necessary to assign to each quark a charge *less* than the electronic charge, viz.:

$$A + \tfrac{2}{3}e, \quad B - \tfrac{1}{3}e, \quad C - \tfrac{1}{3}e.$$

Ideally, it should be possible to reduce the three quarks to one, the other two being excited states of *one* fundamental particle. This may be oversimplifying the problem but *if*, say, quark A proved to be *really* fundamental, many questions would be raised because the basic charge could be 1.067×10^{-19} C instead of 1.602×10^{-19} C. We should have to explain why charges in units of 1.602×10^{-19} C instead of 1.067×10^{-19} C appear in the universe, although this would not violate any known physical principles.

Attempts are now being made to find quarks in the fields of cosmic ray and large accelerator research, but the difficulties are great. The quark mass might be very high and would give birth to the large numbers of other particles. A quark mass of 10 GeV is quite possible and if quarks are to be formed in proton collisions relativistic energies of perhaps 1000 GeV (1 TeV) may be necessary, which means searching for cosmic ray events with incident proton energies of at least 10^{12} eV! The cross-section for each collision will be very low and the flux of cosmic rays of this energy is also very low, so that the probability of getting a free quark on a photographic plate is very small indeed. More intense beams of protons may eventually be available from accelerators when the search for quarks will no doubt be intensified. However, the search for charges $\tfrac{1}{3}e$ and $\tfrac{2}{3}e$ in cosmic radiation still goes on. After a total of many thousands of pictures no events have been found involving either of these fractional charges, and it is estimated that the quark flux is less than about 1.5×10^{-13} quarks m^{-2} s^{-1} sr^{-1} for both charges at altitudes greater than 450 m above sea level, representing about 5 quarks per million years.

It has been suggested that meteorites might have collected quarks during their lifetime in the planetary system of 4.5×10^9 years ($= 4.5$ Ga). Such bits of cosmic material which have arrived on earth would perhaps contain charges less than the electronic charge. Attempts to detect these subelectronic charges in meteorites are now being tried but have yet to be found.

The concentration of quarks in matter is exceedingly small. In meteorites it is estimated to be only 10^{-17} quarks per nucleon, and as low as 10^{-32} quarks per nucleon in air and 10^{-27} quarks per nucleon in sea-water.

The detection and identification of these subelectronic charges obviously calls for a great deal of experimental ingenuity. (See Chapter 28.)

27.9 Conclusions

The great multiplicity of particles make it clear that the words 'fundamental' and 'elementary' have no meaning in their original sense. Indeed it could be argued that a resonance with a lifetime of only 10^{-23} s cannot be regarded as a particle at all! Application of the mathematics of unitary symmetry has given some meaning to the particles and their properties. Unitary symmetry applied to quarks gives rise to certain of the conservation laws we have studied. Unitary symmetry predicts quark conservation in a collision and this leads to the conservation of charge Q of hypercharge Y and of baryon number B and isospin I, but it is only valid for strong interactions.

More sophisticated unitary symmetry methods are being studied which include ordinary spin J as a variable. If all this leads to the conclusion that matter is formed of quarks of very high binding energy, would this be regarded as a new Rutherford–Bohr situation in an entirely new world of physics? The real world is both infinite and infinitesimal and the possibility of quasars at one end of the scale and quarks at the other makes the study of the structure of the universe a fascinating problem. Even so, the most important problem in modern physics is the understanding of structure of the atom as a whole. At present, instead of showing how to build up the material universe from its component atoms we seem to be breaking down the atoms themselves into more and more infinitesimal and ephemeral fragments, demanding bigger and bigger machines to produce more and more particles.

Problems

27.1 Make a list of all the neutral particles and write down the possible products of the high-energy collision process. Apply the conservation laws to find the observed products of this collision. See Fig. 26.10 for photograph.

27.2 Distinguish between a boson, a fermion, a muon, a lepton, a hadron, a baryon, a meson, a nucleon and a hyperon. How would you describe a neutral kaon and the lambda particles?

27.3 A particle is designated $\overline{N^0}$ 1688 MeV. What is this particle? How has it been observed? Write down its quantum numbers.

27.4 Which of the following reactions are possible? Justify your choice.

$$n^0 \to p^+ + e^- + \tilde{v}_e,$$
$$\pi^- + p^+ \to \Xi^- + K^+ + K^0,$$
$$\pi^- \to \mu^- + \tilde{v}_\mu,$$
$$\pi^- + p^+ \to \Sigma^+ + K^-,$$
$$\pi^- \to \mu^- + \tilde{v}_e,$$
$$p^+ \to e^+ + \gamma,$$
$$p^+ + \pi^- \to \Lambda^0 + K^0,$$
$$p^+ + p^+ \to 2n^0 + 2e^+.$$

27.5 Apply the conservation laws to the following reactions; hence pick out those which cannot be observed.

$$\mu^- \to e^- + \nu_e + \tilde{\nu}_e,$$
$$\mu^- \to e^- + \tilde{\nu}_e + \nu_\mu,$$
$$\pi^- + p^+ \to \Xi^0 + K^+ + K^0,$$
$$\pi^- + p^+ \to \Lambda^0 + K^0,$$
$$\pi^- + p^+ \to \overline{\Xi^-} + \Xi^- + n^0,$$
$$\tilde{\nu}_\mu + p^+ \to n^0 + \mu^+,$$
$$K^+ \to \pi^- + 2e^+,$$
$$K^- + n^0 \to \Lambda^0 + \pi^-,$$
$$\pi^- + p^+ \to \Lambda^0 + \pi^-,$$
$$\pi^+ + n^0 \to K^+ + K^0.$$

27.6 Which conservation laws are violated in the following reactions?

$$\tilde{\nu}_e + p^+ \to n^0 + e^+,$$
$$\mu^- \to e^- + \gamma,$$
$$\pi^+ + n^0 \to \Lambda^0 + K^+,$$
$$\pi^+ + \overline{p^+} \to \Sigma^+ + K^-,$$
$$\pi^+ + \overline{p^+} \to \overline{\Sigma^-} + \pi^-,$$
$$\Xi^- \to 2\pi^- + p^+,$$
$$K^- + p^+ \to \Sigma^- + \pi^+,$$
$$\Sigma^- + p^+ \to K^+ + \pi^+,$$
$$\overline{\Sigma^-} + p^+ \to \pi^+ + \pi^+.$$

27.7 Discuss the reality of the resonance particles.

27.8 Draw the meson octet pattern for $J=0$ on rectangular Y, I_3 axes. Hence show the position of the η^0 meson.

27.9 Similarly plot the decuplet of $J=\tfrac{3}{2}$ baryons on a rectangular Y, I_3 plot and show how it led to the prediction of the Ω^- particle and its properties.

27.10 What is meant by 'strangeness' in the strong interactions in nuclear

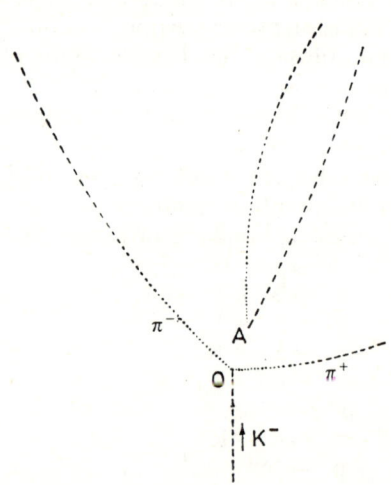

physics? Given that the K^+-meson is assigned a $+1$ unit of strangeness and that the following interactions conserve this quantity
$$\pi^+ + p^+ \rightarrow \Sigma^+ + K^+$$
and
$$K^- + p^+ \rightarrow \Sigma^+ + \pi^-$$
find the strangeness of the K^- and Σ^+ particles.

The figure is taken from a bubble chamber photograph of a K^--meson event. The K^--meson collides with a proton at O and produces two pions and a neutral particle. The neutral decays at the point A into a positive and a negative particle.

(i) Deduce the strangeness of the neutral particle.
(ii) Show that the neutral particle is a baryon and not a meson.
(iii) Explain why the two particles produced at A are not of equal mass and deduce which is the greater.
(iv) Suggest a decay scheme for the neutral particle.

[N]

Chapter 28

Charm and All That

28.1 The Forces of Nature

In the last chapter we saw that the whole of the physical world was controlled by only four forces — gravitational, weak interaction (characteristic decay time 10^{-10} s), electromagnetic (charge interactions) and the strong nuclear force (characteristic decay time 10^{-23} s), in ascending order of strength. Much was known about these forces separately, but there had been little progress towards a unifying theory. However, in the 1960s a deeper interest in the connections between the nuclear forces was shown.

The bubble chamber chamber Gargamelle at CERN was being used in 1970 for neutrino–proton collision experiments. Neutrinos have extremely low collision cross-sections (10^{-48} m²) so that a very high flux of neutrinos is necessary if a neutrino interaction is to be rewarded. The large CERN bubble chamber was ideal for this sort of work. Neutrinos are solely involved in the weak interaction, so that whenever a neutrino event took place the weak interaction was in evidence. The usual reactions were such that in all neutrino collisions the neutrinos changed into charged leptons. Thus the muon neutrino changed to a muon whenever a collision in the bubble chamber took place. The neutrino completely lost its identity. However, some collisions were recorded in which this did not happen — the neutrino collided in the bubble chamber and *still remained a neutrino*. This led to the supposition of a new weak force and a mediating particle not unlike the electromagnetic force and its mediating particle, the photon.

It was Weinberg and Salam in 1967 who suggested the unification of the weak interaction and the electromagnetic force. In the pictorial representation called the Feymann diagram, we can illustrate the transfer of a photon between two electrons via the electromagnetic force, as in Fig. 28.1. Here the photon couples to the electrons at A and B and thus transfers angular momentum. The photon is the mediator or force-carrier.

Similarly the weak interaction can be regarded as the transfer of a massive boson W in analogy with the photon, which is also a boson. The charged bosons $W^{\pm}$ are then the current mediators of the weak interaction.

The case when the boson is a charged W^+ particle is shown in Fig. 28.2. Since the W particles are massive, the uncertainty principle tells us that the range of the interaction must be small, whereas in the case of the massless photon it is large, i.e. of atomic size. For a very heavy boson, the range is almost zero.

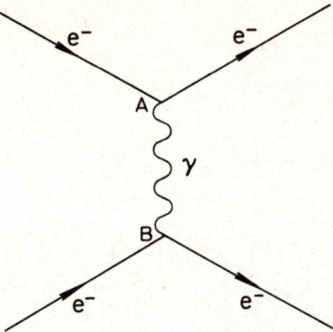

Fig. 28.1 Electromagnetic interaction. Photon mediator.

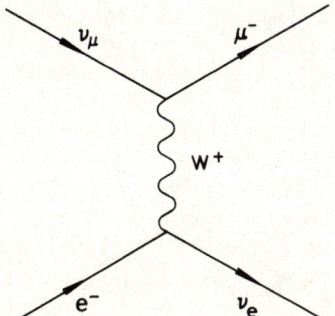

Fig. 28.2 Weak interaction. Positive current mediator.

To explain the fact that some neutrino collisions left the neutrino unchanged, as is the electron in the photon case, the Feynman diagrams of Fig. 28.3 are appropriate. Here the v_μ particle exchanges with the electron via a *neutral* current mediator Z^0 (W^0 in some texts). Similarly, a nucleon may be turned into a number of hadrons, leaving the v_μ particle intact.

This leads to the possibility of three vector bosons W^+, W^- and Z^0 associated with the weak field. A neutrino can therefore generate its W mediator and change into a lepton. The mediator itself can then react with, say, an electron, to change back to a neutrino. The original neutrino could also generate a neutral mediator and *remain* an unchanged neutrino. There was thus a close similarity between the electromagnetic force mediated by the photon and the weak interaction mediated by the $W^\pm$, Z^0 particles. The possibility of a neutral weak interaction mediator Z^0 similar in action to the photon mediator meant dividing the weak interaction into two parts. One mediated by charged bosons $W^\pm$ and the other by the neutral intermediate boson Z^0 as originally proposed by Weinberg. This then suggested that there should be many neutral current interactions among the millions of recorded collisions. Some of these have now been found.

From these experiments it was concluded that there was indeed a distinct similarity between the electromagnetic and the weak interactions. The Weinberg

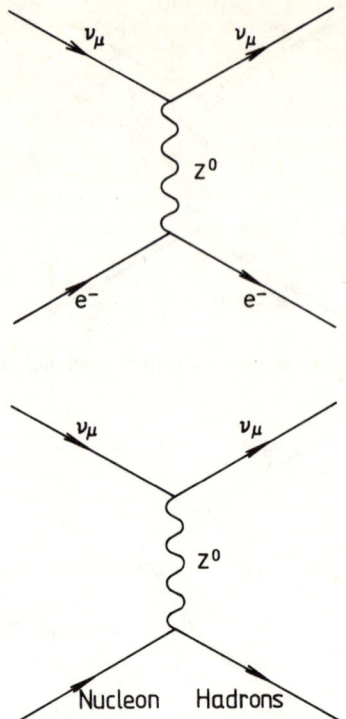

Fig. 28.3 Weak interactions. Neutral current mediator.

and Salam theory indicated that the γ-photon and the Z^0 particle belonged to the same family and were therefore genetically related. They manifest themselves in the electromagnetic and the weak neutral current interaction respectively. The latest (1978) experiments on electron–proton scattering using spin-polarized electrons supports the unified field theory. This was the first step towards a complete unification of the forces of nature.

The $W^\pm$ particles have not yet been found as free particles as their estimated masses are

$$W^\pm \sim 50 \text{ GeV} \quad \text{and} \quad Z^0 \sim 70\text{–}90 \text{ GeV},$$

showing that they are far heavier than the heaviest particle so far found, and must await the development of particle accelerators in the 1980s for their direct observation.

We can now tabulate the new ideas on the forces of nature as in Table 28.1.

The source particles for the quantum mediators are as follows:

	Source particles	*Mediators*
1.	Newtonian masses	Gravitons
2.	Leptons	W particles
3.	Electric charges	Photon
4.	Baryons	Mesons

All the mediators are bosons with spins 0, 1, 2, ..., etc.

TABLE 28.1
Summary of Forces of Nature Proposed in 1978

Group	Force			Boson mediator or quantum carrier	Force strength factor
I			Universal gravitation	Graviton	Very weak, 10^{-39}
II	Weak nuclear interactions involving quarks and leptons	IIA	Weak interaction via charged currents	W^+ \} Intermediate vector bosons W^-	Weak, 10^{-13}
		IIB	Weak interaction via neutral current	Z^0	
III	Electromagnetic, involving electric force acting on charged particles only			γ-photon	10^{-2}
IV	Strong Nuclear Interaction via quarks	IVA	Short range nuclear forces	Meson	1
	Very strong chromodynamic interactions	IVB	Very strong nuclear colour force. Weak at very short range	Gluon	>1

The Weinberg–Salam theory combines IIB and III.

The weak interaction is involved in *all* β-decays and leptonic interactions. It also occurs in a few hadronic interactions such as neutron decay. The strong interaction is *always* involved in hadronic decays and is an essential feature of all high-energy particle interactions.

It will be noticed that in Table 28.1 the subdivision of the strong interaction force IVB introduces the strange terms 'chromodynamics', 'colour force' and 'gluon'. These are the result of intense research in particle physics since 1970, both theoretical and experimental, which we will now describe.

28.2 The Three-Quark Trick

In the last chapter three quarks and their antiquarks were invoked to predict the quantum numbers of the known hadrons. These quarks were labelled A, B and C for convenience, and their various hadronic combinations were shown in Tables 27.6 and 27.7. It is obvious from these tables that the quarks can be combined variously with remarkable success to produce the quantum numbers of all the hadrons. There are no exceptions. Despite this success, and despite vigorous research, free quark particles have not yet been observed and this fact may yet prove to be the cornerstone of quark theory.

The three quarks A, B and C have now been labelled u (up) d (down) and s (sideways, or strange) and are called quark 'flavours' in quark theory jargon. In the three-quark model all mesons ($B=0$) are formed of any quark and any antiquark combinations, i.e.

$$\text{mesons:} \quad q_1 \bar{q}_2 \quad \text{or} \quad q_1 \bar{q}_1,$$

baryons ($B=1$) are formed of any three quark combinations, i.e.

$$\text{baryons:} \quad q_1 q_2 q_3,$$

and antibaryons from any three antibaryon combination, i.e.

$$\text{antibaryons:} \quad \bar{q}_1 \bar{q}_2 \bar{q}_3.$$

Quarks and antiquarks do not mix in baryons. Quarks never appear in isolation. The quantum numbers of u, d and s are based on a triplet symmetry of Y and I_3 (hypercharge and third isospin component), as shown in Fig. 28.4(a) and (b) and in Table 28.2. We also show on Fig. 28.4 the Q coordinates which depend on I_3 and Y through the equation

$$Q = I_3 + \tfrac{1}{2} Y$$

Table 28.2 gives the fundamental quantum numbers of the three quarks and their antiquarks. For all the hadronic quark combinations these quantum numbers are added as in atomic physics. Each quark flavour is a unique combination of Y and I_3, from which all hadron quantum numbers can be derived. Thus, since $B=0$ for mesons ($q_1 \bar{q}_2$) and $B = \pm 1$ for baryons ($q_1 q_2 q_3$; $\bar{q}_1 \bar{q}_2 \bar{q}_3$) it follows that each quark has the same baryon number $B_q = \tfrac{1}{3}$ and each antiquark has $B_{\bar{q}} = -\tfrac{1}{3}$. Quark quantum numbers are assembled in Table 28.3.

All quarks are fermions with $J = \tfrac{1}{2}$, which means that all mesons ($q_1 \bar{q}_2$) are bosons and all baryons ($q_1 q_2 q_3$; $\bar{q}_1 \bar{q}_2 \bar{q}_3$) are also fermions with $J = \tfrac{1}{2}, \tfrac{3}{2}$, etc. For the antiquarks ū, d̄, s̄ and c̄ we invert Q, I_3, B, S, Y and C as in Table 28.3.

To identify a hadron with a given quark combination is relatively simple. For example, both the proton and the π^+ pion have $Q=1$, but whereas the proton is a baryon the pion is a meson. Now the only quark combinations giving $Q=1$ for a

TABLE 28.2
Basic Quantum Numbers of Three Quarks

Quark	I_3	Y	$Q = I_3 + \tfrac{1}{2} Y$
u, ū	$\tfrac{1}{2}, -\tfrac{1}{2}$	$\tfrac{1}{3}, -\tfrac{1}{3}$	$\tfrac{2}{3}, -\tfrac{2}{3}$
d, d̄	$-\tfrac{1}{2}, \tfrac{1}{2}$	$\tfrac{1}{3}, -\tfrac{1}{3}$	$-\tfrac{1}{3}, \tfrac{1}{3}$
s, s̄	0, 0	$-\tfrac{2}{3}, \tfrac{2}{3}$	$-\tfrac{1}{3}, \tfrac{1}{3}$

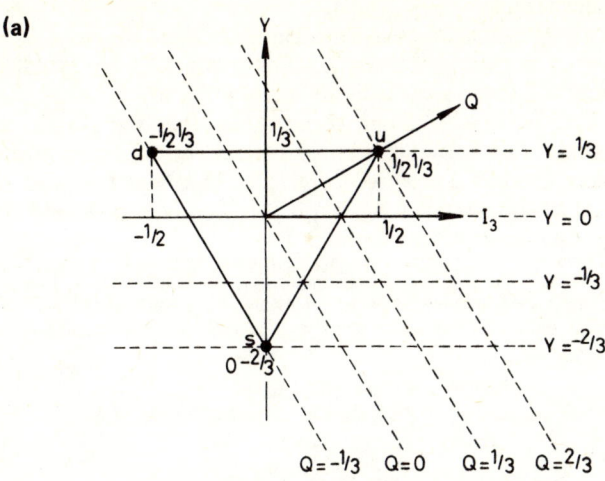

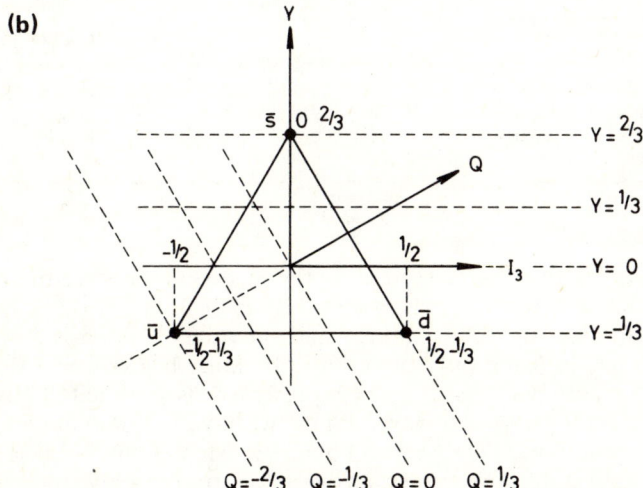

Fig. 28.4 Quark and antiquark triplet symmetries on Q, Y, I_3 plots of three-quark model. (a) The three quarks; (b) the three antiquarks.

baryon are uud and uus and the only way a meson combination can give $Q = 1$ is either $u\bar{d}$ or $u\bar{s}$. In the case of the pion we have $Y = 0$ and we see that $u\bar{d}$ gives $Y = 0$ whereas $u\bar{s}$ gives $Y = 1$, so that $\pi^+ \equiv u\bar{d}$. Similarly, the proton has $Y = 1$ so that uud, giving $Y = 1$, is the correct designation, rather than uus which gives $Y = 0$. In this way *all* the hadrons can be given a quark designation.

Thus the baryon decuplet $J = \frac{3}{2}$ described in the last chapter in connection with the Ω^- particle (Fig. 27.5) can now be rearranged as in Table 28.4.

This is a symmetrical decuplet ranging from ddd to sss including all

TABLE 28.3
Quark Quantum Numbers

Quark	Q	I	I_3	Y	B	S	C
u	$\frac{2}{3}$	$\frac{1}{2}$	$\frac{1}{2}$	$\frac{1}{3}$	$\frac{1}{3}$	0	0
d	$-\frac{1}{3}$	$\frac{1}{2}$	$-\frac{1}{2}$	$\frac{1}{3}$	$\frac{1}{3}$	0	0
s	$-\frac{1}{3}$	0	0	$-\frac{2}{3}$	$\frac{1}{3}$	-1	0
c	$\frac{2}{3}$	0	0	$\frac{1}{3}$	$\frac{1}{3}$	0	1

The quantum numbers Q, I_3, Y, B, S and C change sign for antiquarks. (For convenience we have included the charmed quark c and the charm quantum number C which are discussed in the next section.)

TABLE 28.4
Baryon $J = \frac{3}{2}$ Decuplet Quark Combinations

Baryon	Q = -1	Q = 0	Q = 1	Q = 2	Y	$I_3 = Q - \frac{1}{2}Y$
Delta	Δ^- ddd	Δ^0 udd	Δ^+ uud	Δ^{++} uuu	1	$-\frac{3}{2}, -\frac{1}{2}, \frac{1}{2}, \frac{3}{2}$
Sigma	Σ^- dds	Σ^0 uds	Σ^+ uus		0	$-1, 0, 1$
Xi	Ξ^- dds	Ξ^0 uss			-1	$-\frac{1}{2}, \frac{1}{2}$
Omega	Ω^- sss				-2	0

combinations, and is given as an example of the success of the three-quark model. All quantum numbers can be derived from Table 28.3.

One of the fascinating things about hadron quark synthesis is the great economy of nature in using only three basic building bricks. Even the excited states of mesons and baryons are still represented as outlined. Thus the Δ^+ particle is uud *as is the proton*. However, for Δ^+ we have $J = \frac{3}{2}$ and mass 1238 MeV whereas for the proton we have $J = \frac{1}{2}$ and mass 938 MeV. Thus, because there are many possible spin states, the same quark combination can give more than one hadron with different angular momentum, energy and therefore mass. They are quite distinct states of matter of different masses even though they have the same quark constituents.

We now return to the subject of the previous section — leptons. Quarks can interact with leptons through the weak and electromagnetic forces and theory showed that there was an intimate connection between quarks and leptons. Both were point-like (hadrons were ~1 fm in size) and were regarded as fundamental, but whereas leptons had a neat four-particle symmetry (eν_e; $\mu\nu_\mu$) the quarks had only three, (u, d and s). Somehow the quark production was lagging by one. A new quark was required to rectify this omission. It was expected to be related more to s than either u or d since u and d form a natural pair.

The new quark was called 'charm' and this is the subject of the next section.

28.3 The New Quark — Charm

The four leptons v_e, e, v_μ, μ and their antiparticles are regarded as truly fundamental. They show no structure, unlike the proton, and they have never been subdivided. The three quarks are *also* regarded as truly fundamental, as they also have not been subdivided, but whereas the leptons go in pairs, as above, there is no obvious pairing of the three quarks. Even if u and d are paired the s quark is still a lone quark. Furthermore leptons are free particles whereas quarks are bound particles.

Symmetry of pairing for quarks, in line with the leptons, can be achieved by adding a fourth quark, the charmed quark c. There are now two pairs of quarks, with charges as shown:

$$\text{u, d} \quad \tfrac{2}{3}, -\tfrac{1}{3} \quad \text{and} \quad \text{s, c} \quad -\tfrac{1}{3}, \tfrac{2}{3}$$

in units of e. Here u, d make up a nuclear pair and s, c make up the strange–charmed partnership. The new charmed quark has the identifying quantum number $C=1$. Thus there are eight fundamental quarks and antiquarks based on the quark tetrahedral CYI_3 pattern of Fig. 28.5, which is an extension of the triangular pattern of Fig. 28.4(a). The parameter Q is not included.

The quantum numbers of c are included in Table 28.3. Charm was originally suggested, without any experimental observation, to complete the quark spectrum, and it was based on theoretical speculation prompted by consideration of the weak interaction.

According to the theory, the neutral current boson Z^0 of the weak interaction should mediate the break-up of hadrons containing strange quarks or antiquarks, e.g. the $d\bar{s}$ meson K^0. Now K^0 decays by many modes such as

$$K^0 \to \pi^0 + \pi^0,$$
$$K^0 \to \pi^+ + \pi^- + \pi^0,$$
$$K^0 \to \pi^- + \mu^+ + v_\mu,$$

but the decay $K^0 \to \mu^+ + \mu^-$ is rarely observed. (Although shown in Fig. 26.11 it is suppressed to about 10^{-6} of the other decay processes.) The relative absence of this dimuon decay, so intimately involved with Z^0, cast some doubt on the existence of this mediator. It was postulated that the Z^0 meson was made up of a quark and its own antiquark and in the $K^0 \to \mu^+ \mu^-$ decay process could change from $u\bar{u}$ to $d\bar{d}$ or to $s\bar{s}$. In order to explain the suppression of the $K^0 \to \mu^+ \mu^-$ decay the new quark charm was introduced giving a fourth Z^0 combination of $c\bar{c}$. The theory demanded that the new quark should have the same charge as the u quark and should combine with its antiquark to give an antiphase pairing so that some feasible three-quark decays are removed by destructive interference. Thus the addition of $c\bar{c}$ as a possible Z^0 combination gave theoretical reasons for the non-existence of the $K^0 \to \mu^+ \mu^-$ and other decays. Charm was therefore introduced as a quenching effect to explain experimentally unobserved neutral current weak interactions involving changes in strangeness. Looked at another way, in the $K^0 \to \mu^+ \mu^-$ decay (unobserved) the Z^0 meson changes from a higher to a lower $q_1 \bar{q}_1$ representation but leaves insufficient energy for the production of the $\mu^+ \mu^-$ muon pair. It remains a $q_2 \bar{q}_2$ formation irrespective of the pair with which it started. The effect of a $c\bar{c}$ pair is to explain this and is to suppress some otherwise expected decays such as $K^+ \to \pi^+ e^+ e^-$.

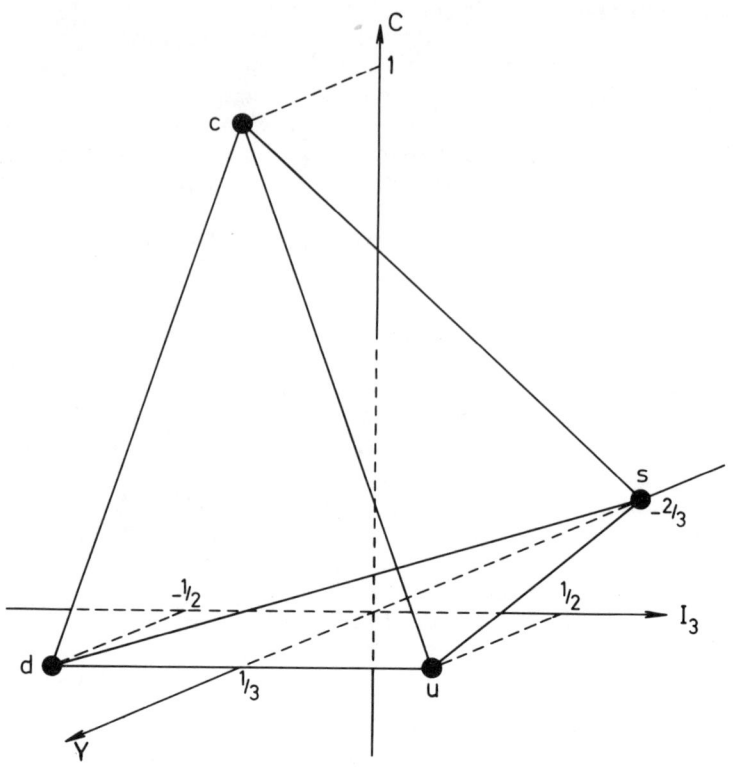

Fig. 28.5 Fundamental tetrahedron for four-quark model.

Quark	I_3	Y	C	$Q = I_3 + \tfrac{1}{2}(Y+C)$
u	$\tfrac{1}{2}$	$\tfrac{1}{3}$	0	$\tfrac{2}{3}$
d	$-\tfrac{1}{2}$	$\tfrac{1}{3}$	0	$-\tfrac{1}{3}$
s	0	$-\tfrac{2}{3}$	0	$-\tfrac{1}{3}$
c	0	$-\tfrac{1}{3}$	1	$\tfrac{2}{3}$

The four quarks have quantum numbers shown in Table 28.3. The charmed quark has $C=1$ and the antiquark $C=-1$, all other quarks having $C=0$. Charm C, like strangeness S, is conserved in strong and electromagnetic interactions but not in weak interactions. The synthesis of the already known hadron properties is not altered since for them $C=0$, but now the possibility of charmed hadrons arises. Before 1974 these had not been found but it was realised that charmed mesons (say, $c\bar{q}$) and charmed baryons (say, cq_2q_3) must be found if the charm theory was to be vindicated. It opened up an entirely new field of hadron possibilities.

The charmed quark is thought to be the heaviest quark and charmed hadrons were expected to decay to lighter quark combinations via the weak interaction. Thus the charmed analogue of the proton is udc ($Q=1$) whereas on the three-quark model the proton is uud, both having $Q=1$.

Thus the four quark flavours and their antiflavours are involved in all the possible hadronic transformations (many undiscovered). The introduction of the charmed quark was made in order to explain the absence of certain weak interaction decays. But does it exist?

28.4 The November Revolution — the J/ψ particle

The reality of the charmed quark was demonstrated experimentally in November 1974. One of the important aspects of quark theory is that it deals with symmetries, so that new particles with precise masses and quantum numbers can be predicted, as in the case of the Ω^- baryon. The experimentalists can then take this hypothetical situation and set the beam energy, target and detector to optimize the search. The probability of the event may be very small but this is offset by the many millions of events recorded and measured.

The first particle discovered containing a charmed quark was the famous J/ψ particle — doubly named because it was discovered almost simultaneously in November 1974 by two American groups working from opposite ends, as it were. At Brookhaven they were using the big proton accelerator to give the simultaneous production of e^+, e^- pairs. On scanning the energy spectrum of the products for resonances, an unknown peak was found at 3·1 GeV representing a new particle. This was a very heavy particle with a relatively long lifetime of 10^{-20} s compared with the 10^{-23} s of the strong interaction decay. This was entirely unexpected. The leader of this group was Samuel Ting. The particle was named the J-particle.

The second group was the Stanford group working with the linear accelerator ring called SPEAR. In this arrangement the products of e^+, e^- *annihilation* were studied. On collision, high-energy γ-rays were produced from which many other particles could emerge. Scans at higher and higher energies were made and some inconsistencies were found at about 3 GeV. After some refinements had been introduced the peak energy was found at 3·1 GeV, and this was a new particle named the ψ particle since the computer display showed tracks shaped like the Greek letter ψ, (see frontispiece for similar example). The leader of the Stanford group was Burton Richter.

The new particle J/ψ (gypsy?) was thus born twice, and was identified as the $c\bar{c}$ meson, confirming the existence of the charmed quark and earning Ting and Richter a Nobel Prize in 1976. As a new particle it could not be formed of a two- or three-quark combination because the permutations of u, d and s had accounted for *all* the known hadrons. Hence it was a new hadron, containing the new quark c, and now identified as the $c\bar{c}$ meson with $J=1$, which seemed to have a 'slowed down' strong interaction rather like the known φ meson which has $s\bar{s}$ and $S=0$. Its decay lifetime was a thousand times longer than that of the strong interaction.

Within 10 days the Stanford group had a heavier particle at 3·68 GeV, identified as an *excited* $c\bar{c}^*$ (J/ψ)* particle decaying to J/ψ by γ-emission whereas J/ψ decays by pions, electrons and muons. The J/ψ particle has no overt charm, i.e. has hidden charm $C=0$, but it supports the idea of a charmed quark and opens up the whole area of charmed hadrons as free particles.

As a $c\bar{c}$ combination it is not unlike the p^+e^- combination of the hydrogen atom and as in the hydrogen atom the higher $c\bar{c}$ energy states could be calculated. One was predicted at 3·67 GeV, exactly as found at Stanford. This was later confirmed by the German team at Hamburg. The charmed quark was now well authenticated.

The discovery of a charmed hadron (i.e. with so-called naked charm) came in July 1976. Particles with naked charm were very heavy and were expected to break up into many smaller particles. The observed $e^+ e^-$ annihilations and decay processes are:

$$e^+ e^- \to \psi''$$

(an excited ψ state), followed by

$$\psi'' \to D^0 + \overline{D^0}$$

where D^0 and $\overline{D^0}$ are new particles with naked charm, followed by the decays

$$D^0 \to K^- + \pi^+,$$
$$\overline{D^0} \to K^0 + \pi^+ + \pi^- + \pi^+ + \pi^-.$$

The existence of such charmed particles could be deduced from the energy—momentum analysis of photographs of collision events. Theory showed that the lightest charmed hadron would be expected at about 1·9 GeV as $c\bar{u}$. This was found at 1·87 GeV and labelled the D^0 meson. The charmed proton at 2·26 GeV was found in August 1976 and identified as udc. Note that as with other quarks, the charmed quark is bound and is (as yet) unobserved. As before, the rule that mesons are $q_1 \bar{q}_2$ and baryons $q_1 q_2 q_3$ still holds, but any q in these combinations may now be c.

Perhaps the most significant result of the four-quark model is that it provides a deeper understanding of how nuclear matter is derived from its building bricks. It provides a similar pattern to the electron model building in *atomic* matter. Hadrons are atom-like systems of quarks and the J/ψ system is a bound system just like the $p^+ e^-$ system of the hydrogen atom. Perhaps a better comparison would be the bound positronium system of e^-, e^+ (i.e. e, ē). By analogy, the $c\bar{c}$ system is called charmonium and leads to energy states like its electronic equivalent. Photon transitions between these states are the common feature. This predicts a subnuclear chemistry with quarks moving in substates analogous to the atomic electron states. The discovery of a fifth quark (Section 28.9) confirms this idea.

28.5 Quark Multiplet Representation

The introduction of the charmed quark requires the two-dimensional plots of Figs 27.4 and 27.5 to be extended to three dimensions. And just as these plots enabled the quantum properties of η^0 and Ω^- to be predicted so the new diagrams enable the properties of many undiscovered hadrons to be predicted. The C, Y, I_3 plot for the meson family for $J=0$ which includes J/ψ written as $c\bar{c}$ is shown in Fig. 28.6. The known hadrons are situated in the hexagon in the $C=0$ plane, the unknowns have $C=1$ or $C=-1$. Some estimated masses are

$$\left. \begin{array}{l} D^0 = 1·863 \\ D^+ = 1·874 \end{array} \right\} \text{corresponding to } K^0 \text{ and } K^+$$
$$D^{0*} = 2·006$$
$$D^{+*} = 2·008$$
$$F^+ = 2·030$$
$$F^{+*} = 2·140 \text{ GeV}$$

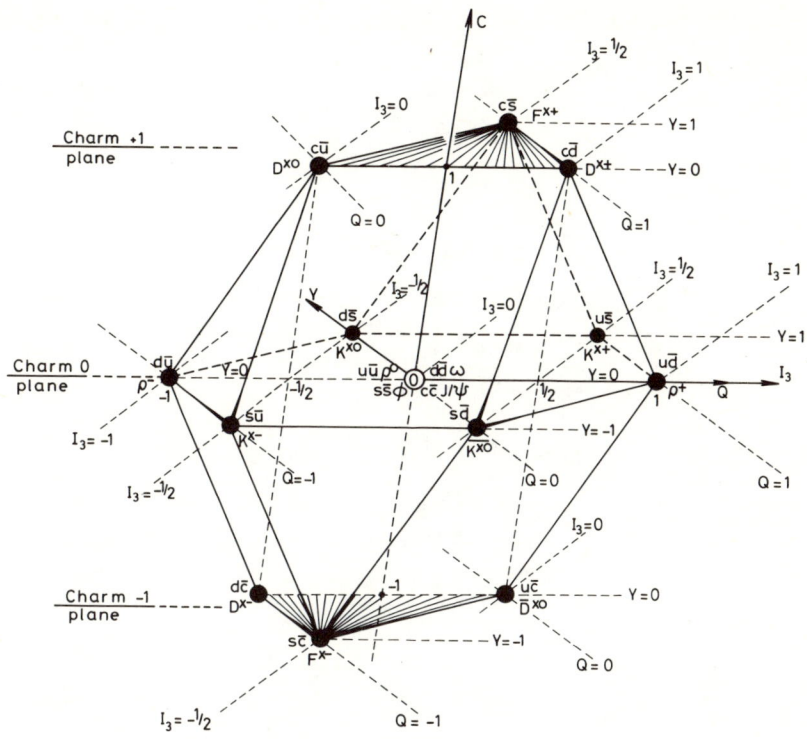

Fig. 28.6 Quantum numbers of charmed excited states of basic meson family with $J = 1$, as predicted by four-quark model.

in energy units, of which the first four have been 'seen'. It is obvious from this that we are on the threshold of a new charm spectroscopy with the higher states decaying to lower states via high-energy γ-emission.

Figure 28.7 is the charm extension of the Ω^- decuplet plot. Here we have tried to plot Q, Y, I_3 and C to show their connection and from it the new charmed hadrons are seen to be those in the $C = 1$, 2 and 3 planes. Thus the charmed proton at cud is predicted and has been found experimentally. Its mass is 2·26 GeV and this figure agrees with the theoretical prediction. The masses and quantum properties of all the charmed hadrons of Fig. 28.7 are predictable and this helps enormously when setting the experimental conditions to look for a new particle. Thus the charge formula for all hadrons, including charm is now, $Q = I_3 + \frac{1}{2}(Y + C)$.

Having established the charm theory as an experimental reality the question now arises — what holds these quarks together within the hadron nucleus? It must be some very strong attractive force to prevent the quarks escaping (free quarks have never been observed) and it turns out to be the strongest possible interquark glue, appropriately called the gluon force.

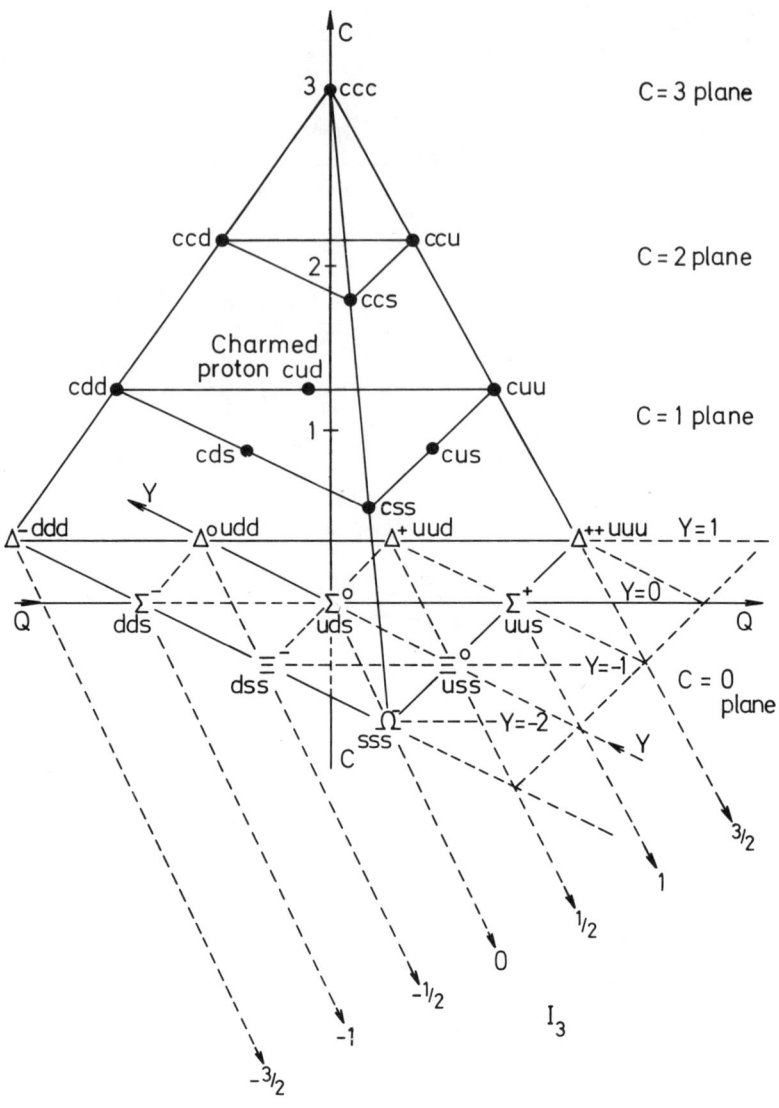

Fig. 28.7 Quantum numbers of charmed baryon $J = \tfrac{3}{2}$ family.

28.6 Gluons and Colour

The long history of particle physics has been one in which imaginative systems of particles have been suggested by the theorists and ultimately proved to be correct by the experimentalists. The simple neutrino is a case in point. Predicted by Pauli almost fifty years ago, it had no mass, no charge, but spin $\tfrac{1}{2}$ to conserve angular momentum in β^--emission. No wonder it had to wait until 1957 before it was proved to be a free particle — there was nothing to measure! But the experiment of Reines and Cowan, who knew what they were looking for, proved beyond doubt that this elusive particle existed and could be used as a bombarding particle, albeit with an extremely low collision cross-section. Neutrinos are now commonplace in particle physics. The discovery of the Ω^- particle is another example of a theoretical prediction being found.

The quark particle model is also a piece of elegant theory. It represents a great simplification of the building processes of nuclear matter. But why have free quarks not been found? With powerful accelerating machines available and tens of millions of particle tracks to analyse one would have expected the odd quark to have shown up by now on some nuclear plate somewhere. Either quarks do not exist or their non-appearance has to be explained.

From time to time we have compared particle physics with atomic physics and we now come to another point of comparison. There are four hadrons each with three identical quarks, viz.:

$$\Delta^+ = \text{uuu}, \quad \Delta^- = \text{ddd}, \quad \Omega^- = \text{sss},$$

and the hadron (ccc), not yet found. Since all quarks are fermions, $J = \frac{1}{2}$, and the three hadrons quoted have $J = \frac{3}{2}$, it follows that their spins are all parallel and the quarks are in the same quantum state. If we extend the Pauli exclusion principle to quarks in hadronic orbitals we could say that no two quarks in a hadron can have the same set of quantum numbers. Clearly the $q_1 q_1 q$ baryons violate this since these baryons must have at least *two* quark spins parallel and sometimes all three. We therefore require a fourth state, different for the three quarks, which will differentiate between them. The fourth quantum number is called 'colour', and because each quark in a $q_1 q_1 q_1$ baryon must be different, it has been customary to use *three* colour states labelled red, green and blue (or sometimes red, white and blue). These are the primary colours, of course, so that the anticolours are the complementary colours i.e. turquoise (cyan), mauve (magenta) and yellow. The introduction of colour must now triple the number of quarks because any quark can be associated with *any* colour, i.e. there are 12 coloured quarks

	R	G	B
u	×	×	×
d	×	×	×
s	×	×	×
c	×	×	×

None of these is observable and the number of hadron possibilities remains the same. Now the theory says that all hadrons are 'colourless' i.e. their quark colours add up to 'neutral white'. This is easily seen for mesons, $q\bar{q}$, i.e. a colour and its complementary colour. In the case of baryons, each quark carries a different colour, so that *all* baryons of whatever quark combination, have $q_1(R)q_2(B)q_3(G)$, giving a neutral white combination. These are sometimes called 'neutral molecules'. We see, then, that colour is a symmetry introduced to give colourless hadrons from coloured quarks.

The strong force of attraction between coloured quarks must be greater than that experienced between mesons and baryons and the theory of the colour force, called appropriately quantum chromodynamics (QCD) in analogy with quantum electrodynamics (QED), shows that there are eight and only eight particles carrying the colour force. These are called coloured gluons. Gluons are the mediators of the colour force, which is the strongest of all known physical forces. Gluons are bosons of $J = 1$, like photons, and can be regarded as the quanta of the colour force, just as photons are the quanta of the electromagnetic force.

We have seen that all hadrons are composed of coloured quarks, but although all three colours are present in a baryon it is impossible to say which colour belongs to which quark. All colour combinations are equally probable. This is a sort of 'colour uncertainty principle', and the only particles observed are those in which the colour forces are saturated in neutral white. The crux of the theory is that all baryons are RGB and all mesons are one of $R\bar{R}$, $G\bar{G}$ and $B\bar{B}$, irrespective of their quark structure. This raises interesting points of structure in the case of the proton. From a completely structureless particle it has gone through many phases, showing internal structure, until it can now be regarded as a closed system of three point-like quarks together with a virtual quark–antiquark pair all exchanging coloured gluons at will. This is not unlike the Rutherford model of the atom with its point-like nucleus.

28.7 The Confinement of Quarks

Quark theory shows that at normal nuclear distances of 10^{-15} m quarks are tightly bound and cannot be shifted, but at much smaller distances they can move about freely. This situation has been called 'infra-red slavery' and 'ultra-violet freedom'. Hadrons thus consist of tightly bound quarks bonded by the colour force. Theoretical reasoning predicts that quarks are absolutely inaccessible to matter probes so that the question arises — if they are inaccessible do they really exist? If leptons are observable, why not quarks?

There have been several attempts to explain this paradoxical situation — short-range freedom and long-range slavery — leading to inaccessibility. One such is the string model, in which the string is composed of gluons and the colour forces holding the quarks together in the hadrons are considered to be concentrated along strings tying the quarks together. Breaking the string between two quarks by a high-energy bombarding particle does not free the two quarks but merely reproduces the original condition by creating a new colourless particle, i.e. a $q_1\bar{q}$ meson, from the break, while the two exposed ends join up again to reproduce the original quark arrangement. It is rather like breaking a bar magnet hoping to isolate the two polarities: an isolated NS [i.e. $N\bar{N}$] pair can never be created by breaking up the magnet.

Another model is the so-called bag model in which the quarks are supposed to be confined to a bag whose membrane is impenetrable. The bag is kept inflated by the quarks, which can only be separated by applying energy and blowing the bag up like a balloon. In equilibrium the pressure of the quark 'molecules' just balances the natural tendency of the bag to shrink, like a bubble's surface tension arrangement.

It is obvious that much inventive thought is being given to the problem of quark confinement. The original three quarks, the charmed quark, the concept of colour and gluons were all ideas born of theory. Experimentalists then found evidence of them. The theory of quark confinement tells us that free quarks cannot exist and are permanently frozen in the hadrons. Some physicists believe that free quarks do exist, but rarely. If not, it is possible that quark confinement is an absolute law of nature.

28.8 The Hunting of the Quark

In spite of all that the theorists say about the inaccessibility of the quark there is intensive research going on for their discovery. Claims have been made for their existence in cosmic ray tracks, but never substantiated. One idea is to look for a particle track with a fractional charge either in ancient cosmic ray events in rocks of great age which have been bombarded for billions of years or in lunar soils. None has been found.

And yet the search goes on. The most direct method is a replica of the Millikan e experiment, as shown in Fig. 28.8.

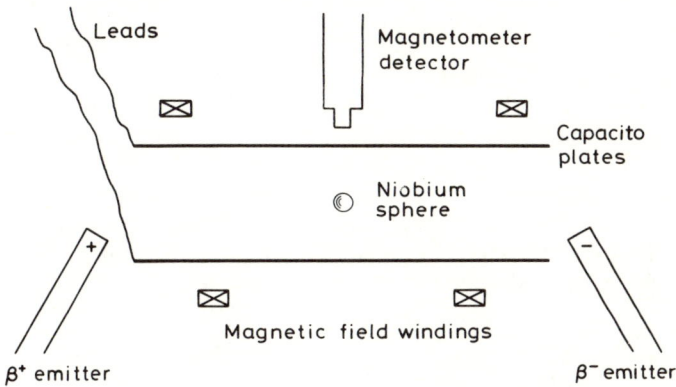

Fig. 28.8 Scheme for free quark Millikan-type experiment.

The diagram shows the essential features of an experiment in which a niobium sphere of 0·25 mm size is kept at liquid helium temperature so that it becomes superconducting. It is 'floated' in a vacuum between two copper capacitor plates by suitable magnetic fields and the movement of the sphere can be detected by means of a specially designed superconducting magnetometer. The whole apparatus is inside a sealed container surrounded by liquid helium. Thus any magnetic field due to charge on the sphere is nearly permanent. Should any free quarks attach themselves to the sphere they should be detectable by the magnetometer.

With no electric field between the plates (they were 1 cm apart) the vertical frequency of the 'bouncing ball' was about 1 per second. When an alternating square wave field was applied to the plates at the resonant frequency of the ball the magnetometer picks up the change in amplitude and in fact can detect changes of 10^{-6} m.

To get rid of the normal extraneous electric charges on the ball, β^- and β^+ sources are used to increase or decrease the number of electrons on the sphere until the net electric charge on it is zero. It was estimated that a charge of 1–2% of e could be measured.

When the niobium spheres were annealed on a niobium substrate prior to insertion between the plates the results were not the same as when the sphere was annealed on a tungsten substrate. Tests have been made on eight suitable niobium spheres. Five were annealed on niobium and three on tungsten. Of the three tungsten-annealed spheres *two* gave charges of $\frac{1}{3}e$ while one gave zero charge. The figures were 0.337 ± 0.009, 0.331 ± 0.024 and -0.001 ± 0.025 in units of e. The niobium-annealed spheres gave no charge. It is curious that the fractional charge was only picked up from the tungsten substrate as if tungsten was a quark catcher.

These experiments have not been confirmed by other workers although many attempts have been made.

If free quarks *are* found, it means quark containment is only approximate or at any rate a first-order effect. If free quarks are never observed we have a new law of nature — the confinement of quarks.

28.9 Latest News: New Quarks

The researches at CERN, Brookhaven, Stanford and Hamburg had completely vindicated the four-quark model and the idea of coloured gluons. There was a satisfactory symmetry between leptons and quarks. But this was soon removed by the discovery of the τ (tau) lepton in America in 1975 with mass 1.8 GeV. This was a very heavy lepton and was confirmed by electron–positron colliding-beam experiments at Hamburg. If we presume that the τ-particle (or tauon) has an associated neutrino v_τ, we are now into the third generation of lepton fermions:

(1) $e, v_e: m_e = 0.5$ MeV
(2) $\mu, v_\mu: m_\mu = 105$ MeV
(3) $\tau, v_\tau: m_\tau = 1800$ MeV

in ascending order of mass. Each lepton has also an antilepton, of course.

The total number of leptons is now six with six further antileptons. The reality of v_τ is based on the fact that it cannot be found by bombarding targets with v_μ particles, hence it has a separate entity. The mass of v_τ has not yet been determined and it will be interesting to see whether or not it has zero mass like v_e and v_μ. At present, theory cannot tell us whether or not there are more leptons to be discovered.

There is another complication. Allowing that the symmetry between quarks and leptons is fundamental, and that the weak and electromagnetic forces are successfully unified, the third pair of leptons requires a third pair of quarks. (The quarks always seem to lag behind the leptons). Two extra quarks were therefore postulated, named top (t) and bottom (b), or truth and beauty, so that the quark list now reads u d, s c and t b in pairs, together with their antiparticles. There are now three generations of fundamental fermions:

(1) $e, v_e:$ u, d
(2) $\mu, v_\mu:$ s, c
(3) $\tau, v_\tau:$ t, b

in ascending order of mass. As before, the new quarks cannot be found in isolation but evidence of their existence was sought in the 10 GeV range since heavy t, b quarks imply heavy hadrons — heavier than anything hitherto found.

Working on the principle that the J/ψ particle was a bound state of c$\bar{\text{c}}$ with hidden charm, a hidden truth (t$\bar{\text{t}}$) or hidden beauty (b$\bar{\text{b}}$) particle was sought and experiments at Fermilab scanning through the 10 GeV range came up with new resonance peaks at 9·4 GeV, 10·0 GeV and possibly at 10 GeV in 1977. These were called upsilon particles Υ. The first two have been confirmed in Germany and are interpreted as a low-lying state of q$\bar{\text{q}}$ and its first excited state, i.e. it is a bound state of either t$\bar{\text{t}}$ or b$\bar{\text{b}}$ with similar properties as J/ψ. The latest work on the width of the Υ resonance indicate that it is a hidden bottom particle, b$\bar{\text{b}}$, where b = $-\frac{1}{3}e$ and t = $\frac{2}{3}e$ to preserve the symmetry of charge. The upsilon particle is therefore a real particle which is similar to the J/ψ particle. The energy gap to the first excited state is the same as for J/ψ, viz.:

$$\Upsilon' = \Upsilon = \psi' - \psi = 0.6 \text{ GeV},$$
$$10.0 - 9.4 = 3.7 - 3.1 = 0.6 \text{ GeV}$$

suggesting that there are 'bottomonium' states of Υ analogous to the 'charmonium' states of J/ψ, and just as naked charm mesons such as D^0 and D^+ have been found in the debris of e^+, e^- annihilations so we could expect an upsilon meson equivalent of D^0 to be found — with naked bottom — i.e. a possible q$\bar{\text{b}}$ or b$\bar{\text{q}}$ particle, with mass greater than half the mass of Υ''. Extending this, one can see that diagrams such as Figs 28.6 and 28.7 would have multidimensional problems of presentation.

By analogy with the atomic case one could designate the charmonium states as 1S, 1P, 2P and 2S (four particles have been found) in which the heavy charmonium states decay to lighter ones with the emission of high-energy photons. Similarly the upsilon bottomonium states could be regarded as 1S, 2S, (3S) with energies of 9·4, 10·0 and 10·4 GeV. Although the 10·4 GeV resonance has not yet been satisfactorily resolved, the latest storage ring facility at Hamburg, named PETRA, is being tuned to this high-energy region. It can generate both particles (e^-, e^+) to 19 GeV so that at 38 GeV it is the most advanced machine of its kind and it is hoped that the higher states of bottomonium represented by Υ, Υ', Υ'', Υ''' will soon be discovered.

It must be remembered that these are extremely high energies and a particle of mass 10 GeV is approaching the mass of a carbon atom.

28.10 Conclusion

We have described three sets of leptons and quarks and shown how these two types of fundamental particle run in parallel. The theory cannot yet say how many of these there should be and cannot explain why the muon and the tauon are so much heavier than the electron. There is no way, as yet, of understanding the properties of leptons. Perhaps leptons and quarks are the real stuff of the universe.

The strong colour force between quarks, carried by the gluons, governs all meson–baryon interactions, but there is no corresponding lepton colour force.

However, it is conjectured that this might not be true. The leptons *may* be involved with a colour force such as

$$\begin{aligned} e, v_e \quad &\text{red,} \\ \mu, v_\mu \quad &\text{blue,} \\ \tau, v_\tau \quad &\text{green,} \end{aligned}$$

giving a lepton colour quantum number in addition to the usual quantum numbers. Thus we might soon have 'lepton flavours'. Strong interactions are generated by quark colour, so weak interactions are expected to be generated by lepton colour and the two never mix. The essential difference is that gluons of quark colour are massless whereas leptonic colours are massive. That is why R(e), B(μ) and G(τ) have been 'seen' but the quarks R(q), B(q) and G(q) have not. Experiments to find the lepton colour force are now in progress; theory says it should be there.

On the theoretical side the prediction of strong interactions has been good. The crux of the theory is the confinement of quarks, as demonstrated by the absence of free quarks, so that if the niobium sphere experiments come up with definite fractional charges, i.e. isolated quarks, the theory must be drastically revised or new parameters introduced. Theory cannot predict the total number of possible quarks, although it is certain that there are only eight gluons and no more.

We have discussed the forces of nature. With the Weinberg–Salam model predicting the unification of the weak and electromagnetic forces it may be that there are now only three basic forces of nature:

(1) gravitational force,
(2) weak/electromagnetic force,
(3) strong/colour force.

At present we know of no other forces. But it is our belief that there is one and only one universe, and that there is one and only one set of physical laws — known and unknown — so that theoreticians are working towards the ultimate goal — the unification of all these universal forces.

There is still much to be done. Man's knowledge of the universe is still infinitesimal, and his researches provide answers which merely beget more questions. Thus quantum chromodynamics (QCD) and quark theory have given answers to some problems but provoked many fresh questions. How many quark or lepton flavours are there? At present, theoreticians have no idea how many flavours there should be. Are quarks, leptons truly fundamental? QCD assumes that quarks do not have a structure. Would one be revealed in collision experiments at much higher energies than those available today? Likewise, QCD is so dependent on quark confinement. Suppose an isolated quark *is* found, what then? And where does gravity fit into the unified field theory? Is the graviton truly a boson mediator of the gravitational field? The integration of astronomy, cosmology and particle physics into a single theory seems very far away, but nevertheless so important that many theoreticians are already working on it. This problem is the 'Holy Grail' of theoretical physics.

It is said that physics is an exact science. With the common usage of terms such as 'magic numbers', 'quarks', 'strange', 'charm', 'truth' and 'beauty' is nuclear physics losing its exactness? Not so. Behind all these whimsical expressions there is an underlying symmetry and simplicity which belies their superficial meanings.

We conclude with a quotation:

> 'Now the smallest Particles of Matter may cohere by the strongest attractions, and compose bigger Particles of weaker Virtue, and many of these may cohere and compose bigger Particles whose Virtue is still weaker and so on for divers successions, until the Progression end in the biggest Particles on which the operations in Chymistry and the Colours of the natural Bodies depend, and which by cohering compose Bodies of a sensible Magnitude.
>
> 'There are therefore agents in Nature able to make the Particles of Bodies stick together by very strong Attractions. And it is the Business of experimental Philosophy to find them out.'†

Who said that? Why, Isaac Newton, of course!
And this 'Business' could perhaps go on for ever!

Problems

28.1

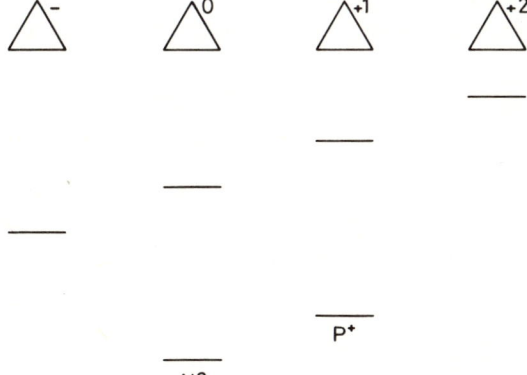

The diagram depicts the energy levels of the $\Delta \to N$ decay system. Insert appropriate values of I, I_3 and Q for each level. Using the selection rule $\Delta I_3 = \pm 1, 0$ show that there are only six possible transitions. Label each transition with the symbol of the particle emitted.

28.2 The following table gives the masses in GeV of some important subnuclear particles.

† Quoted by W. F. Weisskopf in *Science* **149**, p. 1181, 10th September, 1965, from Newton's *Opticks*, p. 394, Dover, New York (1952).

Mass (GeV)	0	0	0·0005	0·105	0·135	0·140	0·494	0·938	0·940	1·020
Particle										

Mass (GeV)	1·116	1·800	1·863	1·868	2·260	3·098	3·684	9·4	10·0	10·4
Particle										

Identify each particle from the text of Chapters 26, 27 and 28 and so complete the table.

28.3 Of the particles identified in Problem 28.2 state which are (a) leptons, (b) mesons, (c) hadrons, (d) bosons and (e) fermions.

28.4 In the above table (Problem 28.2) identify the charmed particles and give their quark constituents. What are their decay products?

28.5 Which has the greater rest mass, the J/ψ particle or the α-particle?

28.6 In the light of the proliferation of quarks, discuss the problem of whether or not they are fundamental particles.

28.7 If quarks seven and eight are ever postulated what will be their probable masses compared with those of the first three generations of quarks?

28.8 What would be the consequences of the discovery of free quarks, i.e. if the niobium sphere experiments prove successful and can be repeated in other laboratories?

28.9 The universe is both infinite and infinitesimal. Discuss.

Appendix A

Relativity Theory

The relativity theory will only be described in sufficient detail to indicate the nature of the evidence which led to the establishment of a relationship between mass and velocity and the implication that mass and energy are equivalent. Many experiments were devised and carried out to try to detect the motion of the earth through the ether, and it was the failure of such experiments which culminated in the relativity theory of Einstein.

Perhaps the most celebrated experiment was that carried out by Michelson and Morley in 1897 using the Michelson interferometer. The experiment amounted to comparing the times required by two light waves to travel equal distances, along paths which were oriented parallel and perpendicular to the direction of motion of the earth in space.

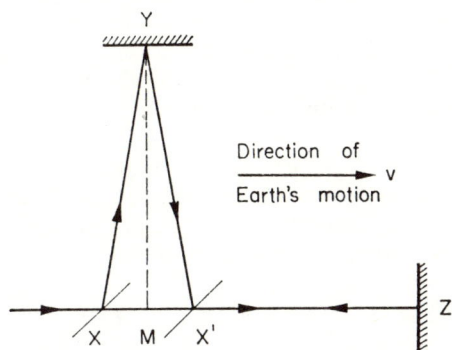

Fig. A.1 The Michelson–Morley experiment.

The light is divided into two beams by a half-silvered mirror at X (Fig. A1). One beam proceeds across the direction of motion of the earth by the path XYX', the mirror having moved the distance XX' during the same time. The other half of the beam proceeds to the mirror Z travelling parallel to the earth's motion. The times taken by the two beams can be compared by observing the interference fringes formed when the beams are reunited. A time difference between the two paths would be revealed as a fringe displacement when the apparatus was continuously rotated. The time taken for the path XZX' is given by

$$t_1 = \frac{d}{c+v} + \frac{d}{c-v},$$

where $d = MZ = MY$ and v is the velocity of the earth relative to the ether. This becomes $2dc/(c^2 - v^2)$ on simplification. The time taken for the light to travel XYX' is given by $t_2 = 2d/(c^2 - v^2)^{1/2}$, which differs from t_1 the time taken for the XZX' path. The difference is given approximately by

$$t_1 - t_2 = \frac{2d}{c}\left(1 + \frac{v^2}{c^2}\right) - \frac{2d}{c}\left(1 + \frac{v^2}{2c^2}\right) = \frac{dv^2}{c^3}.$$

To their surprise, no displacement of the fringes was observed, although the equipment was capable of detecting a displacement of less than a tenth of the expected amount.

Similar experiments, both optical and electrical, failed to reveal any positive effects of the earth's motion in the ether. Following the explanations of Fitzgerald and Lorentz, Einstein, in 1905, formulated his two postulates upon which the special theory of relativity is based. These are:

(1) The laws of physics are the same for all systems having uniform motion of translation with respect to one another. An observer cannot therefore detect the motion of that system by observations confined to the system.
(2) The velocity of light in any given frame of reference is independent of the velocity of the source.

The implications of these postulates are very far-reaching and some indication of their importance will now be given in so far as they affect our ideas relating to the atom. It will be realized that the special theory of relativity is limited to systems moving with uniform velocity relative to one another. Systems moving with acceleration require the general theory of relativity which came ten years later. Our considerations will be limited to the special theory in which systems move with uniform relative velocity.

Consider two sets of axes S, S' (Fig. A.2), the former being fixed and the latter moving with respect to it with uniform velocity v along the direction of the x axis. Further let the origins O, O' coincide at a particular instant to $t = t' = 0$. Consider first a point P whose coordinates in the two systems are (x, y, z, t) and (x', y', z', t').

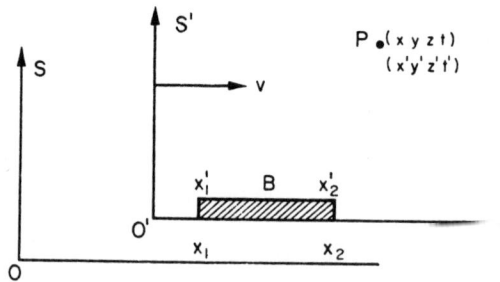

Fig. A.2 Relative motion of two sets of axes.

After a time interval t or t' we may transform from one system S to the other S' by a set of equations $x' = x - vt$, $y' = y$, $z' = z$ and $t' = t$. This is the 'common sense' or Newtonian connection between the two systems. Suppose now that at the instant when the origins coincide a light wave originates at O and O'. In the fixed system S the wave-front will lie upon a sphere defined by $x^2 + y^2 + z^2 = c^2 t^2$, where c is the velocity of light. We may now rewrite this equation in the S' system of coordinates as $(x' + vt)^2 + y'^2 + z'^2 = c^2 t'^2$. This, however, is no longer a sphere as implied by the second postulate of the relativity theory. Moreover it suggests that by measuring the velocity of light in a number of directions it should be possible in principle to determine the velocity of the S' system. The relativity theory requires that the wave-front in the S' system shall also be a sphere given by $x'^2 + y'^2 + z'^2 = c^2 t'^2$ and this is clearly impossible if we insist upon using the Newtonian transformation equations. To give a spherical wave-front in the S' system as well, a new set of equations is required, namely $x' = \beta(x - vt)$, $y' = y$, $z' = z$ and $t' = \beta(t - vx/c^2)$, where $\beta = 1/(1 - v^2/c^2)^{1/2}$. This should be verified by the reader. Moreover, if one wishes to transform from S' back to S the equations become $x = \beta(x' + vt')$, $y = y'$, $z = z'$ and $t = \beta(t' + x'v/c^2)$. These are known as the Lorentz transformation equations.

Certain important consequences follow at once from the Lorentz equations. For example, two events which are simultaneous in the S system, so that $t_1 = t_2$, take place at $t'_1 = \beta(t_1 - vx_1/c^2)$ and at $t'_2 = \beta(t_2 - vx_2/c^2)$ in the S' system. They are therefore simultaneous in the S' system only when $x_1 = x_2$. Furthermore if two events occur at times t_1 and t_2 in the same place (x) in the S system, the time interval between these events measured in the S' system is given by $t'_1 - t'_2 = \beta(t_1 - vx/c^2) - \beta(t_2 - vx/c^2) = \beta(t_1 - t_2)$. Since β is slightly greater than unity the time interval $t'_1 - t'_2$ in S' is slightly longer than in the S system. Thus time appears to have slowed down in the S' system when seen by an observer at rest in the S system. It can similarly be shown that the time in the S system appears to be dilated to the same extent to an observer moving with the S' system.

Suppose a bar B is at rest in system S' and moves with a velocity v relative to S as shown in Fig. A.2. The length of the bar is given by $x'_2 - x'_1$ in the S' system and $x_2 - x_1$ in the S system. How are these related? We must imagine an observer in the S system who has a measuring rod, and as B moves past him he notes the positions x_1, x_2 of the ends of the bar at the same instant. Using the transformation equations we have

$$x_2 - x_1 = \beta[x'_2 - x'_1) + v(t'_2 - t'_1)]$$

and $t'_2 - t'_1 = \beta(t_2 - t_1) - \beta v(x_2 - x_1)/c^2$. But since readings were simultaneous in S, $(t_2 - t_1) = 0$ and therefore $x_2 - x_1 = \beta[(x'_2 - x'_1) - \beta v^2(x_2 - x_1)/c^2]$. This simplifies to $x_2 - x_1 = (x'_2 - x'_1)/\beta$ and as $\beta > 1$ this implies that $x_2 - x_1 < x'_2 - x'_1$, showing that the length of the bar B as seen by an observer in S will be shortened in the ratio $\beta:1$.

If now we return for a moment to the Michelson–Morley experiment we see at once that this is just the amount by which the arm of the interferometer must contract in order to give the null result. Recalling that the time for the light to travel along the arm d_1 perpendicular to the motion is $t_1 = 2d_1/(c^2 - v^2)^{1/2}$ and along the arm d_2 parallel to the motion is $t_2 = 2d_2 c/(c^2 - v^2)$ and that the experiment gives $t_1 = t_2$ we obtain $d_1 = d_2/(1 - v^2/c^2)^{1/2}$. This is just the amount required by the relativity theory as outlined above.

In Newtonian mechanics a body having a velocity u' in the S' system would simply have a velocity $u = u' + v$ in the S system. Although this is very nearly true, important discrepancies appear when the velocities become comparable with the velocity of light. Remembering that $u = dx/dt$ and $u' = dx'/dt'$ and using the Lorentz transformations we have

$$u = \frac{\delta x}{\delta t} = \frac{\beta(\delta x' + v\delta t')}{\beta(\delta t' + v\delta x'/c^2)}$$

$$= \frac{\frac{\delta x'}{\delta t'} + v}{1 + \frac{v \cdot \delta x'/\delta t'}{c^2}}$$

$$= \frac{u' + v}{1 + u'v/c^2}.$$

It is clear that when u' and v are small, the Newtonian method of adding velocities is sufficiently accurate. It is interesting to put $v = c$ when it will be found that u cannot be made to exceed c. This represents the upper limit of velocity for material particles and for radiation.

The above method of adding velocities has an important bearing upon the conservation of mass and the conservation of momentum. If these are *universal* laws, they must be true for whatever system of axes one selects. Expressed mathematically we have in the S' system $\Sigma m = K_1$ for the conservation of mass and $\Sigma mu' = K_2$ for the conservation of momentum, K_1, K_2 being constants and m a typical mass. In the S system, is the conservation of momentum Σmu still true? Using the Newtonian method of adding velocities we can write $\Sigma mu = \Sigma m(u' + v) = \Sigma mu' + v\Sigma m = K_2 + vK_1$. This is still constant, and therefore the conservation of momentum is true for both systems. When, however, one uses the more refined relativity expression obtained earlier we get

$$\Sigma mu = \Sigma m \frac{(u' + v)}{1 + u'v/c^2} = \frac{\Sigma mu'}{1 + u'v/c^2} + \frac{v\Sigma m}{1 + u'v/c^2}.$$

This is no longer constant and we are at once faced with the choice of abandoning either the conservation of mass, the conservation of momentum or of modifying our definition of mass in order to meet the above requirement. This is not such a revolutionary step as it might have seemed earlier. We have already seen how our familiar concepts of length and time have been modified by the relativity theory. If we let m, m' be the masses of a body as observed in the two systems of coordinates S, S' the new definition of mass is given by $m' = \beta m = m/(1 - v^2/c^2)^{1/2}$. Experimental confirmation of this comes from a study of the high-speed particles emitted by radioactive substances (see Section 3.3).

In the problems we shall encounter it is more convenient to refer to m_0 the mass of a body at rest on the earth's surface and to m the mass of the same body in motion with a velocity v relative to the earth's surface. The equation must then be written as $m = m_0/(1 - v^2/c^2)^{1/2}$. Expansion by the binomial theorem yields

$$m = m_0 \left(1 + \frac{v^2}{2c^2} + \frac{3}{8}\frac{v^4}{c^4} + \cdots\right),$$

which may be rewritten as $mc^2 = m_0c^2 + \frac{1}{2}m_0v^2 + \ldots$. The term $\frac{1}{2}m_0v^2$ is already familiar to us as the kinetic energy of a body moving with a velocity v. The above is therefore an energy equation and it would seem that mc^2 is the total energy of a body moving with velocity v, m_0c^2 is the energy associated with a mass m_0 at rest. It now appears that we have to regard mass as a form of energy with c^2 as the conversion factor. The complete expression for the total energy E is given by $E = mc^2 = m_0c^2/(1-v^2/c^2)^{1/2}$ and the momentum $p = mv = m_0v/1-v^2/c^2)^{1/2}$. We can now establish a general relationship between total energy and momentum based upon these two equations. Using the first we have

$$\left(\frac{E}{c}\right)^2 = \frac{m_0^2 c^2}{\left(1 - \frac{v^2}{c^2}\right)},$$

from which, using the second,

$$m_0^2 c^2 = \left(\frac{E}{c}\right)^2 - \left(\frac{mc^2}{c}\right)^2 \frac{v^2}{c^2} = \left(\frac{E}{c}\right)^2 - p^2$$

so that

$$m_0^2 c^2 + p^2 = \left(\frac{E}{c}\right)^2.$$

Thus for a mass at rest $p=0$ and $E = m_0c^2$ and if a quantity of matter Δm_0 disappears, a corresponding amount of energy ΔE is produced according to the equation $\Delta E = \Delta m_0 c^2$. A particle with no rest mass ($m_0 = 0$) can still have momentum given by $p = E/c = mc$ as in the case of a photon in the Compton effect described in Chapter 6.

We now have the relativity equation $E^2 = m_0^2 c^4 + p^2 c^2$ for the energy and momentum of a moving particle. This is the expression used in the discussion on antiparticles on p. 379.

Appendix B

The Dangers of Atomic Radiations

B.1 Introduction

Throughout this book ionization has been shown to be one of the main properties of the atomic and nuclear radiations which have revealed so much about the structure of the atom. Most nuclear particles cause primary ionization of a gas through which they pass while others, such as neutrons, can cause ionization as a result of projection of recoil nuclei. In all cases the passage of penetrating radiation through a gas causes a change in the electrical and chemical behaviour of the gaseous molecules.

Although nuclear and X-radiations have been used for many years, and although there were serious exposure accidents in the very early days to individual workers, it is only recently that this danger has become applicable to the population as a whole. With the building of innumerable particle accelerator machines, nuclear reactors and nuclear power stations, with the vast amount of radiosotope work now being done and with the tests of nuclear bombs in the air causing world-wide fall-out, more and more people are coming into contact with ionizing radiations. This has led to the development of a new subject, 'Health Physics', which teaches the dangers of these radiations, how to handle radioactive sources and how to take the necessary protective measures against the harmful effects of ionizing radiations.

All ionizing radiations are harmful to the human body and even the most minute quantity of a radioactive source should be treated with great circumspection. All personnel in contact with nuclear or X-radiation of any sort should know exactly what precautions to take and know something about the permissible levels of radiation and how to monitor the sources used.

B.2 Biological Effects of Nuclear and Electromagnetic Radiations

The Geiger counter, the ionization chamber and the discharge of a gold leaf electroscope are examples of ionization by radiation passing through a gas. The passage of radiation through a liquid has the same general effect, as shown by the use of liquid scintillation counters, and, within the short range of penetration, ionization also takes place in solids. Ionization of the molecules of living cells constitutes an added biological hazard and must certainly occur when α, β, γ or X-rays pass through living tissue. For neutrons it is possible that various (n, γ) reactions take place in which the γ-rays cause secondary ionization. Fast neutrons give knock-on protons which subsequently cause ionization along their paths.

All living cells consist of an active nucleus surrounded by a fluid called the cytoplasm, and within the nucleus are found the chromosomes which carry all the hereditary factors. The behaviour of the living cell is governed by extremely complex chemical changes in the constituent protein molecules in which the proper functioning of the whole body is controlled by the delicate balance between enzymes and cell molecules. When penetrating radiations pass through such cells it is reasonable to suppose that ionization takes place just as it does in an inanimate liquid. The consequences of the ionization of protein molecules are not fully known on the molecular scale. They are, however, well known in so far as they affect the health of the whole body. The normal chemical action of various proteins is often totally destroyed, and even the whole cell can be destroyed. Some cells have the biological property of self-repair whereas others are irreparably damaged.

The tissues of the human body contain millions of cells which die and are replaced daily by new cells. Ionizing radiations cause a comparatively small number of extra cells to die but this is not noticeable unless the total number of cells of a particular type is small or for some reason replacement is impossible. Examples of this class are the germ cells in the ovary, early embryo cells and brain cells. Other cells have the build-in facility for self-replacement, as in skin cells, or in the case of the male germ cells.

If the exposure to radiation is not lethal to the cell it will nevertheless cause damage which may be carried by the chromosomes when the cell divides and so be transmitted to subsequent generations of the same tissue, or if the germ cells are irradiated the damage can be transmitted to later generations of the same species.

Chromosomes are especially sensitive to ionizing radiations at the moment of cell division and the gene arrangement in the chromosomes can be seriously modified. The normal gene mutation rate can be increased by extra doses of ionizing radiations, so producing abnormalities in the succeeding generations.

The biological effects of ionizing radiations can be superficial, affecting skin and hair, or deep within the body inducing blood disorders, tumours and damage to the bone marrow. The hazard can be external, from a source of radiation some distance away giving whole body irradiation, or it can also arise internally, either from the ingestion of radioactive substances in contaminated food or from the inhalation of radioactive dust in the air. Some cells cannot recover from the radiation damage and the effects on the tissues are cumulative with a characteristic latent period before the radiation effects are manifest. Other cells can recover in time, without long-term deleterious effects. In all cases there is no immediate direct evidence to the individual that excessive radiation has been received.

The biological effects can be divided into three groups: (i) short-term recoverable effects; (ii) long-term irrecoverable effects and (iii) the genetic effect. Groups (i) and (ii) are limited to the individuals who have actually received the radiation while the effects in group (iii) only appear in later generations.

Exposures can be acute, as in an accidental burst of radiation from an unshielded source, or chronic, as in the occupational exposure of a professional radiographer. In all cases there are maximum acceptable levels of radiation which we shall discuss and with which all workers should be familiar. When excessive doses are absorbed the first noticeable disorder is a drop in the white blood cell count, which becomes evident in the first few hours after exposure as shown in Fig. B.1. This is followed by a sickness pattern of diarrhoea, vomiting and fever

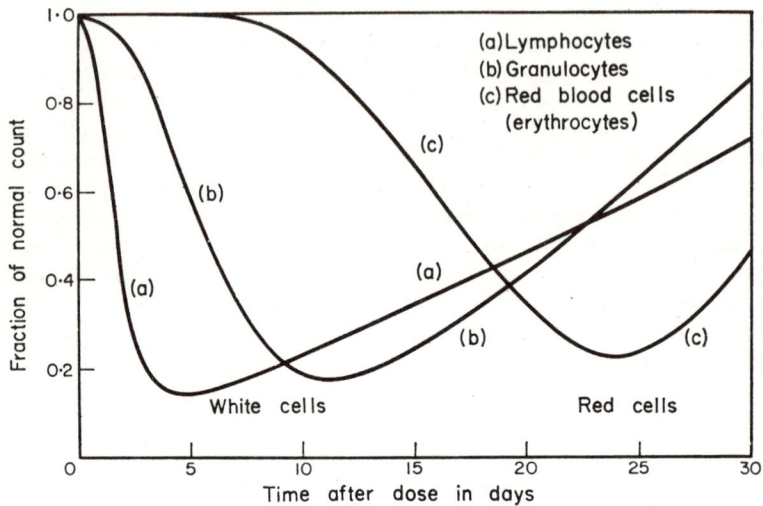

Fig. B.1   Time variation of blood count after single dose of X-rays.

which is now called radiation sickness. Recovery is possible from small acute doses, the time of recovery being weeks or months depending on the dose. Large doses are lethal within a few weeks. Smaller doses produce short-term effects, such as skin disorders and loss of hair, which are generally recoverable. More serious is the damage done to the bone marrow and to other cells which may not have the ability to repair the damage. This leads to leukaemia and to the production of cancerous cells and finally malignant tumours. Unfortunately there is often a time lag of 6–20 years before this is a sufficient accumulation of cell damage to cause the disease to appear with fatal results. This latent period is typical of the long-term effects.

The effects of the third type of damage only appear in the future generations of those irradiated. It is probably true that in the field of radiobiology we know least about the genetic effects of radiation. Changes in gene characteristics, or mutations, are brought about by delicate chemical alterations in the chromosomes and genes which occur at a rate corresponding to the natural rate of evolution of the species. Ionizing radiations increase this natural mutation rate and, since chromosomes in the nuclei of germ cells are most easily damaged, it follows that the reproductive organs are particularly sensitive to radiations when genetic effects are considered. Furthermore, there is no level of radiation below which these genetic effects do not take place to *some* extent. There is therefore no absolute threshold of safety for ionizing radiations.

Experiments on animals show that the genetic effects to be expected from breeding from radiation-dosed human individuals include an increase in mental deficiencies, an increase in the number of monsters born and a general deterioration of the species in quality and population number. These are only some of the world-wide problems arising from any increase of nuclear bomb fall-out.

B.3 Maximum Permissible Radiation Levels for Safety

Medical surveys of individuals affected by ionizing radiations in laboratories, nuclear reactor accidents, the Japanese after Hiroshima and Nagasaki in 1945, and the Marshall Islanders after the U.S.A. thermonuclear explosion of 1st March, 1954, and also innumerable animal experiments have gradually accumulated data of the dose–effect relationship for human beings and so have enabled us to state some tolerance limits for radiation. These are expressed in units of activity and dose rate.

The unit of activity is the becquerel (Bq), being the amount of radioactivity which gives off 1 particle per second. The older 'curie' is therefore 37 GBq, but is still in use.

Dosage is measured in grays (Gy) where 1 gray = absorbed energy in joule per kilogram. The older 'rad' is therefore 10 mGy. The rad is also in current use.

Since the biological effects of some radiations are more pronounced than others a degree of 'relative biological effectiveness' of various radiations has been assumed in the definition of the *rem* as a dose unit. The dose in rem is the product of the dose in rad and the relative biological effectiveness (R.B.E.) of the radiation being used. Thus the R.B.E. of X-rays, γ-rays and β-rays is taken as 1 whereas α-particles have an R.B.E. factor of 10 due to their greater ionization effect.

Having taken into account short- and long-term effects on the individual and genetic effects on future generations, the following MAXIMUM permissible levels of radiation have been recommended by the International Commission on Radiological Protection (1959). Occupational exposure over the whole body should not accumulate a dose over several years exceeding that given by the formula

$$D = 5(N-18),$$

where D is the does in rem and N is the age in years. This formula implies an average dose rate of 5 rem per year or 0·1 rem per week, or 235 rems of 'whole body' radiation for persons occupationally exposed from the age of 18 to the age of 65 years. Workers between the ages of 16 and 18 years should be restricted to 1·5 rem a^{-1}, i.e. to 3 rem before the age of 18. The above formula applies particularly to the gonads and blood-forming organs. For shorter periods of exposure, for the other internal organs, the average dose should not exceed 0·3 rem per week for a working day of 8 hours. The accumulated dose should not be more than 3 rem in 13 weeks. The health tolerance unit for occupational exposure is taken to be 0·1 rem per week and is called an M.P.L. (Maximum Permissible Level). This is 2·5 millirem per hour over a 40-h working week, and is the maximum possible permitted dose for occupational exposure.

It is seen that these rates are comparatively low, and when we compare them with the natural radioactive background to which we are all subject we can see why. The natural background of radiation is made up of two parts, one internal, from the radioactive substances contained in the body, and one external, from the radiation coming from an environment. Each of them is made up of different components which act differently upon different organs. Thus the dose to the gonads is different from that received by the bone marrow. Table B.1 shows the

TABLE B.1
Natural Sources of Background Radioactivity in Millirads Per Year

Source of radiation		Dose to gonads	Dose to bones
External radiation	Cosmic rays	24	24
	Terrestrial γ-rays	47	47
Internal radiation	Potassium 40	21	15
	Carbon 14	4	4
	Radium	2	38
	Total	98	128

radiations received in the U.K. averaged out over the whole area. The external radiations are cosmic rays from high altitudes and local terrestrial γ-radiations from the radioactive rocks in the earth's crust. The internal sources are radioactive potassium and carbon, and radium which does not affect the reproductive organs since it is centred largely in the bones. This natural background is therefore abour 2 mrad per week, i.e. 0·002 rad per week, so that the permitted weekly occupational dose is about 50 backgrounds for a few weeks only. It makes very little difference whether this dose is absorbed in a continuous low-level field or is made up of a number of short high-level exposures, so long as the M.P.L. is not exceeded.

B.4 Precautions against Radiation Hazards

Geiger counter equipment can be modified very easily to give direct readings of dose rate — usually in rads per hour for X- or γ-rays. Portable monitors or dose rate meters are available which can be used in the laboratory during experiments. It is essential to measure the dose rate regularly in a radiation field.

Film badges or pocket dosimeters should be worn by all workers in a permanent γ-radiation field of 10 backgrounds (say 0·02 rad per week) in order to register the accumulated dose over a long period of time. Film badges are developed and compared with standard blackening — dose charts and dosimeters are gradually discharged by the radiations and have calibrated scales

Radioactive sources should be stored in well protected rooms and suitably labelled 'Radioactive Material'. Laboratories should be capable of being washed down thoroughly so that there should be no cracks on the benches, the wall–floor joins should be radiussed, the bench tops, floors and walls painted with hard gloss paint which can easily be cleaned. Experiments with γ-rays particularly should be suitably shielded with lead blocks or with barytes concrete and radioactive waste should only be disposed under supervision. If necessary, decontamination features such as changing from laboratory to outdoor clothing should be enforced. When dealing with unsealed radioactive sources in the laboratory the operator must know what activity can be safely handled in a single batch. This depends on the ease and speed with which any possible contamination can be removed from the laboratory, and on the toxicity of the material. Successful decontamination depends on the design of the laboratory surfaces as already mentioned, while the toxicity depends on the biological half-life of the substance, on the type and energy of the emitted radiations and on the critical organ concerned with each isotope.

Isotopes are classified as Very Highly Toxic, Highly Toxic, Moderately Toxic and Slightly Toxic. Thus ^{90}Sr is classified as Very Highly Toxic because it is a long-lived bone seeker when ingested, ^{45}Ca as only Highly Toxic because it has a much shorter half-life than ^{90}Sr even though it is also a bone seeker, ^{24}Na as Moderately Toxic since although it has only a half-life of about 14 h it is a β^-/γ-emitter and ^{14}C is classified as Slightly Toxic because it is a very soft β^--emitter with no associated critical organ.

A laboratory might then be allowed to accumulate the following maximum activities:

1 millicurie	^{90}Sr
10 millicuries	^{45}Ca
1000 millicuries	^{24}Na
1 curie	^{14}C

This shows that a change of isotopes from ^{14}C to ^{90}Sr in a tracer investigation would have disastrous results if the same total activity of ^{90}Sr as ^{14}C were used. Hence a knowledge of the radiobiological toxicity of an isotope is extremely important.

Rubber gloves should be worn whenever possible, food and drink should never be taken in the radioisotope laboratory, hands should be washed more often than usual, active areas should be roped off and shielded where necessary. As an example, intensity of the γ-radiation from ^{60}Co can be reduced to one-tenth by shielding it with about 50 mm lead, 85 mm steel or by about a 300 mm thickness of ordinary brick wall. The inverse square law for distance attenuation must also be borne in mind.

If the precautions mentioned in this section are taken there is nothing to be feared when dealing with radioactive and X-ray sources, and only foolishness can lead to over exposure.

Further information on the subject of this chapter can be obtained from: *The Hazards to Man of Nuclear and Allied Radiations*, H.M.S.O., 1960; *Code of Practice for the Protection of Persons Exposed to Ionizing Radiations in University Laboratories*, Association of Universities of the British Commonwealth, 1961; *Code of Practice for the Protection of Persons Exposed to Ionizing Radiations in Research and Teaching*, H.M.S.O., 1968.

Appendix C

Complete List of Nuclides of the Elements

By using various modes of excitation, over 1000 isotopes of the elements have been created. It is frequently desirable to know the half-life, the atomic mass, or the type of radiation emitted by certain of these, without too much searching in the literature. This information is assembled in the following table. It refers only to the ground states of nuclides.

The masses are expressed in μu on the $^{12}C = 12 \cdot 000\ 000$ scale as the mass excess $(M - A)$ for each nuclide and are taken from the mass table of J. H. E. Mattuch, W. Thiele and A. H. Wapstra in *Nuclear Physics* **67**, p. 1 (1965).

In the table, the energy of each type of radiation is shown in the accompanying parentheses. The symbol K is used to denote the 'K' electron capture process by the nucleus. The number of γ-rays for many isotopes is too numerous to present completely, and it is then represented by the symbol γs. The designations γ, e^-, and e^+ are for γ-rays, electrons, and positrons, respectively. The percentage abundance of the stable isotopes is presented in the column headed as %. Half-lives are designated as s, min, h, d and a for second, minutes, hours, days, and years, and the energies are in MeV.

Z	Element name	A	Mass excess (μu)	% Abundance	Half-life	Radiation (MeV)
0	Neutron	1	8 665		12·8 min	
1	Hydrogen	1	7 825	99·98		
		2	14 102	0·0156		
		3	16 050		12·4 a	e^-(0·018)
2	Helium	3	16 030	$\sim 10^{-5}$		
		4	2 603	~ 100		
		5	12 297		10^{-22}	α; n
		6	18 893		0·85 s	e^-(3·7)
3	Lithium	6	15 125	7·4		
		7	16 004	92·6		
		8	22 487		0·88s	e^-(12·0)
		9	26 802		0·17 s	

Z	Element name	A	Mass excess ($\mu\mu$)	% Abundance	Half-life	Radiation (MeV)
4	Beryllium	7	16 929		54 d	γ (0·485); K
		8	5 308		10^{-22} s	2α
		9	12 186	100		
		10	13 354		$2\cdot7 \times 10^6$ a	e^-(0·56)
5	Boron	9	13 332		10^{-18} s	
		10	12 939	18·8		
		11	9 305	81·2		
		12	14 354		0·027 s	e^-(13·4)
6	Carbon	10	16 810		19·1 s	e^+(2·2)
		11	11 432		20·5 min	e^+(0·95)
		12	0 000	98·9		
		13	3 354	1·1		
		14	3 242		5730 a	e^-(0·155)
7	Nitrogen	12	18 641		0·0125 s	e^+(16·6)
		13	5 738		10·0 min	e^+(0·92, 1·24)
		14	3 074	99·62		
		15	108	0·38		
		16	6 103		7·4 s	e^-(10·0); γ (6·4)
		17	8 450		4·14 s	e^-(3·7)
8	Oxygen	14	8 597		76·5 s	e^+(1·8); γ (2·3)
		15	3 070		126 s	e^+(1·7)
		16	$-$ 5 085	99·76		
		17	$-$ 867	0·04		
		18	$-$ 840	0·20		
		19	3 578		29·5 s	e^-(3·0, 4·5); γ
9	Fluorine	17	2 095		66 s	e^+(1·7)
		18	937		112 min	e^+(0·6)
		19	$-$ 1 595	100		
		20	$-$ 13		12 s	e^-(5·0); γ (2·2)
10	Neon	19	1 881		18·2 s	e^+(2·20)
		20	$-$ 7 560	90·52		
		21	$-$ 6 151	0·27		
		22	$-$ 8 615	9·21		
		23	$-$ 5 527		40 s	e^-(4·1)
11	Sodium	21	$-$ 2 345		23 s	e^+(2·53)
		22	$-$ 5 563		2·6 a	e^+(1·8, 0·54); γ(1·28)
		23	$-$10 229	100		
		24	$-$ 9 038		14·8 h	e^-(1·39); γ(2·758, 1·38)
		25	$-$10 045		62 s	e^-(3·7, 2·7); γ
12	Magnesium	23	$-$ 5 875		11·9 s	e^+(2·82)
		24	$-$14 958	78·6		
		25	$-$14 161	10·1		
		26	$-$17 407	11·3		
		27	$-$15 655		9·6 min	e^-(1·8, 0·8); γs
13	Aluminium	25	$-$ 9 588		7·3 s	e^+
		26	$-$13 109		6·3 s	e^+(2·99)
		27	$-$18 461	100		
		28	$-$18 095		2·4 min	e^-(3·0); γ(1·8)
		29	$-$19 558		6·7 min	e^-(2·5, 1·4); γ(2·3, 1·2)

Z	Element name	A	Mass excess (μu)	% Abundance	Half-life	Radiation (MeV)
14	Silicon	27	−13 297		4·92 s	e^+(3·64)
		28	−23 070	92·3		
		29	−23 504	4·7		
		30	−26 237	3·0		
		31	−24 651		170 min	e^-(1·486)
15	Phosphorus	29	−18 192		4·6 s	e^+(3·63)
		30	−21 683		2·55 min	e^+(3·5)
		31	−26 235	100		
		32	−26 091		14·3 d	e^-(1·718)
		33	−28 272		25 d	e^-(0·26)
16	Sulphur	31	−20 389		2·8 s	e^+(3·9)
		32	−27 926	95·1		
		33	−28 538	0·74		
		34	−32 135	4·2		
		35	−30 469		87·1 d	e^-(0·17)
		36	−32 910	0·0136		
		37	−28 990		5·0 min	e^-(4·3, 1·6); γ(2·7)
17	Chlorine	33	−22 560		2·8 s	e^+(4·13)
		34	−26 250		33 min	e^+(4·5, 2·6, 1·3); γs
		35	−31 149	75·4		
		36	−31 691		4×10^5 a	e^+; e^-(0·66); K
		37	−34 102	24·6		
		38	−31 995		38 min	e^-(4·81, 2·77, 1·11); γs
		39	−31 992		56 min	e^-(2·5)
18	Argon	35	−24 746		1·86 s	e^+(4·38)
		36	−32 456	0·307		
		37	−33 228		34 d	K
		38	−37 272	0·061		
		39	−35 683		2·4 min; 265 a	e^-(2·1, 0·565)
		40	−37 616	99·632		
		41	−35 500		110 min	e^-(2·55, 1·25); γ(1·3)
19	Potassium	37	−26 635		1·3 s	e^+(4·57)
		38	−30 903		7·5 min	e^+(2·53); γ(2·1)
		39	−36 290	93·1		
		40	−36 000	0·011	1.25×10^9 a	e^-(1·33); γ(1·46); K
		41	−38 168	6·9		
		42	−37 594		12·4 h	e^-(3·55, 1·99); γ(1·53, 0·31)
		43	−37 960		22 h	e^-(1·84, 1·22); γs
		44	−37 960		22 min	e^-(4·9, 1·5); γ(1·16)
20	Calcium	39	−29 309		1·1 s	e^+(5·13)
		40	−37 411	96·97		
		41	−37 725		1.1×10^5 a	K
		42	−41 375	0·64		
		43	−41 220	0·145		
		44	−44 510	2·06		
		45	−43 811		164 d	e^-(0·256)
		46	−46 311	0·0033		
		47	−45 462		5·35 d	e^-(1·4, 0·46); γs

Z	Element name	A	Mass excess (μu)	% Abundance	Half-life	Radiation (MeV)
		48	−47 469	0·185		
		49	−44 325		8·5 min	$e^-(2·0)$; $\gamma(3·0)$
21	Scandium	40	−22 430		0·3 s	$e^-(9·0)$; $\gamma(3·75)$
		41	−30 753		0·87 s	$e^+(4·94)$
		42	−34 505		13·5 d	$e^+(1·4)$
		43	−38 835		4 h	$e^+(1·18, 0·80, 0·39)$; γs
		44	−40 594		3·9 h	$e^+(1·47)$; $\gamma(2·54, 1·17)$
		45	−44 081	100		
		46	−44 827		85 d	$e^-(0·36)$; γ; K
		47	−47 587		80 h	$e^-(0·61, 0·45)$; $\gamma(0·159)$
		48	−47 779		44 h	$e^-(0·64)$; γs
		49	−49 974		57 min	$e^-(2·1)$
22	Titanium	43	−31 500		0·58 s	
		44	−40 28			
		45	−41 871		3·0 h	$e^+(1·02)$
		46	−47 368	7·94		
		47	−48 231	7·75		
		48	−52 049	73·45		
		49	−52 129	5·52		
		50	−55 214	5·34		
		51	−53 397		5·9 min	$e^-(2·13, 1·5)$; γs
23	Vanadium	46	−39 786		0·4 s	$e^+(6·2)$
		47	−45 101		33 min	$e^+(1·9)$
		48	−47 741		16 d	$e^+(0·7)$; $\gamma(1·04)$; K
		49	−51 477		330 d	K
		50	−52 836	0·23		
		51	−56 038	99·77		
		52	−55 220		3·8 min	$e^-(2·73)$; $\gamma(1·45)$
24	Chromium	49	−48 729		42 min	$e^+(1·54, 1·45, 1·39)$; γs
		50	−53 945	4·49		
		51	−55 231		27·5 d	$e^+(0·75, 0·43)$; $\gamma(0·32)$
		52	−59 486	83·78		
		53	−59 347	9·43		
		54	−61 118	2·30		
		55	−59 167		3·5 min	$e^-(2·85)$
25	Manganese	50	−45 785		0·28 s	$e^+(6·5)$
		51	−51 810		46 min	$e^+(2·0)$
		52	−54 432		21 min; 5·7 d	$e^+(2·6)$; $\gamma(1·4)$
		53	−58 705		140 a	$e^+(0·57)$
		54	−59 638		290 d	K; $\gamma(0·84)$
		55	−61 949	100		
		56	−61 089		2·59 h	$e^-(0·7, 1·09, 2·88)$; γs
		57	−59 740		1·7 min	$e^-(2·6)$; γs
26	Iron	52	−51 833		8·3 h	$e^+(0·80)$
		53	−54 428		8·9 min	$e^+(2·7)$; γs
		54	−60 383	5·81		
		55	−61 701		2·9 a	K
		56	−65 063	91·64		
		57	−64 602	2·21		

Z	Element name	A	Mass excess (μu)	% Abundance	Half-life	Radiation (MeV)
		58	−66 718	0·34		
		59	−65 122		46 d	e^-(1·56, 0·46, 0·27); γs
27	Cobalt	54	−51 525		0·18 s	e^+(7·7)
		55	−57 987		18·2 h	e^+(1·5, 1·03, 0·53); γs
		56	−60 153		77 d	e^+(1·5, 0·98, 0·32); γs
		57	−63 704		270 d	K; γ(0·137, 0·123, 0·014)
		58	−64 239		72 d	e^+(0·47); γ(0·805)
		59	−66 810	100		
		60	−66 186		5·3 a	e^-(1·48, 0·31); γ(1·332, 1·172)
		60	−66 186		10·5 min	e^-(1·54); γ(1·332)
		61	−67 560		1·7 h	e^-(1·42, 1·0); γ
		62	−66 054		13·8 min	e^-(2·5); γ(1·17)
28	Nickel	57	−60 231		36 h	e^+(0·84); γs
		58	−64 458	67·7		
		59	−65 657		$75 \cdot \times 10^4$ a	K
		60	−69 213	26·2		
		61	−68 944	1·2		
		62	−71 658	3·7		
		63	−70 336		85 a	e^-(0·062)
		64	−72 042	1·2		
		65	−69 928		2·6 h	e^-(2·10, 1·01, 0·60); γs
		66	−70 915		56·0 h	e^-(0·3)
29	Copper	58	−55 459		3 s	e^+(8·2)
		59	−60 504		81 s	e^+
		60	−62 638		24·6 min	e^+(3·92, 3·0, 2·0); γs
		61	−66 543		3·4 h	e^+(1·21, 1·11, 0·55, 0·23); γs
		62	−67 434		9·8 min	e^+(2·9)
		63	−70 408	69·1		
		64	−70 241		12·8 h	e^-(0·573); e^+(0·565); K
		65	−72 214	30·9		
		66	−71 129		5 min	e^-(2·63), 1·65); γ(1·05)
		67	−72 241		59 h	e^-(0·577, 0·484, 0·395); γ s
		68	−70 230		38 s	e^-(3·0)
30	Zinc	62	−65 620		9·3 h	e^+(0·69, 0·64); γ(0·042)
		63	−66 794		38 min	e^+(2·32); γ
		64	−70 855	48·9		
		65	−70 766		250 d	e^+(0·33); K; γ(1·11)
		66	−73 948	27·8		
		67	−72 855	4·0		
		68	−75 143	18·6		
		69	−73 459		51 min	e^-(0·90)
		70	−74 666	0·7		
		71	−72 490		2·2 min; 3 h	e^-(2·4, 1·5); γs
		72	−73 157		49 h	e^-(1·6, 0·3)
31	Gallium	64	−63 263		2·6 min	e^+
		65	−67 267		15 min	e^+(2·1); γs
		66	−68 393		9·4 h	e^+(4·15, 1·38, 0·90, 0·40); γs
		67	−71 784		78 h	K; γs
		68	−72 008		68 min	e^+(1·9)
		69	−74 426	60·2		

Z	Element name	A	Mass excess (μu)	% Abundance	Half-life	Radiation (MeV)
		70	−73 965		21 min	e^-(1·65, 0·6, 0·4); γs
		71	−75 294	39·8		
		72	−73 628		14·2 h	e^-(3·16, 2·53, 1·51, 0·96); γs
		73	−74 874		5 h	e^-(1·4); γs
32	Germanium	66	−65 200		140 min	e^+
		67	−67 060		20 min	e^+(3·4)
		68	−71 470		250 d	e^+
		69	−72 036		40 h	e^+(1·22, 0·61, 0·22); γs
		70	−75 748	20·6		
		71	−75 044		11 d	K
		72	−77 918	27·4		
		73	−76 537	7·6		
		74	−78 819	36·8		
		75	−77 117		82 min	e^-(1·19, 0·98, 0·92, 0·55); γs
		76	−78 594	7·6		
		77	−76 400		12 h	e^-(2·9, 2·7, 2·2, 0·74); γs
		78			86 min	e^-(0·9)
33	Arsenic	71	−72 887		62 h	e^+(0·81)
		72	−73 237		26 h	e^+(3·34, 2·5, 1·84, 0·67); γs
		73	−76 139		76 d	K; γ(0·052)
		74	−76 067		17 d	e^-(1·36, 0·69); e^+(1·53, 0·92); γs
		75	−78 403	100		
		76	−77 603		26·8 h	e^-(2·97, 2·41, 1·76, 0·36); γs
		77	−79 354		40 h	e^-(0·68, 0·43, 0·16); γs
		78	−78 100		90 min	e^-(4·1, 1·4); γs
		79	−79 110		9 min	e^-(2·2)
34	Selenium	70			44 min	e^+
		72	−72 590		9·5 d	K
		73	−73 186		7·1 h	e^-(1·68, 1·32); γ
		74	−77 524	0·9		
		75	−77 475		125 d	K; γs
		76	−80 793	9·0		
		77	−80 089	7·6	17·5 s	γ(0·14)
		78	−82 686	23·5		
		79	−81 505		$6·5 \times 10^4$ a	e^-(0·16)
		80	−83 472	49·8		
		81	−82 016		19 min	e^-(1·38)
		82	−83 293	9·2		
		83			25 min	e^-(1·5); γs
		84			3 min	e^-
35	Bromine	75	−74 553		1·7 h	e^+(1·7, 0·8, 0·6, 0·3)
		76	−75 820		17 h	e^+(3·5, 1·7, 1·1, 0·8, 0·6); γs
		77	−78 624		57 h	e^+(0·36); K; γs
		78	−78 850		6·4 min	e^+(2·3)
		79	−81 670	50·6		
		80	−81 464		4·4 h	γ(0·048, 0·036)
		80	−81 464		18 min	e^-(2·0, 1·38); e^+(0·886)
		81	−83 708	49·4		
		82	−83 198		36 h	e^-(0·465); γs
		83	−84 832		140 min	e^-(0·96); γ(0·051)
		84	−83 450		33 min	e^-(4·68, 3·56, 2·53, 1·72); γs

Z	Element name	A	Mass excess (μu)	% Abundance	Half-life	Radiation (MeV)
		85	−84 470		3·0 min	e^-(2·5)
		87			55 s	e^-(8·0, 2·6); γs
		88			16·0 s	e^-(2·7, 0·52); γs
36	Krypton	77	−75 520		1·1 h	e^+(1·86, 1·67, 0·85); γs
		78	−79 597	0·35		
		79	−79 932		34 h	e^+(0·60, 0·34); K; γs
		80	−83 620	2·22		
		81	−83 390		2×10^5 a	K
		82	−86 518	11·53		
		83	−85 868	11·55	1·89 h	γ(0·032, 0·009)
		84	−88 496	57·00		
		85	−87 477		4 h; 10·6 h	e^-(0·85); (0·672, 0·15); γs
		86	−89 384	17·37		
		87	−86 635		78 min	e^-(3·8, 3·3 1·3); γs
		88	−85 730		2·8 h	e^-(2·7, 0·9, 0·52); γs
		89	−83 400		3·2 min	e^-(4·0, 2·0)
		90	−80 280		0·5 min	e^-(3·2)
		91			9·3 s	e^-(3·6)
		92			2·3 s	e^-
		93			2·2 s	e^-
		94			1·4 s	e^-
37	Rubidium	81	−80 980		4·7 h	e^+(0·9); γ(0·95)
		82	−82 041		1·3 min	e^+(3·15)
		82	−82 041		6·4 h	e^+(0·775); γs
		83	−85 270		83 d	K; γ(0·525)
		84	−85 619		34 d	e^+(1·63, 0·82); e^-(0·44)
		85	−88 200	72·3		
		86	−88 807		19·5 d	e^-(1·77, 0·68); γ(1·0)
		87	−90 813	27·7	5×10^{10} a	e^-(0·275)
		88	−88 730		18 min	e^-(5·3, 3·6, 2·5); γs
		89	−88 350		15 min	e^-(3·8); γ
		90	−85 180		2·7 min	e^-(5·7)
		91	−83 930		1·7 min	e^-(4·6, 3·0)
38	Strontium	83	−82 800		38 h	e^+(1·15)
		84	−86 569	0·56		
		85	−87 011		65 d	K; γ(0·513)
		85	−87 011		70 min	K; γs
		86	−90 715	9·86		
		87	−91 107	7·02	2·7 h	γ(0·386)
		88	−94 359	82·56		
		89	−92 558		51 d	e^-(1·48)
		90	−92 253		28 a	e^-(0·6)
		91	−89 839		9·8 h	e^-(2·67, 2·03, 1·36, 1·09, 0·61)
		92	−89 020		2·7 h	e^-(0·55)
		93	−85 290		7 min	e^-
39	Yttrium	86	−85 054		15 h	e^+(1·8, 1·19); γs
		87	−89 260		14 h	K; γ(0·38)
		87	−89 260		80 h	e^+(0·7)
		88	−90 472		105 d	e^+(0·83); γs
		89	−94 128	100		
		90	−92 837		64 h	e^-(2·4)

Z	Element name	A	Mass excess (μu)	% Abundance	Half-life	Radiation (MeV)
		91	−92 705		61 d; 50 min	e^-(1·6); γ(1·22, 0·55)
		92	−91 074		3·5 h	e^-(3·6, 2·68, 1·3); γs
		93	−90 448		10·5 h	e^-(3·1)
		94	−88 320		16·5 min	e^-(5·4); γ(1·4)
40	Zirconium	87	−85 510		94 min	e^+(2·10); γ(3·81)
		88	−89 940		85 d	K
		89	−91 086		4·5 min; 79 h	K; e^+(0·90); γs
		90	−95 300	51·4		
		91	−94 358	11·2		
		92	−94 969	17·1		
		93	−93 550		5×10^6 a	e^-(0·060)
		94	−93 686	17·4		
		95	−91 965		63 d	e^-(1·0); γs
		96	−91 714	2·8		
		97	−89 034		17 h	e^-(2·5); γs
41	Niobium	89	−86 920		1·9 h	e^+(2·9)
		90	−88 741		15·6 h	e^+(1·5, 0·86, ·055); γs
		91	−93 140		62 d	K
		92	−92 789		11 d	K; γs
		93	−93 618	100	42 d	γ(0·15)
		94	−92 697		6·6 min	e^-(1·4, 0·4)
		95	−93 168		35 d; 90 h	e^-(0·163); γs
		96	−91 944		22·9 h	e^-(0·75, 0·37); γs
		97	−91 904		74 min	e^-(1·267); γ(0·665)
		98	−89 650		30 min	e^-
42	Molybdenum	91	−88 350		15.5 min; 75 s	e^+(3·7, 2·6); γs
		92	−93 189	15·9		
		93	−93 170		2 a; 6·8 h	K; γs
		94	−94 909	9·1		
		95	−94 161	15·7		
		96	−93 326	16·5		
		97	−93 978	9·5		
		98	−94 591	23·8		
		99	−92 280		68·3 h	e^-(1·23, 0·445); γs
		100	−92 525	9·5		
		101	−89 647		15 min	e^-(2·2, 1·0); γs
		102	−89 750		12 min	e^-
43	Technetium	92	−84 540		4·5 min	K; γ(1·51)
		93	−89 749		44 min	e^+(0·8); γ(1·32)
		94	−90 357		53 min	e^+(2·41); γs
		95	−92 380		62 d; 20 h	e^+(0·4); K; γs
		96	−92 170		4·3 d	K; γs
		97	−93 660		93 d	γ(0·097)
		98	−92 890		2·7 d	e^-; e^+; γs
		99	−93 751		5×10^5 a	e^-(0·32)
		100	−92 160		16 s	e^-(2·8); γ
		101	−92 674		14 min	e^-(1·3); γ
		102	−90 820		1 min	e^-(3·7); γ

Z	Element name	A	Mass excess (μu)	% Abundance	Half-life	Radiation (MeV)
44	Ruthenium	94			57 min	e^+
		95	−90 199		1.6 h	$e^+(1.1)$; $\gamma(0.95)$
		96	−92 402	5.68		
		97	−92 370		2.9 d	K; $\gamma(0.22)$
		98	−94 711	2.22		
		99	−94 064	12.81		
		100	−95 782	12.70		
		101	−94 423	16.98		
		102	−95 652	31.34		
		103	−93 694		40 d	$e^-(6.84, 0.22)$; γs
		104	−94 570	18.27		
		105	−92 321		4.4 h	$e^-(1.15)$; γs
		106	−92 678		1 a	$e^-(0.03)$
		107	−89 870		4 min	$e^-(4.0)$
45	Rhodium	100	−91 874		21 h	$e^+(3.0, 1.3)$; $\gamma(1.8)$
		101	−93 822	0.08		
		102	−93 158		210 d	$e^-(1.1)$; e^+; γ
		103	−94 489	99.92	57 min	γ
		104	−93 341		4.2 min	$\gamma(0.055, 0.080)$
		104	−93 341		44 s	$e^-(2.3)$; γs
		105	−94 329		36 h	$e^-(0.57, 0.25)$; γs
		106	−92 721		30 s	$e^-(3.5)$
		107	−93 247		24 min	$e^-(1.2)$
46	Palladium	100	−91 230		4.0 d	$\gamma(1.8)$; K
		101	−91 930		9 h	$e^+(2.3)$
		102	−94 391	0.8		
		103	−93 893		17 d	
		104	−95 989	9.3		
		105	−94 936	22.6		
		106	−96 521	27.2		
		107	−94 868		7×10^6 a	$e^-(0.035)$
		108	−96 109	26.8		
		109	−94 046		13 h	$e^-(1.03)$
		110	−94 836	13.5		
		111	−92 330		22 min	$e^-(2.15)$; γs
		112	−92 614		21 h	$e^-(0.2)$
47	Silver	102	−88 700		73 min	e^+
		104	−91 404		16.3 min	e^+
		105	−95 540		40 d	K; γs
		106	−93 339		24.5 min	$e^+(2.04)$
		106	−93 339		8.6 d	K; γs
		107	−94 906	51.9	42 s	$\gamma(0.093)$
		108	−94 051		2.3 min	$e^-(1.49, 0.83)$; γs
		109	−95 244	48.1	40 s	$\gamma(0.0875)$
		110	−93 905		24 s	$e^-(2.8)$
		110	−93 905		282 d	$e^-(0.53, 0.087)$; γs
		111	−94 684		7.5 d	$e^-(1.04, 0.80, 0.70)$; γs
		112	−92 936		3.2 h	$e^-(3.5)$; $\gamma(0.86)$
		113	−93 444		5.3 h	$e^-(2.1)$
		115	−91 070		20 min	$e^-(3.0)$

Z	Element name	A	Mass excess (μu)	% Abundance	Half-life	Radiation (MeV)
48	Cadmium	105	−90 530		38 min	$e^+(1.5)$; γ
		106	−93 537	1.21		
		107	−93 385		6.7 h	$e^+(0.32)$; $\gamma(0.847)$
		108	−95 813	0.87		
		109	−95 072		330 d	K; $\gamma(0.080)$
		110	−96 988	12.40		
		111	−95 811	12.75	48.7 min	$\gamma(0.243)$
		112	−97 237	24.07		
		113	−95 591	12.26		
		114	−96 639	28.86		
		115	−94 569		2.3 d	$e^-(1.11)$; γs
		115	−95 569		43 d	$e^-(1.41)$; $\gamma(1.10)$
		116	−95 238	7.58		
		117	−92 761		2.8 h	$e^-(1.5)$
49	Indium	107	−89 640		33 min	e^+
		109	−92 904		4.2 h	$e^+(0.75)$; $\gamma(0.5)$
		110	−92 769		65 min	$e^+(2.25)$; $\gamma(0.65)$
		111	−94 640		2.8 d	K; $\gamma(0.33, 0.246, 0.172)$
		112	−94 456		9 min	$e^+(1.5)$; $e^-(1.0)$
		112	−94 456		23 min	$\gamma(0.095)$
		113	−95 911	4.3	105 min	$\gamma(0.39)$
		114	−95 095		48 d	$\gamma(0.1909)$
		114	−95 095		72 s	$e^-(1.98)$; $e^+(0.65)$; γs
		115	−96 129	95.7	4.5 h	$\gamma(0.34)$
		116	−94 683		13 s	$e^-(2.8)$
		116	−94 683		54 min	$e^-(1.0, 0.87, 0.6)$; γs
		117	−95 466		117 min	$e^-(1.95)$
		118	−93 890		4.5 min	$e^-(1.5)$
50	Tin	111	−91 940		35 min	$e^+(1.51)$
		112	−95 168	0.9		
		113	−94 813		118 d	K; $\gamma(0.401, 0.255)$
		114	−97 227	0.6		
		115	−96 654	0.4		
		116	−98 255	14.1		
		117	−97 041	7.5	14.5 d	$\gamma(0.162)$
		118	−98 394	24.0		
		119	−96 686	8.6	245 d	γ
		120	−97 801	33.0		
		121	−95 773		27 h	$e^-(0.4)$
		122	−96 558	4.8		
		123	−94 262		125 d; 40 min	$e^-(1.42, 1.26)$; γs
		124	−94 728	6.1		
		125	−92 254		10 min; 9.5 d	$e^-(2.3, 2.0, 1.17)$; γs
		126	−92 360		70 min	$e^-(2.7, 0.7)$; $\gamma(1.2)$
51	Antimony	116	−93 370		60 min	$e^+(1.45)$; γ
		117	−95 088		2.9 h	e^+, γ
		118	−94 426		3.5 h	$e^+(3.1)$
		119	−96 065		39 h	K
		120	−94 919		17 min; 6 d	$e^+(1.7)$; K; γs
		121	−96 182	57.2		

Z	Element name	A	Mass excess (μu)	% Abundance	Half-life	Radiation (MeV)
		122	−94 817		2·8 d; 3·5 min	e^-(1·94, 1·36); γs
		123	−95 787	42·8		
		124	−94 027		60 d; 21 min	e^-(3·2, 2·5, 0·60); γs
		125	−94 768		2·7 a	e^-(0·616, 0·299, 0·128); γs
		126	−92 680		9 h	e^-(1·0); γs
		127	−93 023		90 h	e^-(1·2); γs
		129	−90 746		4·2 h	e^-
		132			2·2 min	e^-
		133			4·5 min	e^-
52	Tellurium	118	−94 100		6·0 d	K
		119	−93 602		4·5 d	K; γ(1·6)
		120	−95 977	0·1		
		121	−94 801		17 d; 154 d	K; γ(0·575, 0·213)
		122	−96 934	2·5		
		123	−95 723	0·9	121 d	γ
		124	−97 158	4·6		
		125	−95 582	7·0	58 d	γ
		126	−96 678	18·7		
		127	−94 791		113 d	e^-; γ(0·086)
		127	−94 791		9·3 h	e^-(0·70)
		128	−95 524	31·7		
		129	−93 425		32 d	e^-; γ
		129	−93 425		72 min	e^-(1·8); γ(0·3)
		130	−93 762	34·5		
		131	−91 425		30 h	e^-; γ
		131	−91 425		25 min	e^-(2·0); γ
		132	−91 477		77 h	e^-(0·35); γ(0·22)
		133			60 min; 2 min	e^-(2·4, 1·3); γs
		134			43 min	e^-
		135			2 min	e^-
53	Iodine	122	−92 989		3·6 min	e^-(3·1)
		124	−93 754		4·0 d	e^+(2·2, 1·5, 0·67); γs
		125	−95 422		56 d	K; γ
		126	−94 369		13·0 d	e^-(1·24, 0·85); e^+; γs
		127	−95 530	100		
		128	−94 162		25·0 min	e^-(2·02, 1·59); γ(0·428)
		129	−95 013		$1·7 \times 10^7$ a	e^-(0·13); γ(0·039)
		130	−93 324		12·6 h	e^-(0·61, 1·03); γs
		131	−93 872		8·0 d	e^-(0·60, 0·32); γs
		132	−92 019		2·4 h	e^-(2·2, 1·5); γs
		133	−92 250		22 h	e^-(1·4, 0·5); γs
		134	−90 150		54 min	e^-(3·9, 1·6); γs
		135	−89 980		6·6 h	e^-(1·4, 1·0, 0·47); γs
		136	−85 260		1·5 min	e^-(6·5); γ
		137			22 s	e^-
		138			5·9 s	e^-
54	Xenon	124	−93 880	0·094		
		125	−93 380		18 h	K; γs

Z	Element name	A	Mass excess (μu)	% Abundance	Half-life	Radiation (MeV)
		126	−95 712	0·088		
		127	−94 780		75 s	γ(0·175, 0·125)
		127	−94 780		34 d	K; γs
		128	−96 460	1·90		
		129	−95 216	25·23		
		130	−96 491	4·07		
		131	−94 914	21·17	12 d	γ
		132	−95 839	26·96		
		133	−94 185		5·3 d; 2·3 d	e⁻(0·346); γs
		134	−94 602	10·54		
		135	−92 980		9·4 h	e⁻(0·9)
		135	−92 980		15·6 min	e⁻(0·65); γ
		136	−92 779	8·95		
		137	−88 900		3·9 min	e⁻(4·0)
		138	−86 190		17 min	e⁻
		139	−82 160		40 s	e⁻
		140			16 s	e⁻
		141			3 s	e⁻
		143			1·3 s	e⁻
		145			0·8 s	e⁻
55	Caesium	128	−92 241		3·1 min	e⁺(3·0)
		129	−94 040		31 h	K; γ
		130	−93 280		30 min	e⁺; K
		131	−94 534		10·2 d	e⁺; K
		132	−93 607		7·1 d	K; γ(0·62)
		133	−94 645	100		
		134	−93 177		3 h	e⁻(2·4); γ(0·7)
		134	−93 177		2·3 a	e⁻(0·648, 0·092); γs
		135	−94 230		2·9 × 10⁶ a	e⁻(0·19)
		136	−92 660		13·3 d	e⁻(0·35, 0·28); γs
		137	−93 230		35 a	e⁻(0·17, 0·518); γ(0·662)
		138	−89 200		33 min	e⁻(2·6); γ
		139	−87 100		9·5 min	e⁻
		140	−82 890		66 s	e⁻
56	Barium	129	−91 410		2 h	e⁺
		130	−93 755	0·10		
		131	−93 284		13 d	K; γs
		132	−94 880	0·09		
		133	−94 121		38·8 h	γ(0·276)
		133	−94 121		10 a	K; γ(0·32, 0·085)
		134	−95 388	2·42		
		135	−94 450	6·59		
		136	−95 700	7·81		
		137	−94 500	11·32	2·6 min	γ
		138	−95 000	71·66		
		139	−91 400		85 min	e⁻(2·27); γs
		140	−89 425		12·8 d	e⁻(1·02, 0·48); γs
		141	−85 950		18 min	e⁻(2·8); γ
		142	−83 650		30 s	e⁻
57	Lanthanum	134	−91 340		6·5 min	e⁺(2·7); K
		135	−93 110		18·5 h	K; γ(0·76)
		136	−92 620		9·5 min	e⁺(2·1); K

Z	Element name	A	Mass excess (μu)	% Abundance	Half-life	Radiation (MeV)
		137	−93 960		400 a	e^-
		138	−93 090	0·09		
		139	−93 860	99·91		
		140	−90 562		41·4 h	e^-(2·26, 1·67, 1·32); γs
		141	−89 172		3·6 h	e^-(2·8)
		142	−86 020		74 min	e^-; γ
		143	−84 130		20 min	e^-
		144	−80 400		1 s	e^-
58	Cerium	135	−90 860		22 h	e^+(0·8)
		136	−92 900	0·19		
		137	−92 670		36 h	K; γ(0·75, 0·28)
		138	−94 170	0·25		
		139	−93 570		140 d	K; γ(0·137, 0·144)
		140	−94 608	88·49		
		141	−91 781		32 d	e^-(0·56, 0·41); γ(1·41)
		142	−90 860	11·07		
		143	−87 673		33 h	e^-(1·36); γs
		144	−86 409		290 d	e^-(0·45, 0·307); γs
		145	−82 730		1·8 h	e^-
		146	−81 330		14·6 min	e^-
59	Praseodymium	138	−89 540		2·0 h	e^+(1·4); γs
		140	−90 993		3·5 min	e^+(2·40); γ
		141	−92 404	100		
		142	−90 022		19·2 h	e^-(2·14); γs
		143	−89 219		13·5 d	e^-(0·95)
		144	−86 752		17 min	e^-(2·87); γs
		145	−85 524		4·5 h	e^-(3·2)
		146	−82 410		24·7 min	e^-(3); γ
60	Neodymium	140	−90 670		3·3 d	K; γ
		141	−90 472		2·5 h	e^+(0·78); γ(1·05)
		142	−92 337	27·13		
		143	−90 221	12·20		
		144	−89 961	23·87		
		145	−87 462	8·30		
		146	−86 914	17·18		
		147	−83 926		11·8 d	e^-(0·78, 0·35); γs
		148	−83 131	5·72		
		149	−79 878		18 h	e^-(1·5, 1·1, 0·95); γs
		150	−79 085	5·60		
		151	−76 230		12 min	e^-(1·93); γs
61	Promethium	141	−86 590		20 min	e^+(2·6)
		142	−87 180		2 min	e^+(3·8); γ(1·6)
		143	−89 010		285 d	K; γ
		146	−85 368		2·7 h	e^-(2·0); γ
		147	−84 892		2·7 a	e^-(0·2)
		148	−82 579		5·3 d	e^-(2·5); γ(0·8)
		149	−81 670		47 h	e^-(1·1); γ(0·25)
		150	−79 040		2·7 h	e^-(3·0, 2·01); γs
62	Samarium	143	−85 450		8 min	e^+
		144	−88 011	3·16		

Z	Element name	A	Mass excess (μμ)	% Abundance	Half-life	Radiation (MeV)
		145	−86 606		410 d	K; γ(0·061)
		147	−85 133	15·07	6·7 × 10¹¹ a	α(2·1)
		148	−85 209	11·27		
		149	−82 820	13·84		
		150	−82 724	7·47		
		151	−80 081		70 a	e⁻(0·75)
		152	−80 244	26·63		
		153	−77 898		50 h	e⁻(0·82); γs
		154	−77 718	22·53		
		155	−75 299		23 min	e⁻(1·8); γs
		156	−74 431		10 h	e⁻(0·8)
63	Europium	146	−82 862		38 h	e⁺(0·4)
		147	−83 200		24 d	e⁺(1·0, 0·4); K; γ
		148	−81 890		58·3 d	K; γ(0·69)
		149	−82 000		14 d	K; γ(1·0)
		150	−80 311		13·1 h	e⁺(1·8); e⁻(1·1, 0·8)
		151	−80 162	47·8		
		152	−78 251		9·2 h; 5·3 a	e⁻(1·8, 0·9, 0·36); γs
		153	−78 758	52·2		
		154	−76 947		5·4 a	e⁻(1·9, 0·7, 0·3); γs
		155	−77 070		1·7 a	e⁻(2·23, 0·24); γs
		156	−75 198		15·4 d	e⁻; γ
		157	−74 610		15·4 h	e⁻; γ
64	Gadolinium	149	−80 700		9 d	K; γ(3·0)
		150	−81 395		10⁴ a	α(2·7)
		151	−79 730		150 d	K; γ(2·65)
		152	−80 206	0·20		
		153	−78 497		236 d	K; e⁻; γ
		154	−79 071	2·15		
		155	−77 336	14·78		
		156	−77 825	20·59		
		157	−75 975	15·71		
		158	−75 822	24·78		
		159	−73 632		18 h	e⁻(1·1, 1·9); γs
		160	−72 885	21·79		
		161	−70 280		3·7 min	e⁻(1·6); γs
65	Terbium	152	−75 720		4·5 h	K
		153	−76 510		5·1 d	e⁺
		154	−75 420		17·2 h	e⁺(2·6); γ
		155	−76 370		190 d	K; γ
		157	−75 910		4·7 d	K; γ(1·4)
		159	−74 649	100		
		160	−72 854		76 d	e⁻(0·59)
		161	−72 248		6·8 d	e⁻(0·5); γ(0·049)
66	Dysprosium	156	−76 070	0·05		
		158	−75 551	0·09		
		159	−74 241		134 d	e⁺; K
		160	−74 798	2·29		
		161	−73 055	18·88		
		162	−73 197	25·53		

Z	Element name	A	Mass excess ($\mu\mu$)	% Abundance	Half-life	Radiation (MeV)
		163	−71 245	24·97		
		164	−70 800	28·19		
		165	−68 184		145 min	e^-(1·25, 0·88, 0·42); γs
		166	−67 193		81 h	e^-(0·4); γ
67	Holmium	160	−71 260		20 min	e^+(1·3); γ(1·2)
		161	−72 200		4·6 h	K; γ
		162	−70 878		65 d	e^+(0·8); γ
		163	−71 234		5·2 d	K; γs
		164	−69 610		34 min	e^-(0·95)
		165	−69 579	100		
		166	−67 711		28 h	e^-(1·8, 0·55); γs
68	Erbium	162	−71 260	0·1		
		163	−69 935		11·2 h	K; γ(1·1)
		164	−70 713	1·5		
		165	−69 181		10 h	K
		166	−69 693	32·9		
		167	−67 940	24·4		
		168	−67 617	26·9		
		169	−65 390		9 d	e^-(0·33)
		170	−64 440	14·2		
		171	−61 870		7·6 h	e^-(1·49, 1·05, 0·67); γs
69	Thulium	166	−66 490		7·7 h	e^+(2·1); γ
		167	−66 970		9 d	K; γ(0·95, 0·22)
		168	−65 770		85 d	e^-(0·5); K; γs
		169	−65 755	100		
		170	−63 940		120 d	e^-(0·99, 0·87); γ
70	Ytterbium	168	−65 840	0·14		
		169	−64 470		30·6 d	K; γs
		170	−64 980	3·03		
		171	−63 570	14·24		
		172	−63 640	21·68		
		173	−61 940	16·18		
		174	−61 260	31·77		
		175	−58 860		4·2 d	e^-(0·47, 0·37); γs
		176	−57 320	12·65		
		177	−54 590		1·88 h	e^-(1·3); γs
71	Lutecium	170	−61 170		1·72 d	e^+; K; γ
		171	−61 860		8·5 d	K; γs
		172	−60 740		6·7 d	K
		175	−59 360	97·5		
		176	−57 340	2·5	$7\cdot3 \times 10^{10}$ a	e^-(0·40); γ(0·26)
		176	−57 340		3·7 h	e^-(1·15)
		177	−56 070		6·7 d	e^-(0·44); γs
72	Hafnium	173			23·6 h	K; γ(1·0)
		174	−59 640	0·18		
		176	−58 430	5·30		
		177	−56 600	18·47		
		178	−56 120	27·13		
		179	−53 970	13·85	19 s	γ(0·215)

Z	Element name	A	Mass excess (μu)	% Abundance	Half-life	Radiation (MeV)
		180	−53 180	35·14	5·5 h	e^-(0·44, 0·33); γ
		181	−50 895		43 d	e^-(0·41); γs
73	Tantalum	176			8·0 h	e^+; K; γ(1·3)
		177	−55 350		2·2 d	K; γ
		178	−54 070		9·3 min; 2·1 h	K; γs
		180	−52 456		8·2 h	e^-(0·7, 0·6); γs
		181	−51 993	100		
		182	−49 833		115 d	e^-(0·53); γs
		182	−49 833		16·2 min	e^-(0·2); γ
		183	−48 530		4·8 d	e^-(0·6)
74	Wolfram	178			21·5 d	e^+; γ
		179			5·2 min; 39 min	K; e^-
		180	−53 000	0·13		
		182	−51 699	26·3		
		183	−49 676	14·2		
		184	−48 975	30·6		
		185	−46 481		74 d	e^-(0·43); γ(0·134)
		186	−45 560	28·6		
		187	−42 756		24·1 h	e^-(1·32, 0·63); γs
		188	−41 184		65 d	e^-
75	Rhenium	182	−48 628		14 h	K; γs
		183	−48 740		240 d	K; γs
		183	−48 740		67 h	K; γs
		184	−47 220		52 d; 2 d	K; γs
		185	−46 941	37·1		
		186	−44 980		91 h	e^-(1·07, 0·93, 0·3); γs
		187	−44 167	62·9	4×10^{12} a	e^-(0·043)
		188	−41 647		16 h	e^-(2·1); γs
76	Osmium	184	−47 250	0·018		
		185	−45 887		96 d	K; γ
		186	−46 130	1·59		
		187	−44 168	1·64		
		188	−43 919	13·3		
		189	−41 700	16·1		
		190	−41 370	26·4		
		191	−39 030		16 d; 14 h	e^-(0·15); γs
		192	−38 550	41·0		
		193	−35 773		32 h	e^-(1·10); γ(1·5)
77	Iridium	188	−40 878		41·5 h	e^+(2·0); γ(1·8)
		190	−39 170		12·6 d; 3 h	K; γ
		191	−39 360	38·5		
		192	−37 300			e^-; γs
		193	−36 988	61·5		
		194	−34 875		19 h	e^-(2·18); γ(1·35)
		195	−34 110		2·3 h	e^-(1·0)
78	Platinum	190	−40 050	0·01		
		191	−38 550		3·0 d	K; γs
		192	−38 850	0·8		

Z	Element name	A	Mass excess (μμ)	% Abundance	Half-life	Radiation (MeV)
		193	−36 940		4·3 d	K; γs
		194	−37 275	32·8		
		195	−35 187	33·7	3·8 d	γ(0·337, 0·126)
		196	−35 033	25·4		
		197	−32 653		17·4 h	e⁻(0·68); γs
		197	−32 653		82 min	γs
		198	−32 105	7·3		
		199	−29 420		31 min	e⁻(1·8)
79	Gold	190	−35 290		4·3 min	e⁺
		191	−36 450		18 h	e⁺
		192	−35 380		4·0 h	e⁻(1·9); K; γ
		193	−35 760		15·8 h	K; γ
		194	−34 582		39·2 h	K; γ
		195	−34 949		182 d	K; γ
		196	−33 445		14 h	K; γ
		196	−33 445		5·6 d	e⁻(0·36); γs
		197	−33 459	100		
		198	−31 769		2·7 d	e⁻(1·38, 0·96, 0·29); γs
		199	−31 227		3·3 d	e⁻(0·46, 0·30, 0·25); γs
		200	−29 300		48 min	e⁻(2·5)
		201	−28 080			
80	Mercury	196	−34 180	0·15		
		197	−32 640		23 h	γ(0·275); e⁻(0·165); K
		197	−32 640		64 h	γ(0·191, 0·077); K
		198	−33 244	10·1		
		199	−31 721	17·0	43 min	γ(0·53)
		200	−31 673	23·1		
		201	−29 692	13·2		
		202	−29 358	29·7		
		203	−27 120		46·5 d	e⁻(0·208); γ(0·279)
		204	−26 505	6·8		
		205	−23 790		5·5 min	e⁻(1·75)
81	Thallium	198	−29 530		1·8 h	K; γ; e⁻
		199	−30 540		7·2 h	K; γ; e⁻
		200	−29 038		27 h	e⁻(0·40); γs
		201	−29 250		72 h	K; γ
		202	−28 050		11·4 d	K; γ(0·435); e⁻(0·35)
		203	−27 647	29·5		
		204	−26 135		2·7 a	e⁻(0·76); K
		205	−25 558	70·5		
		206	−23 896		4·2 min	e⁻(1·8)
	(Ac C″)	207	−22 550		4·76 min	e⁻(1·47); γ
	(Th C″)	208	−17 987		3·1 min	e⁻(1·82); γ(2·62)
		209	−14 704		2·2 min	e⁻(1·8)
	(Ra C″)	210	−9 946		1·32 min	e⁻(1·80)
82	Lead	199	−27 140		1·5 h	K
		200	−28 040		18 h	K
		201	−27 140		7 h	e⁺; K
		202	−27 997		500 a	K
		203	−26 771		52 h	K; γs
		204	−26 956	13	68 min	γ(0·90)
		205	−25 520		10 min	e⁺

Z	Element name	A	Mass excess (μu)	% Abundance	Half-life	Radiation (MeV)
		206	−25 532	26.3		
		207	−24 097	20.8		
		208	−23 350	51.5		
		209	−18 918		3.3 h	$e^-(0.72)$
	(Ra D)	210	−15 813		25 a	$e^-(0.025)$; γs
	(Ac B)	211	−11 258		36.1 min	$e^-(0.5, 1.4)$; $\gamma(0.83)$
	(Th B)	212	− 8 095		10.6 h	$e^-(0.59, 0.36)\gamma$
	(Ra B)	214	− 234		26.8 min	$e^-(0.65)$; γs
83	Bismuth	197			2 min	$\alpha(6.2)$
		198			7 min	$\alpha(5.8)$; K
		199	−21 560		27 min	$\alpha(5.47)$; K
		200	−21 060		90 min	$\alpha(5.1)$
		204	−22 190		12 h	K
		206	−21 611		6.4 d	$\gamma(0.93)$; K
		207	−21 562		50 a	γs; K
		208	−20 269			K
		209	−19 606	100		
	(Ra E)	210	−15 879		4.8 d	$\alpha(5.0)$; $e^-(1.65, 1.080)$; γ
	(Ac C)	211	−12 700		2.16 min	$\alpha(6.619)$; e^-; γ
	(Th C)	212	− 8 721		60.5 min	$\alpha(6.054)$; $e^-(2.25)$; γ
		213	− 5 683		46.5 min	α; $e^-(1.2)$
	(Ra C)	214	− 1 314		19.7 min	$\alpha(5.502)$; $e^-(3.15)$
		215	1 830		8 min	$\alpha(8.3)$
		216	6 330			
84	Polonium	203	−18 530		47 min	α; K
		205	−18 800		4 h	$\alpha(5.17)$; K
		206	−19 676		9 d	$\alpha(5.2)$; K; γs
		207	−18 442		5.7 h	$\alpha(5.1)$; γ; K
		208	−18 757		3 a	$\alpha(5.1)$
		209	−17 574		200 a	$\alpha(4.09)$
		210	−17 124		140 d	$\alpha(5.298)$; γ
	(Ac C′)	211	−13 343		0.5 s	$\alpha(7.434)$
	(Th C′)	212	−11 134		3×10^{-7} s	$\alpha(8.776)$
		213	− 7 175		3.2×10^{-6} s	$\alpha(8.3)$
	(Ra C′)	214	− 4 799		1.6×10^{-4} s	$\alpha(7.68)$
	(Ac A)	215	− 577		1.83×10^{-9} s	$\alpha(7.365)$
	(Th A)	216	1 922		1.58×10^{-1} s	$\alpha(6.774)$; e^-
		217	6 064		1 s	$\alpha(6.5)$
	(Ra A)	218	8 930		3.05 min	$\alpha(5.998)$
		219				
		220				
85	Astatine	207	−14 440		1.7 h	$\alpha(5.7)$
		208	−13 390		6.5 h	$\alpha(5.6)$
		210	−12 964		8.3 h	K
		211	−12 538		7.5 h	$\alpha(5.94)$; γ; K
		212	− 9 276		0.25 s	α
		213	− 6 930			
		214	− 3 660			$\alpha(8.78)$
		215	− 1 337		10^{-4} s	$\alpha(8.04)$

Z	Element name	A	Mass excess (μu)	Abundance	Half-life	Radiation (MeV)
		216	2 411		3×10^{-4} s	$\alpha(7\cdot64)$
		217	4 648		0·02 s	$\alpha(7\cdot0)$
		218	8 607		2 s	$\alpha(6\cdot63)$
		219	11 290		54 s	$\alpha(6\cdot27)$; e^-
		220	15 370			
86	Radon	212	− 9 293		23 min	$\alpha(6\cdot17)$; K
		214	− 4 310		1 s	$\alpha(8\cdot6)$
		215	− 1 310		1 s	$\alpha(8\cdot0)$
		216	272		1 s	$\alpha(8\cdot1)$
		217	3 896		1×10^{-3} s	$\alpha(7\cdot7)$
		218	5 603		0·019 s	$\alpha(7\cdot1)$
	(An)	219	9 481		3·92 s	$\alpha(6\cdot824)$
	(Tn)	220	11 401		54·5 s	$\alpha(6\cdot282)$
		221	15 230		24 min	α; e^-
	(Rn)	222	17 531		3·825 d	$\alpha(5\cdot486)$
		223				
		224				
87	Francium	219	9 257		0·02 s	$\alpha(7\cdot3)$
		220	12 337		27·5 s	$\alpha(6\cdot7)$
		221	14 183		4·8 min	$\alpha(6\cdot3)$
		222	17 630		14·8 min	e^-
		223	19 736		21 min	$e^-(1\cdot2)$; γ
		224	23 590			
		225				
		226				
88	Radium	218	7 170			
		219	10 050			$\alpha(8\cdot0)$
		220	11 029			$\alpha(7\cdot49)$
		221	13 892		30 s	$\alpha(6\cdot71)$
		222	15 376		38 s	$\alpha(6\cdot51)$
		223	18 501		11·2 d	$\alpha(5\cdot719, 5\cdot607)$; γs
		224	20 218		3·6 d	$\alpha(5\cdot66, 5\cdot44)$; $\gamma(0\cdot25)$
		225	23 528		14·8 d	$e^-(0\cdot2)$
		226	23 360		1620 a	$\alpha(4\cdot79, 4\cdot61, 4\cdot21)$; $\gamma(0\cdot188)$
		227	29 159		41·2 min	e^-
		228	31 138		6·7 a	$e^-(0\cdot030)$
		229				
		230			1 h	$e^-(1\cdot2)$
89	Actinium	221	15 680			
		222	17 760		5·5 s	$\alpha(6\cdot96)$
		223	19 144		2·2 min	$\alpha(6\cdot64)$; K
		224	21 690		2·9 h	$\alpha(6\cdot17)$; K
		225	25 153		10·0 d	$\alpha(5\cdot8)$
		226	26 160		22 h	e^-
		227	27 753		27·7 a	$\alpha(4\cdot94)$; $e^-(0\cdot02)$; γ
		228	31 080		6·13 h	$\alpha(4\cdot54)$; $e^-(1\cdot5, 2\cdot0)$; γ
		229	32 800		66 min	α
		230	36 210		1 min	$e^-(2)$
		231	38 550			
		232				

Z	Element name	A	Mass excess (μu)	% Abundance	Half-life	Radiation (MeV)
90	Thorium	222				
		223	20 920		0.1 s	α(7·55)
		224	21 477		1 s	α(7·13)
		225	23 927		8·0 min	α(6·57); K
		226	24 901		30·9 min	α(6·30)
		227	27 706		18·6 d	α(6·05, 5·67); γ
		228	28 750		1·9 a	α(5·42, 5·34); γ
		229	31 652		7340 a	α(5·02, 4·94, 4·85)
		230	33 087		8×10^4 a	α(4·68, 4·61); γ
		231	36 291		25·6 a	e^-(0·302, 0·216); γ
		232	38 124	100	$1{\cdot}39 \times 10^{10}$ a	α(3·98); γ(0·055)
		233	41 469		23·5 min	e^-(1·23)
		234	43 583		24·1 d	e^-(0·192, 0·104); γ
91	Protactinium	225	26 230		2·0 s	α
		226	27 810		1·8 min	α(6·81)
		227	28 811		38·3 min	α(6·46); K
		228	31 010		22 h	α(6·09, 5·85); K
		229	32 022		1·5 d	α(5·69); K
		230	34 433		17·7 d	α; e^-(1·1); K
		231	35 877		34 300 a	α(4·66, 5·04)
		232	38 612		1·32 d	e^-(0·28)
		233	40 132		27·4 d	e^-(0·53)
		234	43 298		1·14 min	e^-(2·32, 0·8)
		235	45 420		23·7 min	e^-(1·4)
		236	49 230			
92	Uranium	226				
		227				
		228	31 387		9·3 min	α(6·72)
		229	33 481		58 min	α(6·42)
		230	33 937		20·8 d	α(5·85)
		231	36 270		4·2 d	K
		232	37 168		70 a	α(5·31)
		233	39 522		$1{\cdot}63 \times 10^5$ a	α(4·83); e^-; γ; K
		234	40 904	0·006	$2{\cdot}5 \times 10^5$ a	α(4·76)
		235	43 915	0·71	$7{\cdot}1 \times 10^8$ a	α(4·52)
		236	45 637		$2{\cdot}46 \times 10^7$ a	α(4·49); γ(0·05)
		237	48 608		6·7 d	e^-(0·26); γ(0·5)
		238	50 770	99·28	$4{\cdot}50 \times 10^9$ a	α(4·18); γ(0·045)
		239	54 300		23·5 min	e^-(0·56, 1·2); γ
		240	56 594		18 h	e^-
		241				
		242				
93	Neptunium	229				
		230	37 680			
		231	38 280		50 min	α(6·2); K
		232	39 860		13 min	γ; K
		233	40 670		35 min	α(5·53); γ; K
		234	42 860		4·4 d	γ(1·9); K

Z	Element name	A	Mass excess (μu)	% Abundance	Half-life	Radiation (MeV)
		235	44 049		435 d	α(5·06); K
		236	46 624		22 h	e^-(0·51, 0·36); γ(0·15)
		237	48 056		$2 \cdot 2 \times 10^7$ a	α(4·77); γ(0·065)
		238	50 896		2·1 d	e^-(0·258, 1·272); γs
		239	52 924		2·33 d	e^-(0·676); γ(0·023)
		240	56 080		7·3 min	e^-(1·30)
		241	58 200		60 min	e^-(0·89); γ
		242	61 840			
94	Plutonium	232	41 180		36 min	α(6·6); K
		233	42 972			
		234	43 315		8·5 h	α(6·2); K
		235	45 270		26 min	α(5·85); K
		236	46 071		2·7 a	α(5·75); γ(0·045)
		237	48 298		40·0 d	K
		238	49 511		90 a	γ(0·0436, 0·100)
		239	52 146		$2 \cdot 4 \times 10^4$ a	α(5·1); γ(0·035, 0·050)
		240	53 882		6700 a	α(5·16)
		241	56 737		14 a	α(4·91); e^-(0·01)
		242	58 725		5×10^5 a	α(4·88)
		243	61 972		5 h	e^-(0·39); γ(0·095, 0·12)
		244	64 100			
		245	67 830			
		246	70 090			
95	Americium	236	49 160			
		237	49 840		1·3 h	α(6·01); K
		238	51 940		1·9 h	γ(1·35); K
		239	53 016		12 h	α(5·77); γ(0·28); K
		240	55 280		51 h	γ(1·4); K
		241	56 714		460 a	α(5·48); γ(0·06)
		242	59 502		16·0 h; 100 a	e^-(0·63); γ(0·04)
		243	61 367		10^4 a	α(5·27); γ(0·75)
		244	64 355		10 h	e^-(0·39); γ(0·4)
		245	66 340		2 h	e^-(0·91); γ(0·36)
		246	69 660		25 min	e^-(2·10); γ(0·10)
		247	72 090			
		248				
96	Curium	236				
		237				
		238	53 036		2·5 h	α(6·5); K
		239	54 880		2·8 h	γ(0·19); K
		240	55 545		27 d	α(6·26); K
		241	57 542		35 d	α(5·95); γ(1·45); K
		242	58 788		162 d	α(6·11); γ(0·044)
		243	61 370		32 a	α(5·78); γ(0·28); K
		244	62 821		18 a	α(5·8); γ(0·43)
		245	65 371		920 a	α(5·15); γ(0·17)
		246	67 202		550 a	α(5·37)
97	Berkelium	240				
		241	60 100			
		242	61 790			

Z	Element name	A	Mass excess ($\mu\mu$)	% Abundance	Half-life	Radiation (MeV)
		243	62 965		4·5 h	α(6·7); γ(0·84)
		244	65 170		4·5 h	α(6·67); γ(1·06); K
		245	66 272		4·9 d	α(6·37); γ(0·38); K
		246	68 770		45 h	γ(1·09); K
		247	70 260		950 a	α(5·6); γ(0·27); K
		248	72 960		2·3 h	e^-(0·65); K
		249	74 883		310 d	e^-(0·125); γ(0·32); α(5·4)
		250	78 270		3·2 h	e^-(1·76)
98	Californium	242				
		243	65 310			
		244	65 969		25 min	α(7·17)
		245	67 905		45 min	α(7·11); K
		246	68 766		56 h	α(6·75); γ(0·42)
		247	71 070		2·4 h	γ(0·3); K
		248	72 262		350 d	α(6·26)
		249	74 749		360 a	α(6·2); γ(0·39)
		250	76 384		10 a	α(6·0); γ(0·43)
		253	85 020		19 d	e^-(0·27)
99	Einsteinium	244				
		245	71 060		1·2 min	α(7·7)
		246	72 430		7 min	α(7·35)
		247	73 580			
		248	75 280		25 min	α(6·87); K
		249	76 258		2 h	α(6·76); K
		250	78 610		7·7 h	K
		253	84 730		200 d	α(6·6); γ(0·5)
		254	87 900		1·3 a	α(6·42); γ(0·6)
100	Fermium	246				
		247				
		248	77 092		36 s	α
		249	79 140		150 s	α(7·9)
		250	79 490		30 min	α(7·43)
		251	81 190		7 h	α(6·89); K
		252	82 562		23 h	α(7·04)
		254	86 839		3·2 h	α(7·2); γ(0·04)
		255	89 640		20 h	α(7·02); γ(0·58)
101	Mendelevium	250				
		251	84 620			
		252	86 120			
		253	86 940			
		254	89 470			
		255	90 550		30 min	α(7·30); K
		256			90 min	K
102	Nobelium	253	91 340			
		254	91 140		3 s	α(8·8)
		255	92 730		15 s	α(8·2)
103	Lawrencium	257	98 940		8 s	α(8·6)
104	Kurchatovium					

Z	Element name	A	Mass excess (μu)	% Abundance	Half-life	Radiation (MeV)
105	Hahnium					
106						
107						
108						

Index

Absorption
    α-particles, 213
    β^--particles, 221
    γ-rays, 223
    neutrons, 227
Absorption edge, X-ray, 114–116
Accelerating machines, 258ff
  growth and future, 269
  table, 273
Actinide series, 345
Actinium series, 54
Activity, measure of, 226
Advances Gas-cooled Reactor (AGR), 335
Age measurements
  by C-14 method, 314
  by tree-ring counting, 315
Alkali metal spectra, 93
Alpha particles
  absorption, 213
  Blackett cloud chamber photographs, 233
  Bragg curve, 213
  charge, 35
  charge-mass ratio, 33
  E/M ratio, 34
  emitters, 212
  identification, 36
  Geiger–Nuttall rule, 213
  range, 214
  scattering, Rutherford, 38, 231ff
  straggling, Bragg curve, 213
  theory of, emission, 214ff
  transuranic elements, 340ff
Alternating gradient proton synchrotron, 268
Americium, 343
Ampere's theorem, 151
Anderson, 298, 377
Andrews, 13
Angular momentum, 86
Angular momentum vectors, 146
  summary, 150
Anomalous Zeeman effect, 188
Antineutrino, 219
Antineutron, 385
Antiparticles, 379
Antiproton, 383
Archaeological dating, 314
Artificial radioactivity, 298ff
Aston, 55
Atom
  models, Thomson, Rutherford, 37
  one-electron, Schrodinger solution, 139
Atomic bomb, 333
Atomic mass number, 46
  relative atomic mass, 202

Atomic mass unit (unified), 56, 202
Atomic number, 46, 231
Atomic spectra, 63, 85ff
Atomic theory of matter, Greek, 1
Atomicity, 10
Avogadro, 1
Avogadro constant, 7
Avogadro's hypothesis, 1, 6

Background radiation, table, 456
Bainbridge mass spectrograph, 55
Balmer series formula, 86
Band spectra, 63
Barn, 278
Baryon
  conservation, 409
  decuplet, 419
  definition, 392
  groups, 415
  number, 409
Beauty (quark), 442
Becquerel (unit), 226
Berkelium, 343
Beryllium neutron source, 310
Beta particles
  absorption and range, 221
  and neutrinos, 218
  e/m ratio, 31, 32
  energy spectrum, 219
  from fission decay chains, 330
  origins in nucleus, 302
Betatron, 265
Bethe carbon-nitrogen cycle, 349
Bethe proton–proton cycle, 349
Bevatron, 265
Binding energy of nuclide, 203
  and liquid drop model, 282
  per nucleon, 205
Biological effects of radiation, 452ff
Black body radiation spectrum, 69
Blackett cloud chamber photographs, 233
Bohr, A., collective model of nucleus, 290
Bohr, N.,
  compound nucleus theory, 236
  correspondence principle, 98
  liquid drop theory, 281
  theory of hydrogen atom, 86
Bohr–Wheeler theory of fission, 323
Boltzmann constant, 10, 70
Bombarding experiments, 232ff
Born interpretation of waves and particles, 127
Born probability function, 133, 134
Boson (definition), 394
Bottom (quark), 442

Boyle's law, 6
Boyle temperature, 14
Brackett series, 89
Bragg curve, α-particles, 213
Bragg law, X-ray diffraction, 77
Bragg X-ray spectrometer, 78
Breeder reactor, 335
Bremsstrahlung, 222
Brownian motion, 2
Bubble chamber, 241
Bucherer, e/m for β-particles, 32
Build-up factor, 224

Cadmium/boron reactor control rods, 334
Calibration of C-14 scale by tree-ring counting, 316
Californium, 343
Carbon-12 mass scale, 202
Carbon-14 age-determination method, 314
Carbon-nitrogen cycle, 349
Cathode rays, 22
Cerenkov counter, 254
C.E.R.N. accelerator, 268
Chadwick estimation of mass of neutron, 43
Charm, 426ff
 quark, 433
Charmed excited states, quantum numbers, 437
Charmed particles, 436
Classification of elementary particles, 392, 413
Classification of radioisotopes, 457
Cloud chamber, 240
Cockcroft–Walton proton accelerator, 258
 reactions, 235
Collective model of the nucleus, 290
Colour and gluons, 438
Coloured quarks, 440
Compound nucleus, 234, 236
Compton effect, 80
Conduction of electricity,
 in glass, 20
 in liquids, 19
Conservation laws for particles, 409
 table, 414
Constant composition, law of, 1
Constant density of nuclei, 280
Controlled fission, reactor, 331
Correspondence, principle of, 98
Cosmic rays, 364ff
 air showers, 371
 altitude effect, 364
 cascade showers, 367
 composition, table, 368
 detection, 371
 discovery, 364
 energy spectrum, 366
 geomagnetic effect, 368

Cosmic rays—*contd.*
 heavy nuclides in, 365
 maximum energy of, 373
 mesons in, 369
 origin, 360, 373
 primary particles in, 364
 protons in, 365
 secondary particles, 366
 stars, 366
 telescope, 373
Cosmotron, 268
Critical energy of fission, 323
Critical potentials, 99
Crookes' dark space, 21
Cross-sections, neutron, 277
Curie (definition), 226
Curie-Joliot, 42
Curium, 343
Cusp fusion machine, 354
Cyclotron, 263

Dalton, 1
Davisson and Germer experiment, 122
Dead Sea Scrolls, age, 315
de Broglie's law, 120
Degeneracy, exchange, 164
Degrees of freedom, 11
Delayed neutrons, 330
Deuterium, discovery, 93
 in fusion reactions, 350
Deuteron reactions, 237
Diffraction,
 electron, 122
 neutron, 312
 optical, 61
 X-ray, 75
Diffuse series, 94, 111
Diffusion cloud chamber, 241
Dirac theory of electron, 383
Disintegration constant, 47
Dulong and Petit's law, 12
Dunnington e/m for electrons, 23

Eigenfunction in wave mechanics, 136
 normalization, 134, 140
Einstein relativity theory (special), 447ff
 energy-mass relation, 204, 451
 energy-momentum relation, 379, 451
 mass-velocity relation, 450
 theory of photoelectric effect, 72
 two postulates of special relativity, 448
Einsteinium, 344
Electrochemical equivalent, 19
Electrolysis, 20
Electromagnatic spectrum, 59ff
Electromagnatic theory, Maxwell, 63

Electron properties, 19ff
  charge (Millikan), 25
  mass, 27
Electron capture, 301
  density distribution, 128, 133
  diffraction, 122
  microscope, 125
  neutrino, 389
  spin angular momentum, 111ff
  spin quantum number, 112
Electron synchrotron, 266
Elementary particles, 402
  classification, 392, 413
e/m for electrons
  Thomson method, 22
  Duddington method, 24
e/m for β^- rays
  Bucherer method, 33
  Kaufmann method, 31
E/M for α-particles
  Rutherford and Robinson method, 33
Endoergic reactions, 234
Energy levels,
  γ-rays, 225
  helium, 171
  hydrogen, 90
  nuclear, 287
  sodium, 96
  superheavy elements, 292
  wave-mechanical, 141
  X-rays, 117
Equipartition of energy, 11
Eta meson, discovery of η^0, 417
Even/odd distribution of stable nuclides, 201
Exchange degeneracy, 164
Excitation potentials, 99
Exoergic reactions, 234
Expansion chamber, 240

Faraday constant, 20
Faraday dark space, 21
Faraday laws of electrolysis, 19
Fast breeder reactor, 335
Feather rule for β^- absorption, 222
Fermi, 221, 320, 405
Fermilab, tevatron, 268
Fermion(definition), 392
Fermium, 344
Filling order
  of electron levels, 178
  of nuclear levels, 289
Fine structure
  and electron spin, 109ff
  H-α line, 110, 159
  sodium D-lines, 112, 113
  X-ray levels, 117
Fissile nuclides, table, 325

Fission, 320ff
  beta chains in, 330
  Bohr–Wheeler theory of, 323
  cross-sections, 322, 332
  delayed neutrons in, 330
  distribution of fission products, 327
  energy of, 325
  fast reactor, 336
  neutron spectrum, 328
  power reactor, 334
  products, $At^{-1.2}$ decay law, 331
  prospects for fusion power, 335
  reactivity, 333
  reactor poisons, 331, 332
  sustained fission, 331
Forces of nature, 402, 426
  table, 429
Franck and Hertz experiment, 99
Free quark experiment, 441
Frequency modulated cyclotron, 265
Friedrich and Knipping experiment, 75
Frisch, 320
Fundamental particles, 403
Fundamental series, 94
Fusion, 348ff
  conditions for reaction, 348
  future of fusion power, 360
  Lawson's criterion, 365
  plasma, 349
  reactions and cross-sections, 350, 351
  reactor, 355
  source of stellar energy, 348
  Tokomak systems, 357

g-factor (Landé), 188
Gamma-rays,
  absorption, 223
  as bombarding particle, 237
  build-up factor, 224
  energy levels, 225
  half-value thickness, 224
  in fission, 327
  interaction with matter, 223
  photoelectric effect, 301
  radiation hazards, 452
  tenth-value thickness, 224
Gas multiplication factor, 246
Geiger law, 213, 235
Geiger-Muller counter, 248
  characteristics, 249
  dead-time, 251
  gas fillings, 250
  quenching vapour, 249
  threshold voltage, 250
Geiger–Nuttall rule, 214
Genetic effects of ionizing radiations, 454
Genetically related isomers, 303
Gerlach–Stern experiment, 153

Gluons, 438
Gross fission product decay law, 331
Group velocity, 121

H-α line, 85, 109
Hadrons, 395
  quark synthesis, 420
Hahnium, 345
Half-life
  in radioactive decay, 49
  of neutron decay, 311
Half-value thickness, 224
Heavy ion bombardment, 344
Heavy ion fusion, 361
Heisenberg uncertainty principle, 125, 127
  quantum mechanics, 132
  operations, 137
Helium spectrum, 167–170
  energy level diagram, 171
Hertz radiation experiment, 65
Huygens' principle, 61
Hydrogen atomic spectrum, 85
  Bohr theory, 86
  quantum mechanics of, 143
Hypercharge, 411
Hyperfine structure, 180
Hyperons, 392

Inner (total) quantum number, j, 112, 150
Interference, 59
Internal conversion, 301
Intersecting ring accelerator, 269
Ionic conduction, 20
Ionization chamber, 244
Islands of isomerism, 305
Isobars, 197, 199
Isomers (nuclear), 303
Isospin (I), 410
  third component of (I_3), 411
Isotopes, 52, 197
  artificial radioactive, 298ff
Isotope effect in line spectra, 90

Jeans (Rayleigh–Jeans law), 70
J/ψ particle, 435
  excited states of, 436
jj coupling, 170
Joliot, 42, 300

K-electron capture, 301
K-series, X-rays, 79, 116
Kaons, 389
Kaufmann e/m for $β^-$-rays experiment, 31
Kinetic theory of gases, 3
Kurchatovium, 345

LS (Russell–Saunders) coupling, 165
L-series, X-rays, 79, 116
Lamb–Rutherford experiment, 159
Lanthanide series, 345
Larmon precession, Zeeman effect, 185
Laue, V., X-ray diffraction, 75
Lawrence cyclotron, 263
Lawrencium, 344
Lawson's criterion, fusion, 356
Lenard dynamids, 37
Lennard–Jones molecular forces, 14, 15
Lepton (definition), 392
  colour, 444
  conservation, 410
Leukaemia, 454
Lifetime, average, mean
  of decay process, 48
  of fission reactor neutrons, 333
Linear accelerator, 261
Liquid drop model of nucleus, 281
  in theory of fission, 323
Lorentz theory of Zeeman effect, 183
  transformation in relativity, 449
Lyman series, 88

Macroscopic neutron cross-section, 278
Magic numbers
  evidence, 285
  derivation from spin-orbit coupling, 288
Magnetic moments,
  atoms, 151
  in Zeeman effect, 185
  nuclei, 180
Magnetic quantum number, 145, 150, 186
Magnetic vector coupling, 153
Magneton,
  Bohr μ_B, 152
  nuclear μ_N, 180
Mass defect, 203
Mass-energy conversion factor, 204
Mass excess, 201
Mass formula from liquid drop model, 284
Mass number, atomic, 46
Mass unit, unified atomic, 56, 202
Maximum permissible radiation levels, 455
Maxwell, 2, 63
Mean free path
  molecules, 7
  neutrons, 279
Mean lifetime
  radioactive decay, 48
  reactor neutrons, 333
Mendelevium, 344
Mesic atoms, 394
Meson, 377, 385, 392
  generated from quarks, 422
  groups, 413

Meson—contd.
 octet, 417
 resonances, 407
MeV (definition), 208
Michelson–Morley experiment, 426 & 47
Microscopic neutron cross-section, 278
Millikan,
 cosmic ray experiments, 364
 oil drop experiment, 25
 photoelectric experiment, 73
Mirror machines (fusion), 354
Missing elements, X-ray spectra, 103
Models
 atom, 37
 nucleus, 281–291
Moderator, reactor, 333
Molecular heat capacities, 12
 sizes, 16
 speeds, 6
Moon, early history and superheavy elements, 294
Moseley's experiments, 102
Multiple proportions, law of, 1
Multiplication factor, reactors, 331
Multiplicity
 atomic term, 113, 151
 gas, Geiger counter, 246
 particle, 411, 413
Muon (μ-meson), 377
 decay, 387ff
Muon neutrinos, 389
Muonium atom, 397

Negative energy states (Dirac), 382
Neptunium, 34
 series, 50, 54
Neutrino in β emissions, 219
 Reines and Cowan experiment, 220
Neutron, 42, 236
 absorption, 277
 bombarding reactions, 313
 counter, 255
 cross-sections, 278
  energy dependence, 281
  table, 332
 decay and half-life, 311
 diffraction, 312
 energy classification, 311
 induced reactions, 313
 mass of, 310
 mean free path, 279
 multiplication factor (reactors), 331
 properties, 310ff
 radii and $A^{1/3}$ rule, 280
 to proton ratio, 212
 transition to proton in nucleus, 302
 wave nature of, 311

Nitrogen transmutation (Rutherford), 232
Nobelium, 344
Non-penetrating orbits, 97
Normalization of wave function, 140
Normal Zeeman effect, 183, 187
n-p junction ionization detector, 251
Nuclear bombs, fission and fusion, 327, 333
Nuclear charge, 41, 102
Nuclear diameter, 41
Nuclear emulsion plate, 255
Nuclear energy levels (shell model), 286ff
Nuclear fission, 320ff
Nuclear forces, 402, 406
Nuclear, fusion, 348ff
Nuclear isomerism, 303
Nuclear magnetic moment, 180
Nuclear magneton, 180
Nuclear models, 277ff
Nuclear moderator, 330
Nuclear power reactors, 236
Nuclear reactor energies, Q values, 234, 236
Nuclear reactor, 331
Nuclear shells and magic numbers, 284
Nuclear shell model, theory of, 286–290
Nuclear size, 41
Nuclear spin angular momentum, 180
 quantum number, 180
Nuclear stability curve, 200
Nuclear transformations, 236
Nucleon, 46
 quark structure of, 420, 440
Nuclide, 199
 table of (Table C), 458

Ω^- **particle**, 418, 435
Odd-even distribution of stable nuclides, 201
Oil-drop experiment (Millikan), 25
Omega-minus particle,
 prediction, 418, 419
 discovery, 420
 decuplet structure, 419
  in charm family, 438
 photograph, 421
One-electron atom model
 Schrodinger solution, 139
Orbital electron capture, 301
 quantum number l, 97, 109
Oxygen-16 mass scale, 202

Packing fraction, 203
Pair production, 223, 298
Paralysis time(counters), 251
Particle symmetries, 416ff
Particles, table of stable and semi-stable, 391
Paschen series, 89
Paschen–Back effect, 192
Pauli exclusion principle, 162, 165

485

Pauli neutrino, 218
Penetrating orbits, 97
Periodic table, 173, 179
  empiricle rules, 176
Permeability of free space, μ_0, 65
Permittivity of free space, ε_0, 65
Pfund series, 89
Photoelectric effect, 71
Photon, 72
Physical constants table, xvi
π-meson, pion, 385
Pinch effect (fusion), 353
Planck quantum theory, 70
Planck constant, 70
Plasma (fusion), 349ff
  reactions in hydrogen, 350
  containment of, 352
  instabilities in, 353
  reaction times, 353
Plutonium, 341
Positive rays, Thomson, 52–54
Positron, 298, 379
Positronium, 383, 436
Powell, discovery of π-meson, 378
Power reactor, 334
Precautions against ionizing radiations, 456
Pressure of a gas, 4
Principal quantum number (n), 87, 150
Principal series, 94
Principal specific heat capacities, 9
Probability function, 133, 134
Proton-neutron transitions in nuclei, 302
Proton-proton collisions, 383
Proton-proton cycle in stellar energy, 349
Proton reactions, 235
Proton synchrotron, 267

Quantization of angular momentum, 88, 89, 98
Quantum defects, 95
Quantum (wave) mechanics, 131ff
Quantum numbers, summary of atomic, 150
Quantum of strange particles, 409ff
Quantum theory, 69ff
Quantum chromodynamics (Q.C.D.), 439
Quarks, 420
  baryons generated from three, 420
  experiment to find free, 441
  mesons generated from three, 422
  summary of quantum numbers
    three quarks, 430
    four quarks, 432
Quark colour, 439
Quark confinement, 440
Quenching agent in Geiger-Muller tubes, 250

Rad, 455
Radiation hazards, precautions, 456
Radioactivity, 46ff

Radioactive series, 49, 50
Radioactive equilibrium, 50, 51
Radioactive decay law, 47
Radioisotopes,
  production, 306
  uses, 307
Radon, 34, 51
Rare earth elements, 176
Ratio of principal specific heat capacities of gases, 9
Rayleigh–Jeans law, 70
Reactivity (fission reactor), 33
Reduced mass, 92
Relative biological effectiveness, 455
Relativity, theory of (special), 447
Rem, 455
Resonance potentials (atomic), 99
Resonance particles (nuclear), 403, 406, 407
Rock salt (NaCl) lattice, 78
Rod contraction (relativity), 449
Root mean square speed, 5
Russell–Saunders coupling (**LS**), 165
Rutherford model of the atom, 37
Rutherford and Geiger charge on α-particle, 35
Rutherford and Robinson E/M ratio for α-particles, 34
Rutherford and Royds identification of α-particles, 36
Rutherford single α-particle scattering theory, 38
  in hydrogen, 231
  in nitrogen, 232–235
Rydberg constant, 88, 92

Schrodinger equation, wave mechanics, 134ff
Scintillation counter, 251
Selection rules, 97
  for l, 97, 151
  for j, 113, 151
Semiconductor counter, 251
Semi-empiricle mass equation, 284
Sharp series, 94
Shell model of nucleus, 284–290
Short lived resonance states, 401ff
Single α-particle scattering, 38
Sodium spectrum, 93
  D-lines, 113
  Zeeman pattern, 191
Spark chamber, 252
Spatial quantization,
  l, 145
  s, 156
  j, 158
Specific heat capacities, 9
Spectra, 62

Spin-orbit coupling
 in atoms, 113
 in nuclei, 288
Stable and semi-stable nuclides, 377ff
Stable nuclides, 199–201
Stable particles, classification, 392
 table, 391
Stationary states (Bohr), 86
Statistical interpretation of waves and particles (Born), 127
Stellar energy, 348
Stern–Gerlach experiment, 153–156
Stonehenge (date), 315–316
Straggling of α-particles, 213
Strong interaction, 402
Superheavy elements, 291
Synchrocyclotron, 265

Tandem van de Graaff generator, 260
Tau lepton, 442
Temperature of a gas, 6
Tevatron, Fermilab machine, 268
Thermal conductivity of a gas, 8
Thermal neutron cross-sections, 332
Thomson, J. J.
 e/m of electron experiment, 22
 model of atom, 37
 positive ray apparatus, 54
Thomson, G. P., and Reid electron diffraction experiment, 124
Thorium series, 52
Time dilation (relativity), 449
Tokomak fusion systems, 357–360
Top and bottom quarks, 442
Total (atomic) quantum number (j), 112, 150
Transformations, Lorentz, 450
Transuranic elements, 340ff
Tree-ring dating, 315
Tritium, 237
 in fusion reactors, 350
Truth, beauty quarks, 442

Uncertainty principle, Heisenberg, 125, 127
Unification of forces of nature, 444
Unified mass unit (u), 56, 202
Uranium series, 51
Uranium isotopes in fission, 320ff
Uranium cross-sections, 322

Van Allen radiation belts, 370
Van de Graaff generator, 260
Van der Waals' equation, 12
Van der Waals' forces, 13
Vector bosons, 426
Vector coupling, 153, 157, 165–168
Vector model of atom, 150ff
Vector model, Zeeman effect, 184
Viscosity of a gas, kinetic theory, 8, 9

Wave associated with electron, 131
Wave function, 133
 of two electrons, 165
Wave mechanics, 131
Wave theory of light, 59
Wave velocity, 121
Weak interaction mediators, 426
Wien black body radiation, 69
Wien displacement law, 69
Wilson cloud chamber, 240

X-rays,
 absorption spectrum, 114
 absorption edges, 115, 116
 characteristic, 101
 continuous (white) spectrum, 79
 diffraction, 75
 discovery, 74
 energy levels, 105
 fine structure, 116, 117
 wavelengths, 77
X-ray spectrometer, 78

Young's interferometer experiment, 60
Yukawa, nuclear particle of, 377

Zeeman effect, 183ff
 anomalous, 188
 in strong field, 192
 Lorentz theory, 183
 normal, 183–187
 of sodium D-lines, 191
 vector model of, 184